GAMES OF STRATEGY

■

GAMES OF STRATEGY

Avinash Dixit

■

Susan Skeath

W · W · Norton & Company

New York · London

The text of this book is composed in Utopia
with the display set in Machine and Myriad
Composition by Gina Webster
Manufacturing by Courier
Book design by Jack Meserole

Editor: Ed Parsons
Associate Managing Editor: Jane Carter
Production Manager: Roy Tedoff
Project Editor: Kate Barry
Editorial Assistant: Mark Henderson

Library of Congress Cataloging-in-Publication Data
Dixit, Avinash K.
Games of Strategy / Avinash Dixit, Susan Skeath.
p.   cm.
Includes bibliographical references and index.
ISBN 0-393-97421-9
1. Game theory. 2. Policy sciences. 3. Decision making. I. Skeath, Susan. II. Title.
HB144.DS9   1999
519.3—dc21   98–49972

W. W. Norton & Company, Inc., 500 Fifth Avenue, New York, N.Y. 10110
www.wwnorton.com
W. W. Norton & Company Ltd., Castle House, 75/76 Wells Street, London W1T 3QT
4 5 6 7 8 9 0

■

To the memories of our fathers,

Kamalakar Ramachandra Dixit
and
James Edward Skeath

Contents

PART TWO
Concepts and Techniques

4 Games with Simultaneous Moves   79

PART THREE
Some Broad Classes of Games and Strategies

10 Evolutionary Games   320

PART FOUR
Applications to Specific Strategic Situations

13 Brinkmanship: The Cuban Missile Crisis 435

14 Strategy and Voting 462

15 Bidding Strategy and Auction Design 494

16 Bargaining 521

17 Markets and Competition 550

Preface

Game theory is a relative newcomer to the family of academic fields of thought. Its first substantial text, *Theory of Games and Economic Behavior* by John von Neumann and Oskar Morgenstern (Princeton, N.J.: Princeton University Press, 1943), was published less than 60 years ago. But it has rapidly become an important subject. Its basic theory developed through the 1950s and 1960s, with important contributions from John Nash, Thomas Schelling, and others. Then the pace accelerated as the theory began to find applications to issues in such diverse fields as international relations in the 1960s; economics, business, and evolutionary biology in the 1970s and 1980s; and political science in the 1980s and 1990s. Now we are at a point where terms from game theory, such as "zero-sum games" and "the prisoners' dilemma," have become a part of the language. As Paul Samuelson says, "To be literate in the modern age, you need to have a general understanding of game theory."

Colleges and universities should attempt to impart such understanding to all of their students. Courses and textbooks on game theory are indeed proliferating rapidly. However, most of them suffer from two severe limitations—too narrow an orientation and too many prerequisites.

Most game theory courses take their approach from some particular discipline, whether it be economics, politics, business, or biology. These courses are taught with the assumption that students already know the concepts and jargon of their particular science. Many courses also assume a lot of mathematical background; they assume substantial knowledge of, and routinely use, calculus and probability theory. Such disciplinary and mathematical prerequisites restrict these courses to the more quantitatively oriented juniors and seniors majoring in a few subjects.

Game-theoretic concepts are actually much more basic, and can be conveyed with benefit to much larger groups of freshmen and sophomores without any prerequisites in any of the fields of application, and with only high school mathematics. Both of us have taught courses of this kind for several years. We have found that students take to the concepts of game theory with ease. Indeed, in economics the game-theoretic view of competition—strategic interactions between firms as they try to outdo each other in their attempts to attract customers—is much more natural and appealing than the standard textbook story of balancing supply and demand in impersonal markets. There are similar examples from other fields.

We believe that there is a strong case for reversing the usual order whereby general introductory courses in each subject are followed by advanced subject-specific courses in game theory. In the more natural progression, all students interested in the social and biological sciences would complete a freshman course in elementary game theory before going on to more detailed study of one of the specialized fields. Students intending to specialize in the natural sciences would also find game theory a more interesting and useful way to satisfy their distribution requirements than many introductory courses in particular social sciences. This book is the product of our experiences teaching game theory at such an elementary level and with such aims, and we hope it will enable others to develop and teach similar courses for beginners.

We emphasize that our book assumes no prior knowledge of economics, political science, or biology. We explain from first principles those concepts we use from these disciplines. Similarly, we do not require any knowledge of college mathematics or statistics. Of course the subject is inherently quantitative; totally nonnumerate students will not flourish in it. But the most basic high school algebra—for example, the ability to solve two linear equations in two unknowns—suffices for almost everything we do. We explain from first principles, and using familiar examples, the few simple rules for manipulating probabilities. In a few places where the use of elementary calculus (taking derivatives of simple functions) actually makes the analysis simpler, we do offer it, but we also offer a noncalculus alternative.

We also recognize that a few students with the best quantitative background and ability will find their intellectual curiosity whetted and will want more. We have therefore gathered some of the more technical and advanced material into one chapter—Chapter 7—for such readers; it can be omitted by others without loss of continuity.

The book covers a lot of material, and teachers can select from it to suit the length and special emphases of their courses. The ideas introduced in Chapters 1 and 2, and the theory and techniques developed in Chapters 3 to 5, are basic to any course. Chapter 6 extends and combines the techniques of Chapters 3 through 5; it can be covered in a more superficial or selective manner. Chapter 7 is technical and optional as we said earlier.

The next five chapters examine specific classes of games, and therefore offer more scope for choice. Chapter 8 on the prisoners' dilemma is important to any treatment of game theory. Chapters 9 on strategic moves, 10 on evolutionary games, 11 on collective-action games, and 12 on uncertainty and asymmetric information are largely independent of one another. Some of these can be studied in depth while others are only skimmed.

The final group of chapters is on applications—Chapters 13 on brinkmanship, 14 on voting, 15 on auctions, 16 on bargaining, and 17 on markets—and allows the greatest freedom of choice in course design. In most of these chapters, we begin with one or two simple numerical examples and then develop the ideas in greater generality. Teachers may choose to cover some of the topics in depth and convey only a flavor of the rest using the examples.

In our own courses we first used Avinash Dixit and Barry Nalebuff's *Thinking Strategically* (Norton, 1991) as the main text. But that book was directed at more of a general "trade" readership, and so lacked the kind of pedagogic material a course textbook needs: repetition and summaries for reinforcement, precise definitions and highlighted statements of key concepts, problems, and questions to allow students to test their understanding, and so on. Therefore we had to supplement the book with lecture notes, problem sets, and other handouts or course packets. In the process of creating these tools, we gradually improved and replaced the examples and cases used to convey the ideas in *Thinking Strategically,* thus effectively replacing almost all of that book with new material. Here we have assembled and organized all of these things into one package, and written a proper textbook at the Dixit–Nalebuff level. However, we have retained a very important feature of that book, namely its reliance on examples and cases. Our ultimate aim is to convey the concepts of game theory, but through the medium of examples and cases instead of setting out the theory by itself in an abstract mathematical way.

We have tried out successive drafts of this manuscript on our own classes for the past two years. Several teachers at other universities and colleges used the manuscript in their courses; some even developed new courses based on the manuscript. This list includes Amanda Bayer (Swarthmore College), Larry Evans (The College of William and Mary), Kimberly Katz (Mount Holyoke College), Greg Trandel (University of Georgia), and Randall Waldron (University of South Dakota). They, along with their students and our own students, gave us extremely valuable feedback. The result is not only the correction of many errors—typographical as well as substantive—but also major improvements in exposition, the introduction of new material, and the deletion of material found to be too difficult or unsuitable. We are grateful to all of these teachers and students—for their willingness to venture into a new type of course using a draft manuscript; for the encouragement they gave us based on their experience; and for their friendly, perceptive, and constructive criticisms of the manuscript drafts.

In addition, many people read the whole manuscript: Vincent Crawford (University of California, San Diego), Hiroyuki Kawakatsu (University of California, Irvine), Barry Nalebuff (Yale), Ed Parsons (our editor at Norton), and two anonymous readers commissioned by him. Others read individual chapters: Dilip Abreu and Pierpaolo Battigalli (Princeton), Frank Milne (Queen's University, Canada), and Sylvia Nasar (*New York Times*). All of these people gave us many helpful comments concerning the substance as well as the writing, and we are happy to acknowledge our debt to them all. We are also grateful to Kate Barry (our project editor at Norton) for expertly guiding the book through the production process and for being tolerant of our requests for many last-minute changes.

Susan Skeath would also like to thank her husband, Paul van Mulbregt, for his support and encouragement, as well as for reading and commenting on early drafts of various chapters. In addition, her children deserve praise for their patience and tolerance, if not for learning how to push the computer's reset button.

Finally, we were fortunate to have an outstanding copy editor, Susan Middleton. She not only improved our writing at numerous points, but also understood the substance, and caught an embarrassingly large number of slips and errors. If the book is judged to be well written and relatively error-free, she deserves a great deal of the credit. Of course, the two authors retain joint responsibility for any errors and shortcomings that remain, in the sense that each of us blames the other.

A. D.
S. S.
April 1999

PART ONE

▪

Introduction and General Principles

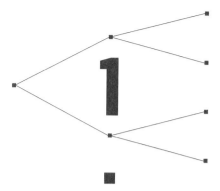

Basic Ideas and Examples

ALL INTRODUCTORY TEXTBOOKS begin by attempting to convince the student readers that the subject is of great importance in the world, and therefore merits their attention. The physical sciences and engineering claim to be the basis of modern technology and therefore of modern life; the social sciences discuss big issues of governance, for example, democracy and taxation; the humanities claim that they revive your soul after it has been deadened by exposure to the physical and social sciences and to engineering. Where does the subject games of strategy, often also called game theory, fit into this picture, and why should you study it?

We offer a practical motivation much more individual and closer to your personal concerns than most other subjects. You play games of strategy all the time: with your parents, siblings, friends, enemies, even with your professors. You have probably acquired a lot of instinctive expertise, and we hope you will recognize in what follows some of the lessons you have already learned. We will build on this experience, systematize it, and develop it to the point where you will be able to improve your strategic skills and use them more methodically. Opportunities for such uses will appear throughout the rest of your life; you will go on playing such games with your employers, employees, spouses, children, and even strangers.

Not that the subject lacks wider importance. Similar games are played in business, politics, diplomacy, wars—in fact, whenever people interact to strike mutually agreeable deals or to resolve conflicts. Being able to recognize such games will enrich your understanding of the world around you, and will make you a better participant in all its affairs.

It will also have a more immediate payoff in your study of many other subjects. Economics and business courses already use a great deal of game-theoretic thinking. Political science is rapidly catching up. Biology has been importantly influenced by the concepts of evolutionary games, and has in turn exported these ideas to economics. Psychology and philosophy also interact with the study of games of strategy. Game theory has become a provider of concepts and techniques of analysis for many disciplines, one might say all disciplines except those dealing with completely inanimate objects.

1 WHAT IS A GAME OF STRATEGY?

The word *game* may convey an impression that the subject is frivolous or unimportant in the larger scheme of things—that it deals with trivial pursuits like gambling and sports when the world is full of more weighty issues like war and business and your education and career and relationships. Actually, games of strategy are "not just a game"; all of these weighty issues are instances of games, and game theory helps us understand them all. But it will not hurt to start with gambling or sports.

Most games involve chance, skill, and strategy in varying proportions. Playing double or nothing on the toss of a coin is a game of pure chance, unless you have exceptional skill in doctoring or tossing coins. A hundred-yard dash is a game of pure skill, although some chance elements can creep in; for example, a runner may simply have a slightly off day for no clear reason.

Strategy is a skill of a different kind. In the context of sports it is a part of the mental skill needed to play well; it is the calculation of how best to use your physical skill. For example, in tennis you develop physical skill by practicing your serves (first serves hard and flat, second serves with spin or kick) and passing shots (hard, low, and accurate). The strategic skill is knowing where to put your serve (wide, or on the T) or passing shot (crosscourt, or down the line). In football, you develop such physical skills as blocking and tackling, running and catching, and throwing. Then the coach, knowing the physical skills of his[1] own team and those of the opposing team, calls the plays that best exploit his team's skills and the other team's weaknesses. The coach's calculation constitutes the strategy. The physical game of football is played on the gridiron by jocks; the strategic game is

[1]In this book, generic pronouns will be male in the chapters for which Dixit had the primary responsibility, and female in the chapters for which Skeath was primarily responsible. This device achieves a gender balance for the book as a whole, avoids the usual awkward constructions of he/she, and tells the reader whom to blame for errors or flaws in any chapter. Of course, when we refer to particular persons (for example, Seles and Hingis in the tennis games) or construct examples with players of particular gender (such as James and Dean in the chicken game), we refer to them by pronouns of the appropriate gender no matter in which chapter they appear.

played in the offices and on the sidelines by coaches and by nerdy assistants.

A hundred-yard dash is a matter of exercising your physical skill as best you can; it offers no opportunities to observe and react to what other runners in the race are doing, and therefore no scope for strategy. Longer races do involve strategy—whether you should lead to set the pace, how soon before the finish should you try to break away, and so on.

Strategic thinking is essentially about your interactions with others: someone else is also doing similar thinking at the same time and about the same situation. Your opponents in a marathon may try to frustrate or facilitate your attempts to lead, as they think best suits their interests. Your opponent in tennis tries to guess where you will put your serve or passing shot; the opposing coach in football calls the play that will best counter what he thinks you will call. Of course, just as you must take into account what the other player is thinking, he is taking into account what you are thinking. Game theory is the analysis, or science, if you like, of such interactive decision making.

When you think carefully before you act—when you are aware of your objectives or preferences and of any limitations or constraints on your actions, and choose your actions in a calculated way to do the best according to your own criteria—you are said to be behaving rationally. Game theory adds another dimension to rational behavior, namely interaction with other equally rational decision makers. In other words, game theory is the science of rational behavior in interactive situations.

We do not claim that game theory will teach you the secrets of perfect play or ensure that you will never lose. For one thing, your opponent can read the same book, and both of you cannot win all the time. More importantly, many games are complex and subtle enough, and most actual situations involve enough idiosyncratic or chance elements, that game theory cannot hope to offer surefire recipes for action. What it does is to provide some general principles for thinking about strategic interactions. You have to supplement these ideas and some methods of calculation with many details specific to your situation before you can devise a successful strategy for it. Good strategists mix the science of game theory with their own experience; one might say that game playing is as much art as science. We will develop the general ideas of the science, but will also point out its limitations and tell you when the art is more important.

You may think that you have already acquired the art from your experience or instinct, but you will find the study of the science useful nonetheless. The science systematizes many general principles that are common to several contexts or applications. Without general principles, you would have to figure out from scratch each new situation that requires strategic thinking. That would be especially difficult to do in new areas of application; for example, if you learned your art by playing games against parents and siblings and must now practice strategy against business competitors. The general principles of game theory provide you with a ready reference point. With this foundation in place, you can proceed much more quickly and confidently to acquire and add the situation-specific features or elements of the art to your thinking and action.

2 SOME EXAMPLES AND STORIES OF STRATEGIC GAMES

With the aims announced in the previous section, we will begin by offering you some simple examples, many of them taken from situations you have probably encountered in your own lives, where strategy is of the essence. In each case we will point out the crucial strategic principle. Each of these principles will be discussed more fully in a later chapter, and after each example we will tell you where the details can be found. But don't jump to them right away; for a while, just read all the examples to get a preliminary idea of the whole scope of strategy and of strategic games.

A. Which Passing Shot?

Tennis at its best consists of memorable duels between top players: Martina Navratilova versus Chris Evert, John McEnroe versus Ivan Lendl, Pete Sampras versus Andre Agassi. The current hot rivalry is between Monica Seles and Martina Hingis. Therefore picture a future U.S. Open final. Hingis at the net has just volleyed to Seles on the baseline. Seles is about to hit a passing shot. Should she go down the line, or crosscourt? And should Hingis expect a down-the-line shot and lean slightly that way, or expect a crosscourt shot and lean the other way?

Conventional wisdom favors the down-the-line shot. The ball has a shorter distance to travel to the net, so the other player has less time to react. But this does not mean that Seles should use that shot all of the time. If she did, Hingis would confidently come to expect it and prepare for it, and the shot would not be so successful. To improve the success of the down-the-line passing shot, Seles has to use the crosscourt shot often enough to keep Hingis guessing on any single instance.

Similarly in football, with a yard to go on third down, a run up the middle is the percentage play, that is, the one used most often, but the offense must throw a pass occasionally in such situations "to keep the defense honest."

Thus the most important general principle of such situations is not what Seles *should* do but what she *should not* do: she should not do the same thing all the time or systematically. If she did, then Hingis would learn to cover that, and Seles's chances of success would fall.

Not doing any one thing systematically means more than not playing the same shot in every situation of this kind. Seles should not even mechanically switch back and forth between the two shots. Hingis would spot and exploit this *pattern*, or indeed any other detectable system. Seles must make the choice on each particular occasion *at random* to prevent this guessing.

This general idea of "mixing one's plays" is well known, even to sports commentators on television. But there is more to the idea, and these further aspects require analysis in greater depth. Why is down-the-line the percentage shot? Should one play it 80% of the time or 90% or 99%? Does it make any difference if the occasion is particularly big; for example, does one throw that pass on third down during the regular season but not in the Super Bowl? In actual practice, just how does one mix one's plays? What happens when a third possibility (the lob) is introduced? We will be able to examine and answer such questions in Chapter 5.

B. The GPA Trap

You are enrolled in a course that is graded on a curve. No matter how well you do in absolute terms, only 40% of the students will get As, and only 40% will get Bs. Therefore you must work hard, not just in absolute terms, but relative to how hard your classmates (actually, "class enemies" seems a more fitting term in this context) work. All of you recognize this, and after the first lecture you hold an impromptu meeting in which all students agree not to work too hard. As weeks pass by, the temptation to get an edge on the rest of the class by working just that little bit harder becomes overwhelming. After all, the others are not able to observe your work in any detail, nor do they have any real hold over you. And the benefits of an improvement in your grade point average are substantial. So you hit the library more often, and stay up a little longer.

The trouble is, everyone else is doing the same. Therefore your grade is no better than it would have been if you and everyone else had abided by the agreement. The only difference is that all of you have spent more time working than you would have liked.

This is an example of the prisoners' dilemma. In the original story, two suspects are being separately interrogated and invited to confess. One of them, say A, is told, "If the other suspect, B, does not confess, then you can cut a very good deal for yourself by confessing. But if B does confess, then you would do well to confess, too; otherwise the court will be especially tough on you. So you should confess no matter what the other does." B is told to confess, using similar reasoning. Faced with this choice, both A and B confess. But it would have been better for both if neither had confessed, because the police had no really compelling evidence against them.

Your situation is similar. If the others slacken, then you can get a much better grade by working hard; if the others work hard, then you had better do the same, else you will get a very bad grade. You may even think that the label "prisoner" is very fitting for a group of students trapped in a required course.

There is a prisoners' dilemma for professors and schools, too. Each professor can make his course look good or attractive by grading it slightly more liberally, and each school can place its students in better jobs or attract better applicants by grading all of its courses a little more liberally. Of course, when all do this, none has any advantage over the others; the only result is rampant grade inflation, which compresses the spectrum of grades, and therefore makes it difficult to distinguish abilities.

People often think that in every game there must be a winner and a loser. The prisoners' dilemma is different—both or all players can come out losers. People play (and lose) such games every day, and the losses can range from minor inconvenience to potential disaster. Spectators at a sports event stand up to get a better view, but when all stand, no one has a better view than when they were all sitting. Superpowers acquire more weapons to get an edge over their rivals, but when both do so, the balance of power is unchanged; all that has happened is that both have spent economic resources that they could have used for better purposes, and the risk of accidental war has escalated. The magnitude of the potential cost of such games to all players makes it important to understand the ways in which mutually beneficial cooperation can be achieved and sustained. We will devote the whole of Chapter 8 to the study of this game.

Just as the prisoners' dilemma is potentially a lose-lose game, there are win-win games too. International trade is an example; when each country produces more of what it can do relatively best, all get to share in the fruits of this international division of labor. But successful bargaining about the division of the pie is needed if the full potential of trade is to be realized. The same applies to many other bargaining situations. We will study these in Chapter 16.

C. "We Can't Take the Exam Because We Had a Flat Tire"

Here is a story, probably apocryphal, that circulates on the undergraduate e-mail networks; each of us independently received it from our students:

> There were two friends taking Chemistry at Duke. Both had done pretty well on all of the quizzes, the labs, and the midterm, so that going into the final they had a solid A. They were so confident the weekend before the final that they decided to go to a party at the University of Virginia. The party was so good that they overslept all day Sunday, and got back too late to study for the Chemistry final that was scheduled for Monday morning. Rather than take the final unprepared, they went to the professor with a sob story. They said they had gone up to UVA and had planned to come back in good time to study for the final but had had a flat tire on the way back. Because they didn't have a spare, they had spent most of the night looking for help. Now they

were really too tired, so could they please have a makeup final the next day? The professor thought it over and agreed.

The two studied all of Monday evening and came well prepared on Tuesday morning. The professor placed them in separate rooms and handed the test to each. The first question on the first page, worth 10 points, was very easy. Each of them wrote a good answer, and greatly relieved, turned the page. It had just one question, worth 90 points. It was: "Which tire?"

The story has two important strategic lessons for future partygoers. The first is to recognize that the professor may be an intelligent game player. He may suspect some trickery on the part of the students, and may use some device to catch them. Given their excuse, the question was the likeliest such device. They should have foreseen it, and prepared their answer in advance. This idea that one should look ahead to future moves in the game, and then reason backward to calculate one's best current action, is a very general principle of strategy, which we will discuss and elaborate upon in Chapter 3. We will also use it, most notably in Chapter 9.

But it may not be possible to foresee all such professorial countertricks; after all, professors have much more experience of seeing through students' excuses than students have of making up such excuses. If the pair are unprepared, can they independently produce a mutually consistent lie? If each picks a tire at random, the chances are only 25% that the two will pick the same one. (Why?) Can they do better?

You may think that the front tire on the passenger side is the one most likely to suffer a flat, because a nail or a shard of glass is more likely to lie closer to the side of the road than the middle, and the front tire on that side will encounter it first. You may think this is good logic, but that is not enough to make it a good choice. What matters is not the logic of the choice, but making the same choice as your friend does. Therefore you have to think about whether your friend would use the same logic and would consider that choice equally obvious. But even that is not the end of the chain of reasoning. Would your friend think that the choice would be equally obvious to you? And so on. The point is not whether a choice is obvious or logical, but whether it is obvious to the other that it is obvious to you that it is obvious to the other. . . . In other words, what is needed is a convergence of expectations about what should be chosen in such circumstances. Such a commonly expected strategy on which the players can successfully coordinate is called a focal point.

There is nothing general or intrinsic to the structure of all such games that creates such convergence. In some games, a focal point may exist because of chance circumstances about the labeling of strategies, or some experience or knowledge shared by the players. For example, if the passenger's front side of a car were for some reason called the Duke's side, then two Duke students would be very likely to choose it without any need for explicit prior understanding. Or if the driver's front side of all cars were painted orange (for safety, to be easily visible to oncoming cars), then two Princeton students would be very likely to choose that tire, because orange is the Princeton color. But without some such clue, tacit coordination might not be possible at all.

We will study focal points in more detail in Chapter 4. Here in closing we merely point out that when asked in classrooms, over 50% of students choose the driver's front side. They are generally unable to explain why, except to say that it seems the obvious choice.

D. Why Are Professors So Mean?

Many professors have an inflexible rule not to give makeup exams, and never to accept late submission of problem sets or term papers. Students think the professors must be really hardhearted to behave in this way. The true strategic reason is often exactly the opposite. Most professors are kindhearted, and would like to give their students every reasonable break and accept any reasonable excuse. The trouble lies in judging what is reasonable. It is hard to distinguish between similar excuses and almost impossible to verify their truth. The professor knows that on each occasion he will end up by giving the student the benefit of the doubt. But the professor also knows that this is a slippery slope. As the students come to know that the professor is a soft touch, they will procrastinate more and produce ever flimsier excuses. Deadlines will cease to mean anything, and examinations will become a chaotic mix of postponements and makeup tests.

Often the only way to avoid this slippery slope is to refuse to take even the first step down it. Refusal to accept any excuses at all is the only realistic alternative to accepting them all. By making an advance commitment to the "no excuses" strategy, the professor avoids the temptation to give in to all.

But how can a softhearted professor maintain such a hardhearted commitment? He must find some way to make a refusal firm and credible. The simplest way is to hide behind an administrative procedure or university-wide policy. "I wish I could accept your excuse, but the university won't let me" not only puts the professor in a nicer light, but removes the temptation by genuinely leaving him no choice in the matter. Of course, the rules may be made by the same collectivity of professors as hides behind them, but once made, no individual professor can unmake the rule in any particular instance.

If the university does not provide such a general shield, then the professor can try to make up commitment devices of his own. For example, he can make a clear and firm announcement of the policy at the beginning of the course. Any time an individual student asks for an exception, he can invoke a fairness principle, saying, "If I do this for you, I would have to do it for everyone." Or the professor can acquire a reputation for toughness by acting tough a few times. This may be an unpleasant thing for him to do, and it may run against his true inclination, but it helps in the long run over his whole career. If a professor is believed to be tough, few students will try excuses on him, so he will actually suffer less pain in denying them.

We will study commitments, and related strategies, like threats and promises, in considerable detail in Chapter 9.

E. Roommates and Families on the Brink

You are sharing an apartment with one or more other students. You notice that the apartment is nearly out of dishwasher detergent, paper towels, cereal, beer, and other items. You have an agreement to share the actual expenses, but the trip to the store takes time. Do you spend your time, or do you hope that someone else will spend his, leaving you more time to study or relax? Do you go and buy the soap, or stay in and watch TV to catch up on the soap operas?[2]

In many situations of this kind, the waiting game goes on for quite a while before someone who is really impatient for one of the items (usually beer) gives in and spends the time for the shopping trip. Things may deteriorate to the point of serious quarrels or even breakups among the roommates.

This game of strategy can be viewed from two perspectives. In one, each of the roommates is regarded as having a simple binary choice—to do the shopping, or not. The best outcome for you is where someone else does the shopping and you stay at home; the worst is where you do the shopping while the others get to use their time better. If both do the shopping (unknown to each other, on the way home from school or work) there is unnecessary duplication and perhaps some waste of perishables; if neither does, there can be serious inconvenience or even disaster if the toilet paper runs out at a crucial time.

This is analogous to the game of chicken that used to be played by American teenagers. Two of them drove their cars toward each other. The first to swerve to avoid a collision was the loser (chicken); the one who kept driving straight was the winner. We will analyze the game of chicken further in Chapter 4, and also in Chapters 5 and 10.

A more interesting dynamic perspective on the same situation regards it as a "war of attrition," where each roommate tries to wait out the others, hoping that someone else's patience will run out first. In the meantime the risk escalates that the apartment will run out of something critical, leading to serious inconvenience or a blowup. Each player lets the risk escalate to the point of his own tolerance; the one revealed to have the least tolerance loses. Each sees how close to the brink of disaster the others will let the situation go. Hence the name

---

[2] This example comes from Michael Grunwald's "At Home" column, "A Game of Chicken," in the *Boston Globe Magazine*, April 28, 1996.

"brinkmanship" for this strategy and this game. It is a dynamic version of chicken, offering richer and more interesting possibilities.

One of us (Dixit) was privileged to observe a brilliant example of brinkmanship at a dinner party one Saturday evening. The company was sitting in the living room before the dinner around 7:30, when the host's 15-year-old daughter appeared at the door and said, "Bye, Dad." The father asked, "Where are you going?" and the daughter replied, "Out." After a pause that was only a couple of seconds but seemed much longer, the host said, "All right, bye."

Your strategic observer of this scene was left thinking how it might have gone differently. The host might have asked, "With whom?" and the daughter might have replied, "Friends." The father could have refused permission unless the daughter told him exactly where and with whom she would be. One or the other might have capitulated at some such later stage of this exchange, or it could have led to a blowup.

This was a risky game for both the father and the daughter to play. The daughter might have been punished or humiliated in front of strangers; an argument could have ruined the father's evening with his friends. Each had to judge how far to push the process, without being fully sure whether and when the other might give in, or whether there would be an unpleasant scene. The risk of an explosion would increase as the father tried harder to force the daughter to answer, and as she defied each successive demand.

In this respect the game played by the father and the daughter was just like that between a union and a company's management who are negotiating a labor contract, or superpowers who are encroaching on each other's sphere of influence in the world. Neither side can be fully sure of the other's intentions, so each side explores these through a succession of small incremental steps, each of which escalates the risk of mutual disaster. The daughter in our story was exploring previously untested limits of her freedom; the father was exploring previously untested—and perhaps unclear even to himself—limits of his authority.

This was an example of brinkmanship, a game of escalating mutual risk, par excellence. Such games can end in one of two ways. In the first way, one of the players reaches the limit of his own tolerance for risk and concedes. (The father in our story conceded quickly, at the very first step. Other fathers might be more successful strict disciplinarians, and their daughters might not even initiate a game like this.) In the second way, before either has conceded, the risk they both fear comes about, and the blowup (the strike or the war) occurs. The feud in our host's family ended "happily"; although the father conceded and the daughter won, a blowup would have been much worse for both.

We will analyze the strategy of brinkmanship more fully in Chapter 9, and in Chapter 13 examine a particularly important instance of it, namely the Cuban missile crisis of 1962.

F. The Dating Game

When you are dating, you want to show off the best attributes of your personality to your date, and to conceal the worst ones. Of course, you cannot hope to conceal them forever if the relationship progresses; but you are resolved to improve, or hope that by that stage the other person will accept the bad things about you with the good ones. And you know that the relationship will not progress at all unless you make a good first impression; you won't get a second chance to do so.

Of course, you want to find out everything, good and bad, about the other person. But you know that if the other is as good at the dating game as you are, he or she will similarly try to show the best side and hide the worst. You will think and examine the situation more carefully, and try to figure out which signs of good qualities are real and which ones can easily be put on for the sake of making a good impression. Even the worst slob can easily appear well groomed for a big date; ingrained habits of courtesy and manners that are revealed in a hundred minor details may be harder to simulate for a whole evening. Flowers are relatively cheap; more expensive gifts may have value, not for intrinsic reasons, but as credible evidence of how much the other person is willing to sacrifice for you. And the "currency" in which the gift is given may have different significance depending on the context; from a millionaire, a diamond may be worth less in this regard than the act of giving up valuable time for your company or time devoted to some activity at your request.

You should also recognize that your date will similarly scrutinize your actions for their information content. Therefore you should take actions that are credible signals of your true good qualities, and not just the ones that anyone can imitate.

This is important not just on a first date; revealing, concealing, and eliciting information about the other person's deepest intentions remain important throughout a relationship. Here is a story to illustrate that.

Once upon a time in New York City there lived a man and a woman who had separate rent-controlled apartments, but whose relationship had reached the point where they were using only one of these. The woman suggested to the man that they give up the other apartment. The man, an economist, explained to her a fundamental principle of his subject: it is always better to have more choice available. The probability of their splitting up might be small, but given even a small risk, it would be useful to retain the second low-rent apartment. The woman took this very badly and promptly ended the relationship!

Economists who hear this story say that it just confirms their principle that greater choice is better. But strategic thinking offers a very different and more compelling explanation. The woman was not sure of the man's commitment to the relationship, and her suggestion was a brilliant strategic device to elicit the truth. Words are cheap; anyone can say, "I love you." If the man had put his property where his mouth was, and given up his rent-controlled apartment, that would have been concrete evidence of his love. The fact that he refused to do so constituted hard evidence of the opposite, and the woman did right to end the relationship.

These are examples, designed to appeal to your immediate experience, of a very important class of games, namely those where the real strategic issue is manipulation of information. Strategies that convey good information about yourself are called signals; strategies that induce others to act in ways that will credibly reveal their private information, good or bad, are called screening devices. Thus the woman's suggestion of giving up one of the apartments was a screening device, which put the man in the situation of offering to give up his apartment, or else revealing his lack of commitment. We will study games of information, and signaling and screening, in Chapter 12.

3 OUR STRATEGY FOR STUDYING GAMES OF STRATEGY

We have chosen several examples that relate to your experiences as amateur strategists in real life, to illustrate some basic concepts of strategic thinking and strategic games. We could continue, building a whole stock of dozens of similar stories. The hope would be that when you face an actual strategic situation, you might recognize a parallel with one of these stories, which would help you decide the appropriate strategy for your own situation. This is the *case study* approach taken by most business schools. It offers a concrete and memorable vehicle for the underlying concepts. However, each new strategic situation typically involves a unique combination of so many variables that an intolerably large stock of cases is needed to cover all of them.

An alternative approach focuses on the general principles behind the examples, and so constructs a *theory* of strategic action, namely formal game theory. The hope here is that, facing an actual strategic situation, you might recognize which principle or principles apply to it. This is the route taken by the more academic disciplines, such as economics and political science. A drawback to this approach is that the theory is presented in a very abstract and mathematical manner, without enough cases or examples. This makes it difficult for most beginners to understand or remember the theory and to connect the theory with reality afterward.

But knowing some general theory has an overwhelming compensating advantage. It gives you a deeper understanding of games, and of *why* they have the outcomes they do. This helps you play better than you would if you merely read some cases and knew the recipes for *how* to play some specific games. With the knowledge of why, you can think through new and unexpected situations where a mechanical follower of a "how" recipe would be lost. A world champion of checkers, Tom Wiswell, has expressed this beautifully: "The player who knows how will usually draw, the player who knows why will usually win."[3] This is not to be taken literally for all games; some games may be hopeless situations for one of the players no matter how knowledgeable he may be. But the statement contains the germ of an important general truth—knowing why gives you an advantage beyond what you can get if you merely know how. For example, knowing the why of a game can help you foresee a hopeless situation and avoid getting into such a game in the first place.

Therefore we will take an intermediate route that combines some of the advantages of both approaches—case studies (how) and theory (why). We will organize the subject around its general principles, generally one in each of the chapters to follow. Therefore you don't have to figure these out on your own from the cases. But we will develop the general principles through illustrative cases rather than abstractly, so the context and scope of each idea will be clear and evident. In other words, we will focus on theory, but build it up through cases, not abstractly.

Of course such an approach requires some compromises of its own. Most importantly, you should remember that each of our examples serves the purpose of conveying some general idea or principle of game theory. Therefore we will leave out many details of each case that are incidental to the principle at stake. If some examples seem somewhat artificial, please bear with us; we have generally considered the omitted details and left them out for good reasons.

A word of reassurance. Although the examples that motivate the development of our conceptual or theoretical frameworks are deliberately selected for that purpose—even at the cost of leaving out some other features of reality—once the theory has been constructed, we pay a lot of attention to its connection with reality. Throughout the book we examine factual and experimental evidence as to how well the theory explains reality. The frequent answer—very well in some respects and less well in others—should give you cautious confidence in using the theory, and a spur to contributing to the formulation of better theories. In appropriate places, we examine in great detail how institutions evolve in practice to solve some problems pointed out by the theories; note in particular our discussion in Chapter 8 of how prisoners' dilemmas arise and are solved in reality, and a similar discussion of more general collective-action problems in Chapter 11. Finally, in Chapter 13 we will examine the use of brinkmanship in the Cuban missile crisis. Such theory-based case studies, which take rich factual details of a situation and subject them to an equally detailed theoretical analysis, are becoming common in such diverse fields as business studies, political science, and economic history; we hope our original study of an important episode

[3] Quoted in Victor Niederhoffer, *The Education of a Speculator* (New York: Wiley, 1997), p. 169. We thank Austin Jaffe of Pennsylvania State University for bringing this aphorism to our attention.

in the diplomatic and military areas will give you an interesting introduction to this genre.

To pursue our approach in which examples lead to general theories that are then tested against reality and used to interpret reality, we must first identify the general principles that serve to organize our discussion. We will do this in Chapter 2 by classifying or dichotomizing games along several key dimensions of different strategic issues or concepts. Along each dimension, we will identify two extreme pure types. For example, one such dimension concerns the order of moves, and the two pure types are ones where the players take turns making moves (sequential games) and those where all players act at once (simultaneous games). Actual games rarely correspond to exactly one of these conceptual categories; most partake of some features of each extreme type. But each game can be located in our classification by considering which issues or dimensions bear on it and how it mixes the two pure types in each dimension. To decide how to act in a specific situation, one then combines in appropriate ways the lessons learned for the pure types.

Once this general framework has been constructed in Chapter 2, the chapters that follow will build on it, developing several general ideas and principles for each player's strategic choice and the interaction of all players' strategies in games.

SUMMARY

Businesspeople, politicians, diplomats, and others in the performance of their duties, as well as ordinary individuals in their family and social relationships, interact strategically with others virtually every day. Such interactions are called strategic games, and how to think rationally and choose good actions or strategies in strategic games is the subject of game theory. This book provides an introduction to the general principles and methods of calculation needed to be good game players.

Strategy is of the essence in interactions where the actions of participants alter one another's outcomes. Individuals attempt to "win" games through such strategies as the random mixing of actions, coordination, commitment to a particular move, the gradual escalation of mutual risk, and the use of signals to provide information or screens to obtain it. The general principles illustrated will be developed in context, using examples, so as to be less abstract than pure theory.

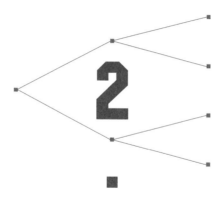

How to Think About
Strategic Games

CHAPTER 1 GAVE SOME simple examples of strategic games and strategic thinking. In this chapter we begin a more systematic and analytical approach to the subject. We choose some crucial conceptual categories or dimensions, in each of which there is a dichotomy of types of strategic interactions. For example, one such dimension concerns the timing of the players' actions, and the two pure types are games where the players act in strict turns (sequential moves), and where they act at the same time (simultaneous moves). We discuss some issues that arise in thinking about each pure type in this dichotomy, and in similar dichotomies, with respect to other matters, such as whether the game is played only once or repeatedly, and what the players know about each other.

In the chapters that follow, we will examine each of these categories or dimensions in more detail, and show how the analysis can be used in several specific applications. Of course, most actual applications are not of a pure type but involve a mixture. Moreover, in each application, two or more of the categories have some relevance. The lessons learned from the study of the pure types must therefore be combined in appropriate ways. We will show how to do this, using the context of our applications.

In this chapter we also state some basic concepts and terminology—such as strategies, payoffs, and equilibrium—that are used in the analysis, and briefly discuss solution methods. We also provide a brief discussion of the uses of game theory and an overview of the structure of the remainder of the book.

1 DECISIONS VERSUS GAMES

When a person (or team or firm or government) decides how to act in dealings with other people (or teams or firms or governments), there must be some cross-effect of their actions; what one does must affect the outcome for the other. When George Pickett (of Pickett's Charge at the battle of Gettysburg) was asked to explain the Confederacy's defeat in the Civil War, he responded, "I think the Yankees had something to do with it."[1]

For the interaction to become a strategic game, however, we need something more, namely the participants' mutual awareness of this cross-effect. What the other person does affects you; if you know this, you can react to his actions, or take advance actions to forestall the bad effects his future actions may have on you and to facilitate any good effects, or even take advance actions so as to alter his future reactions to your advantage. If you know that the other person knows that what you do affects him, you know that he will be taking similar actions. And so on. It is this *mutual awareness* of the cross-effects of actions, and the actions taken as a result of this awareness, that constitute the most interesting aspects of strategy.

This distinction is captured by reserving the label **strategic games** (or sometimes just **games,** since we are not concerned with other types of games, like those of pure chance or pure skill) for interactions between mutually aware players, and **decisions** for action situations where each person can choose without concern for reaction or response from others. If Robert E. Lee (who ordered Pickett to lead the ill-fated Pickett's Charge) had thought that the Yankees had been weakened by his earlier artillery barrage to the point that they no longer had any ability to resist, his choice to attack would be a decision; if he was aware that the Yankees were prepared and waiting for his attack, then the choice became a part of a (deadly) game. The simple rule is that unless there are two or more players, each of whom responds to what others do (or what each thinks the others might do), it is not a game.

Strategic games arise most prominently in head-to-head confrontations of two participants: the arms race between the United States and the Soviet Union from the 1950s through the 1980s, wage negotiations between General Motors and the United Auto Workers, or a Super Bowl matchup between the San Francisco 49ers and the Buffalo Bills. By contrast, interactions involving a large number of participants seem less prone to the issues raised by mutual awareness. Because each farmer's output is an insignificant part of the whole nation's or the world's output, the decision of one farmer to grow more or less corn has almost

---

[1] James M. McPherson, "American Victory, American Defeat," in *Why the Confederacy Lost*, ed. Gabor S. Boritt (New York: Oxford University Press, 1993), p. 19.

no effect on the market price, and not much appears to hinge on thinking of agriculture as a strategic game. This was indeed the view prevalent in economics for many years. A few confrontations between large companies, as in the U.S. auto market that was once dominated by GM, Ford, and Chrysler, were usefully thought of as strategic games, but most economic interactions were supposed to be governed by the impersonal forces of supply and demand.

In fact, game theory has a much greater scope. Many situations that start out as impersonal markets with thousands of participants turn into strategic interactions of two or just a few. This happens for one of two broad classes of reasons—mutual commitments or private information.

Consider commitment first. When you are contemplating building a house, you can choose one of several dozen contractors in your area; the contractor can similarly choose from several potential customers. There appears to be an impersonal market. Once each side has made a choice, however, the customer pays an initial installment, and the builder buys some materials for the plan of this particular house. The two become tied to each other, separately from the market. Their relationship becomes *bilateral*. The builder can try to get away with a somewhat sloppy job or can procrastinate, and the client can try to delay payment of the next installment. Strategy enters the picture. Their initial contract has to anticipate their individual incentives, and specify a schedule of installments of payments that are tied to successive steps in the completion of the project. Even then, some adjustments have to be made after the fact, and these bring in new elements of strategy.

Next consider private information. Thousands of farmers seek to borrow money for their initial expenditures on machinery, seed, fertilizer, and so forth, and hundreds of banks exist to lend to them. Yet the market for such loans is not impersonal. A borrower with good farming skills who puts in a lot of effort will be more likely to be successful and will repay the loan; a less-skilled or lazy borrower may fail at farming and default on the loan. The risk of default is highly personalized. It is not a vague entity called "the market" that defaults, but individual borrowers who do so. Therefore each bank will have to view its lending relationship with each individual as a separate game. It will seek collateral from, or investigate the creditworthiness of, each borrower. The farmer will look for ways to convince the bank of his quality as a borrower; the bank will look for effective ways to ascertain the truth of the farmer's claims.

Similarly, an insurance company will make some efforts to determine the health of individual applicants, and will check for any evidence of arson when a claim for a fire is made; an employer will inquire into the qualifications of, and monitor the performance of, individual employees. More generally, when participants in a transaction possess some private information bearing on the outcome, each bilateral deal becomes a game of strategy, even though the larger picture may have thousands of very similar deals going on.

To sum up, when each participant is significant in the interaction, either because each is a large player to start with, or because commitments or private information narrow the scope of the relationship to a point where each is an important player *within* the relationship, we must think of the interaction as a strategic game. Such situations are the rule rather than the exception in business, in politics, and even in social interactions. Therefore the study of strategic games forms an important part of all fields that analyze these issues.

2 CLASSIFYING GAMES

Games of strategy arise in many different contexts, and accordingly have many different features that require study. This task can be simplified by grouping these features into a few categories or dimensions, along each of which we can identify two pure types of games, and then recognize any actual game as a mixture of the pure types. We develop this classification by asking a few questions that will be pertinent for thinking about the actual game you are playing or studying.

A. Are the Moves in the Game Sequential or Simultaneous?

Moves in chess are sequential: White moves first, then Black, then White again, and so on. By contrast, participants in an auction for an oil-drilling lease or a portion of the airwave spectrum make their bids simultaneously, in ignorance of competitors' bids. Most actual games combine aspects of both. In a race to research and develop a new product, the firms act simultaneously, but each competitor has partial information about the others' progress, and can respond. During one play in football, the opposing offensive and defensive coaches simultaneously send out teams with the expectation of carrying out certain plays, but after seeing how the defense has set up, the quarterback can change the play at the line of scrimmage, or call a time-out so the coach can change the play.

The distinction between **sequential** and **simultaneous moves** is important because the two types of games require different types of interactive thinking. In a sequential-move game, each player must think: if I do this, how will my opponent react? Your current move is governed by your calculation of its *future* consequences. With simultaneous moves, you have the trickier task of trying to figure out what your opponent is going to do *right now*. But you must recognize that, in making his own calculation, the opponent is also trying to figure out your current move, while at the same time recognizing that you are doing the same with him . . . . Both of you have to think your way out of this circle.

In Chapters 3 and 4 we will study the two pure cases: sequential-move games, where you must look ahead to act now, and simultaneous-move games, where you must square the circle of "He thinks that I think that he thinks . . ." In each case we will devise some simple tools for such thinking—trees and payoff tables—and obtain some simple rules to guide actions.

Study of sequential games also tells us when it is an advantage to move first and when second. Roughly speaking, this depends on the relative importance of commitment and flexibility in the game in question. For example, the game of economic competition among rival firms in a market has a first-mover advantage if one firm, by making a firm commitment to compete aggressively, can get its rivals to back off. But in political competition, a candidate who has taken a firm stand on an issue may give his rivals a clear focus for their attack ads, and the game has a second-mover advantage.

The knowledge of the balance of these considerations can also help you devise ways to manipulate the order of moves to your own advantage. That in turn leads to the study of strategic moves, like threats and promises, which we will take up in Chapter 9.

B. Are the Players' Interests in Total Conflict, or Is There Some Commonality?

In simple games like chess or football, there is a winner and a loser. One player's gain is the other's loss. Similarly, in gambling games, one player's winnings are the others' losses, so the total is zero. This motivates the name **zero-sum games** for such situations. More generally, the idea is that the players' interests are in complete conflict. Such conflict occurs when players are dividing up any fixed amount of possible gain, whether it be measured in yards, dollars, acres, or scoops of ice cream. Because the available gain need not always be exactly zero, the term **constant-sum game** is often substituted for zero-sum; we will use the two interchangeably.

Most economic and social games are not zero-sum. Trade, or economic activity more generally, offers scope for deals that benefit everyone. Joint ventures can combine the participants' different skills and generate synergy to produce more than the sum of what they could have produced separately. But the interests are not completely aligned either; the partners can cooperate to create a larger total pie, but they will clash when it comes to deciding how to split this among them.

Even wars and strikes are not zero-sum games. A nuclear war is the most striking example of a situation where there can be only losers, but the concept is far older. Pyrrhus, the king of Epirus, defeated the Romans at Heraclea in 280 B.C., but at such great cost to his own army that he exclaimed: "Another such victory and we are lost." Hence the phrase "Pyrrhic victory." In the 1980s, at the height of the frenzy of business takeovers, the battles among rival bidders led to such costly escalation that the successful bidder's victory was often similarly Pyrrhic.

Most games in reality have this tension between conflict and cooperation, and many of the most interesting analyses in game theory come from the need to handle it. The players' attempts to resolve their conflict—distribution of territory or profit—is influenced by the knowledge that if they fail to agree, the outcome will be bad for all of them. One side's threat of a war or a strike is its attempt to frighten the other side into conceding its demands.

Even when a game is constant-sum for all players, when there are three (or more) players, we have the possibility that two of them will cooperate at the expense of the third; this leads to the study of alliances and coalitions. We will discuss and illustrate these ideas later, especially in Chapter 16 on bargaining.

C. Is the Game Played Once or Repeatedly, and with the Same or Changing Opponents?

A game played just once is in some respects simpler and in others more complicated than one with a longer interaction. You can think about a one-shot game without worrying about its repercussions on other games you might play in the future against the same person, or against others who might hear of your actions in this one. Therefore actions in one-shot games are more likely to be unscrupulous or ruthless. For example, an automobile repair shop is much more likely to overcharge a transient motorist than a regular customer.

In one-shot encounters, each player doesn't know much about the others; for example, what their capabilities and priorities are, whether they are good at calculating their best strategies or have any weaknesses that one can exploit, and so on. Therefore in one-shot games, secrecy or surprise are likely to be important components of good strategy.

Games with ongoing relationships involve the opposite considerations. You have the opportunity to build a reputation (for toughness, fairness, honesty, reliability, and so forth, depending on the circumstances) and to find out more about your opponent. The players together can better exploit mutually beneficial prospects by arranging to divide the spoils over time (taking turns to "win") or to punish a cheater in future plays (eye-for-an-eye or tit-for-tat). We will discuss these possibilities in Chapter 8, on the prisoners' dilemma.

More generally, a game may be zero-sum in the short run but have scope for mutual benefit in the long run. For example, each football team likes to win, but they all recognize that close competition generates more spectator interest, which benefits all teams in the long run. That is why they agree to a drafting scheme where teams get to pick players in reverse order of their current standing, thereby reducing the inequality of talent. In long-distance races, the runners or cyclists often develop a lot of cooperation; two or more of them can help one another by taking turns to follow in one another's slipstream. Near the end of the race, the cooperation collapses as all of them dash for the finish line.

Here is a useful rule of thumb for your own strategic actions in life. In a game that has some conflict and some scope for cooperation, you will often think up a great strategy for winning big and grinding a rival into dust, but have a nagging worry at the back of your mind that you are behaving like the worst 1980s yuppie. In such a situation, the chances are that the game has a repeated or ongoing aspect that you have overlooked. Your aggressive strategy may gain you a short-run advantage, but its long-run side effects will cost you even more. Therefore you should dig deeper and recognize the cooperative element, and alter your strategy accordingly. You will be surprised how often niceness, integrity, and the golden rule of doing to others as you would have them do to you turn out to be not just old nostrums, but good strategies as well, when you consider the whole complex of games you will be playing over the course of your life.

D. Do the Players Have Full or Equal Information?

In chess, each player knows exactly the current situation and all the moves that led to it, and each knows that the other aims to win. This situation is exceptional; in most other games, there is some piece of **information** that one player has but others don't. Then the players' attempts to infer, conceal, or sometimes convey this information become an important part of the game and the strategies.

In bridge or poker each player has only partial knowledge of the cards the others hold. Their actions (bidding and play in bridge, the number of cards taken and the betting behavior in poker) give information to opponents. Each player tries to manipulate his actions to mislead the opponents (and in bridge, to inform one's partner truthfully), but in doing so each must be aware that the opponents know this, and that they will use strategic thinking to interpret one's actions.

You may think that if you have superior information, you should always conceal it from other players. But that is not true. For example, suppose you are the CEO of a pharmaceutical firm that is engaged in an R&D competition to develop a new drug. If your scientists make a discovery that is a big step forward, you may want to let your competitors know, in the hope that they will give up their own searches. In war, each side wants to keep its tactics and troop deployments secret; but in diplomacy, if your intentions are peaceful, then you desperately want other countries to know and believe this.

The general principle here is that you want to release your information selectively. You want to reveal the good information (the kind that will draw responses from the other players that work to your advantage) and conceal the bad (the kind that may work to your disadvantage).

This raises a problem. Your opponents in a strategic game are purposive rational players and know that you are one too. They will recognize your incentive to exaggerate or even to lie. Therefore they are not going to accept your unsupported declarations about your progress or capabilities. They can be convinced only by objective evidence or by actions that are credible proof of your information. Such actions on the part of the more-informed player are called **signals,** and strategies that use them are called **signaling.** Conversely, the less-informed party can create situations in which the more-informed player will have to take some action that credibly reveals his information; such strategies are called **screening,** and the methods they use are called **screening devices.** The word *screening* is used here in the sense of testing in order to sift or separate, not in the sense of concealing. Recall that in the dating game in Section 2.F of Chapter 1, the woman was screening the man to test his commitment to their relationship, and her suggestion that the pair give up one of their two rent-controlled apartments was the screening device. If the man had been committed to the relationship, he might have signaled this by acting first and volunteering to give up his apartment; this action would have been a signal of his commitment.

Now we see how, when different players have different information, the manipulation of information itself becomes a game, perhaps more important than the game that will be played after the information stage. Such information games are ubiquitous, and to play them well is essential for success in life. We will study more games of this kind in greater detail in Chapter 12.

E. Are the Rules of the Game Fixed or Manipulable?

The rules of chess, card games, or sports are given, and every player must follow them, no matter how arbitrary or strange they seem. But in games of business, politics, and ordinary life, the players can make their own rules to a greater or lesser extent. For example, in the home, parents constantly try to make the rules, and children constantly look for ways to manipulate or circumvent those rules. In legislatures, rules for the progress of a bill (including the order in which amendments and main motions are voted on) are fixed, but the game that sets the agenda—*which* amendments are brought to vote first—can be manipulated; that is where political skill and power have the most scope, and we will address these issues in detail in Chapter 14.

In such situations, the real game is the "pregame" where rules are made, and your strategic skill must be deployed at that point. The actual playing out of the subsequent game can be more mechanical; you could even delegate it to someone else. However, if you "sleep" through the pregame, you might find that you have lost the game before it ever began. For many years, American firms ignored the rise of foreign competition in just this way, and ultimately paid the price. Others, like oil magnate John D. Rockefeller, Sr., benefitted from an unwillingness to participate in any "game" in which they could not also participate in making the rules.[2]

[2] For more on the methods used in Rockefeller's rise to power, see Ron Chernow (*Titan*, New York: Random House, 1998).

The distinction between changing rules and acting within the chosen rules will be most important for us in our study of strategic moves, such as threats and promises. Questions of how you can make your own threats and promises credible, or how you can reduce the credibility of your opponent's threats, basically have to do with a pregame that involves manipulating the rules of the subsequent game in which the promises or threats may have to be carried out. More generally, the strategic moves that we will study in Chapter 9 are essentially ploys for such manipulation of rules.

But if the pregame of rule manipulation is the real game, what fixes the rules of the pregame? Usually these depend on some hard facts related to the players' innate abilities. In business competition, one firm can take preemptive actions that alter subsequent games between it and its rivals; for example, it can expand its factory or advertise in a way that twists the results of subsequent price competition more favorably to itself. Which firm can do this best or most easily depends on which one has the managerial or organizational resources to make the investments or to launch the advertising campaigns.

Of course, players may also be unsure of their rivals' abilities. This often makes the pregame one of unequal information, requiring more subtle strategies and occasionally resulting in some big surprises. We will comment on all these issues in the appropriate places in the chapters that follow.

F. Are Agreements to Cooperate Enforceable?

We saw that most strategic interactions involve a mixture of conflict and common interest. Then there is a case to be made that all participants should get together and reach an agreement about what everyone should do, balancing their mutual interest in maximizing the total benefit and their conflicting interests in the division of gains. Such negotiations can take several rounds, in which agreements are made on a tentative basis, better alternatives are explored, and the deal is finalized only when no group of players can find anything better. The concept of the core in Chapter 17 embodies such a process and its outcome. However, even after the completion of such a process, additional difficulties often arise in putting the final agreement into practice. For instance, all the players must perform, in the end, the actions that were stipulated for them in the agreement. When all others do what they are supposed to do, any one participant can typically get a better outcome for himself by doing something different. And if each one suspects that the others may cheat in this way, he would be foolish to adhere to his stipulated cooperative action.

Agreements to cooperate can succeed if all players act immediately and in the presence of the whole group, but agreements with such immediate implementation are quite rare. More often the participants disperse after the agreement is reached, and then take their actions in private. Still, if these actions are observable to the others, and a third party, for example a court of law, can enforce compliance, then the agreement of joint action can prevail.

However, in many other instances individual actions are neither directly observable nor enforceable by external forces. Without enforceability, agreements will only stand if it is in all participants' individual interests to abide by them. Games among sovereign countries are of this kind, as are many games with private information, or games where the actions are either outside the law or too trivial or too costly to enforce in a court of law. In fact, games where agreements for joint action are not enforceable constitute a vast majority of strategic interactions.

Game theory uses a special terminology to capture the distinction between situations in which agreements are enforceable and those in which they are not. Games in which joint-action agreements are enforceable are called **cooperative,** and those in which such enforcement is not possible, and individuals must be allowed to act in their own interests, are called **noncooperative.** This has become standard terminology, but it is somewhat unfortunate because it gives the impression that the former will produce cooperative outcomes and the latter will not. In fact, individual action can be compatible with the achievement of a lot of mutual gain, especially in repeated interactions. The important distinction is that in so-called noncooperative games, cooperation will emerge only if it is in the participants' separate and individual interests to continue to take the prescribed actions. This emergence of cooperative outcomes from noncooperative behavior is one of the most interesting findings of game theory, and we will develop the idea in Chapters 8 and 10.

We will adhere to the standard usage, but emphasize that the terms *cooperative* and *noncooperative* refer to the way actions are implemented or enforced— collectively in the former mode and individually in the latter—and not to the nature of the outcomes.

As we said above, most games in practice do not have adequate mechanisms for external enforcement of joint-action agreements. Therefore most of our analytical development will proceed in the noncooperative mode. The few exceptions include our discussion of bargaining in Chapter 16, and a brief treatment of markets and competition in Chapter 17.

3 SOME TERMINOLOGY AND BACKGROUND ASSUMPTIONS

When one thinks about a strategic game, the logical place to begin is by specifying its structure. This includes all the strategies available to all the players, their information, and their objectives. The first two aspects will differ from one game to another along the dimensions discussed in the previous section, and one must locate one's particular game within that framework. The objectives raise some new and interesting considerations. Here we discuss aspects of all these matters.

A. Strategies

Strategies are simply the choices available to the players, but even this basic notion requires some further study and elaboration. If a game has purely simultaneous moves made only once, then each player's strategy is just the action taken on that single occasion. But if a game has sequential moves, then the actions of a player who moves later in the game can respond to what other players (or he himself) have done at earlier points. Therefore each such player must make a complete plan of action, for example: "If the other does A, then I will do X, but if the other does B, then I will do Y." This complete plan of action constitutes the strategy in such a game.

There is a very simple test to determine whether your strategy is complete. It should specify how you would play the game in such full detail—describing your action in every contingency—that if you were to write this all down, hand it to someone else, and go on vacation, this other person acting as your representative could play the game just as you would have played it. He would know what to do on each occasion that could conceivably arise during the course of play, without ever needing to disturb your vacation for instructions on how to deal with some situation you had not foreseen.

This test will become clearer in Chapter 3, when we develop and apply it in some specific contexts. For now, you should simply remember that a strategy is a complete plan of action.

This notion is similar to the common usage of the word *strategy* to denote a longer-term or larger-scale plan of action, as distinct from tactics that pertain to a shorter term or a smaller scale. For example, the military makes strategic plans for a war or a large-scale battle, while tactics for a smaller skirmish or a particular theater of battle are often left to be devised by lower-level officers to suit the local conditions. But game theory does not use the term *tactics* at all. The term *strategy* covers all the situations, meaning a complete plan of action when necessary, and a single move if this is all that is needed in the particular game being studied.

The word *strategy* is also commonly used to describe the decisions of an individual over a fairly long time span and a sequence of choices, even though there is no game in our sense of purposive and aware interaction with other individuals. Thus you have probably already chosen a career strategy. When you start earning an income, you will make saving and investment strategies and eventually plan a retirement strategy. This usage of the term *strategy* has the same sense as ours—a plan for a succession of actions in response to evolving circumstances. The only difference is that we are reserving it for a situation, namely a game, where the circumstances evolve because of actions taken by other purposive players.

B. Payoffs

When asked what a player's objective in a game is, most newcomers to strategic thinking respond that it is "to win"; but matters are not always so simple. Sometimes the margin of victory matters; for example, in R&D competition, if your product is only slightly better than the nearest rival's, your patent may be more open to challenge. Sometimes there may be smaller prizes for several participants, so winning isn't everything. Most importantly, very few games of strategy are purely zero-sum or win-lose; they combine some common interest and some conflict among the players. Thinking about such mixed-motive games requires more refined calculations than the simple dichotomy of winning and losing, for example, comparisons of the gains from cooperating versus cheating.

We will give each player a complete numerical scale with which to compare all logically conceivable outcomes of the game, corresponding to each available combination of choices of strategies by all the players. The number associated with each possible outcome will be called that player's **payoff** for that outcome. Higher payoff numbers attach to outcomes that are better in this player's rating system.

Sometimes the payoffs will be simple numerical ratings of the outcomes, the worst labeled 1, the next worst 2, and so on all the way to the best. In other games there may be more natural numerical scales, for example, money income or profit for firms, viewer-share ratings for television networks, and so on. In many situations, the payoff numbers are only educated guesses; then we should do some sensitivity tests by checking that the results of our analysis do not change significantly if we vary these guesses within some reasonable margin of error.

Two important points about the payoffs need to be understood clearly. First, the payoffs for one player capture everything in the outcomes of the game that he cares about. In particular, the player need not be selfish, but his concern about others should be already reflected in his numerical payoff scale. Second, we will suppose that if the player faces a random prospect of outcomes, then the number associated with this prospect is the average of the payoffs associated with each component outcome, each weighted by its probability. Thus if in one player's ranking, outcome A has payoff 0 and outcome B has payoff 100, then the prospect of a 75% probability of A and a 25% probability of B should have the payoff $0.75 \times 0 + 0.25 \times 100 = 25$. This is often called the **expected payoff** from the random prospect. The word *expected* has a special connotation in the jargon of probability theory. It does not mean what you think you will get, or expect to get—it is the mathematical or probabilistic or statistical expectation, meaning an average of all possible outcomes, where each is given a weight proportional to its probability.

The second point creates a potential difficulty. Consider a game where players get or lose money and payoffs are measured simply in money amounts. Using the above example, if a player has a 75% chance of getting nothing and a 25% chance of getting $100, then the expected payoff as calculated above is $25. That is also the payoff the player would get from a simple nonrandom outcome of $25. In other words, this way of calculating payoffs entails that a person should be indifferent to whether he receives $25 for sure or faces a risky prospect of which the average is $25. One would think that most people would be averse to risk, preferring a sure $25 to a gamble that yields only $25 on the average.

A very simple modification of our payoff calculation gets around this difficulty. We measure payoffs not in money sums, but using a nonlinear rescaling of the dollar amounts. This is called the expected utility approach, and we will discuss it in detail in the Appendix to Chapter 5. For now, please take our word that incorporating differing attitudes toward risk into our framework is a manageable task. Almost all of game theory is based on the expected utility approach, and it is indeed very useful, although not without flaws. We will adopt it in this book, but we also indicate some of the difficulties that it leaves unresolved, using a simple example in Chapter 7, Section 1.

C. Rationality

Each player's aim in the game will be to achieve as high a payoff for himself as possible. But how good is each player at pursuing this aim? This question is not about whether and how other players pursuing their own interests will impede him; that is in the very nature of a game of strategic interaction. We mean how good is each player at calculating the strategy that is in his own best interests, and at following this strategy during the actual course of play.

Much of game theory assumes that players are perfect calculators and flawless followers of their best strategies. This is the assumption of **rational behavior.** Observe the precise sense in which the term *rational* is being used. It means that each has a consistent set of rankings (values or payoffs) over all the logically possible outcomes, and calculates the strategy that best serves these interests. Thus rationality has two essential ingredients: complete knowledge of one's own interests, and flawless calculation of what actions will best serve those interests.

It is equally important to understand what is *not* included in this concept of rational behavior. It does not mean that players are selfish; a player may rate highly the well-being of some other, and incorporate this into his payoffs. It does not mean that players are short-run; in fact calculation of future consequences is an important part of strategic thinking, and actions that seem irrational from the immediate perspective may have valuable long-term strategic roles. Most importantly, being rational does not mean sharing the value system that other players, or sensible people, or ethical or moral people, would use; it means merely pursuing one's own value system consistently. Therefore, when one player carries out the analysis of how other players will respond (in a game with sequential moves) or of the successive rounds of thinking about thinking (in a game with simultaneous moves), he must recognize that the other players calculate the consequences of their choices using their own value or rating system. You must not impute your own value systems or standards of rationality to others, and assume that they would act as you would in that situation. Thus many "experts" commenting on the Persian Gulf conflict in late 1990 predicted that Saddam Hussein would back down "because he is rational"; they failed to recognize that Saddam's value system was different from the one held by most Western governments and by the Western experts.

Of course, in general each player does not really know the other players' value systems; this is part of the reason why in reality many games have incomplete and asymmetric information. In such games, trying to find out the values of others and trying to conceal or convey one's own become important components of strategy.

Game theory assumes that all players are rational. How good is this assumption, and therefore how good is the theory that employs it? At one level, it is obvious that the assumption cannot be literally true. People often don't even have full advance knowledge of their own value systems; they don't think ahead about how they would rank hypothetical alternatives, and then remember these rankings until they are actually confronted with a concrete choice. Therefore they find it very difficult to perform the logical feat of tracing all possible consequences of their and other players' conceivable strategic choices and ranking the outcomes in advance in order to choose which strategy to follow. Even if they knew their preferences, the calculation would remain far from easy. Most games in real life are very complex, and most real players are limited in their thinking and computational abilities. In some games like chess, it is known that the calculation for the best strategy can be performed in a finite number of steps, but no one has succeeded in performing it, and good play remains largely an art.

The assumption of rationality may be closer to reality when the players are regulars who play the game quite often. Then they benefit from having experienced the different possible outcomes. They understand how the strategic choices of various players lead to the outcomes, and how well or badly they themselves fare. Then we as analysts of the game can hope that their choices, even if not made with full and conscious computations, closely approximate the results of such computations. We can think of the players as implicitly choosing the optimal strategy, or behaving as if they were perfect calculators. We will offer some experimental evidence in Chapter 7 that the experience of playing the game generates more rational behavior.

The advantage of making a complete calculation of your best strategy, taking into account the corresponding calculations of a similar strategically calculating rival, is that then you are not making mistakes that the rival can exploit. In many actual situations, you may have specific knowledge of the way in which the other players fall short of this standard of rationality, and you can exploit this in devising your own strategy. We will say something about such calculations, but very often this is a part of the "art" of game playing, not easily codifiable in rules to be followed. Of course, one must always beware of the danger that the others are merely pretending to have poor skills or strategy, losing small sums through bad play, hoping that you will then raise the stakes, when they can raise the level of their play and exploit your gullibility. When this risk is real, the safer advice to a player may be to assume that the rivals are perfect and rational calculators, and to choose his own best response to them. In other words, one should play to the opponents' capabilities instead of their limitations.

D. Common Knowledge of Rules

We suppose that, at some level, the players have a common understanding of the rules of the game. In a *Peanuts* cartoon, Lucy thought that body checking was allowed in golf, and decked Charlie Brown just as he was about to take his swing. We do not allow this.

The qualification "at some level" is important. We saw how the rules of the immediate game could be manipulated. But this merely admits that there is another game being played at a deeper level, namely where the players choose the rules of the superficial game. Then the question is whether the rules of this deeper game are fixed. For example, in the legislative context, what are the rules of the agenda-setting game? These may be that the committee chairs have the power. Then how are the committees and their chairs elected? And so on. At some basic level, the rules are fixed by the constitution, or by the technology of campaigning, or by general social norms of behavior. We ask that all players recognize the given rules of this basic game, and that is the focus of the analysis. Of course, that is an ideal; in practice we may not be able to proceed to a deep enough level of analysis.

Strictly speaking, the rules of the game consist of (1) the list of players, (2) the strategies available to each player, (3) the payoffs of each player for all possible combinations of strategies pursued by all the players, and (4) the assumption that each player is a rational maximizer.

Game theory cannot properly analyze a situation where one player does not know whether another player is participating in the game, or what the entire sets of actions available to the other players are from which they can choose, or what their value systems are, or whether they are conscious maximizers of their own payoffs. But in actual strategic interactions, some of the biggest gains are to be made by taking advantage of the element of surprise and doing something your rivals never thought you capable of. Several vivid examples can be found in historic military conflicts. For example, in 1967 Israel launched a preemptive attack that destroyed the Egyptian air force on the ground; in 1973 it was Egypt's turn to create a surprise by launching a tank attack across the Suez Canal.

It would seem, then, that the strict definition of game theory leaves out a very important aspect of strategic behavior, but in fact matters are not that bad. The theory can be reformulated so that each player attaches some small probability to the situation where such dramatically different strategies are available to the other players. Of course, each player knows his own set of available strategies. Therefore the game becomes one of asymmetric information, and can be handled using the methods we will develop in Chapter 12.

The concept of common knowledge itself requires some explanation. For some fact or situation X to be common knowledge between two people, A and B, it is not enough for each of them separately to know X. Each should also know that the other knows X; otherwise, for example, A might think that B does not know X, and might act under this misapprehension in the midst of a game. But then A should also know that B knows that A knows X, and the other way around, otherwise A might mistakenly try to exploit B's supposed ignorance of A's knowledge. Of course, it doesn't even stop there. A should know that B knows that A knows that B knows, and so on ad infinitum. Philosophers have a lot of fun exploring the fine points of this infinite regress and the intellectual paradoxes it can generate. For us, the general notion that the players have a common understanding of the rules of their game will suffice.

E. Equilibrium

Finally, what happens when rational players' strategies interact? Our answer will generally be in the framework of **equilibrium.** This simply means that each player is using the strategy that is the best response to the strategies of the other players. We will develop game-theoretic concepts of equilibrium in Chapters 3 through 5, and then use them in subsequent chapters.

Equilibrium does not mean that things don't change; in sequential-move games the players' strategies are the complete plans of action and reaction, and the position evolves all the time as the successive moves are made and responded to. Nor does equilibrium mean that everything is for the best; the interaction of rational strategic choices by all players can lead to bad outcomes for all, as in the prisoners' dilemma. But we will generally find that the idea of equilibrium is a useful descriptive tool and organizing concept for our analysis. We will discuss this in greater detail later, in connection with specific equilibrium concepts. We will also see how the concept of equilibrium can be augmented or modified to remove some of its flaws and to incorporate behavior that falls short of full calculating rationality.

Just as the rational behavior of individual players can be the result of experience in playing the game, the fitting of their choices into an overall equilibrium can come about after some plays that involve trial and error and nonequilibrium outcomes. We will look at this issue in Chapters 4 and 7.

Defining an equilibrium is not hard; actually finding an equilibrium in a particular game—that is, solving the game—can be a lot harder. Throughout this book we will solve many simple games involving two or three players, each of them having two or three strategies or one move each in turn. Many people believe this to be the limit of the reach of game theory, and therefore believe that the theory is useless for the more complex games that occur in reality. That is not true.

Of course, humans are severely limited in their speed of calculation and in their patience for performing long calculations. Therefore humans can easily solve only the simple games with two or three players and strategies. But computers are very good at speedy and lengthy calculations. Many games that are far beyond the power of human calculators are easy games for computers. The level of complexity that occurs in many games in business and politics is already within the powers of computers. Even in games like chess that are far too complex to solve completely, computers have reached a level of ability comparable to that of the best humans; we discuss chess in more detail in Chapter 3.

Computer programs for solving quite complex games exist, and more are appearing rapidly. *Mathematica* and similar program packages contain routines for finding mixed-strategy equilibria in simultaneous-move games. Gambit, a National Science Foundation project led by Professors Richard D. McKelvey of the California Institute of Technology and Andrew McLennan of the University of Minnesota, is producing a comprehensive set of routines for finding equilibria in sequential- and simultaneous-move games, in pure and mixed strategies, and with varying degrees of uncertainty and incomplete information. We will refer to this project again in Chapters 3 to 5 and 7. The biggest advantage of the project is that its programs are in the public domain and can easily be obtained from its web site with the URL *http://www.hss.Caltech.edu/~gambit/Gambit.html.*

Why then do we set up and solve several simple games in detail in this book? The reason is that understanding the concepts is an important prerequisite for making good use of the mechanical solutions that computers can deliver, and understanding comes from doing simple cases yourself. This is exactly how you learned and now use arithmetic. You came to understand the ideas of addition, subtraction, multiplication, and division by doing many simple problems mentally or using paper and pencil. With this grasp of basic concepts, you can now use calculators and computers to do far more complicated sums than you would ever have the time or patience to do manually. If you did not understand the concepts, you would make errors using calculators; for example you might solve $3 + 4 \times 5$ by grouping additions and multiplications incorrectly as $(3 + 4) \times 5 = 35$ instead of correctly as $3 + (4 \times 5) = 23$.

Thus the first step of understanding the concepts and tools is essential. Without it, you would never learn to set up correctly the games you ask the computer to solve. You would not be able to inspect the solution with any feeling for whether it was reasonable, and if it was not, would not be able to go back to your original specification, improve it, and solve it again until the specification and the calculation correctly capture the strategic situation you want to study. Therefore please pay serious attention to the simple examples we solve, and the drill exercises we ask you to solve, especially in Chapters 3 through 5.

F. Dynamics and Evolutionary Games

The theory of games based on assumptions of rationality and equilibrium has proved very useful, but it would be a mistake to rely on it totally. When games are played by novices who do not have the necessary experience to perform the calculations to choose their best strategies, explicitly or implicitly, their choices, and therefore the outcome of the game, can differ significantly from the predictions of analysis based on the concept of equilibrium.

However, we should not abandon all notions of good choice; we should recognize the fact that even poor calculators are motivated to do better for their own sakes and will learn from experience and by observing others. We should allow for a dynamic process in which strategies that proved to be better in previous plays of the game are more likely to be chosen in later plays.

The **evolutionary** approach to games does just this. It is derived from the idea of evolution in biology. Any individual animal's genes strongly influence its behavior. Some behaviors succeed better in the prevailing environment, in the sense that the animals exhibiting those behaviors are more likely to reproduce successfully and pass their genes to their progeny. An evolutionary stable state, relative to a given environment, is the ultimate outcome of this process over several generations.

The analogy in games would be to suppose that strategies are not chosen by conscious rational maximizers, but instead that each player comes to the game with a particular strategy "hardwired" or "programmed" in. The players then confront other players who may be programmed to play the same or different strategies. The payoffs to all the players in such games are then obtained. The strategies that fare better—in the sense that the players programmed to play them get higher payoffs in the games—multiply faster, while the worse strategies decline. In biology the mechanism of this growth or decay is purely genetic transmission through reproduction. In the context of strategic games in business and society, the mechanism is much more likely to be social or cultural—observation and imitation, teaching and learning, greater availability of capital for the more successful ventures, and so on.

The object of study is the dynamics of this process. Does it converge to an evolutionary stable state? Does just one strategy prevail over all others in the end, or can a few strategies coexist? Interestingly, in many games the evolutionary stable limit is the same as the equilibrium that would result if the players were consciously rational calculators. Therefore the evolutionary approach gives us a backdoor justification for equilibrium analysis.

The concept of evolutionary games has thus imported biological ideas into game theory; there has been an influence in the opposite direction, too. Biologists have recognized that significant portions of animal behavior consist of strategic interactions with other animals. Members of a given species compete with one another for space or mates; members of different species relate to each other as predators and prey along a food chain. The payoff in such games in turn contributes to reproductive success and therefore to biological evolution. Just as game theory has benefitted by importing ideas from biological evolution for its analysis of choice and dynamics, biology has benefitted by importing game-theoretic ideas of strategies and payoffs for its characterization of basic interactions between animals. We have truly an instance of synergy or symbiosis.

G. Observation and Experiment

All of Section 3 to this point has concerned how to think about games, or how to analyze strategic interactions. This constitutes theory. This book will give an extremely simple level of theory, developed through cases and illustrations instead of formal mathematics or theorems, but it will be theory just the same. All theory should relate to reality in two ways. Reality should help structure the theory, and reality should provide a check on the results of the theory.

We can find out the reality of strategic interactions in two ways, one by observing them as they occur naturally, and the other by conducting special experiments that help us pin down the effects of particular conditions. Both methods have been used, and we will mention several of each in the proper contexts.

Many people have studied strategic interactions—the participants' behavior and the outcomes—under experimental conditions, in classrooms among "captive" players, or in special laboratories with volunteers. Auctions, bargaining, prisoners' dilemmas, and several other games have been studied in this way. The results are a mixture. Some conclusions of the theoretical analysis are borne out; for example, in games involving buying and selling, the participants generally settle quickly on the economic equilibrium. In other contexts the outcomes differ significantly from the theoretical predictions; for example, prisoners' dilemmas and bargaining games show more cooperation than theory based on the assumption of selfish maximizing behavior would lead us to expect, while auctions show some gross overbidding.

At several points in the chapters that follow, we will review the knowledge that has been gained by observation and experiments, discuss how it relates to the theory, and consider what reinterpretations, extensions, modifications of the theory have been made, or should be made, in the light of this knowledge.

4 THE USES OF GAME THEORY

We began Chapter 1 by saying that games of strategy are everywhere—in your personal and working life; in the functioning of the economy, society, and polity around you; in sports and other serious pursuits; in war; and in peace. This should be motivation enough to study such games systematically, and that is what game theory is about, but your study can be better directed if you have a clearer idea of just how you can put game theory to use. We suggest a threefold method.

The first use is in *explanation*. Many events and outcomes prompt us to ask: why did that happen? When the situation involves interaction of decision makers with different aims, game theory often supplies the key to understanding the situation. For example, cutthroat competition in business is the result of the rivals being trapped in a prisoners' dilemma. At several points in the book we will mention actual cases where game theory helps us to understand how and why the events unfolded as they did. This includes the detailed case study of the Cuban missile crisis from the perspective of game theory.

The other two uses evolve naturally from the first. The second is in *prediction*. When looking ahead to situations where multiple decision makers will interact strategically, we can use game theory to foresee what actions they will take and what outcomes will result. Of course, prediction for a particular context depends on its details, but we will prepare you to do this by analyzing several broad classes of games that arise in many applications.

The third use is in *advice* or *prescription:* we can act in the service of one participant in the future interaction, and tell him which strategies are likely to yield good results and which ones are liable to lead to disaster. Once again such work is context-specific, and we equip you with several general principles and techniques and show you how to apply them to some general types of contexts. For example, in Chapter 5 we will show how to mix moves, in Chapter 8 we will examine alternative ways of overcoming prisoners' dilemmas, and in Chapter 9 we will discuss how to make your commitments, threats, and promises credible.

Of course, the theory is far from perfect in performing any of the three functions. To explain an outcome one must first have a correct understanding of the motives and behavior of the participants. As we saw earlier, most of game theory takes a specific approach to these matters, namely the framework of rational choice of individual players and the equilibrium of their interaction. Actual players and interactions in a game might not conform to this framework. But the proof of the pudding is in the eating. Game-theoretic analysis has greatly improved our understanding of many phenomena, as reading this book should convince you. The theory continues to evolve and improve as the result of ongoing research. This book will equip you with the basics so you can more easily learn and profit from the new advances as they appear.

When explaining a past event, we can often use historical records to get a good idea of the motives and the behavior of the players in the game. When attempting prediction or advice, there is the additional problem of determining what motives will drive the players' actions, what informational or other limitations they will face, and sometimes even who the players will be. Most importantly, if game-theoretic analysis assumes that the other player is a rational maximizer of his own objectives when in fact he is unable to do the calculations or is a clueless person acting at random, the advice based on that analysis may prove wrong. This risk is reduced as more and more players recognize the importance of strategic interaction and think through their strategic choices or get expert advice on the matter, but some risk remains. Even then, the systematic thinking made possible by the framework of game theory helps keep the errors down to this irreducible minimum, by eliminating the errors that arise from faulty logical thinking about the strategic interaction. Also, game theory can take into account many kinds of uncertainty and incomplete information, including that about the strategic possibilities and rationality of the opponent. We will discuss a few examples of this in the chapters to come.

5 THE STRUCTURE OF THE CHAPTERS TO FOLLOW

In this chapter we introduced several considerations that arise in almost every game in reality. To understand or predict the outcome of any game, we must know in greater detail all of these ideas. We also introduced some basic concepts that will prove useful in such analysis. However, trying to cope with all of the issues at once merely leads to confusion and a failure to grasp any of them. Therefore we will build up the theory one issue at a time. We will develop the appropriate technique for analyzing that issue, and illustrate it.

In the first group of chapters, from 3 to 7, we will construct and illustrate the most important of these concepts and techniques. We will examine purely sequential-move games in Chapter 3, and introduce the techniques—game trees and rollback reasoning—that are used to analyze and solve such games. In Chapter 4 we will turn to games with simultaneous moves, and develop for them another set of concepts—payoff tables, dominance, and Nash equilibrium. In Chapter 5 we will turn to simultaneous-move games that require the use of randomization or mixed strategies. In Chapter 6 we will show how games that have some sequential moves and some simultaneous moves can be studied by combining the techniques developed in Chapters 3 through 5. Chapter 7 will then examine some deeper conceptual issues, provide a method for analyzing games of dynamic competition, discuss some more complex examples of earlier material, and develop a little general theory of mixed strategies; this is a harder chapter that can be omitted without loss of continuity.

The ideas and techniques developed in Chapters 3 through 7 are the most basic ones: correct forward-looking reasoning for sequential-move games, and equilibrium strategies—pure and mixed—for simultaneous-move games. Equipped with these concepts and tools, we can proceed to apply them to study some broad classes of games and strategies. We will do this in the block of chapters from 8 through 12. Chapter 8 is about the most famous game of them all— the prisoners' dilemma. We will study whether and how cooperation can be sustained, most importantly in a repeated or ongoing relationship. Chapter 9 is about strategies that manipulate the rules of a game, by seizing a first-mover advantage and making a strategic move. Such moves are of three kinds—commitments, threats, and promises. In each case, credibility is essential to the success of the move, and we will outline some ways of making such moves credible.

All these theories and applications are based on the supposition that the players in a game fully understand the nature of the game and deploy calculated strategies that best serve their objective in the game. Such rationally optimal behavior is sometimes too demanding of information and calculating power to be believable as a good description of how people really act. Therefore Chapter 10 will look at games from a very different perspective. Here the players are not calculating and do not pursue optimal strategies. Instead each player is tied, as if genetically preordained, to a particular strategy. The population is diverse, and different players have different predetermined strategies. When players from such a population meet and act out their strategies, which strategies perform better? And if the more successful strategies proliferate better in the population, whether through inheritance or imitation, then what will the eventual structure of the population look like? It turns out that such evolutionary dynamics often favor exactly those strategies that would be used by rational optimizing players. Thus our study of evolutionary games lends useful indirect support to the theories of optimal strategic choice and equilibrium we studied in the previous chapters.

In most of the work in the earlier chapters, each game involves two or perhaps three players. Chapter 11 will turn to situations where large populations interact strategically, games that concern problems of collective action. Each person's actions have an effect—in some instances beneficial, in others, harmful—on the others. The outcomes are generally not the best from the aggregate perspective of the society as a whole. We will clarify the nature of these outcomes and discuss some simple policies that can lead to better outcomes.

Chapter 12 studies games with uncertainty and asymmetric information. We will examine strategies for coping with risk, and even for using risk strategically. We will also study the important strategies of signaling and screening that are used for manipulating and eliciting information.

In the final group of five chapters, 13 to 17, we will take up specific applications to situations involving strategic interactions. Here we will use as needed the ideas and methods from all the earlier chapters. We will start with a particularly interesting dynamic version of a threat, known as the strategy of brinkmanship. We will elucidate its nature in Chapter 13 and apply the idea to study the Cuban missile crisis of 1962. Following brinkmanship, Chapter 14 is about voting in committees and elections. We will look at the variety of voting rules available and some paradoxical results that can arise. In addition, we will address the potential for strategic behavior not only by voters but also by candidates in a variety of election types.

Chapters 15 through 17 will look at mechanisms for the allocation of valuable economic resources: Chapter 15 will treat auctions, Chapter 16 will consider bargaining processes, and Chapter 17 will look at markets. In our discussion of auctions, we will emphasize the roles of information and attitudes toward risk in the formulation of optimal strategies for both bidders and sellers. We will also take the opportunity to apply the theory to a well-publicized set of very recent auctions, the Federal Communication Commission's electromagnetic spectrum (airwave) auctions. Chapter 16 will discuss bargaining in both cooperative and noncooperative settings. Finally, Chapter 17 will consider games of market exchange, building on some of the concepts used in bargaining theory and including some theory of the core.

All this adds up to a lot of material; how might readers or teachers with more specialized interests choose from it? Chapters 3 through 6 constitute the core theoretical ideas that are needed throughout the rest of the book. Chapters 8 and 9 are likewise important for the general classes of games and strategies they discuss. Beyond that, there is a lot from which to pick and choose. Chapter 7 goes somewhat deeper into theory and mathematics. This will appeal to those with more scientific and quantitative backgrounds and interests, but those who come from the social sciences or humanities and have less quantitative background can omit it without loss of continuity. Chapter 10 will be of greatest interest to those with interests in biology, but similar themes are emerging in the social sciences too, and students from that background should aim to get the gist of the ideas even if they skip the details. The opposite is true of Chapter 11—collective-action games occur most frequently in social interactions but have biological parallels. Chapter 12 deals with an important topic because most games in practice have incomplete and asymmetric information, and the players' attempts to manipulate information is a vital aspect of many strategic interactions. However, the concepts and techniques for analyzing information games are inherently somewhat more complex. Therefore some readers and teachers may choose to study just the examples that convey the basic ideas of signaling and screening and leave out the rest. On the other hand, courses where students have stronger quantitative backgrounds may do this chapter earlier, say immediately after Chapter 6, in view of the importance of the subject. Chapters 13 and 14 present topics from political science—international diplomacy and elections respectively—and Chapters 15 through 17 cover topics from economics—auctions, bargaining, and markets. Courses with more specialized audiences may choose a subset, and indeed expand upon the ideas we discuss.

Whether you come from mathematics, biology, economics, politics, other sciences, or from history or sociology, the theory and examples of strategic games will stimulate and challenge your intellect. We urge you to enjoy the subject even as you are studying or teaching it.

SUMMARY

Strategic games situations are distinguished from individual decision-making situations by the presence of significant interactions among the players. Games can be classified according to a variety of categories including the timing of play, the common or conflicting interests of players, the number of times an interaction occurs, the amount of information available to the players, the type of rules, and the feasibility of coordinated action.

Learning the terminology for a game's structure is crucial for analysis. Players have *strategies* that lead to different *outcomes* with different associated *payoffs*. Payoffs incorporate everything that is important to a player about a game, and are calculated using probabilistic averages or *expectations* if outcomes are random or involve some risk. *Rationality*, or consistent, behavior is assumed of all players who must also be aware of all of the relevant rules of conduct. *Equilibrium* arises when all players use strategies that are best responses to others' strategies; some classes of games allow learning from experience and the study of dynamic movements toward equilibrium. The study of behavior in actual game situations provides additional information about the performance of the theory.

Game theory may be used for explanation, prediction, or prescription in various circumstances. While not perfect in any of these roles, the theory continues to evolve; the importance of strategic interaction and strategic thinking has also become more widely understood and accepted.

KEY TERMS[3]

constant-sum game (19)
cooperative game (24)
decision (16)
equilibrium (30)
evolutionary game (32)
expected payoff (26)
game (16)
information (21)
noncooperative game (24)
payoff (26)

rational behavior (27)
screening (22)
screening devices (22)
sequential moves (18)
signaling (22)
signals (22)
simultaneous moves (18)
strategies (25)
zero-sum game (19)

EXERCISES

1. Determine which of the following situations describe games and which describe decisions. In each case, indicate what specific features of the situation caused you to classify it as you did.

   (a) A group of grocery shoppers choosing what flavor of yogurt to purchase

   (b) A pair of teenage girls choosing dresses for their prom

   (c) A college student considering what type of postgraduate education to pursue

   (d) Microsoft and Netscape choosing prices for their Internet browsers

   (e) A state gubernatorial candidate picking a running mate

2. Consider the strategic games described below. In each case, state how you would classify the game according to the six dimensions outlined in the text: whether moves are sequential or simultaneous, whether the game is zero-sum, whether the game is repeated, whether there is full information, whether the rules are fixed, and whether cooperative agreements are possible. If you do not have enough information to classify a game in a particular dimension, explain why not.

   (a) *Rock–Paper–Scissors:* On the count of three, each player makes the shape of one of the three items with his hand. Rock beats scissors, scissors beats paper, and paper beats rock.

   (b) *Roll-call voting:* Voters cast their votes orally as their names are called. In a two-candidate election, the candidate with the most votes wins.

   (c) *Sealed-bid auction:* Bidders seal their bids in envelopes for a bottle of wine. The highest bidder wins the item and pays his bid.

---

[3] The number in parentheses after each key term represents the page where that term is defined or discussed.

3. "A game player would never prefer an outcome in which every player gets a little profit to an outcome in which he gets all the available profit." Is this statement true or false? Explain why in one or two sentences.

4. You and a rival are engaged in a game in which there are three possible outcomes: you win, your rival wins (you lose), and the two of you tie. You get a payoff of 50 if you win, a payoff of 20 if you tie, and a payoff of zero if you lose. What is your expected payoff in each of the following situations:

   (a) There is a 50% chance that the game ends in a tie, but only a 10% chance that you win. (There is thus a 40% chance that you lose.)

   (b) There is a 50-50 chance that you win or lose. There are no ties.

   (c) There is an 80% chance that you lose and a 10% chance that you win or that you tie.

5. Explain the difference between game theory's use as a predictive tool and its use as a prescriptive tool. In what types of real-world settings might these two uses be most important?

PART TWO

▪

Concepts and Techniques

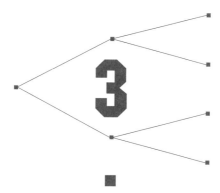

Games with Sequential Moves

SEQUENTIAL-MOVE GAMES involve strategic situations in which there is a strict order of play. Players take turns making their moves, and they know what players who have gone before them have done. To play well in such a game, participants must use a particular type of interactive thinking. Each player must consider: if I make this move, how will my opponent respond? Whenever actions are taken, players need to think about how their current actions will influence future actions, both for their rivals and for themselves. Players thus decide their current moves based on calculations of future consequences.

Most actual games combine aspects of both sequential- and simultaneous-move situations. But the concepts and methods of analysis are more easily understood if they are first developed separately for the two pure cases. Therefore in this chapter we study purely sequential games. Chapter 4 will be devoted to purely simultaneous games, and Chapters 6 and then 7 will show how to combine the two types of analysis in more realistic mixed situations. The analysis presented here can be used whenever a game includes sequential decision making. Analysis of sequential games also provides information about when it is to a player's advantage to move first and when it is better to move second. Players can then devise ways to manipulate the order of moves to their advantage.

1 A SENATE RACE AS A SEQUENTIAL-MOVE GAME

As we said earlier, when game players take turns making moves and can observe what others have done before them, then the game between them is said to be sequential. Many interactions in the business and political world have some sequential aspects to them. One common situation that arises in politics involves the entry decision of a challenger into a race—for a Senate seat, for example—and the posture taken by the incumbent Senator. In many cases, incumbents know that challengers are considering entering the race, and they make decisions—about which policies to promote or advocate against, what their campaign advertising budgets will be, and so on—in anticipation of a challenger's response. Similarly, the challengers carefully watch the behavior of incumbents in order to assess their chances for winning if they enter the race. We can tell a simple story about two such political candidates for office, one incumbent and one potential challenger, that can then be analyzed as a sequential-move game using the concepts and tools of game theory.

Suppose a particular Senate seat is currently occupied by Gray (the incumbent). It is generally known that the only likely challenger for Gray's seat in an upcoming election is Congresswoman Green (the potential challenger). In this race, Gray must determine whether to launch a preemptive advertising campaign for her seat, and Green has to decide whether to enter the race. Gray can advertise (and thereby convey to her constituents her own stellar qualities or her opponent's despicable ones) or not (and reach out to fewer constituents but have lower costs). Green will either come into the race or stay out. We assume that the order of play entails the incumbent, Gray, making a choice regarding her advertising campaign first and the potential challenger, Green, making an entry decision after observing Gray's advertising strategy.

Clearly there will be different outcomes for the candidates under different combinations of moves by Gray and Green. In this particular story, there are four possible combinations of moves: (1) Gray does not advertise and Green stays out; (2) Gray advertises and Green stays out; (3) Gray does not advertise and Green then enters; (4) Gray advertises and Green then enters. How would each candidate fare in each case?

Given Gray's incumbency, she is quite likely to retain her seat regardless of her advertising choice, and such advertising is also costly. Green's main purpose if she enters this race is to be seen to do well and to get name recognition in her party for the future. Then it follows that if Gray chooses not to advertise and Green stays out of the race, Gray's showing will be the best possible and come at the least cost, while Green has no showing and gets no recognition. (Green also makes herself look weak in this situation, by staying out when the incumbent takes a nonaggressive stance.) If Green stays out when Gray has chosen to advertise, Green again gets no reward (but she can justifiably stay out of such a hopeless situation without looking weak), while Gray's overall outcome is somewhat worse than it would have been without the advertising expense. Of course, Green may decide to enter the race. If Gray chooses not to advertise and then Green decides to enter, they each have a respectable showing, although Gray still wins the seat. For Gray, her outcome in this case is worse than when she advertises and Green stays out because she must share some of her spotlight with Green; for Green, doing respectably well in the race earns her recognition as a rising star, and this outcome is best for her. Finally, if Gray chooses to advertise and Green then enters, Green discovers that Gray's large base of loyal constituents makes it impossible to garner any votes; Green loses very badly and loses any chance at a future in politics, a very bad outcome. Even Gray is hurt by the need to compete against Green while paying for her ads (especially if they are negative); her outcome in this final situation is also the worst of all.

This information tells us how each of our players would rate the four possible outcomes. The easiest way to turn their ratings into payoffs is to associate numerical values with each outcome and to have those values increase with the attractiveness of the outcomes. Because there are four outcomes here, we use the numbers 1 through 4 to indicate how each outcome is rated, and we focus only on the orderings of those outcomes; the orderings are all that matter in the present context. This means that all similar games in which the payoffs associated with the outcomes have the same orderings as those used here will yield the same predictions for behavior. Thus our analysis is very general.

Let us consider how to order the outcomes from the perspective of each of the players. Take Gray first. The immediately preceding discussion indicated that she likes best the outcome "Gray doesn't advertise/Green stays out"; in her rating this outcome is associated with the payoff of 4. The situation "Gray advertises/ Green enters" is the worst for Gray, with a payoff of 1. Of the remaining two outcomes, Gray prefers "Gray advertises/Green stays out" to "Gray doesn't advertise/Green enters" because running unopposed (even with the cost of advertising) increases her reputation more than winning a tightly contested race in which she does not advertise. Therefore the former outcome yields a payoff of 3 in Gray's rating, and the latter outcome gets a payoff of 2.

For Green, the best situation (a payoff of 4) is "Gray doesn't advertise/Green enters." The outcome "Gray advertises/Green stays out" is next, with a payoff of 3, because Green can say to the party leaders that Gray's ads made the campaign hopeless. Then comes the outcome "Gray doesn't advertise/Green stays out." In this situation, the party leaders regard Green as a wimp and may not support her in the future, so Green associates a payoff of 2 with this outcome. Finally, "Gray advertises/Green enters" is the worst (a payoff of 1) for Green because she gets trounced in the election and irreparably damages her reputation with the public.

We can now analyze this sequential-move game. First, we show how to illustrate such a game in a usable schematic format. Then we move on to specify each of the component parts of the game that we introduced in Chapter 2. Finally, we explain how to find an equilibrium and to predict behavior in this type of game.

2 ILLUSTRATING SEQUENTIAL-MOVE GAMES

Sequential-move games, like our market-entry game, are most easily illustrated using **game trees;** these trees are generally referred to as the **extensive form** of a game. Game trees are joint decision trees for all of the players in a game. The trees illustrate all of the possible actions that can be taken by all of the players and also indicate all of the possible outcomes from the game. They are made up of **nodes** and **branches** and are often drawn from left to right across a page. However, game trees can be drawn in any orientation that best suits the game at hand: bottom up, sideways, top down, or even radially outward from a center. The tree is a metaphor, and the important feature is the idea of successive branching, as decisions are made at the tree nodes.

Nodes are connected to each other by the branches and come in two types. The first node type is called a **decision node.** Decision nodes represent specific points in the game at which decisions are made. Each decision node is associated with the player who chooses an action at that node. In addition, every tree will have one decision node that is the game's **initial node,** the starting point of the game. The second type of node is called a **terminal node.** Terminal nodes represent end points of the game. Each terminal node has associated with it a set of outcomes for the players involved in the game. These outcomes represent the payoffs received by each player if the game has followed the branches that lead to this particular terminal node. The more players in a game and the more opportunities they have to make choices, the more possible terminal nodes in the game tree.

The branches of a game tree represent the possible actions that can be taken from any decision node. Each branch leads from a decision node on the tree either to another decision node, generally for a different player, or to a terminal node. The tree must account for *all* of the possible choices that could be made by a player at each node, so some game trees include branches associated with the choice "do nothing." There must be at least one branch leading from each decision node, but there is no maximum. Every decision node can have only one branch leading to it, however.

To illustrate the Senate race game using a game tree, you need to consider the components of the game. The two players are Gray and Green. Gray is the first to choose an action in the game; she decides between no advertising (No

Ads) and advertising (Ads). Green makes her decision after Gray; Green decides between coming into the race (In) and staying out (Out). The tree for this game must then have three decision nodes, one for Gray to start the game and two for Green following upon each of the two actions that could be taken by Gray. Each of these decision nodes will have two branches leading from it, representing the choices made at each node. Then, with Green's two possible actions at each of her decision nodes, the tree will have four terminal nodes. Each terminal node will indicate the payoffs for each player.

The tree for this game is illustrated in Figure 3.1. The initial node, node *a*, is labeled to indicate that Gray makes her decision there. Similarly, nodes *b* and *c* are labeled Green. Each branch is also labeled with an action, Ads or No Ads at node *a*, and In or Out at nodes *b* and *c*. The terminal nodes show the payoffs to Gray and Green, in that order. The payoffs in this case indicate how each player rates the outcome associated with each terminal node. Figure 3.1 includes a column heading above the payoffs to indicate the order in which they are listed. Generally, the payoff to the player who moves first is listed first. In a game with many players, payoffs are usually listed in the order in which players made their first moves.

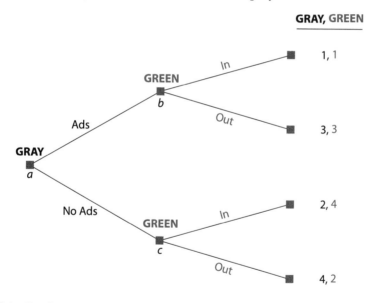

FIGURE 3.1   Tree for Senate Race Game

3 DESCRIBING STRATEGIES

Section 3 of Chapter 2 discussed the component descriptors of a game of strategy. These descriptors include the players, the information they possess about the play of the game, their actions, strategies and payoffs, and the different possible outcomes of the game. The tree in Figure 3.1 illustrates a game with two *players*, each of whom has full and complete *information* about what she can do and what her opponent has done (or can do). In addition, each player has two possible *actions* at each node, and *payoffs* are listed at the four terminal nodes that show the four possible *outcomes* of the game. The only information that is not contained in the game tree in Figure 3.1 pertains to *strategies* for the two players in the game.

Recall that strategies are rules telling players which action to choose at each possible step of the game. A strategy, in general, maps out a plan of attack for each player under all eventualities. We define a **pure strategy** as a rule that tells a player what choice to make at each of her possible decision nodes in the game. A pure strategy is then a detailed plan telling each player what action to take under every contingency. In a sequential-move game, describing all of the pure strategies available to the player who moves first is a relatively simple task. However, describing all of the pure strategies available to players moving later in the game or for those who move more than once gets more complicated. The complication arises from the fact that each pure strategy for a player must identify the actions made by other players who have moved previously and must describe a set of actions for a player's move as contingent on those previous moves.

Consider, again, the game described in Figure 3.1. Gray has one decision node in this game at which she can choose either No Ads or Ads. She has two pure strategies from which to choose. Gray's first pure strategy is to choose not to advertise; her second pure strategy is to choose to advertise.

The description of pure strategies for Green is not as simple. Green makes only one move in the game when it is actually played, but she has two decision nodes that might be reached. Green's set of pure strategies must include rules for her behavior at both of her decision nodes. Each rule must specify an action for Green at both of her nodes, given the previous action by Gray *and* there must be enough rules to cover all of the possible combinations of play on the part of both players. There are then four possible pure strategies for Green. She can choose, at the start of the game, to pursue any of the following decision rules: (1) if Gray chooses Ads, then choose In, and if Gray chooses No Ads, then choose In; (2) if Gray chooses Ads, then choose Out, but if Gray chooses No Ads, then choose In; (3) if Gray chooses Ads, then choose In, but if Gray chooses No Ads, then choose Out; (4) if Gray chooses Ads, then choose Out, and if Gray chooses No Ads, then choose Out. These four pure strategies for Green can also be described in shorthand as four sets of actions, one action per decision node, starting with the decision node at the top of the tree and working down. (For trees with no obvious "top," one must clearly state the order in which the actions are listed.) In shorthand, Green's four strategies are: (1) In, In; (2) Out, In; (3) In, Out; and (4) Out, Out. Think of these as detailed instructions given by Green to her administrative assistant. In Chapter 2, Section 3.A, we said that a strategy had to be a plan of action so complete that you could hand it to someone else and instruct them to carry it out on your behalf; this illustrates the idea.

Obviously, the description of pure strategies becomes more complex when players make decisions at a number of nodes over the course of a game or when there are multiple players in a game. We will consider such games, and the pure strategies of the players in them, later in this chapter.

Now, however, we need to turn our attention to determining equilibrium outcomes in sequential-move games. How do we expect players to behave in sequential games? What is the best strategy for each to employ? In the case of the Senate race game, which pure strategy is chosen by Gray and which pure strategy is chosen by Green? To answer such questions about the actual play of a sequential-move game, we need to think carefully about how to predict behavior in such games. To reach an *equilibrium outcome* requires that players choose actions that are best responses to the actions of the other players. In a sequential-move game, players choose pure strategies that are contingent on opponents' behavior. The contingent nature of the sequential-move game means that players must look ahead and reason back to determine which action is best in any given situation or at any given node of the game tree.

4 ROLLBACK EQUILIBRIUM

The concept of looking ahead and reasoning back is captured in the methodology used to determine behavior in sequential-move games. This methodology is known as **rollback.** As the name suggests, using rollback involves starting to think about what will happen at all the terminal nodes, and literally "rolling back" through the tree to the initial node as you do your analysis. As this reasoning involves working backward one step at a time, the method is also called **backward induction.** But *rollback* is simpler and is becoming more widely used. We will follow this usage, but in other sources on game theory you will sometimes find the older *backward induction.* Just remember that the two are equivalent.

The best way to learn how to use rollback is to look at a game tree like the one in Figure 3.1. There, the top two terminal nodes show the outcomes for Green's choice of entering the race (In) or not (Out) following Gray's choice regarding whether to advertise. If Green chooses In, she achieves her worst outcome (1); if she chooses Out, she achieves her second-best outcome (3). Clearly it is better for Green to choose Out at node *b*. Rollback tells us to identify Out as Green's optimal action at this node and to identify In as an action that would never be taken in equilibrium. Similarly, Green's optimal action at node *c* is In. Choosing In when Gray does not advertise gives Green her highest payoff of 4 from node *c*, while choosing Out gives her only 2.

(a) Pruning at terminal nodes

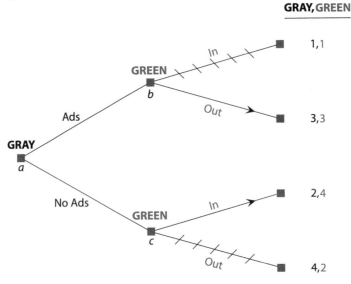

(b) Fully pruned tree

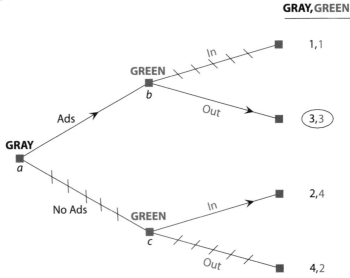

FIGURE 3.2 Using Rollback

Many game theorists identify the optimal actions at each node by drawing an arrow along the branch from that action, as shown in Figure 3.2a. Branches that are not marked with arrows are considered to have been **pruned** from the tree. Our analysis of Green's options shows that the bottom branch from node *b* should be marked with an arrow, while the top branch at that node should be pruned from the tree. From node *c*, the top branch should be marked with an arrow while the bottom branch is pruned. In general, players are not expected to take actions associated with pruned branches. Thus any terminal nodes of the game that are reached by following pruned branches are ruled out as possible end points of the game. Outcomes associated with pruned branches, such as the first and fourth outcomes in Figure 3.1, are expected not to occur.

To continue analyzing this game using rollback, it is necessary to step back to the previous set of decision nodes. In this case, there is only one remaining node, the one at which Gray decides whether she should advertise. Gray must compare the outcomes associated with No Ads and Ads in order to decide which choice gives her a better payoff from the game. Because the tree has already been pruned, Gray knows that choosing Ads will lead to Green choosing Out and leave Gray with a payoff of 3. Choosing No Ads leaves Gray with a payoff of 2, because Green is sure to choose In from node *c*. The optimal action for Gray at node *a*, then, is to choose Ads. The top branch from node *a* should be marked with an arrow and the bottom branch should be pruned from the tree. The fully pruned tree is illustrated in Figure 3.2b, and the remaining branches define the optimal strategies for each player, leading us to the **rollback equilibrium** outcome of the game; the equilibrium set of payoffs is shown circled.

The rollback equilibrium is described by the set of strategies that lead to it. In this game, the analysis shows that Green's best pure strategy is "stay out if Gray advertises, and go in if Gray does not" (Out, In) and Gray's best pure strategy, given the predicted actions of Green, is to use Ads. Thus the equilibrium is (Ads; Out, In). This collection of strategies leads to the outcome at the second terminal node of the tree at which Gray and Green both receive their second-best payoffs of 3. Rollback allows us to reach this outcome simply by following the arrows on the tree branches starting from the initial node. At node *a*, we follow the arrow up the Ads branch and at node *b* we follow the arrow down along the Out branch. This simple procedure takes us right to the predicted equilibrium outcome of the game in which Gray advertises and Green does not enter.[1]

The complete set of arrows, on one hand, and the full pruning, on the other, convey the same information in precisely opposite ways. The complete set of arrows shows what *will* be chosen at each node, while the full pruning shows what *will not* be chosen at each node. In Figure 3.2b we showed both, but we could have shown one without the other and not lost any information. But while the two have the same information content, arrows have a twofold practical advantage: (1) they produce a cleaner picture, and, more importantly, (2) they show the outcome of the rollback equilibrium visibly as a continuous link of arrows from the initial node to a terminal node. Therefore in the future we will generally use arrows instead of

---

[1] This example, often referred to as the *entry deterrence game*, arises in the economic context of market entry. Played in its sequential form as described, an incumbent can deter entry by choosing the appropriate strategy.

pruning. When you draw game trees, for a while you should show both; once you have enough practice, you can choose either to suit your taste.

There are three issues that should be addressed here in relation to the equilibrium outcome found by rollback. First, notice that the actual equilibrium play of a sequential-move game traces a path through the game tree that misses most of the branches and nodes. Calculating the best actions that would be taken if these other nodes were reached, however, is an important part of determining the ultimate equilibrium. Choices made early in the game are affected by players' expectations of what would happen if they chose to do something other than their best action and by what would happen if any opposing player chose to do something other than what was best for her. These expectations, based on predicted actions at out-of-equilibrium nodes (nodes associated with branches pruned during the process of rollback), keep players choosing optimal actions at each node.

For instance, Green's optimal pure strategy in the entry game has her choosing to stay out of the market when Gray advertises but to enter when Gray does not. In the equilibrium play of the game, Gray chooses to advertise, so it would seem not to matter what Green would choose to do at node c. This reasoning is faulty, however. It is actually extremely important to specify Green's action at node c because her behavior at that node is critical to Gray's decision regarding her play at node a. Green could have chosen the strategy that entailed staying out at node b *and* at node c—that is, Out, Out.

Recall that a strategy is a complete plan of action. Green might have made such a complete plan, for example, in a discussion with family or friends. If Gray knows that Green is following this strategy, say because of a leak from one of Green's friends, then Gray should anticipate that she will get her best outcome with a payoff of 4 if she does not advertise but a payoff of only 3 if she chooses to advertise. Gray would then choose *not* to advertise, and the outcome of the game would change. The new outcome with Gray using no ads and Green staying out would give Green a lower payoff than she obtained in the original equilibrium (2 rather than 3), primarily because she chose a strategy that did not incorporate her best action at every one of her possible decision nodes in the game. Thus it is vital that players consider all nodes of the game in determining their equilibrium pure strategies.

Second, notice that in order to use rollback successfully, players must be able to identify every possible outcome associated with every possible action available at every possible node. The necessity of such perfect or complete information should be clear. Without such knowledge, a player might not be able to identify her own best choice at a particular node or she might not be able to identify her opponent's best choice(s) at later nodes. In either case, she would be forced to make a decision without being able to predict the outcome of that decision, so there would always be a chance she might choose an action detrimental to her own interests. We assume when using rollback that such situations do not arise because players are completely informed about the order of play, the available actions, and the consequences of those actions.

Using the Senate race example again, we see that Gray's ability to choose her optimal action at node a depends on her knowing what the game tree looks like from nodes b and c. If Gray could not identify the outcomes associated with the various terminal nodes of the tree, she would not be able to predict, using rollback, which actions would be chosen by Green. If Gray could not predict Green's actions, she would be unable to identify her own best course of action at node a and she would be forced to choose blindly between the two alternatives. Under such circumstances, Gray might choose the action that would yield her the lower of the two possible outcomes, a choice that would not be made in the presence of complete information. The assumption that all players possess perfect knowledge of the game is crucial for finding the rollback equilibrium of a game.

Third, when using rollback a player must figure out what other players will do at future nodes using the payoff scale or the value system of those players, and must not impute to them her own value system. The question is not "What would I do if I were in that situation?" but "What is the other player going to do in that situation?" Of course, players often cannot be sure of others' value systems. That is why games often have incomplete or asymmetric information, and we will study some such games in Chapter 12.

5 A SEQUENTIAL-MOVE GAME WITH ONLY "ONE" PLAYER

When we think of a game, we generally expect it to have at least two players interacting with each other. However, many individuals face situations over the course of their lives in which they must choose from a variety of strategic paths. The most obvious, and probably the most important, decision that individuals make is choosing a career. This choice involves several stages, and at each stage some choices open up or are foreclosed. Can such a situation be represented as a game too, even though there is only one player?

It turns out that the career decision and many other similar one-player strategic situations can be described as games if we recognize that future choices are actually made by a different player. That other player is your future self who will be subject to different influences and will have different views about the ideal outcome of the game. By considering your future self as a rival player in the career choice decision-making process, you can turn what looks like a simple decision into a game.

Put yourself in the position of someone at an early stage of this process, a college junior or senior considering your educational choices after graduation.

In principle there are thousands of possibilities, including a doctorate in deconstructionist literary criticism, and a master's in home economics. But for today's ambitious young Americans (and increasingly, Europeans and Asians too) only three choices are worth considering: medical school, law school, and business school. So let us build your game tree with these three branches leading out of the initial node.

You cannot know which of these branches to take without looking ahead to where each will lead eventually. Here the possibilities start to multiply unmanageably. To keep the discussion simple, we show, in Figure 3.3, only a few prominent ones.

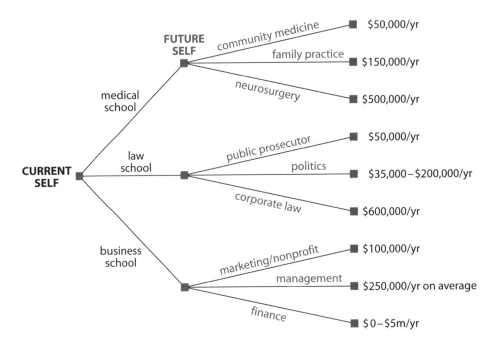

FIGURE 3.3 "One"-Player Game

After medical school you can work in community medicine for $50,000 a year (and a warm glow in your heart), become a family physician earning $150,000, or a neurosurgeon earning $500,000. After law school you can become a public prosecutor at $50,000 a year. Or you can go into politics, take up an entry-level position at $35,000 a year (as governor of Arkansas) and hope to rise to $200,000 (as President of the United States). Or you can become a corporate lawyer at $600,000 a year. There are three possibilities after business school too. You can take your marketing skills to the nonprofit sector and earn $100,000. Or you can go into management, and earn $250,000 on average—zero if the rival company's managers have understood strategic thinking better than you have, and $500,000 if you are the better strategist. Or, you can go into high finance, hoping to earn several million a year, but also risk losing everything. To be concrete, suppose your earnings in this activity can range from $0 to $5 million. These career ideas and earnings estimates help you fill in the full details of the game tree facing you.

You could consider the tree as a decision tree only, in which all the choices made at every node of the tree are made by your current self. When you use rollback to determine your path through the tree, you might find the decisions relatively obvious. Suppose that you are a twenty-year-old idealist, who after medical school would choose community medicine so you can serve the neighborhood in which you grew up. (Such a choice implies that monetary rewards are not your primary goal in choosing a career and that you include your moral beliefs when determining the outcomes associated with your choices.) If you go to law school, your idealism would lead you to the public prosecutor's office in the hopes of contributing to the general reduction of crime on city streets. After business school, your best opportunity to pursue a publicly minded career is through work at a nonprofit corporation. If it has always been your dream to be a community doctor, you would see your equilibrium path through the tree as leading to medical school and then community medicine.

But now consider how the situation looks if you describe it as a game against your future self, which is likely to be a more accurate rendition of reality. You should recognize that the postprofessional school choice is going to be made by a different you: a twenty-six year old with (at least) $150,000 of accumulated debt, to whom a large income has become essential. This future you is going to choose neurosurgery after medical school. Therefore the current you should use rollback and realize that an ultimate career in community medicine is not really an option. You now have to compare neurosurgery with the final outcomes of the other professions, where those final outcomes are chosen by your future self. After law school the last of the three choices may also look the best to the future you, unless you expect to be someone who really, really enjoys hearing "Hail to the Chief" hundreds of times. And finally, after business school the future you may also be most tempted by the final of the three choices, depending on how you assess the risks and rewards.

Now the situation is clear for your initial choice: medical school leading to neurosurgery, law school leading to corporate law, or business school leading to the world of high finance. Of these, law seems better than medicine: more money, and less risk of malpractice suits. Therefore the choice is between corporate law and the business of high finance, and this game comes down to your current preferences between the two. We leave this one open for you to decide for yourself. But the important general point we want you to understand is that, when you consider your career decision from a strategic perspective, it is not a decision but a game. Even though only one player appears to be involved, that one player may have different preferences at different stages of the decision-making process, making her actually two (or more) different players. Thus we can describe many one-player strategic situations as games if we recognize that players can change over time.

Notice that we had to simplify the true problem of career choice to tell our story here. We began by limiting the choices at the initial node to the three popular professional schools. Even within each of these professions, we considered only three alternatives and left out others. For example, after law school we could have allowed a fourth kind of career, namely a private lawyer in an affluent suburb: wills, mortgages, divorces, $175,000 per year, and golf most afternoons. What we used as terminal nodes really aren't: The political career track entails many further decisions and much uncertainty about final outcomes and payoffs.

In most situations you face, you can construct a tree in varying degrees of detail. How much detail is adequate to give you a good understanding of your choice, without making the tree so complex that you can't see your way through it at all, is something that can be learned by experience. This is one aspect of strategy that is more art than science, and we return to the question of complexity and simplification of trees in Sections 7 and 8.

6 ORDER ADVANTAGES

Sequential-move games have the important characteristic that there is always a player who moves first. By default, there is then also always a player who moves second (and third and fourth and so on, if the game is large enough). Many young children are obsessed with making sure that they get to be the player that goes first in any and every game that they play. Is this strategy really a sound one for all strategic games?

In the games we have analyzed so far, the ability to move first does seem to be in a player's best interest—there is evidence of a **first-mover advantage.** The Senate race game, for example, is one in which the ability to *set the stage* for a later player is beneficial to the first mover. Using rollback, Gray can look forward to Green's optimal responses and make a first move that is to her own advantage.

We can see that this ability to set the stage in the Senate race game is to a player's advantage by considering the outcome of the game when Green, rather than Gray, goes first. If Green makes her entry decision and then Gray makes her advertising choice, the game tree and its outcomes would be illustrated as in Figure 3.4. Rollback shows that the equilibrium in this reordered game has Green choosing to enter and then Gray choosing not to advertise. (Gray's full rollback equilibrium strategy is No Ads, No Ads.) The new equilibrium outcome gives Green her best payoff of 4 and Gray only her third-best payoff of 2. In the same way that Gray could guarantee herself one of her two best outcomes when she

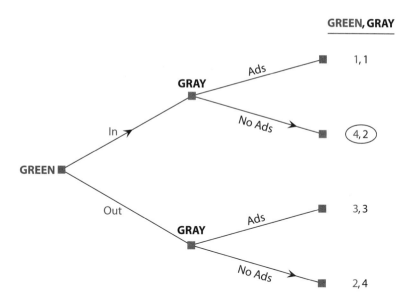

FIGURE 3.4 Change of Move Order in the Senate Race Game

set the stage, so can Green now guarantee herself her best outcome. The first-mover advantage lies in being able to set the stage appropriately.

Many sequential-move games do have first-mover advantages, but there is no rule saying that going first is always better. In some situations, players may benefit from being able to react to an earlier move. When reaction, or moving second, is beneficial, there is a **second-mover advantage.** Suppose, for example, that you and a friend have an entire chocolate cake to share between you and a knife with which to split it. Neither of you is willing (for presumably obvious reasons) to allow the other both to cut the cake and to decide who gets which piece. Therefore you decide that one of you will cut and the other will determine the allocation of the two pieces. In the ensuing game, the cutter moves first and the allocator moves second. It should be clear that the allocator has an advantage in this game; she can carefully observe the cutter's handiwork and make an informed decision about which piece of cake should be hers. The second-mover advantage can be credited to the ability to *react* to the other player's action in this case.

In other forms of competition too, there can be second-mover advantage. Consider two firms who make competing products and are setting the prices. If you knew the price your competitor had set, you could undercut it a little and divert a lot of customers away from the rival's product. Thus the second mover can do better. And in another version of a political campaign game, if you can get your rival candidate to commit to a particular policy position, you can target your negative ads more effectively, and choose your own position to appeal to a majority of voters.

Of course, sometimes which player has the advantage in a game will depend on the specifics of the particular play of a game or the special skills of the players involved. For instance, a game in which the first or second mover could win, and in which neither first nor second is always better, is the game of matchsticks. In matchsticks, two players take turns removing matchsticks from a pile The last player to remove a matchstick wins. Winning this game as the first or second mover depends much less on your strategic skills than on the number of matchsticks in the pile. If there are an even number of matchsticks, the second mover will remove the last stick and win the game. With an odd number of matchsticks, the first mover wins. There is no advantage here in being able to react to your opponent, and there is no advantage to being able to set the stage for later players. It is simply the setup of the game that determines who wins any particular play of the game.

Players may also find that they have special skills (or lack thereof) that make it better for them to go first (or second) in certain games. Consider the cake-cutting game again. Clearly, the first person wants to cut the cake as evenly as possible to ensure she ends up with close to half. If you are playing this game and your surgical skills leave something to be desired, you might do better going second, as we suggested above. However, if you are playing this game and your surgical and spatial skills are excellent, you might prefer to go first. A circle cut out of the middle of the cake must have a radius approximately 70% as big as the radius of the full cake to give it half the area of the full cake. Cut exactly, the center circle creates an optical illusion, looking like more than half the cake. If you could cut a center circle that looked like more than half the cake but was actually less than half, you would do better by going first.

7 ADDING MORE PLAYERS

The analysis of sequential-move games presented in the context of a two-player game in which there are two choices for each player can easily be extended to a variety of more complex strategic situations. In this section, we consider games with more than two players. The most obvious difference between two- and, say, three-player games is the complexity of the game tree. For each player to have the opportunity to move just once, the tree must have at least three levels. As before, the number of available choices at each node adds to the size of the tree as well.

A simple example of a three-player sequential-move game involves three major department stores contemplating whether to locate in a particular metropolitan area. Two new shopping malls are opening. Urban Mall is located close to the large and rich population center of the area; Rural Mall is farther out in a relatively rural and poorer area. Being in a more densely populated area, Urban Mall is smaller in size and can accommodate at most two departments stores as anchors for the mall. Rural mall can accommodate up to three anchor stores.

Three department-store companies—Big Giant, Titan, and Frieda's—are considering locating in one of these two malls. None of these three wants to have its store in both malls because there is sufficient overlap of customers between the malls that locating in both would just be competing with oneself. Each store prefers to be in a mall with one or more other department stores over being alone in the same mall, because a mall with multiple department stores will attract sufficiently many more total customers that each store's profit will be higher. Further, each store prefers Urban Mall to Rural Mall because of the richer customer base. Each store must choose between trying to get a space in Urban Mall (knowing that if the attempt fails they will try for a space in Rural Mall) and trying to get a space in Rural Mall directly (without even attempting to get into Urban Mall).

In this case, the stores rank the five possible outcomes as follows: 5 (best), in Urban Mall with one other department store; 4, in Rural Mall with one or two other department stores; 3, alone in Urban Mall; 2, alone in Rural Mall; and 1 (worst), alone in Rural Mall after having attempted to get into Urban Mall and failed, by which time other nondepartment stores have signed up the best anchor locations in Rural Mall.

The three stores are sufficiently different in their managerial structures that they experience different lags in doing the paperwork required to request an expansion of space in a new mall. Frieda's, by virtue of its relatively small size and efficient managerial staff, can make its request very early, before the other stores are ready to do so. Titan is the most efficient in setting up new stores but the least efficient in readying a location plan, so it moves last. Big Giant makes its move in between the other two, before Titan but after Frieda's. Once all three have made their requests, the malls decide which stores to let in. Because of the name recognition that both Big Giant and Titan have with the potential customers, a mall would take either (or both) of those stores before it took Frieda's.

This structure leads to a game tree for the mall location game as illustrated in Figure 3.5. Note that the final "move" by the malls, which simply follows a particular rule for accepting stores, is not shown explicitly; the malls do not have any real strategic role in the game. What we see in the tree is that, at the initial node of the tree, Frieda's chooses between going for Urban Mall or Rural Mall. Big Giant, which moves second, then decides whether to try for Urban Mall or Rural Mall from either node b or node c, depending on Frieda's initial choice. Finally, Titan moves third and chooses to try for Urban Mall or Rural Mall from node d, e, f, or g, depending on the *combination* of choices made earlier by Frieda's and Big Giant. At the eight terminal nodes of the tree, outcomes show the payoffs to each

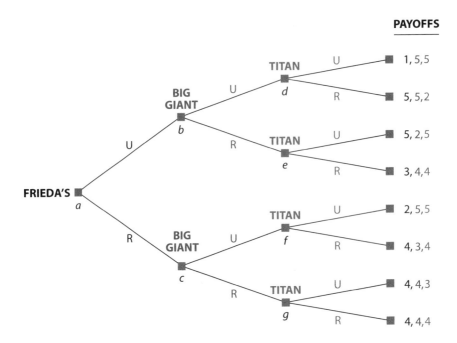

FIGURE 3.5 Three-Player Game Tree

department store, given the malls' decision rule for accepting store location requests, in the order in which they moved.

One aspect of the increased complexity of this three-person game can be seen in the game tree itself. There are more nodes and more branches than in the trees we have previously drawn. But a more interesting, and more intricate, aspect of the game's complexity involves the description of the pure strategies available to the players. Titan, in particular, has a large number of contingencies to account for when setting out all of its pure strategies.

Frieda's and Big Giant have pure-strategy sets similar to those for Gray and Green in the Senate race example from Section 1. Frieda's (like Gray) has two pure strategies: "request a location in Urban Mall" or "request a location in Rural Mall." Big Giant (like Green) has four possible pure strategies, which are expressed as contingent rules based on Frieda's earlier behavior: (1) "try for Urban Mall if Frieda's does, and try for Urban Mall if Frieda's tries for Rural Mall directly" (Urban, Urban, or more simply UU); (2) "try for Urban if Frieda's does, and try for Rural if Frieda's does" (Urban, Rural, or UR); (3) "try for Rural if Frieda's tries for Urban, and try for Urban if Frieda's tries for Rural" (Rural, Urban, or RU); or (4) "try for Rural if Frieda's tries for Urban, and try for Rural if Frieda's does so" (Rural, Rural, or RR).

When we get to Titan, we must add a second level of contingencies to its pure strategies. Not only must Titan's strategies specify what it does, given what Frieda's has done; they must also specify its choices, given what Big Giant has done after Frieda's. Remember also that each of Titan's pure strategies must specify an action for the store at each of its four decision nodes, so each of its pure strategies will have four components. It turns out that there are 16 pure strategies for Titan.

We use the same type of shorthand notation as that just described to set out Titan's strategies. Each strategy is a four-letter sequence. Each of the letters can be either U or R, and represents Titan's planned action at one of the nodes *d* through *g*. The first letter is the planned action at node *d*, the second at node *e*, and so on. For example, URUU is a particular strategy for Titan that consists of choosing R at node *e* and U at each of the nodes *d*, *f*, and *g*. Choosing one of two possibilities for each of the four letters makes a total of 16 combinations, and these are Titan's 16 strategies. Listed in full, they are (1) UUUU, (2) UUUR, (3) UURU, (4) URUU, (5) RUUU, (6) UURR, (7) URRU, (8) RRUU, (9) URUR, (10) RURU, (11) RUUR, (12) URRR, (13) RURR, (14) RRUR, (15) RRRU, and (16) RRRR.[2] You should notice that these 16 strategies include one in which Titan never tries directly for the Rural Mall (strategy 1), four in which it tries for Rural Mall at exactly one node (strategies 2 to 5), six in which it tries for Rural Mall at exactly two nodes (strategies 6 to 11), four in which it tries for Rural Mall at exactly three nodes (strategies 12 to 15), and one in which it always tries for Rural Mall directly (strategy 16). Mathematically, the number of pure strategies for Titan comes from computing 2 to the power 2 to the power 2 (2^{2^2}), which equals 2 to the fourth power (2^4), or 16.[3]

---

[2]Although Titan has 16 distinct strategies, several of them are strategically equivalent in the context of this particular game. What really matters to Titan in the history of previous actions is the number of Us and Rs, not the identity of the players. Thus, it does not matter to Titan whether it is at node (e) or (f) because each is reached by one prior U and one prior R; the store's payoffs from a U and from an R are the same at each node. Given this fact, strategies 3 and 4 are essentially equivalent, as are 6 and 9, 8 and 10, and 13 and 14.

[3]Some general rules can be established for determining the number of pure strategies available to players in a sequential game. These rules depend on how many choices each player has at each node, whether all players have the same number of choices, and whether the number of choices at each node for a single player is always the same. For games that have simple combinations of numbers of choices, the rules are relatively easy to derive.

Consider the rules for two relatively simple types of sequential-move games. The first type of game has *n* players, all of whom have the same number of choices available at each decision node in the game, and we call that number of choices *b*, for branching factor. (The mall location example fits this description. There are three department stores, each of which has two choices at each possible node, so $n = 3$ and $b = 2$.) To determine the number of pure strategies available to player *n* in such a game, you compute *b* to the power *b* to the power $(n-1)$. (Titan has 2 to the power 2 to the power 2 pure strategies as noted above.)

The second type of simple sequential-move game also has *n* players. Each player has the same number of choices available at each of her decision nodes, but that number may differ from the number of choices available to other players at each of their decision nodes. In this case, b_1 equals the number of choices available to the first player at each of her decision nodes, b_2 equals the number of choices available to the second player at each of her decision nodes, and so on. Then the number of pure strategies available to player *n* is computed as b_n to the power that equals the product of b_1, b_2, ... up to b_{n-1}. (If $n = 1$, the power is 1.)

More complex rules can be developed for more complex games.

We can now find the rollback equilibrium and the equilibrium strategies for each player. Despite the complexity of the strategy sets available, the process of rollback is still simple. At node *d*, Titan compares a payoff of 5 when it tries for Urban with a payoff of 2 when it tries directly for Rural, so its best option is Urban. At node *e*, Urban gets Titan a payoff of 5 while Rural gets it 4, so it tries for Urban. At node *f*, Titan again gets a 5 with Urban and a 4 with Rural, so it chooses Urban. Finally, at node *g*, Urban results in a 3 while Rural gets a 4, so Titan chooses Rural. Notice that collecting these four actions shows that Titan's equilibrium pure strategy is number 2: UUUR. Big Giant's optimal choices can now be determined. At node *b*, Urban leads to an ultimate payoff of 5 for Big Giant (given that Titan tries for Urban from node *d*) and Rural leads to a payoff of 2 (given that Titan chooses Urban from node *e*). Therefore Big Giant chooses Urban from node *b*. From node *c*, Big Giant's choices yield it 5 if it chooses Urban and 4 if it chooses Rural, so it tries for Urban from that node also. Big Giant's equilibrium strategy is to always try for Urban—that is, its equilibrium strategy is UU. The last step of rollback shows that Frieda's gets a payoff of 1 if it chooses Urban but a payoff of 2 if it chooses Rural from the start. Frieda's equilibrium strategy is to try Rural first. Following the designated branches through the tree takes us to the fifth terminal node at which Big Giant and Titan have chosen the Urban Mall and Frieda's has tried for Rural Mall directly.

The set of rollback equilibrium strategies used by the players is (Rural, UU, UUUR). Frieda's chooses to take a place in the Rural Mall and the other two stores request (and get) locations in Urban Mall. The payoffs to the three stores, in order of their moves are 2, 5, and 5. The outcome for this game may be surprising; even though Frieda's makes its mall location request first, it actually chooses to request a space in the less attractive location. This is because in its rollback calculation Frieda's looks ahead to see that it will surely be turned down by Urban Mall when the other larger stores both request spaces there too. Thus Frieda's does not bother to request a space in Urban; it does better to go straight to Rural.

The reasoning used here to illustrate the game and to find its rollback equilibrium can be extended to a game with any number of players. We could extend the mall location example to encompass all of the perhaps 300 stores vying for 200 retail spaces in the two (or more) malls, for example. Then the game tree would be huge, with 300 sets of decision nodes, giving the store to move 2^{299} decision nodes. The strategy set descriptions for stores coming after about the first 10 would be horrendously complex but analytically no different from what we have seen here. Rollback would still lead you back through the tree, step by step, to an eventual equilibrium solution.

8 ADDING MORE MOVES

We saw in the last section that adding more players increases the complexity of the analysis of sequential-play games. In this section we consider another type of complexity, which arises from adding additional moves to the game; this is achieved most simply in a two-person game by allowing players to alternate moves more than once. Then the tree is enlarged in the same fashion that a multiple-player game tree would be, but later moves in the tree are made by the same players who made decisions earlier in the same game.

Many common games, like tic-tac-toe, checkers, and chess, are actually two-person strategic games with alternating sequential moves. The use of game trees and rollback should allow us to "solve" such games; we should be able to determine the rollback equilibrium outcome and the equilibrium strategies leading to that outcome. Unfortunately, as the complexity of the game grows and as strategies become more and more intricate, the search for an optimal solution becomes more and more difficult as well. This is when computer routines like Gambit, which we mentioned in Chapter 2, become useful—when manual solution is no longer really feasible.

A. Tic-Tac-Toe

Start with the most simple of the three examples mentioned in the previous paragraph, tic-tac-toe, and consider an easier-than-usual version in which two players (X and O) each try to be the first to get two of their symbols to fill any row, column, or diagonal of a two-by-two game board. The first player has four possible positions in which to put her X. The second player then has three possible actions at each of four decision nodes. When the first player gets to her second turn, she has two possible actions at each of 12 (4×3) decision nodes. As Figure 3.6 shows, even this mini-game of tic-tac-toe has a very complex game tree. And this tree is actually relatively small, because the game is guaranteed to end after the first player moves a second time; however, there are still 24 terminal nodes to consider.

We show this tree merely as an illustration of how complex game trees can become in even simple (or simplified) games. As it turns out, using rollback on the mini-game of tic-tac-toe leads us quickly to an equilibrium. Rollback shows that all of the choices for the first player at her second move lead to the same outcome. There is no optimal action; any move is as good as any other move. Thus when the second player makes her first move, she also sees that each possible move yields the same outcome and she, too, is indifferent among her three choices at each of her four decision nodes. Finally, the same is true for the first

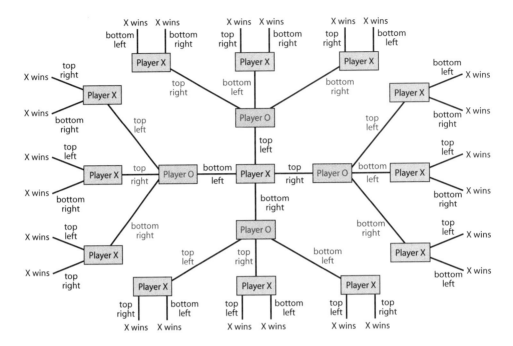

FIGURE 3.6 A More Complex Tree

player on her first move; any choice is as good as any other, so she is guaranteed to win the game.

Although this version of tic-tac-toe has an interesting tree, its solution is not as interesting. The first player always wins, so choices made by either player cannot affect the ultimate outcome. Most of us are more familiar with the three-by-three version of tic-tac-toe. To illustrate that version with a game tree, we would have to show that the first player has nine possible actions at the initial node, the second player has eight possible actions at each of nine decision nodes and then the first player, on her second turn, has seven possible actions at each of 8×9 nodes while the second player, on her second turn, has six possible actions at each of $7 \times 8 \times 9$ nodes. This pattern continues until eventually the tree stops branching so rapidly because certain combinations of moves lead to a win for one player and the game ends. But no win is possible until at least the fifth move. Drawing the complete tree for this game requires a very large piece of paper or very tiny handwriting.

Most of you know, however, how to achieve at worst a tie when you play three-by-three tic-tac-toe. So there is a simple solution to this game that can be found by rollback, and a learned strategic thinker can reduce the complexity of the game considerably in the quest for such a solution. It turns out that, as in the two-by-two version, many of the possible paths through the game tree are strategically identical. Of the nine possible initial moves, there are only three types;

you put your X in either a corner position (of which there are four possibilities), a side position (of which there are also four possibilities), or the (one) middle position. Using this method to simplify the tree can help reduce the complexity of the problem and lead you to the description of an optimal rollback equilibrium strategy. Specifically, we could show that the player who moves second can always guarantee at least a tie with an appropriate first move and then by continually blocking the first player's attempts to get three symbols in a row.[4]

B. Chess

When we consider more complicated games, like checkers, and especially chess, finding a complete solution becomes much more difficult. Students of checkers claim, but have not proven, that the second player can always guarantee a tie. Those who have studied and are still studying chess have even less to report despite the fact that chess is the quintessential game of sequential moves and therefore technically amenable to full rollback analysis.

In chess, White opens with a move, Black responds with one, and so on in turns. All the moves are visible to the other player, and nothing is left to chance, as it would be in card games that involve shuffling and dealing. Therefore the "purest" kind of strategic reasoning in chess involves looking ahead to the consequences of your move in just the way we explained earlier. An example of such reasoning might be: "If I move that pawn now, my opponent will bring up her knight and threaten my rook. I should protect the square to which the knight wants to move with my bishop, before I move the pawn." Moreover, a chess game must end in a finite number of moves. The rules of chess declare that a game is drawn if a given position on the board is repeated three times during the course of play. Since there are only a finite number of ways to place the 32 (or fewer after captures) pieces on 64 squares, a game could not go on infinitely long without running up against this rule.

Therefore in principle chess is amenable to full rollback analysis. We should know what will happen in a rollback equilibrium where both players use the optimal strategies: a win for White, a win for Black, or a draw. Then why has such analysis not been carried out? Why do people continue to play, using different strategies and getting different results in different matchups?

The answer is that for all its simplicity of rules, chess is a bewilderingly com-

---

[4]If the first player puts her first symbol in the middle position, the second player must put her first symbol in a corner position. Then the second player can guarantee a tie by taking the third position in any row, column, or diagonal that the first player tries to fill. If the first player goes to a corner or a side position first, the second player can guarantee a tie by going to the middle first and then following the same blocking technique. Note that, if the first player picks a corner, the second player picks the middle, and the first player then picks the corner opposite from her original play, the second player must not pick one of the remaining corners if she is to ensure at least a tie.

plex game. From the initial set position of the pieces, White can open with any one of 20 moves,[5] and Black can respond with any of 20. Therefore 20 branches emerge from the first node of the tree, each leading to a second node from each of which 20 more emerge. There are already 400 branches, each leading to a node from which many more branches emerge. For a while, at each turn the number of available choices actually increases to a number much bigger than 20. And a typical game goes on for 40 or more moves by each player. It has been estimated that the total number of possible moves in chess is on the order of 10^{120}, or a "one" with 120 zeros after it. That is more billions than Carl Sagan could say in a lifetime of saying "billions and billions and billions." A supercomputer a thousand times faster than your PC, making a billion calculations a second, would need approximately 3×10^{103} years to check out all these moves.[6] Astronomers offer us much less time before the sun turns into a red giant and swallows the earth.

There is clearly no hope of ever drawing the complete tree for chess. Even in the smaller game starting from any specific position, the remaining tree becomes far too complex in just five or six moves. Looking ahead even that far is possible only because just a little experience enables a player to recognize that several of the logically possible moves are clearly bad, and therefore to dismiss them out of consideration.

The general point is that while a game may be amenable in principle to a complete solution by rollback, its complete tree may be too complex to permit such solution in practice. Faced with such a situation, what is a player to do? We can learn a lot about this by reviewing the history of attempts to program computers to play chess.

When computers first started to prove their usefulness for complex calculations in science and business, many mathematicians and computer scientists thought that a chess-playing computer program would soon beat the world champion. It took a lot longer, even though computer technology improved dramatically, while human thought progressed much slower. Finally, in December 1992, a German chess program called Fritz2 beat the world champion Gary Kasparov in some blitz (high-speed) games. Under regular rules, where each player gets two and one-half hours to make 40 moves, humans retained greater superiority for longer. A team sponsored by IBM put a lot of effort and resources into the development of a specialized chess-playing computer and its associated software. In February 1996, this package, called Deep Blue, was pitted against Gary Kasparov in a best-of-six series. It caused a sensation by winning the first

[5] White can move one of eight pawns forward either one square or two, or she can move one of the two knights in one of two ways (to the squares labeled rook-3 or bishop-3).

[6] This calculation need be done only once, because once the game was solved, anyone could use the solution and no one would actually need to play. Everyone would know whether White had a win or Black could force a draw. Players would toss to decide who got which color. They would then know the outcome, shake hands, and go home.

game, but Kasparov quickly figured out its weaknesses, improved his counter-strategies, and won the series handily. Over the next 15 months, the IBM team improved Deep Blue's hardware and software, and the resulting Deeper Blue beat Kasparov in another best-of-six series in May 1997.

To sum up, computers have progressed in a combination of slow patches and some rapid spurts, while humans have held some superiority but have not been able to improve sufficiently fast to keep ahead. Closer examination reveals that the two use quite different approaches to thinking through the very complex game tree of chess.

If you cannot look ahead to the end of the whole game, how about looking part of the way, say five or 10 moves, and working back from there? Of course, the game need not end within this limited horizon. That is, the nodes you reach after five or 10 moves will not generally be terminal nodes. Only terminal nodes have payoffs specified by the rules of the game, from which you can work backward. Therefore you must attach payoffs to the nonterminal nodes at the end of your limited look-ahead, but you cannot know exactly what those payoffs should be. You need some indirect way of assigning plausible payoffs to nonterminal nodes, because you are not able to explicitly roll back from a full look-ahead. A rule that assigns such payoffs is called an **intermediate valuation function.**

In chess, humans and computer programs both use such partial look-ahead in conjunction with an intermediate valuation function. The typical method is as follows. Each piece is assigned a value, say 1 for a pawn, 4 for a bishop, and 9 for a queen. Positional and combinational advantages are assigned their own values. For example, two rooks that are aligned in support of each other are more valuable than they would be when isolated from each other. The sum of all the numerical values attached to pieces and their combinations in a position is the intermediate value of that position. This quantification and trade-off between different good or bad aspects of the position is made, not on the basis of an explicit backward calculation, but from the whole chess-playing community's experience of play in past games starting from such positions or patterns. Then a move is judged by the value of the position to which it is expected to lead after an explicit forward-looking calculation for a certain number, say five or six, of moves.

The evaluation of intermediate positions has progressed furthest with respect to chess openings, that is, the first dozen or so moves of a game. Of course, each opening can lead to any one of a vast multitude of further moves and positions, but experience enables players to sum up certain openings as being more or less likely to favor one player or the other. This knowledge has been written down in massive books of openings, and all top players and computer programs remember and use this information.

At the end stages of a game, when only a few pieces are left on the board, backward reasoning on its own is often simple enough to be doable and complete enough to give the full answer. For example, some endings are known or easy to figure out: If one player has the king and a rook while the other has only her king, then the first will win; if one player has her king and a bishop and the other has only her king, then the second player can achieve a draw. Farther away from the end, however, the position that would be reached at the end of your five- or six-move horizon is no simpler than the current position—it may be even more complex—and players have to use indirect methods to evaluate it.

The hardest stage is the midgame, where positions have evolved into a level of complexity that will not simplify within a few moves. Looking ahead five or even 10 moves will not give you a sure conclusion from where to reason back. If you could spend a lot of computing effort to improve your look-ahead ability from five moves to six, it is not clear that this would always improve your performance. You may do well in situations where disaster would have struck on precisely the sixth move, but at other times after six moves you may end up in a more complex situation that you cannot evaluate. In other words, to find a good move in a midgame position, it may be better to have a good intermediate valuation function than to be able to calculate another few moves farther ahead.

This is where the art of chess playing comes into its own. The best human players develop an intuition or instinct that enables them to sniff out good opportunities and avoid subtle traps in a way that computer programs find hard to match. Computer scientists have found it generally very difficult to teach their machines the skills of pattern recognition that humans acquire and use instinctively, for example, recognizing faces and associating them with names. The art of the midgame in chess is also an exercise in recognizing and evaluating patterns in the same, still mysterious way. This is where Kasparov has his greatest advantage over Fritz2 or Deep Blue. That also explains why computer programs do better against humans at blitz or limited-time games: The humans do not have the time to marshal their art of the midgame.

In other words, the best human players have subtle "chess knowledge," based on experience or ability to recognize patterns, which endows them with a better "intermediate valuation function." Computers have the advantage when it comes to raw or brute-force calculation. For a while, chess programs were designed to imitate human thinking. This was the approach of the artificial intelligence community. But it did not work well. Chess programmers gradually found that it was better to "let computers be computers," and use their computational power to look ahead to more moves in less time. Now, while both human and computer players use a mixture of look-ahead and intermediate valuation, they use them in different proportions: Humans do not look as many moves ahead but have better intermediate valuations based on knowledge; computers have less sophisticated valuation functions but look ahead farther using their superior computational powers.

Most recently, chess computers have begun to acquire more knowledge. When modifying Deep Blue in 1996 to 1997, IBM enlisted the help of human experts to improve the intermediate valuation function in its software. These consultants played repeatedly against the machine, noted its weaknesses, and suggested how the valuation function should be modified to correct the flaws. An episode of this process is particularly instructive.[7] Grandmaster Joel Benjamin "noticed that whenever Deep Blue had a pawn exchange that could open a file for a rook, it would always make the exchange earlier than was strategically correct. Deep Blue had to understand that the rook was already well placed, and that it didn't have to make the exchange right away. It should award evaluation points for the rook being there even before the file is opened." In other words, Deep Blue's intermediate valuation function knew that a well-positioned rook with an open file was good, so it immediately made the captures that opened the file and created this situation. But it failed to recognize that the mere positioning of the rook had "option value"—once the rook was there, it could exercise the option of making the pawn exchange that opens up the file at any time. If the pawn was serving some other good purpose where it was, then the actual exchange would be best postponed. This subtle kind of thinking results from long experience and awareness of complex interconnections among the pieces on the board; it is easier for humans to acquire, often without even knowing that they have it, than it is to codify into the formal steps of a computer program.

If humans can gradually make explicit their subtle knowledge and transmit it to computers, what hope is there for human players who do not get reciprocal help from computers? At times during their 1997 encounter, Kasparov was amazed by the human or even superhuman quality of Deep Blue's play. He even attributed one of the computer's moves to "the hand of God." And matters can only get worse: The brute-force calculating power of computers is increasing rapidly while they are simultaneously, but more slowly, gaining some of the subtlety that constitutes the advantage of humans.

Finally, will the time come when the machine is able to spot its own weaknesses and modify its own intermediate valuation function? In other words, can the machine learn from its own experience and acquire chess knowledge as humans do? If that ever happens, will such programs spread to other areas of strategic interaction, like business? And then what will it mean to be human anyway?

We leave such questions to philosophers, who have thousands of years of experience in discussing the really big questions about life, the universe, and everything, and not finding any answers to any of them. We only point out that the more powerful computers are at the disposal of human users. And after all, humans can do one thing computers cannot do—a human can pull the plug or switch off the power to a computer, but computers have no such ability to terminate humans.

---

[7] This account is taken from Bruce Weber, "What Deep Blue Learned in Chess School," *New York Times*, May 18, 1997.

The abstract theory of chess says that it is a finite game that can be solved by rollback. The practice of chess involves a lot of "art" based on experience, intuition, and subtle judgment. Is this bad news for the use of rollback in sequential-move games? We think not. It is true that theory does not take us all the way to an answer. But it does take us a long way; looking ahead a few moves constitutes an important part of the approach that mixes brute-force calculation of moves with a knowledge-based assessment of intermediate positions. And as computational power increases, the role played by brute-force calculation, and therefore the scope of the rollback theory, will also increase.

We can sum up the issue of complexity in the following way. For really simple games, we can find the rollback equilibrium by verbal reasoning without having to draw the game tree explicitly. For truly complex games, like chess, only a small part of the game tree can be drawn, and one must use a combination of two methods—calculation based on the logic of rollback, and rules of thumb for valuing intermediate positions based on experience. For games having an intermediate range of complexity, verbal reasoning is too hard but a complete tree can be drawn and used for rollback. Sometimes one may enlist the aid of a computer to draw and analyze a moderately complicated game tree.

Thankfully, most of the strategic games that we encounter in economics, politics, sports, business, and daily life are far less complex than chess. The games may have a number of players who move a number of times; they may even have a large number of players or a large number of moves. But we have a chance at being able to draw a reasonable-looking tree for those games that are sequential in nature. The logic of rollback of course remains valid, and it is also often the case that once you understand the idea of rollback, you can carry out the necessary logical thinking and solve the game without explicitly drawing a tree. Moreover, it is precisely at this intermediate level of difficulty, between the simple examples that we solved explicitly in this chapter and the insoluble cases like chess, that computer software like Gambit is most likely to be useful; this is indeed fortunate for the prospect of applying the theory to solve many games in practice.

9 EXPERIMENTS IN STRATEGY

How well do actual participants in sequential-move games perform the calculations of rollback reasoning? There is very little systematic evidence, but classroom and research experiments with some games have yielded outcomes that appear to counter the predictions of the theory. Some of these experiments and their outcomes have interesting implications for the strategic analysis of sequential-move games.

For instance, many experimenters have had subjects play a single-round bargaining game in which two players, designated A and B, are chosen from a class or a group of volunteers. The experimenter provides a dollar (or some known total), which can be divided between them according to the following procedure. Player A proposes a split, for example, "75 to me, 25 to B." If player B accepts this, the dollar is divided as proposed by A. If B rejects the proposal, neither player gets anything.

Rollback in this case predicts that A should propose "99 to me, 1 to B" and B should accept. This particular outcome almost never happens. Most players assigned the A role propose a much more equal split. In fact, 50-50 is the single most common proposal. Furthermore, most players assigned the B role turn down proposals that leave them 25 percent or less of the total, and walk away with nothing; some reject proposals that would give them 40 percent of the pie.[8]

Many game theorists remain unpersuaded that these findings undermine the theory. They counter with some variant of the following argument: "The sums are so small as to make the whole thing trivial in the players' minds. The B players lose 25 or 40 cents, which is almost nothing, and perhaps gain some private satisfaction that they walked away from a humiliatingly small award. If the total were a thousand dollars, so that 25 percent of it amounted to real money, the B players would accept." But this argument does not seem to be valid. Experiments with much larger stakes show similar results. Experiments conducted in Indonesia, with sums that were small in dollars but amounted to as much as three months' earnings for the participants, showed no clear tendency on part of the A players to make less equal offers, or on the part of the B players to accept more unequal offers.[9]

The participants in these experiments typically have no prior knowledge of game theory, and no special computational abilities. But the game is extremely simple; surely even the most naive player can see through the reasoning, and direct questions after the experiment generally show that most participants do. The results show not so much the failure of rollback as the theorist's error in supposing that each player cares only about her own money earnings. Most societies instill in their members a strong sense of fairness, which then causes most A players to offer 50-50 or something close, and the B players to reject anything that is grossly unfair.[10] This argument is supported by the observation that even

---

[8]For detailed accounts of this and related games, read Richard H. Thaler, "Anomalies: The Ultimatum Game," *Journal of Economic Perspectives*, vol. 2, no. 4 (fall 1988), pp. 195–206, and Douglas D. Davis and Charles A. Holt, *Experimental Economics* (Princeton University Press, 1993), pp. 263–269.

[9]Lisa Cameron, "Raising the Stakes in the Ultimatum Game: Experimental Evidence from Indonesia," working paper 345, Princeton University, Industrial Relations Section, Princeton, N.J., July 1995.

[10]One could argue that this social norm of fairness may actually have value in the ongoing evolutionary game being played by the whole of society. Players who are concerned with fairness reduce transaction costs and the costs of fights, which can be beneficial to society in the long run. More in-depth analysis of evolutionary games is provided in Chapter 10.

in a most drastic variant called the *dictator game*, where the A player decides on the split and the B player has no choice at all, many As give significant shares to the Bs.

Some experiments have been conducted to determine the extent to which fairness plays into behavior in ultimatum games like the dictator game. These show that changes in the information available to players and to the experimenter have a significant effect on the final outcome. In particular, when the experimental design is changed so that not even the experimenter can identify who proposed (or accepted) the split, the extent of sharing drops noticeably.

Another experimental game with similarly paradoxical outcomes goes as follows. Two players are chosen and designated A and B. The experimenter puts a dime on the table. Player A can take it or pass. If A takes the dime, the game is over, with A getting the 10 cents and B getting nothing. If A passes, the experimenter adds a dime, and now B has the choice of taking the 20 cents or passing. The turns alternate, and the pile of money grows, until reaching some limit, say a dollar, that is known in advance by both players.

We show the tree for this game in Figure 3.7. Because of the appearance of the tree, this type of game is often called the *centipede game*. You may not even need the tree to use rollback on this game. B is sure to take the dollar at the last stage, so A should take the 90 cents at the penultimate stage, and so on. Thus A should take the very first dime and end the game.

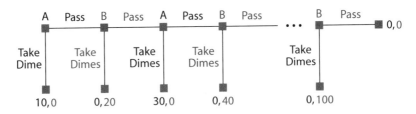

Payoffs all shown as A, B

FIGURE 3.7   The Centipede Game

However, in most classroom or experimental settings, such games go on for at least a few rounds. Remarkably, by behaving "irrationally," the players as a group make more money than they would if they followed the logic of backward reasoning. Sometimes A does better and sometimes B, but sometimes they even solve this conflict or bargaining problem. In a classroom experiment that one of us (Dixit) conducted, one such game went all the way to the end. B collected the dollar, and quite voluntarily gave 50 cents to A. Dixit asked A, "Did you two conspire? Is B a friend of yours?" and A replied, "No, we didn't even know each other before. But he is a friend now."

Once again, what is revealed is not that players cannot calculate and use game-theoretic logic, but that their value systems and payoffs are different than those attributed to them by the theorist who predicted that the game should end with A taking the dime on the first step. We will come across some similar evidence of cooperation that seems to contradict rollback reasoning when we look at finitely repeated prisoners' dilemma games in Chapter 8.[11]

The examples discussed here seem to indicate that apparent violations of strategic logic can be explained by recognizing that people do not care merely about their own money payoffs, but internalize concepts like fairness. But not all observed plays, contrary to the precepts of rollback, have some such explanation. People do fail to look ahead far enough, and they do fail to draw the appropriate conclusions from attempts to look ahead. For example, when issuers of credit cards offer favorable initial interest rates or no fees for the first year, many people fall for them without realizing that they may have to pay much more later. Therefore the game-theoretic analysis of rollback and rollback equilibria serves an advisory or prescriptive role as much as it does a descriptive role. People equipped with the theory of rollback are in a position to make better strategic decisions and get higher payoffs, no matter what is included in their payoff calculation. And game theorists can use their expertise to give valuable advice to those who are placed in complex strategic situations but lack the skill to determine their own best strategies.

SUMMARY

Sequential-move games require players to consider the future consequences of their current moves before choosing their actions. Analysis of pure sequential-move games generally requires the creation of a *game tree*. The tree is made of up *nodes* and *branches* that show all of the possible actions available to each player at each of her opportunities to move, as well as the payoffs associated with all possible outcomes of the game. Strategies for each player are complete plans that describe actions at each of the player's decision nodes contingent on all possible combinations of actions made by players who acted at earlier nodes. The equilibrium concept employed in sequential-move games is that of *rollback equilibrium*, in which players' equilibrium strategies are found using a process known as *rollback*, or *backward induction*.

Different types of games entail advantages for different players, such as *first-mover* advantages. The inclusion of many players or many moves enlarges the game tree of a sequential-move game but does not change the solution process.

---

[11] Once again, one wonders what would happen if the sum added at each step were a thousand dollars instead of a dime.

In some cases, drawing the full tree for a particular game may require more space or time than is feasible. Such games can often be solved by identifying strategic similarities between actions that reduces the size of the tree or by simple logical thinking.

Chess, the ultimate strategic sequential-move game, is theoretically amenable to solution via rollback. The complexity of the game, however, makes it effectively impossible to draw a full game tree and to write out a complete solution. In actual play, elements of both art (identification of patterns and of opportunities versus peril) and science (forward-looking calculations of the possible outcomes arising from certain moves) have a role in determining player moves. Human chess players still maintain an advantage in the use of knowledge and instinct, but computers perform the raw calculations necessary for true rollback much faster and more accurately than humans. As computers increase in power, a more complete solution of chess using rollback comes closer to becoming a reality.

Tests of the theory of sequential-move games seem to suggest that actual play shows irrationality of the players or the failure of the theory to adequately predict behavior. The counterargument points out the complexity of actual preferences for different possible outcomes and the usefulness of strategic theory for identifying optimal actions when actual preferences are known.

KEY TERMS

backward induction (49)

branch (46)

decision node (46)

extensive form (46)

first-mover advantage (56)

game tree (46)

initial node (46)

intermediate valuation function (67)

node (46)

pruning (50)

pure strategy (48)

rollback (49)

rollback equilibrium (51)

second-mover advantage
  advantage (57)

terminal node (46)

EXERCISES

1. Suppose two players, First and Second, are involved in a sequential-move game. First moves first, Second moves second, and each player moves only once.

   (a) Draw a game tree for a game in which First has two possible actions (Up or Down) at each node and Second has three possible actions (Top, Middle, or Bottom) at each node. Label the initial node of the tree "I," each decision node "D," and each terminal node "T." How many of each node type are there?

(b) Draw a game tree for a game in which First and Second each have three possible actions (Sit, Stand, or Jump) at each node. Label the nodes as in part (a). How many of each node type are there?

(c) Draw a game tree for a game in which First has four possible actions (North, South, East, or West) at each node and Second has two possible actions (Stay or Go) at each node. Label the nodes as in part (a). How many of each node type are there?

2. Use rollback to find equilibria for the following games:

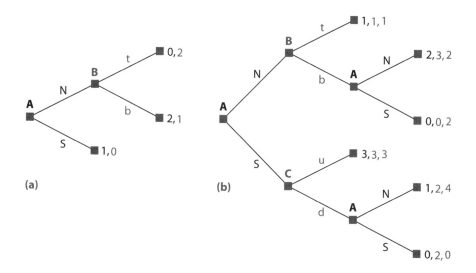

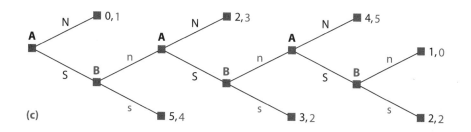

Identify the equilibrium outcome and the complete strategies for each player in each case.

3. In each of the games described in Exercise 2, how many pure strategies (decision rules) are available to each player? For each game, write out all of the pure strategies for each player. (You may use a shorthand notation.)

4. Consider the rivalry between Airbus and Boeing to develop a new commercial jet aircraft. Suppose Boeing is ahead in the development process, and Airbus is considering whether to enter the competition. If Airbus stays out, it earns zero profit while Boeing enjoys a monopoly and earns a profit of $1 billion. If Airbus decides to enter and develop the rival airplane, then Boeing has to decide whether to accommodate Airbus peaceably or to wage a price war. In the event of peaceful competition, each firm will make a profit of $300 million. If there is a price war, each will lose $100 million because the prices of airplanes will fall so low that neither firm will be able to recoup its development costs.

(a) Draw the tree for this game. Find the rollback equilibrium.

5. "In a sequential-move game, the player who moves first is sure to win." Is this statement true or false? State the reason for your answer in a few brief sentences, and give an example of a game that illustrates your answer.

6. Consider the following simplified version of the Massachusetts House of Representatives vote regarding reinstitution of the death penalty in the state: There are four Representatives, of whom three (Representatives A, B, and C) voted for the death penalty and one (Representative D) voted against in a preliminary vote count. Between that vote and the final vote, a crucial legal decision in the state convinces Representatives B and C that they must reconsider their original "for" votes and work to ensure that the death penalty not be reinstated. Ideally, neither Representative wants to have to change her vote (which would show an inability to "stick to one's guns" as well as a willingness to vote against public opinion in their constituencies) but would be willing to do so if necessary. The death penalty will only be reinstated if it gains a *strict majority* of votes in the final tally.

When the final vote is taken, Representatives vote in alphabetical order. A and D, who have not changed their views on the issue, prefer outcomes in which they have cast votes that reflect their views on the issue to outcomes in which they vote contrary to their views; each then prefers the outcome in which her view predominates. Thus, Representative A, who supports the death penalty, most prefers the outcome in which the bill passes and she votes for it; her next preferred outcome is that it fails but she has voted in favor; finally, she prefers the outcome in which it passes but she votes against it, to the outcome in which having it fails with her voting against. Similarly, D, who opposes the death penalty, most prefers the outcome in which it fails and she votes against; her next preferred outcome is that it passes but she has voted against; finally, she prefers the outcome in which it fails and she has voted "for" to the outcome in which it passes with her voting for it.

B and C would like to see the death-penalty bill fail, but each would most prefer that it fail by virtue of someone *else* changing her vote. Each next prefers the outcomes in which the bill fails and she has voted against it. The least-preferred outcome for each is that the bill passes even though they change their votes; slightly better is an outcome in which the bill passes but they vote for it, not changing their original votes.

(a) Draw the game tree for the final round vote on the death-penalty bill. Label all nodes and branches, and fill in payoffs for each voter at each terminal node.

(b) Find the rollback equilibrium of this game. What is each player's full equilibrium strategy?

(c) In the actual Massachusetts election, the Representative whose position most closely resembled that of B in the game in parts (a) and (b) was the one to change his vote. Analysts argued that "if 'The [Changed] Vote' had not been cast by 'B,' some other legislator would have switched to change [the] 81-to-79 decision returning capital punishment to [a] tie. . . . Some said there were about 10 other legislators ready to change their votes if 'B' had not stepped forward."[12] Use your answer to part (b) to explain how game theory can be used to verify this claim.

(d) The same analysts noted: "Ironically, opponents knew that a changed vote would have to come, not from the eight people who had [already publicly] switched from opposing the death penalty [to supporting it in the preliminary vote], but from legislators who had steadfastly maintained their support of it." What does this statement imply about the payoffs for these eight legislators in the final vote? Craft a game-theoretic argument that backs up the claim made here.

7. The centipede game illustrated in Figure 3.7 can be solved using rollback without drawing the complete game tree. In the version of the game discussed in the text, players A and B alternately had the opportunity to claim or pass a growing pile of dimes (to a maximum of 10) placed on the table. Suppose the rules of the games were changed so that:

(a) Player A gets a nickel reward every time she passes, giving her opponent another turn. Find the rollback equilibrium strategies for each player.

(b) Two rounds of the game are played with the same two players, A and B. In the first round, A may not keep more than five dimes, and B may not keep more than nine. Find the rollback equilibrium strategies for each player.

(c) Two rounds of the game are played with the same two players, A and B. In the first round, A may not keep more than five dimes, and B may not keep more than four. Find the rollback equilibrium strategies for each player.

8. Two players, A and B, take turns choosing a number between 1 and 10 (inclusive). A goes first. The cumulative total of all the numbers chosen is calculated

---

[12] Robert A. Jordan, "Slattery Takes Heat for Switching Votes, but Others Were Prepared to Do the Same," *Boston Globe*, November 9, 1997.

as the game progresses. Consider two alternative endings: (i) The player whose choice of number takes the total to exactly 100 is the winner. (ii) The player whose choice of number causes this total to equal or exceed 100 is the loser. For each case, answer the following questions:

(a) Who will win the game?

(b) What are the optimal strategies (complete plans of action) for each player?

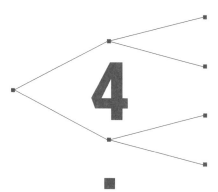

Games with Simultaneous Moves

R ECALL FROM CHAPTER 2 that games are said to have simultaneous moves if players must move without knowledge of what their rivals have chosen to do. Obviously this is so if players choose their actions at exactly the same time. A game is also simultaneous when players choose their actions in isolation, with no information about what other players have done or will do, even if the choices are made at different hours of the clock. (For this reason, simultaneous-move games are often referred to as games of *imperfect information* or *imperfect knowledge*.) This chapter focuses on games that have such purely simultaneous interactions among players. We consider a variety of types of simultaneous games, introduce a solution concept called Nash equilibrium for these games, and study games with one equilibrium, many equilibria, or no equilibrium at all.

Many familiar strategic situations can be described as simultaneous-move games. The various producers of television sets, stereos, or automobiles make decisions about product design and features without knowing what rival firms are doing about their own products. Voters in U.S. elections "simultaneously" cast their individual votes; no voter knows what the others have done when she makes her own decision. Also, the interaction between a soccer goalie and an opposing striker during a penalty kick requires both players to make their decisions simultaneously—the goalie cannot afford to wait until the ball has actually been kicked to decide which way to go, for then it would be far too late.

Strategies for simultaneous-move games cannot be made contingent on another's action, as is possible with sequential-move games. However, in many types of simultaneous-move games, a player can reason through the game from the perspective of her opponent to determine her opponent's best play and therefore her own best play as well. This is exactly what a Nash equilibrium accomplishes. When this solution method fails, players may rely on other mechanisms (like historical or cultural clues) to determine their best actions or, in some instances, they may be forced to choose strategies that randomize their actions. Each of these possibilities is considered in this chapter. Later chapters, especially Chapters 6 and 7, contain more detail on specific issues raised here.

1 ILLUSTRATING SIMULTANEOUS-MOVE GAMES

Simultaneous-move games are most often represented diagrammatically using a **game table** (also called a **game matrix** or **payoff table**). The table is called the **normal form** or the **strategic form** of the game. Games with any number of players can be illustrated using a game table, but its dimension must equal the number of players.

For a two-player game, the table is two-dimensional and appears similar to a spreadsheet. The row and column headings of the table are the strategies available to the first and second players, respectively. The size of the table, then, is determined by the numbers of strategies available to the players. Each cell within the table lists the payoffs to all players that arise under the configuration of strategies that placed players into that cell. Games with three players require three-dimensional tables; we consider them later in this chapter.

A. Zero-Sum Versus Non-Zero-Sum Games

As mentioned, the game table lists payoffs to all players in each of its cells. For some games, **zero-sum** or **constant-sum games** specifically, a shorthand method of payoff description can be used. Because players' interests are wholly opposed in such games, any benefit gained by one player is lost to another. This type of interaction guarantees that payoffs to all players will sum to a constant value, often zero (hence the name), in each cell of the table. In a two-player zero-sum game, then, it is enough to list in each cell the payoff to the first player; that to the second player can be attained by subtracting the first player's payoff from zero, or whatever the constant sum of the game happens to be. This shortcut makes it simpler to write and read the table.

Non-zero-sum games, also called **variable-sum games,** are games in which players have some common interests, so one does not gain strictly as a rival loses and there is no simple relationship between payoffs for different players. Thus the game table must show separately a payoff for every player in each of its cells.

We will use primarily the terms *zero-sum* and *non-zero-sum* in order to be consistent throughout the text; the terms *zero-sum* and *constant-sum* are used interchangeably in the literature, as are the terms *non-zero-sum* and *variable sum*.

B. Constructing an Actual Game Table

We illustrate the idea of a game table using the children's game *rock–paper–scissors*. It is a two-player simultaneous game in which each player has, perhaps not surprisingly, three pure-strategy choices: Rock, Paper, or Scissors. Notice that the description of strategies here is much simpler than for players in sequential-move games. A strategy is still a complete plan of action for a player, specifying her move, given all currently available information. But in a simultaneous-move game in which players cannot formulate strategies as contingent rules, a player's strategy is equivalent to her action or move at any given point in the game.

Rock–paper–scissors requires players to form their hands into one of the three shapes, and to simultaneously show their hands (or call out a choice of item) to each other. The game is a simple one in which a player wins, loses, or ties. Ties occur when both players show the same item. Wins and losses are determined as follows: paper "covers" rock, so a player showing paper wins over a player showing rock; scissors "cut" paper, so a player showing scissors wins over a player showing paper; and rock "breaks" scissors, so a player showing rock wins over a player showing scissors.

The table of payoffs for this game is shown in Figure 4.1a. The table has two dimensions. The first player's three strategies are used to label the rows of the table; the second player's three strategies are used to label the columns. Payoffs are shown in each cell, using W for win, L for lose, and T for tie. The first player's payoff is listed first in each case.

Notice that rock–paper–scissors is a zero-sum game. Whenever the first player wins, the second player must lose, and whenever the first player ties, the

(a) All payoffs shown

		PLAYER 2		
		R	P	S
	R	T, T	L, W	W, L
PLAYER 1	P	W, L	T, T	L, W
	S	L, W	W, L	T, T

(b) Zero-sum shorthand

		PLAYER 2		
		R	P	S
	R	T	L	W
PLAYER 1	P	W	T	L
	S	L	W	T

FIGURE 4.1   Rock–Paper–Scissors Game

second player must also tie. The table for this game could be illustrated as in Figure 4.1b, which shows a table of exactly the same size and dimension as in Figure 4.1a. Only one payoff is shown in each cell, however. That payoff, the payoff to Player 1, indicates exactly what Player 2 receives as well. To reduce crowding in our game tables, we will use this shorthand method of payoff delineation any time we are dealing with a zero-sum game.

2 NASH EQUILIBRIUM

As was the case for sequential-move games, the task following illustration of a simultaneous-move game and description of all the strategies available to the players is to predict the actual behavior when the game is played. In a sequential-move game, each player (after the first move) makes choices contingent on choices made earlier in the game. This aspect of the game allows us to use rollback to find the rollback equilibrium. In simultaneous-move games, we again look for an equilibrium in which each player's action is a best response to the actions of the other players, but we cannot use rollback.

The equilibrium concept we employ here is known as **Nash equilibrium.** This concept is named for mathematician, economist, and Nobel Prize winner John Nash, whose work in the late 1940s and early 1950s established this solution method for noncooperative games.[1] Nash described the equilibrium of a general noncooperative game as a configuration of strategies, one for each player, such that each player's strategy is best for her, given that all the other players are playing their equilibrium strategies. At a Nash equilibrium, each player must be satisfied with the strategy choice she has made, given what other players have chosen. In other words, no player should want to change her strategy once she has seen what her rivals have done. If she would want to change, then she must not have started out by choosing her equilibrium strategy.

This is a natural and appealing criterion in a game in which the players are choosing their actions independently, each motivated by her own payoff (but which, remember, might include altruistic components). Nash equilibrium is indeed the fundamental solution concept for noncooperative games. We will accept it as natural, and demonstrate and explain how it works, in this chapter and the next. We will return to examine its foundations in more detail in Chapter 7.

Nash equilibrium strategies for players in simultaneous-move games can be either pure or mixed. **Pure strategies** specify nonrandom courses of action for

---

[1] Nash also proposed a solution to cooperative games, which we discuss in Chapter 16; he shared the 1994 Nobel Memorial Prize in Economics with John Harsanyi and Reinhard Selten. A biography of Nash has recently been written by *New York Times* columnist Sylvia Nasar: *A Beautiful Mind* (New York: Simon & Schuster, 1998).

players; that is, the move to be made at each time is specified without any uncertainty. **Mixed strategies** specify that an actual move will be chosen randomly from the set of underlying pure strategies with specific probabilities. Thus in a game like rock–paper–scissors, the players each have three pure strategies (Rock, Paper, and Scissors), and a mixed strategy would consist of a rule like "play each pure-strategy one-third of the time." Our focus in this chapter is on the identification of pure-strategy Nash equilibria. Chapter 5 focuses on Nash equilibria with mixed strategies.

There are a number of ways to identify the pure-strategy Nash equilibrium of a simultaneous-move game. These methods can be ranked in order of their level of generality and difficulty. We begin with the method that is least general in that it works for only a small class of games, but when it works, it is the easiest of all methods to apply. Successive sections treat more general but less easy methods. When faced with any particular game, you will often find it best to try out the methods in the same order—first see if the easiest one works, if not try the next one, and so on.

3 DOMINANT STRATEGIES

A player in a simultaneous-move game may have any number of pure strategies available to her. We call one of these strategies her **dominant strategy** if it outperforms all of her other strategies, no matter what any opposing players do. If you have three available strategies, A, B, and C, while your opponent similarly has three available strategies a, b, and c, then C is your dominant strategy if and only if your payoff from playing C against a is higher than your payoffs from playing A or B against a, and similarly C gives you a higher payoff than either A or B when played against b or c. Note that a dominant strategy is *not* a strategy that yields *you* uniformly higher payoffs than *your opponent*. Rather, it always gives *you* higher payoffs than *you* would receive playing any other strategy, against every opposing strategy.

More formally, if your payoff from choosing C when your opponent chooses a is written as $P(C, a)$ and if other payoffs are written in the same way, then C is your dominant strategy if and only if:

$$P(C, a) \geq P(A, a) \text{ and } P(C, a) \geq P(B, a) \text{ and } P(C, b) \geq P(A, b) \ldots$$

for all possible pairings of strategies and with at least one strict inequality (> rather than ≥). If all of the inequalities are strict, then we say that C *strictly dominates* A and B and is a **strictly dominant strategy.**[2] If there are weak inequalities

[2] Some people would say that C *strongly dominates* A and B in this situation and that C is a *strongly dominant strategy*. We will use "strict" rather than "strong," but you should know that they can be used interchangeably.

(≥) and at least one strict inequality, then we say that C *weakly dominates* A and B and is a **weakly dominant strategy.** We also say that A and B are **dominated strategies;** both are dominated by strategy C.

Notice that the delineation of strategies as dominant and dominated is a simple task in a game in which players have only two strategies. In such a case, if one strategy is dominant, then the other strategy must be dominated. (Of course, there may also be no dominant strategy, a possibility we consider later in Section 7.) If there are three or more strategies available to a player, however, dominance is not as cut-and-dried. One strategy (say, A) may be dominated by another (B or C or both), but there may still be no dominant strategy. If B and C are not uniformly ranked against each other, so that B outperforms C against some of the opponent's possible strategies but B is outperformed by C against other strategies, then neither B nor C qualifies as a dominant strategy.

Dominant strategies, when they exist, are extremely useful in helping us solve simultaneous-move games. Because a dominant strategy guarantees you the highest possible payoff of all of your strategies against any possible choice of action made by your opponent, you should play a dominant strategy if you have one. Similarly, you should expect your opponent to use her dominant strategy if she has one. Thus, the *first* thing to do in solving a simultaneous-move game is to look for a dominant strategy. In some games, such as those considered in Section 4, both players will have dominant strategies and solution of the game will be immediate. In other games, such as those considered in Section 5, only one player will have a dominant strategy. Solution of such games is nearly as immediate as those for which both players have dominant strategies, since all but one strategy for one of the players is eliminated as a possible final choice.

Why does this process yield a Nash equilibrium? The player who has the dominant strategy, by choosing it, is obviously choosing her best action no matter what the other player does. The other player is then doing what is best for herself, given this choice of the first player. Thus each player is choosing her best response, given what the other player is doing, and that is the definition of a Nash equilibrium.

If there are no dominant strategies for any player in a game, the *second* thing to look for is a dominated strategy. Just as a player should use a dominant strategy if it exists, players should not use dominated strategies if they exist. It is possible to have games, such as those considered in Section 6, in which there is no dominant strategy but in which successive elimination of dominated strategies leaves only one outcome. By reasoning similar to that in the previous paragraph, it is easy to see that it must be the Nash equilibrium of the game. However, if there are no dominant strategies in a game, and if successive elimination of dominated strategies leaves more than one outcome, we must move on to a new method of solution, as is done in Section 7.

4 WHEN BOTH PLAYERS HAVE DOMINANT STRATEGIES

In this section we consider games in which both (or all) players have dominant strategies. Such games are usually illustrated with an example known as the **prisoners' dilemma.** The prisoners' dilemma can be told in any number of contexts, and we will meet many instances of this game especially in Chapters 8 and 11, but the situation most true to the original story involves two suspected felons, their captors, and their willingness to confess to a crime that, it turns out, they actually did commit.

Consider a story line of the type that appears regularly in the television program *NYPD Blue*. A husband and wife have been brought to the precinct under the suspicion that they were conspirators in the murder of a young woman. Detectives Sipowicz and Simone place the suspects in separate detention rooms and interrogate them one at a time. There is little concrete evidence linking the pair to murder, although there is some evidence they were involved in kidnapping the victim. The detectives explain to each suspect that they are both looking at jail time for the kidnapping charge, probably three years, even if there is no confession from either of them. In addition, the husband and wife are told, individually, that the detectives "know" what happened and "know" how one had been coerced by the other to participate in the crime; it is implied that jail time for a solitary confesser will be significantly reduced if the whole story is committed to paper. (In a scene from virtually every episode, the yellow legal pad and a pencil are produced and placed on the table at this point.) Finally, they are told that if both confess, jail terms could be negotiated down but not as much as in the case of one confession and one denial.

Both husband and wife are then involved in a two-person, simultaneous-move game in which each has to choose between confessing and not confessing to the crime of murder. They both know that no confession leaves them each with a three-year jail sentence for involvement with the kidnapping. They also know that if one of them confesses, he or she will get a short sentence of one year for co-operating with the police, while the other will go to jail for a minimum of 25 years. If both confess, they figure that they can negotiate for jail terms of 10 years each.

The choices and outcomes of the game being played here can be summarized by the game table in Figure 4.2. Payoffs to the player choosing the row, the husband in this case, are listed first (in black), and payoffs to the player choosing the column, the wife in this case, are listed second (in blue). Payoffs here are the lengths of the jail sentences associated with each outcome, so low numbers are better for each player. In that sense, this example differs from most of the games we analyze in which large payoffs are good rather than bad. For example, in the

		WIFE	
		Confess (Defect)	Deny (Cooperate)
HUSBAND	Confess (Defect)	10 yr, 10 yr	1 yr, 25 yr
	Deny (Cooperate)	25 yr, 1 yr	3 yr, 3 yr

FIGURE 4.2 Prisoners' Dilemma

Senate race and mall games of Chapter 3 we used 1 for the worst and worked upward so that large was good even with rankings. You should be aware that "large is good" is not always true, however. When payoffs indicate rankings, people often use 1 for the best alternative and successively higher numbers for successively worse ones. Thus if you ever write a payoff table where large numbers are bad, you should alert the reader by pointing this out clearly. And when reading someone else's example, be aware of the possibility.

Recall that the first step in solving a simultaneous-move game is to look for a dominant strategy. To identify a dominant strategy, it is necessary to compare the outcomes achieved with each strategy. For the husband in Figure 4.2, a confession gets him 10 years in prison if his wife also confesses and only one year in prison if his wife does not confess. The alternative strategy of not confessing gets him 25 years in jail if his wife confesses and three years in jail if she does not. Comparing his two choices, we see that confessing is better for the husband if his wife confesses (10 rather than 25 years behind bars) and confessing is also better if his wife does not confess (one rather than three years behind bars). Thus no matter what the wife chooses to do in this case, the husband does better to confess to the crime. This fact means that Confess is a dominant strategy for the husband.

We can use the same logic to consider the wife's position. She too does better to confess when her husband confesses (10 versus 25 years in jail), and she does better to confess when her husband does not (one versus three years in jail). Thus she too has a dominant strategy to confess to the crime.

Given that players should use dominant strategies if they have them, and given that both players here have dominant strategies, we can now move quickly to a solution for this game. The Nash equilibrium of the game involves both husband and wife choosing to confess. The equilibrium outcome yields each player a payoff of 10 years in jail.

Any game with the same general payoff pattern as that illustrated in Figure 4.2 can be described as a prisoners' dilemma. More specifically, a prisoners' dilemma has three essential features. First, each player has two strategies: to *cooperate* with one's rival (deny any involvement in the crime, in our example) or to *defect* from cooperation (confess to the crime, here). Second, each player also has a dominant strategy (to confess or to defect from cooperation). Finally, the dominance solution equilibrium is worse for both players than the nonequilibrium situation in which each plays the dominated strategy (to cooperate with rivals).

Games of this type are particularly important in the study of game theory for two reasons. The first is that the payoff structure associated with the prisoners' dilemma arises in many quite varied strategic situations in economic, social, political, and even biological competitions. This wide-ranging applicability makes it an important game to study and to understand from a strategic standpoint. We devote the whole of Chapter 8 and sections in several other chapters to its study.

The second reason that prisoners' dilemma games are integral to any discussion of games of strategy is the somewhat curious nature of the equilibrium outcome achieved in such games. Both players follow conventional wisdom in choosing their dominant strategies, but the resulting equilibrium outcome yields them payoffs that are lower than they could have achieved if they had each chosen their dominated strategies. Thus the equilibrium outcome in the prisoners' dilemma is actually a "bad" outcome for the players; they could find another outcome that they both prefer to the equilibrium outcome. This particular feature of the prisoners' dilemma has received considerable attention from game theorists who have asked an obvious question: what can players involved in a prisoners' dilemma do to achieve the better outcome? We leave this question to the reader momentarily, as we continue our discussion of simultaneous games, but return to it in detail in Chapter 8.

5 WHEN ONLY ONE PLAYER HAS A DOMINANT STRATEGY

We have seen that games with two players, each of whom has a dominant strategy, are quite straightforward to solve. Games with two players, in which just one has a dominant strategy, are also relatively simple to solve. Again, finding an equilibrium outcome relies on the assumption that any player with a dominant strategy will use that strategy in equilibrium. Thus if you are playing in a game in which you do not have a dominant strategy but your opponent does, you can assume that she will use her dominant strategy and so you can choose your equilibrium action (your *best response*) accordingly.

We illustrate the possibility that one player has a dominant strategy while the other does not using a famous game actually played in a naval engagement in 1943, during World War II. The game in question, the *battle of the Bismarck Sea*, is named for the body of water in the southwestern Pacific Ocean separating the Bismarck Archipelago from Papua–New Guinea.[3] In 1943, a Japanese admiral was ordered to transport troops and lead a supply convoy to New Guinea. The Japanese had a choice between a rainy northern route and a sunnier southern route. The U.S. air forces knew that the convoy would sail and wanted to send bombers after it but they did not know which route the Japanese would take. The Americans had to send reconnaissance aircraft to scout for the Japanese, but they had only enough planes to explore one route at a time.

The sailing time was three days. If the Japanese convoy was on the route that the Americans explored first, the United States could send its bombers straightaway; if not, a day of bombing was lost by the Americans. In addition, the poor weather on the northern route made it likely that visibility would be too limited for bombing on one day in three. Thus the Americans could anticipate two days of active bombing if they explored the northern route and found the Japanese immediately, and also two days of bombing if they explored the northern route but discovered that the Japanese had gone south. If the Americans explored the southern route first and found the Japanese there, they could get in three days of bombing, but if they found that the Japanese had gone north, they would get only a single day of bombing completed.

Because the Japanese had to choose their route without knowing which direction the Americans would search first, and because the Americans had to choose the direction of their initial reconnaissance without knowing which way the Japanese would go, this game has simultaneous moves. The expected number of days of bombing, in relation to the choices made by the two sides, can be illustrated in a game table as in Figure 4.3.

The game illustrated in Figure 4.3 is zero-sum: The Americans prefer a larger number of bombing days and the Japanese prefer a smaller number. For this reason, we have shown only one payoff in each cell of the game table (as in Figure 4.1b). Higher numbers are simultaneously better outcomes for the row player (U.S.) and worse outcomes for the column player (Japan).

		JAPANESE NAVY	
		North	South
U.S. AIR FORCES	North	2	2
	South	1	3

FIGURE 4.3 Battle of the Bismarck Sea

[3] See O. Haywood, "Military Decisions and Game Theory," *Journal of the Operations Research Society of America* (November 1954), pp. 365–385.

We look now for dominant strategies in this game. The United States does better by choosing North if Japan has chosen North (two days of bombing rather than one), but it does better choosing South if Japan has chosen South. The Americans, then, do not have a dominant strategy. But the Japanese do. If the U.S. reconnaissance planes choose North, it does not matter what the Japanese Navy chooses. (If they choose North, they are spotted at once, but one day out of three is lost to bad weather, so they are bombed for two days; if they choose South, the Americans waste a day looking for them in the wrong place, but then get two days of bombing in clear weather.) If the United States goes south, the Japanese do better when they go north. North is then Japan's *weakly dominant* strategy.

The choice for Japan in the game is now clear; they will take the northern route. The Americans can now take this into account when making their own decision. The United States should choose the best strategy for themselves with the expectation that the Japanese will play North. That means that the United States will send their own reconnaissance to the north and the Nash equilibrium of the game occurs in the top left cell of the game table. Each side goes north, and the Americans achieve two (out of a possible three) days of bombing.

It is not known whether the commanders on the two sides in 1943 thought things through in quite this way. It is known, however, that the outcome was just as game theory predicts. The Japanese convoy took the northern route, the American planes also searched there, and two days of bombing inflicted serious damage on the Japanese ships.

This example illustrates how simple the solution of a simultaneous-move game can be even when both sides do not have dominant strategies. If just one player has a dominant strategy (strict or weak), the other player knows that the dominant strategy will be chosen and can plan accordingly. When looking for a quick solution to a simultaneous-move game, players should check for a dominant strategy for themselves first and lacking one, should then check for a dominant strategy for their opponents.

6 SUCCESSIVE ELIMINATION OF DOMINATED STRATEGIES

The games we have considered so far have had only two pure strategies available to each player. In such games, if one strategy is dominant, the other is dominated, so choosing the dominant strategy is equivalent to eliminating the dominated one. In larger games, as mentioned above, some of a player's strategies may be dominated even though there is no single strategy that dominates all of the others. If players find themselves in a game of this type, they may be able to reach an equilibrium by removing dominated strategies from consideration as possible choices. Removing dominated strategies reduces the size of the game and then the "new" game may have another dominated strategy, for the same player or for her opponent, that can also be removed. Or the "new" game may even have a dominant strategy for one of the players. **Successive or iterated elimination of dominated strategies** uses this process of removal of dominated strategies and reduction in the size of a game until no further reductions can be made. If this process ends in a unique outcome, that outcome is the Nash equilibrium of the game and the strategies that yield it are the equilibrium strategies for each player.

If successive elimination of dominated strategies does yield a unique equilibrium outcome, the game is said to be **dominance solvable.** The process may not yield such a unique outcome, but even then, the reduction in the size of the game brought about by the elimination of even one strategy may make the game easier to solve using one or more of the techniques described in following sections. Thus eliminating dominated strategies is a useful step toward solving a large simultaneous-play game, even when it does not completely solve the game.

We illustrate the idea using a hypothetical but realistic story of economic competition. Consider a small college town with a population of dedicated pizza eaters but able to accommodate only two pizza shops, Donna's Deep Dish and Pierce's Pizza Pies. Each seller has to choose a price for its pizza. To keep things simple we will suppose that only three prices are available: high, medium, and low. If a high price is set, the sellers can achieve a profit margin of $12 per pie. The medium price yields a profit margin of $10 per pie and the low price only $5 profit per pie. Each store has a loyal captive customer base who will buy 3,000 pies per week, no matter what price is charged in either store. There is also a floating demand of 4,000 pies per week. The people who buy these pies are price-conscious and will go to the store with the lower price; if both stores are charging the same price, this demand will be split equally between them.

The game table in Figure 4.4 shows the profits per week going to each pizza seller in thousands of dollars. To see where the profit values come from in each cell, consider the sample situation when Donna's Deep Dish chooses a medium price and Pierce's Pizza Pies chooses a high price. In that case, Donna's will get its 3,000 loyal customers and all of the 4,000 floating customers because of its lower price. Thus Donna's sells a total of 7,000 pies and makes a profit of $10 per pie, giving it a $70,000 profit per week. (Because we are denoting profit levels in thousands in the table, the $70,000 profit for Donna's becomes "70.") Pierce's has only its 3,000 loyal customers to serve but a higher profit margin of $12 per pie. Its total profit per week is then $36,000 ("36" in the table). Similar calculations yield the numbers for the rest of the table.

		PIERCE'S PIZZA PIES		
		High	Medium	Low
DONNA'S DEEP DISH	High	60, 60	36, 70	36, 35
	Medium	70, 36	50, 50	30, 35
	Low	35, 36	35, 30	25, 25

FIGURE 4.4 Successive Elimination of Dominated Strategies ($'000)

Neither player in this game has a dominant strategy. From Donna's perspective we see that a medium price is the best choice when Pierce's chooses High or Medium, but High is best when Pierce's chooses Low. Exactly the same is true for Pierce's Pizza Pies; Medium is best against High or Medium, but High is best against Low. Thus no single strategy is uniformly at least as good as or better than every other strategy (for either player); nothing is dominant.

However, we can find a *dominated* strategy—Low—for each player. Although a strategy need only be dominated by one other strategy, Low happens to be dominated by both Medium and High in this particular game. To check this, note that Donna's payoffs from charging a low price are less than its payoffs from charging a medium price for each possible pricing choice made by Pierce's Pizza Pies. When Pierce's charges a high price, Medium gets Donna's 70 while Low gets it only 35; when Pierce's charges a medium price, Medium gets Donna's 50 while Low gets it only 35; and when Pierce's charges a low price, Medium gets Donna's 30 while Low gets it only 25. Similarly for the comparison of High against Low. The same checking verifies that Low is dominated by both Medium and High for Pierce's Pizza Pies also.

Because Low is a dominated strategy for each player, we can predict that neither store will charge a low price and each player should assume that its rival will not price low. This information allows us to effectively reduce the size of the game to one where each player has only two pricing strategies, High and Medium. The remaining payoff table contains just the top left two rows and columns.

Within the reduced game table, we can continue our search for dominant or dominated strategies. When the only choices are High or Medium, we see that Medium dominates High for each player. For Donna's Deep Dish, Medium yields either 70 (against High) or 50 (against Medium) as opposed to High, which yields 60 (against High) or 36 (against Medium). The same is true for Pierce's Pizza Pies.

It is worthwhile to consider why Medium dominates High after Low has been eliminated but not before. When the rival is playing Low, Medium gives a worse payoff (30) than High (36). But once you have figured out that your rival is not going to play Low, you need to compare Medium against High for only the other two choices of your rival. And, there, Medium outperforms both a rival's high price (yielding 70 rather than 60) and her medium price (yielding 50 rather than 36).

Under these circumstances, we find ourselves with a game in which both players have dominant strategies. If both players use their dominant strategies, the Nash equilibrium of this game entails both sellers pricing at Medium and each earning $50,000 in profit per week. Take note: the two-by-two game with high and medium prices is a prisoners' dilemma. We will return to this example in Chapter 8.

This example also conceals a further useful lesson in economics. Competitive pricing decisions always involve a trade-off between two considerations. If you charge a lower price, that will get you some customers who would not have bought from you otherwise and adds to your profit. But you make less from the customers who would have bought from you even at the higher prices, and this is a sacrifice of profit. Whether you should lower your price depends on whether the addition to your profit from the new demand exceeds the reduction in profit from the captive buyers. Even more importantly, you should look for strategies that will let you treat new and captive demand differently, thereby getting the added profit without making the sacrifice. Many ploys, such as introductory offers and quantity discounts, serve this purpose. We will consider some strategies of this type in later chapters.

It should now be clear why Low is a poor strategy in this example. Each store's captive customer base is large enough to make it unprofitable to give them all low prices. But unless the rival is charging Low, moving from High to Medium *is* profitable. Winning the floating demand (if the rival is charging High) or sharing this demand (if the rival is charging Medium) is worth the sacrifice of $2 (from $12 to $10) in the profit margin on the captive customers.

A. Symmetric Games

Notice that everything that is true for Donna's Deep Dish is also true for Pierce's Pizza Pies. Donna's profit numbers when it prices High against each of Pierce's possible prices were the same as Pierce's profit numbers when it priced High against each of Donna's possible prices. The same strategy is dominated for each player in the full version of the game. The same strategy is dominant for each player in the reduced version of the game.

Games with these features are known as **symmetric games;** both (or all) players face exactly the same choices and exactly the same outcomes are associated with those choices. You can see the symmetry in the game table itself. Payoffs in the cells along the diagonal of the table (the top left corner, the middle, and the bottom right corner) are the same for both players; payoffs in the cells on either side of the diagonal are mirror images of each other when reflected through the diagonal of the table.

Given this symmetry of payoffs, you might expect to find an equilibrium in which the players make the same choices and achieve the same outcomes. This is true in many games. However, symmetry of payoffs does not mean that when the other player chooses some strategy A, your *best response* is to choose A yourself. Indeed, it could be the case that your best response to the other's A is B, and conversely the other's best response to your B is A. Then the game has an *asymmetric equilibrium*. In fact, there are games with symmetric payoffs that have no symmetric equilibria in pure strategies. The game chicken, discussed later in this chapter, is a prominent example. This game does have a symmetric equilibrium in mixed strategies, which we consider in Chapter 5.

B. Weak Versus Strict Dominance

In Section 3 we defined two concepts of dominance: weak and, by contrast, strict. We said that strategy A (strictly) dominates B for a player if, for each choice the other player(s) can make, this player gets a clearly *better* payoff with A than with B. We say that strategy A weakly dominates B for a player if, for each choice the other player(s) can make, A gives *at least as good* a payoff for this player as does B, and A gives her a better payoff than B for at least one choice made by the other(s).

For example, in the Bismarck Sea game, for each of the two American choices, the Japanese got at least as good a payoff from their own choice of North as they did from South; and for one of the two American choices, namely South, the Japanese did definitely better with North than with their South. We called this weak dominance. In the pizza game, each player does strictly better by using Medium than by using Low against each of the three choices the other might make. This is strict dominance.

Successive elimination of dominated strategies is an unambiguous procedure when the dominance is strict. But in the case of weak dominance, the outcome can in general depend on the order in which the elimination is carried out. What is more, successive elimination of weakly dominated strategies may yield some but not all of the Nash equilibria of a game. We will return to this issue later in the chapter.

7 MINIMAX STRATEGIES IN ZERO-SUM GAMES

Many simultaneous-move games will have no dominant strategies and no dominated strategies. Others may have one or several dominated strategies, but iterated elimination of dominated strategies will not yield a unique outcome. In such cases, we need a next step in the process of finding a solution to the game. We are still looking for a Nash equilibrium in which every player does the best she can, given the actions of the other player(s), but we must now rely more heavily on the logic of strategic thinking than on the rules associated with dominant strategies.

One logical approach to solving a zero-sum game with no dominance involves the idea of *minimax*. The minimax method of solution relies on the view that players in such games are pessimistic about their chances of achieving good outcomes. As a player, you figure that your opponent is going to do the best she can; that is, she will pick a strategy that makes you as bad off as she can make you. Because of the nature of the zero-sum game, any choice that yields the best set of outcomes for your opponent does the opposite for you. Similarly, your opponent figures that you will pick a strategy from among your choices that minimizes her payoff from whatever strategy she chooses.

In this situation, what should the players do? Suppose the payoff table shows the row player's payoffs, and Row wants the outcome to be a cell with as high a number as possible. Then Column wants the outcome to be a cell with as low a number as possible. Using the pessimistic logic described in the previous paragraph, Row figures that, for each of her rows, Column will choose the column with the lowest number in that row. Therefore Row should choose the row that gives her the highest among these lowest numbers, or the *maximum* among the *minima*—the **maximin** for short. Similarly, Column reckons that, for each of her columns, Row will choose the row with the largest numbers in that column. Then Column should choose the column with the smallest number among these largest ones, or the *minimum* among the *maxima*—the **minimax**.

If Row's maximin value and Column's minimax value occur in the same cell of the game table, then that outcome is a Nash equilibrium of the game. For example, suppose Row's maximin choice is her third row and Column's minimax choice is her fourth column, and that the value of the payoff is cell $(3, 4)$ represents both the maximin (for Row) and the minimax (for Column). Then in the third row, the cell in the fourth column must have the smallest number in the row; if Row chooses the strategy associated with her third row, then Column's best response is the strategy associated with her fourth column. Conversely, in the fourth column the cell in the third row must have the largest number in the column; if Column chooses the strategy associated with her fourth column, then Row's best response is the strategy associated with her third row. Thus these choices are mutual best responses, and that is a Nash equilibrium.

This method of finding equilibria in zero-sum games should be called the maximin-minimax method, but is called simply the **minimax method** for short. It will lead you to a Nash equilibrium in pure strategies if one exists.

To illustrate the minimax method, we use an example from the game of (American) football. The team on the offense is attempting to move the ball forward to improve its chances of kicking a field goal. It has four possible strategies: a run or one of three different-length passes (Short, Medium, and Long). The defense can adopt one of three strategies to try to keep the offense at bay: a run defense, a pass defense, or a blitz of the quarterback. The game is zero-sum; the offense tries to gain yardage while the defense tries to prevent them from doing so.

Suppose we have enough information about the underlying strengths of the two teams to work out the probabilities of completing different plays and to determine the average gained yardage that could be expected under each combination of strategies. With such information, we could construct a game table as in Figure 4.5. The payoffs in each cell are the expected yards gained by the offense.[4] The offense wants this number to be as large as possible; the defense wants it to be as small as possible.

		DEFENSE			
		Run	Pass	Blitz	
OFFENSE	Run	2	5	13	min = 2
	Short Pass	6	5.6	10.5	min = 5.6
	Medium Pass	6	4.5	1	min = 1
	Long Pass	10	3	−2	min = −2
		max = 10	max = 5.6	max = 13	

FIGURE 4.5 The Minimax Method

Our system for solving simultaneous-move games requires that we check this table first for dominance. For the offense, Run is not dominated by any other strategy; others do better against a Run defense, but none do better against Blitz. Neither is Short Pass nor Long Pass dominated; in both cases, other strategies do better against one or two of the defensive plays but not against all of them. Medium Pass, however, is weakly dominated by Short Pass; Short Pass gets more yards against either a Pass or a Blitz defense and the same number of yards against a Run defense. We can then eliminate Medium Pass as a possible equilibrium strategy for the offense.

Looking now at the defense, remember that it wants to keep the yardage

---

[4]Each of these numbers may itself be an expected value that combines probabilities and outcomes. Thus the −2 in the (Long Pass, Blitz) cell may arise because with a probability of 0.6 the offense's quarterback is sacked, leading to a loss of 10 yards, with a probability of 0.3 a pass is attempted but incomplete, and with a probability of 0.1 a pass of 40 yards is completed. (The numbers were constructed by a team of expert consultants especially convened on a fall Sunday afternoon.)

numbers as low as possible. With that in mind, note that the Run defense is not dominated (since it is the best choice against a Run); neither is the Pass defense (the best choice against a Short Pass) nor the Blitz (the best choice against either the Medium or Long Pass). No further strategies can be eliminated; the game is not dominance solvable.

At this point, we turn to the minimax method. Note that minimax will work on the original table as well as on the reduced table lacking the dominated strategy, Medium Pass. We will check the full table, for practice.

Remember that the offense (the row player) wants to maximize the number of yards it gains, and the defense (the column player) wants to minimize that number. Begin by finding the lowest number in each row (the offense's worst payoff from each strategy) and the highest number in each column (the defense's worst payoff from each strategy).

The offense's worst payoff from Run is 2; its worst payoff from Short Pass, 5.6; its worst payoff from Medium Pass, 1; and its worst payoff from Long Pass, –2. The defense's worst payoff from Run is 10; its worst payoff from Pass, 5.6; and its worst payoff from Blitz, 13. Most beginning game theorists find it easiest to label the row minima at the far right of each row of the table and the column maxima at the bottom of each column, as we have done in Figure 4.5.

The next step is to find the best of each players' worst-possible outcomes, the largest row minimum and the smallest column maximun. The largest of the row minima is 5.6, so the offense can ensure itself a gain of 5.6 yards by playing the Short Pass; this is its maximin. The lowest of the column maxima is 5.6, so the defense can be sure of holding the offense down to a gain of 5.6 yards by deploying its Pass defense. This is the defense's minimax.

Looking at these two strategy choices, we see that the maximin and minimax values are found in the same cell of the game table. Thus the offense's maximin strategy is its best response to the defense's minimax and vice versa; we have found the Nash equilibrium of this game. That equilibrium entails the offense attempting a Short Pass while the defense defends against a Pass. A total of 5.6 yards will be gained by the offense (and given up by the defense).

It is important to realize that the minimax method, by virtue of being a more general approach to the solution of simultaneous-move games, can also be used to find the Nash equilibrium of any dominance-solvable zero-sum game. Recall the battle of the Bismarck Sea from Section 5, whose table is illustrated in Figure 4.3. Applying minimax to that game shows that North is the United States' minimax strategy; the row minima are 2 (for North) and 1 (for South), and 2 is the largest of these. Similarly, North is the minimax strategy for Japan; the column maxima are 2 (for North) and 3 (for South) and 2 is the smallest of these. The maximin and minimax are the same, so (North, North) must be the Nash equilibrium of the game. The minimax method should arrive at the same equilibrium as is attained using the dominance criteria in any zero-sum game.

It is also critical to note that the minimax method may fail to find an equilibrium in some games, just as dominance solvability may fail to do so. In such cases, we move on to consider the possibility that there may be more than one equilibrium or perhaps no equilibrium in pure strategies at all. These cases are considered in the Sections 11 and 12. In addition, the minimax method may fail when applied to non-zero-sum games. In such games, choosing the strategy that makes your minimum payoff as large as possible may not be your best strategy because in a non-zero-sum game the opponent's best need not be your worst, so there is in general no reason to believe that your opponent is out to make your payoff as low as possible.

8 CELL-BY-CELL INSPECTION

There is an alternative to all of the solution methods considered so far in this chapter. This alternative, known as **cell-by-cell inspection,** can be used to find the Nash equilibrium (or Nash equilibria, if there are more than one) of any zero-sum or non-zero-sum game. The technique is not a sophisticated one, but you can always be sure that it will work and you can use it to check an equilibrium found using another solution method. Moreover, it will give you *all* the Nash equilibria of the game, whereas other methods may miss some in games with multiple Nash equilibria.

Cell-by-cell inspection entails a visual inspection of each and every cell of a game table in search of an outcome, or outcomes, that satisfy the Nash equilibrium requirements. Such an outcome must be one in which all players are using strategies that are their best responses to the strategies of the other players. The easiest way to verify that a player is using her best response is just to check, at a particular outcome, whether she would want to change her choice of strategy. If she would, then the strategy she is using cannot be her best response to the strategy(ies) chosen by her rival(s). If she wouldn't want to change, and if no other player would want to change her action either, then you know that the outcome you are considering must be a Nash equilibrium.

You can use cell-by-cell inspection on any of the game tables that have appeared so far in this chapter. For games with few strategies for each player, there are not many cells to inspect. As an example, let us use the pizza pricing game illustrated originally in Figure 4.4. That game involves two pizza parlors choosing among three possible prices; each store wants to end up with the largest-possible profit values.

We begin the cell-by-cell inspection with the cell in the upper left-hand corner of the table, redrawn in Figure 4.6. This cell is associated with both stores choosing to set a high price for their pizzas. We need to check whether this out-

		PIERCE'S PIZZA PIES		
		High	Medium	Low
DONNA'S DEEP DISH	High	60, 60	36, 70	36, 35
	Medium	70, 36	50, 50	30, 35
	Low	35, 36	35, 30	25, 25

FIGURE 4.6 Cell-by-Cell Inspection

come constitutes a Nash equilibrium. To do this, we simply determine whether either Pierce's Pizza Pies or Donna's Deep Dish would want to set a medium or a low price instead of a high price when their rival is setting a high price. For Pierce's Pizza Pies, setting a medium price when Donna's Deep Dish sets a high price is better than also pricing high; Medium yields a profit of 70 against High, while High yields only a profit of 60 against High. Pierce's would want to change its strategy from High to Medium if it found itself in the (High, High) outcome, so both stores pricing high cannot be a Nash equilibrium. We indicate the fact that High is not Pierce's optimal response to Donna's choice of High by drawing a diagonal line through Pierce's payoff in the (High, High) cell.

Because the pizza pricing game is symmetric, we can also show that Donna's would want to change *its* strategy from High to Medium if it found itself in the (High, High) outcome. Therefore a similar analysis from Donna's perspective shows that (High, High) cannot be a Nash equilibrium of this game. We show Donna's desire to move from the (High, High) cell by drawing a diagonal line through its payoff as well. In this case, there are diagonal lines drawn through both payoffs in the (High, High) cell indicating that neither player does best to choose High if their rival is also choosing High. Clearly, only a cell with no diagonal lines drawn through any of its payoffs can be associated with a Nash equilibrium outcome.

We continue our inspection of the cells in the table by moving along the top row. At the (High, Medium) outcome, Pierce's is happy with its choice of Medium but Donna's could increase its profits by changing to a medium pricing policy. In the (High, Medium) cell, then, we put a diagonal line through Donna's payoff; the (High, Medium) outcome cannot be a Nash. You can check the remaining seven cells in the table yourself. We have indicated which player(s) want to move from a given cell with a diagonal line through the payoff of any player that wants to move.

The only cell from which neither player wants to move is the cell associated with the outcome (Medium, Medium). At that outcome, neither Donna's nor

Pierce's can improve their profits by choosing a different pricing policy, given that their rival is also pricing at Medium. Shifting to High lowers each store's profit from 50 to 36; shifting to Low reduces their profits to 35. Notice that this cell is the only one in the table that has no diagonal lines. This cell arises from players using their Nash equilibrium strategies, and our cell-by-cell inspection has led us to the same answer that we found in Section 6. The unique pure strategy equilibrium of this game is for both pizza stores to use a medium pricing policy.

In some games, successive elimination of dominated strategies or the mini-max approach may fail to yield an equilibrium for the game. In such cases, cell-by-cell inspection may be the only method that works to find a pure-strategy Nash equilibrium of the game. It is tedious to check the strategy choices of every player in every cell of the table in a large game; but such inspection is guaranteed to find a pure-strategy equilibrium if there is one. In the future, the mechanical task of finding equilibria may gradually be taken over by computer programs; Gambit, which we have mentioned before, is currently one of the most widely available.

9 PURE STRATEGIES THAT ARE CONTINUOUS VARIABLES

In all of the games we have studied so far, each player had only a finite num-ber—often as few as two or three—pure strategies. But consider games in which each player is choosing a magnitude like a price or an expenditure or a length. Such magnitudes come in multiples of some minimum units, for exam-ple, cents for money or inches for length. Therefore the number of choices available to the player is finite. But these units are usually spaced so closely to-gether that it makes sense to think of the magnitude as varying continuously over a range. In this section we show you how to find the Nash equilibria in such games.

Consider a game in which two car dealers (one for Ford, one for Subaru) want to erect flagpoles from which to fly their huge American flags in order to at-tract customers. Each wants a nice high flagpole, although additional height in-creases the risk of the pole bending; neither wants to have the shorter pole, however. The dealers figure their payoffs in terms of customers attracted into the showroom each month. Each car dealer must order her flagpole without know-ing what the other has done, so the game has simultaneous moves.

If flagpoles came only in 10-foot increments, we could describe this game in terms of the 10 (or so) possible heights that each dealer could reasonably con-sider and draw a (relatively large) game table with which to do our analysis using one of the methods described earlier in the chapter. However, the dealerships know that they can get flagpoles in any length, down to a fraction of an inch.

Thus the set of pure strategies available to each car dealer is enormous. Rather than specifying each possible length as an individual pure strategy, which would require a lot of time and energy on our part as well as a ridiculously large game table, we describe the car dealers' strategies as continuous variables that can take on values anywhere from 20 (feet) to 200 (feet), including fractional units.

The method we will use to find Nash equilibria in such games is known as **best-response analysis.** We know that in equilibrium, each player's strategy must represent its best response to that of the other. Call the strategy of the Ford dealer F and that of the Subaru dealer S. For each value in the range of F from which the Ford dealer can make her choice (20 to 200 feet), we would find the value of S that is best for the Subaru dealer (maximizes her payoff). We would then show this best response as a curve—a **best-response curve,** or a **reaction curve**—graphing S against F. Conversely, for each value of S in the range from which the Subaru dealer makes her choice, we would find the best-response value of F from the Ford dealer's perspective. We would show the resulting graph of F against S in the same figure as the graph of S against F. Nash equilibrium is where the choices are best responses against each other, that is, where the two curves intersect. We will leave the derivation of the reaction curves in this example as an exercise and continue with a more familiar example.

We return now to our two pizza stores, Donna's Deep Dish and Pierce's Pizza Pies. In Section 6 we confined each to choosing just one of three prices, high, medium, or low. But in reality, prices can be any (almost) continuously variable number. So now we will allow the two pizzerias the freedom to choose a specific value for price. Unfortunately it is not possible to keep the arithmetic sufficiently simple and still reproduce our earlier numbers exactly, but we can come close.

In setting its price, each store will have to calculate the consequences for its profit. To keep things relatively simple we will put the stores in a very symmetric relationship, but readers with a little more mathematical skill can do a similar analysis using much more general numbers or even algebraic symbols. Suppose it costs \$3 to make each pizza. Suppose further that experience or market surveys have shown that when Donna's price is P_{Donna} and Pierce's price is P_{Pierce}, their respective sales Q_{Donna} and Q_{Pierce} (measured in thousands of pizzas per week) will be given by

$$Q_{\text{Donna}} = 12 - P_{\text{Donna}} + 0.5\,P_{\text{Pierce}}$$
$$Q_{\text{Pierce}} = 12 - P_{\text{Pierce}} + 0.5\,P_{\text{Donna}}$$

The key idea in these equations is that if one store raises its price by \$1 (say, Donna increases P_{Donna} by \$1), its sales will go down by 1,000 (Q_{Donna} changes by -1×1000) and those of the other store will go up by 500 (Q_{Pierce} changes by 0.5×1000)—presumably 500 of Donna's customers will switch to Pierce's and another 500 will switch to some other kind of food.

Pierce's profit per week (in thousand dollars), call it Y_{Pierce}, is given by the product of its net revenue per pie (selling price less cost or $P_{\text{Pierce}} - 3$), and the number of pies sold:

$$Y_{\text{Pierce}} = (P_{\text{Pierce}} - 3) \, Q_{\text{Pierce}} = (P_{\text{Pierce}} - 3)(12 - P_{\text{Pierce}} + 0.5 P_{\text{Donna}}) \, .$$

Pierce's will set its price P_{Pierce} to maximize this payoff. Doing this for each possible level of Donna's price P_{Donna} will give us Pierce's best response, and then we can graph it.

There are two ways to find the value of P_{Pierce} that maximizes Y_{Pierce} for each P_{Donna}. First, we can multiply out and rearrange the terms on the right-hand side of the above expression for Y_{Pierce} :

$$Y_{\text{Pierce}} = -3(12 + 0.5 P_{\text{Donna}}) + (3 + 12 + 0.5 P_{\text{Donna}}) P_{\text{Pierce}} - (P_{\text{Pierce}})^2$$

$$= -3(12 + 0.5 P_{\text{Donna}}) + \left(\frac{15 + 0.5 P_{\text{Donna}}}{2}\right)^2 - \left(\frac{15 + 0.5 P_{\text{Donna}}}{2} - P_{\text{Pierce}}\right)^2$$

In this final form of the expression, only the last term involves P_{Pierce}. It appears with a negative sign, so what appears after the negative sign should be made as small as possible in order to make the whole expression (Pierce's profit) as large as possible. But that last term is a square, and therefore its smallest value is zero, which is achieved when

$$P_{\text{Pierce}} = 0.5(15 + 0.5 P_{\text{Donna}}) = 7.5 + 0.25 P_{\text{Donna}}.$$

This equation solves for the value of P_{Pierce} that maximizes Pierce's profit, given a particular level value of Donna's price, P_{Donna}. In other words, it is exactly what we want, the rule for Pierce's best response.

The other method to derive a best-response rule is to use calculus. Use the first line of the above rearrangement of the expression for Y_{Pierce}, and take its derivative with respect to P_{Pierce} (holding the other store's price, P_{Donna}, fixed):

$$\frac{dY_{\text{Pierce}}}{dP_{\text{Pierce}}} = (15 + 0.5 P_{\text{Donna}}) - 2 P_{\text{Pierce}}$$

The first-order condition for P_{Pierce} to maximize Y_{Pierce} is that this derivative should be zero. Setting it equal to zero and solving for P_{Pierce} gives the same equation as above for the best-response rule. (The second-order condition is $d^2 Y_{\text{Pierce}} / dP_{\text{Pierce}}^2 < 0$, which is satisfied since the second-order derivative is just -2.)

Donna's best response rule can be found similarly, using either method. Since the costs and sales of the two stores are entirely symmetric, the equation is obviously going to be

$$P_{\text{Donna}} = 0.5(15 + 0.5 P_{\text{Pierce}}) = 7.5 + 0.25 P_{\text{Pierce}}.$$

Both rules are used in the same way to develop best-response graphs. If Pierce's sets a price of 16, for example, then Donna's plugs this value into its best-response rule to find $P_{\text{Donna}} = 7.5 + 0.25\,(16) = 11.5$; similarly, Pierce's best response to Donna's 16 is also 11.5, and each store's best response to the other's price of 4 (or 8) is 8.5 (or 9.5).

Figure 4.7 shows the full graphs of these two best-response relations. Owing to the special features of our example, namely the linear relation between quantity sold and prices charged, and the constant cost of producing each pizza, each of the two best-response curves is a straight line. For other specifications of demands and costs the curves can be other than straight, but the method of obtaining them is the same, namely first holding one store's price (say P_{Donna}) fixed and finding the value of the other store's price (say P_{Pierce}) that maximizes the second store's profit, and then the other way around.

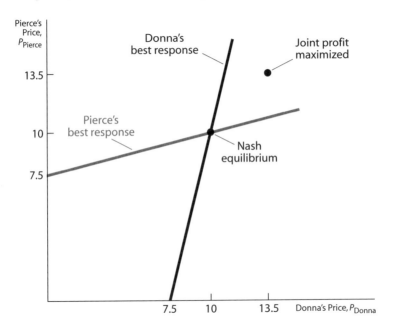

FIGURE 4.7 Best-Response Curves and Equilibrium in the Pizza Pricing Game

The point of intersection of the two best-response curves is the Nash equilibrium of the pricing game between the two stores. The specific values for each store's pricing strategy in equilibrium can be found algebraically by solving the two best-response equations jointly for P_{Donna} and P_{Pierce}. We deliberately chose our example to make the equations linear, and the solution is easy. In this case, we simply substitute the expression for P_{Pierce} into the expression for P_{Donna} to find

$$\begin{aligned}
P_{\text{Donna}} &= 7.5 + 0.25\,P_{\text{Pierce}} \\
&= 7.5 + 0.25\,(7.5 + 0.25\,P_{\text{Donna}}) \\
&= 9.375 + 0.0625\,P_{\text{Donna}}
\end{aligned}$$

This last equation simplifies to $P_{Donna} = 10$; given the symmetry of the problem, it is simple to determine that $P_{Pierce} = 10$ also. Thus, in equilibrium, each store charges $10 and makes a profit of $7 on each of the 7,000 pizzas [$7,000 = (12 - 10 + 0.5 \times 10) \times 1,000$] it sells each week, for a total profit of $49,000.

Our main purpose here is to illustrate how the Nash equilibrium can be found in a game where the strategies are continuous variables, like prices. But it is interesting to take a further look into this example and connect it with what we learned from the pizza example earlier in this chapter.

Begin by observing that each best-response curve slopes upward. Specifically, when one store raises its price by $1, the other's best response is to raise its own price by 0.25, or 25 cents. When one store raises its price, some of its customers switch to the other store, and its rival can then best profit from them by raising its price part of the way. Thus a store that raises its price is helping increase the other's profit. In Nash equilibrium, where each store chooses its price independently and out of concern for its own profit, it does not take into account this benefit it conveys to the other. Could they get together and cooperatively agree to raise their prices, thereby raising both profits? Yes. Suppose the two stores charged $12 each. Then each would make a profit of $9 on each of the 6,000 pizzas [$6,000 = (12 - 12 + 0.5 \times 12) \times 1,000$] it would sell each week, for a total profit of $54,000.

This is exactly like the prisoners' dilemma that the two stores faced when each of them had the choice between just two prices, High and Medium. Their profits were higher when both charged the high price, but each was tempted to cut its own price to divert some sales from its rival toward itself, and when they both did so, both ended up with lower profits. In the same way, the more profitable price of $12 is not a Nash equilibrium. The separate calculations of the two stores will lead them to undercut such a price. Suppose that Donna's somehow starts by charging $12. Using the best-response formula, we see that Pierce's will then charge $7.5 + 0.25 \times 12 = 10.5$, or $10.50. Then Donna's will come back with its best response to that: $7.5 + 0.25 \times 10.5 = 10.125$, or about $10.13. Continuing this process, the prices of both will converge toward the Nash equilibrium price of $10.

But what price is jointly best for the two stores? Given the symmetry, suppose both charge the same price P. Then the profit of each will be

$$Y_{Donna} = Y_{Pierce} = (P-3)(12-P+0.5\,P) = (P-3)(12-0.5\,P)$$

The two can choose P to maximize this expression. Using either of the above methods, it is easy to see that the solution is $P = 13.5$, or $13.50. The resulting profit for each store is $55,125 per week.

In the jargon of economics, such collusion to raise prices to the jointly optimal level is called a *cartel*. The high prices hurt consumers, and regulatory agencies of the U.S. government often try to prevent the formation of cartels and to make firms compete with each other. Explicit collusion over price is illegal, but it may be possible to maintain tacit collusion in a repeated prisoners' dilemma, as we have hinted before and will study in detail in Chapter 8. (In some other countries, governments are themselves in collusion with producers and actually foster cartels.)

10 THREE PLAYERS

So far we have analyzed only games between two players. All of the methods of analysis that we have discussed, however, can be used to find the pure-strategy Nash equilibria of any simultaneous-play game among any number of players. When a game is played between more than two players, each of whom has a relatively small number of pure strategies, the analysis can be done with a game table as we did in the first eight sections of this chapter.

Let us consider a game among three players, each of whom has two pure strategies. The three players, Emily, Nina, and Talia, all live on the same small street. They have each been asked to contribute toward the creation of a flower garden at the intersection of their small street with the main highway. The town has promised to match the contributions made by any, or all, of the three in order to provide a very pleasant garden. Of course, the ultimate size and splendor of the garden will depend on how many of them contribute. Furthermore, while each player is happy to have the garden—and happier as its size and splendor increase—each is reluctant to contribute because of the cost she must incur to do so.

Suppose Emily is contemplating the possible outcomes of the street-garden game. There are six possibilities to consider. Emily can choose either to contribute or not to contribute when both Nina and Talia contribute, or when neither of them contribute, or when just one of them contributes. From her perspective, the best-possible outcome, with a rating of 6, would be to take advantage of her good-hearted neighbors and to have both Nina and Talia contribute while she did not. Emily could then enjoy a medium-size garden without having had to put up her own hard-earned cash. If both of the others have contributed and Emily also contributes, she gets to enjoy a large, very splendid garden but at the cost of her own contribution; she rates this outcome second-best, or 5.

At the other end of the spectrum are the outcomes that arise when neither Nina nor Talia contributes to the garden. If that were the case, Emily would again prefer not to contribute because she would be footing the bill for a public garden that everyone could enjoy; she would rather have the flowers in her own yard. Thus when neither other player is contributing, Emily ranks the outcome in which she contributes as a 1 and the outcome in which she does not as a 2.

In between these cases are the situations in which either Nina and Talia contributes to the flower garden but not both. When one of them contributes, Emily knows that she can enjoy a small garden without contributing; she also feels that the cost of her contribution outweighs the increase in benefit that she gets from being able to increase the size of the garden. Thus she ranks the outcome in which she does not contribute, but still enjoys the small garden, as a 4 and the outcome in which she does contribute, to provide a medium garden, as a 3. Because Nina and Talia have the same views as Emily on the costs and benefits of contributions and garden size, each of them orders the different outcomes in the same way—the worst outcome being the one in which oneself contributes and the other two do not, and so on.

If all three women decide whether to contribute to the garden without knowing what their neighbors will do, we have a three-person simultaneous-move game. To find the Nash equilibrium of the game, we then need a game table. For a three-player game, the table would need to be three-dimensional and the third player's strategies would correspond to the new dimension. The easiest way to add a third dimension to a two-dimensional game table is to add pages: the first page of the table shows payoffs for the third player's first strategy, the second page shows payoffs for the third player's second strategy, and so on.

We show the three-dimensional table for the street-garden game in Figure 4.8. It has two rows for Emily's two strategies, two columns for Nina's two strategies, and two pages for Talia's two strategies. We show the pages side by side so you can see everything at the same time. In each cell, payoffs are listed for the row player first, the column player second, and the page player third; in this case, the order is Emily, Nina, Talia.

TALIA chooses:

Contribute

		NINA	
		Contribute	Don't
EMILY	Contribute	5, 5, 5	3, 6, 3
	Don't	6, 3, 3	4, 4, 1

Don't Contribute

		NINA	
		Contribute	Don't
EMILY	Contribute	3, 3, 6	1, 4, 4
	Don't	4, 1, 4	2, 2, 2

FIGURE 4.8   Street-Garden Game

Our first test should be to determine whether there are dominant strategies for any of the players. In one-page game tables, we found this test to be simple; we just compared the outcomes associated with one of a player's strategies with the outcomes associated with another of her strategies. In practice this required, for the row player, a simple check within columns of the single page of the table, and vice versa for the column player. Here, we must check in both pages of the table to determine whether any player has a dominant strategy.

For Emily, we compare the two rows of both pages of the table and note that when Talia contributes, Emily has a dominant strategy not to contribute, *and* when Talia does not contribute, Emily also has a dominant strategy not to contribute. Thus the best thing for Emily to do, regardless of what either of the other player's does, is not to contribute. Similarly, we see that Nina's dominant strategy—in both pages of the table—is not to contribute. When we check for a dominant strategy for Talia, we have to be a bit more careful. We must compare outcomes that keep Emily and Nina's behavior constant, checking Talia's payoffs from choosing Contribute versus Don't. That is, we compare cells *across* pages of the table—the top left cell in the first page (on the left) with the top left cell in the second page (on the right), and so on. As for the first two players, this process indicates that Talia also has a dominant strategy not to contribute.

Each player in this game has a dominant strategy, which must therefore be her equilibrium pure strategy. The Nash equilibrium of the street-garden game entails all three players choosing not to contribute to the street garden and getting their second-worst payoffs; the garden is not planted, but no one has to contribute either.

Notice that this game is yet another example of a prisoners' dilemma. There is a unique Nash equilibrium in which all players receive a payoff of 2. Yet there is another outcome in the game—in which all three neighbors contribute to the garden—that for all three players yields higher payoffs of 5. Even though it would be beneficial to each of them for all to pitch in to build the garden, no one has the individual incentive to do so. As a result, gardens of this type are either not planted at all or paid for through tax dollars—because the town government can require its citizens to pay such taxes. In Chapter 11 we will encounter more such dilemmas of collective action and study some methods for resolving them.

Before we leave the street-garden game, you should note that we could have approached the search for a Nash equilibrium of the game using the cell-by-cell

TALIA chooses:

Contribute

		NINA	
		Contribute	Don't
EMILY	Contribute	$\cancel{5},\cancel{5},\cancel{5}$	$\cancel{3},6,\cancel{3}$
	Don't	6,$\cancel{3}$,$\cancel{3}$	4,4,$\cancel{1}$

Don't Contribute

		NINA	
		Contribute	Don't
EMILY	Contribute	$\cancel{3},\cancel{3}$,6	$\cancel{1}$,4,4
	Don't	4,$\cancel{1}$,4	2,2,2

FIGURE 4.9 Cell-by-Cell Inspection in the Street-Garden Game

inspection method. Because you may encounter multiple-person games in which you will need this method, Figure 4.9 shows you the same game table as it would look after conducting such an inspection. Diagonal lines through individual payoffs indicate which player would want to change her strategy.

Note that when Emily considers changing her strategy, as the row player she can change only the row position of the game's outcome. Emily can move the outcome only from a given cell in a given row, column, and page to another cell in a different row but the same column and same page of the table. Similarly, Nina can change only the column position of the outcome, moving it to a cell in another column but in the same row and same page of the table. Finally, Talia can change only the page position of the game's outcome. She can move the outcome to a different page, but the row and column positions must remain the same.

11 MULTIPLE EQUILIBRIA IN PURE STRATEGIES

Although the many examples provided so far in this chapter give the impression that each game has only a single, unique Nash equilibrium, it is possible for there to be situations in which there are two (or more) possible equilibria within a single game. Such multiple equilibria arise in several different contexts. The equilibria may involve no conflict in payoffs among the players of the game so that all players prefer the same equilibrium. These are games of **assurance** and are discussed in Section 11.A. Another possibility is that, in a game with two players and two pure-strategy Nash equilibria, each player strictly prefers one of the equilibria and prefers another outcome to the second equilibrium. Such situations arise in games of **chicken,** which are considered in Section 11.B. A third possibility is that a game might have two equilibria, one of which is strictly preferred by each player, as in chicken, but players might prefer both equilibria over the other possible outcomes of the game. If the payoff pattern of a game matches this description, we have what we call a **battle of the two cultures,** described in section 11.C. Finally, we consider in Section 11.D the consequences of iterated elimination of dominated strategies on a game in which some of the eliminated strategies are only weakly dominated and in which multiple Nash equilibria exist.

A. Assurance

Our example of this class of games is the nuclear arms race between the two superpowers after World War II; this interpretation of that situation comes from political scientist Robert Jervis.[5] Each of the two players, the United States and the

---

[5] Robert Jervis, "Cooperation Under the Security Dilemma," *World Politics*, vol. 30,(1978) pp. 167–214.

U.S.S.R., has the choice either to build more weapons or to refrain from doing so. The best outcome for each is when neither builds—the balance of power is maintained, and the resources that would have gone to armaments can instead be deployed to improve the standard of living of the citizens. This outcome rates a 4 for each player. The worst outcome for each country, rated 1, is the one in which it refrains but the other builds; this endangers its security and forces it to offer concessions in many situations involving negotiations with the other power. The outcome in which both build is rated 2 by each. Each gives a rating of 3 to the outcome in which it builds and the other refrains. Figure 4.10 shows the game table. Cell-by-cell inspection quickly reveals that the game has two Nash equilibria—one in which both refrain and the other in which both build.

		U.S.S.R.	
		Refrain	Build
U.S.	Refrain	4,4	1,3
	Build	3,1	2,2

FIGURE 4.10 The Arms Race as an Assurance Game

Both countries would clearly prefer the equilibrium in which both refrain, but that does not automatically ensure that each will choose the Refrain strategy when playing this game. If one player thought that the other would for some reason choose Build, then it should choose Build too. To reach one of the game's Nash equilibria, the two have to coordinate their choices, but the coordination can be tacit. To be induced to choose Refrain, what each really needs is the *assurance* that the other will also choose Refrain. Hence the name for this type of game. How might the two achieve such assurance and thereby coordination?[6] There are two likely approaches, one based on the use of strategic moves, the other on the existence or creation of a focal point.

An individual player can affect the choice of equilibrium in an assurance game by deciding which equilibrium she prefers and making a preemptive move aimed at achieving that equilibrium. In the arms-race example, the United States could simply pass a law or a constitutional amendment by which it committed itself never to build nuclear weapons. With this assurance, according to the payoff table of Figure 4.10, the U.S.S.R. would choose to refrain too. Actions like those of the United States in this example are called **strategic moves** and are used by players to influence the outcome of a game. There is much more to the

[6]The need for coordination to get to equilibrium means that games of this type are also often called *coordination games.*

use of strategic moves than the simple attainment of the mutually preferred outcome in an assurance game, however, as we will see in Chapter 9.

Coordination can also arise if one of the multiple equilibria is a **focal point,** namely an outcome in which players have a common understanding that, of all the possible equilibria of the game, this one is the obvious one to choose. Focal points thus require a **convergence of expectations** on the part of the players in the game. It is not enough that the United States thinks that countries in this situation should play Refrain. It must believe that the U.S.S.R. thinks likewise, and that the U.S.S.R. believes that the United States thinks likewise, and so on.

Before we examine in more generality how such a convergence of expectations can be brought about and how the preferred outcome in assurance games can be obtained, we make two remarks about the example in Figure 4.10. Jervis compares this game to the *stag hunt* described by the 18th-century French philosopher Jean-Jacques Rousseau. If all the hunters in a group cooperate, they can encircle and trap a stag, but if one of them goes off to chase a rabbit, the circle is broken and the stag escapes. However, from Jervis's description it is not clear whether each hunter will find it in its individual interest to cooperate if all the others are doing so. If not, the payoffs of 3 and 4 in Figure 4.10 must be interchanged, and the game becomes a prisoners' dilemma. In fact, other people have interpreted the stag hunt as a prisoners' dilemma; we will see an example in Chapter 11.

In the specific example of the nuclear arms race, we know that from the 1950s through the 1980s this game had the (Build, Build) outcome, not the (Refrain, Refrain) outcome. Perhaps the payoffs for the United States and the U.S.S.R. were not as described in Figure 4.10, and they were playing a prisoners' dilemma rather than an assurance game. But there is another possibility. If they did not know each other's payoffs, then each might suspect the other of wanting to Build regardless, or each might suspect that the other might suspect this in turn, and so on. Indeed, when we mentioned the possibility of the United States unilaterally giving up nuclear weapons, you probably thought that would be a foolish thing to do, and in thinking this you probably had in mind some such suspicion about the U.S.S.R. In Chapter 12 we will see how such lack of information can lead to a bad outcome for players.

Back to focal points. If one equilibrium assumes higher payoffs for both players than another, as in an assurance game, then that higher payoff by itself may achieve the necessary convergence of all players' expectations. But that is not the only consideration. For example, if a two-player game has three Nash equilibria, two of them yielding payoffs of 7 to each player and the third yielding only 6, then the players acting separately, unable to think which of the two equilibria yielding 7 they could coordinate on, might regard the equilibrium yielding 6 to each as the focal point because of its uniqueness.

In *all* games with multiple equilibria, (not just in the assurance game), one of the equilibria can emerge as the outcome if it becomes a focal point on which the expectations of the players converge. One player may dislike the outcome that emerges, but once it has become focal, she cannot do any better than to play her strategy in that equilibrium. Unlike in the assurance game, however, one player may get a higher payoff in one equilibrium and the other player a higher payoff in another equilibrium. Therefore rankings of payoffs by themselves cannot achieve such convergence. Furthermore, in a game of "pure" coordination—for example, a game in which the players must independently choose one of two possible places to meet, and it doesn't matter what the choice is so long as everyone makes the same choice—both equilibria may entail identical payoffs; again, the payoff rankings will not help players to coordinate their moves. This makes the creation of the common understanding and the emergence of a focal point much harder. Successful convergence of expectation on one equilibrium often depends on facts beyond the mere mathematical description of strategies and payoffs in the game, for example, on the players' backgrounds. Are they of the same nationality? Do they share a similar history, culture, or language? Are they acquainted with each other, or with the conventions or norms of behavior of each other's identifying group? In some cases, such cultural norms will make one equilibrium the obvious choice while in other cases it may be impossible to attain the common understanding necessary for coordination.

Consider the example of a game in which two players are being held in separate rooms. Each is asked to pick one of the following colors: gold, green, or orange. If the two pick the same color, each will be given $10. If they pick different colors, neither will get anything. If such an experiment is conducted with Dartmouth ("Big Green") students, they invariably choose green; at Montana State (home of the "Blue and Gold"), students invariably choose gold; and at the University of Illinois (known for the "Orange Krush"), the choice is orange. If the experiment is done with a mixed group, when one player doesn't know where the other comes from, chances are that each will pick the color of his or her own alma mater, and two-thirds of the time they will end up with nothing. Other similar examples of focal-point equilibria were made famous by Thomas Schelling.[7]

B. Chicken

Choosing among two or more Nash equilibria sometimes involves more than the convergence of expectations needed in assurance games. The well-known game of chicken is an example in which new strategic considerations arise. Again, the general pattern of payoffs associated with chicken can arise in many different circumstances, but the most famous example revolves around a game that was supposedly played by American teenagers in the 1950s. Two teenagers take their

[7] Thomas C. Schelling, *The Strategy of Conflict* (Cambridge, Mass.: Harvard University Press, 1960), pp. 54–58.

cars to opposite ends of Main Street, Middle-of-Nowhere, USA, at midnight and start to drive toward each other. The one who swerves to avoid a collision is the "chicken," and the one who keeps going straight is the winner. If both maintain a straight course, there is a collision in which both cars are damaged and both players injured.[8]

The payoffs for chicken depend on how negatively one rates the "bad" outcome—being hurt and damaging your car in this case—against being labeled chicken. As long as words hurt less than crunching metal, a reasonable payoff table for the 1950s version of chicken is found in Figure 4.11. Each player most prefers to win, having the other be chicken, and each least prefers the crash of the two cars. In between these two extremes, it is better to have your rival be chicken with you (to save face) than to be chicken by yourself.

		DEAN	
		Swerve (Chicken)	Straight (Tough)
JAMES	Swerve (Chicken)	0,0	−1,1
	Straight (Tough)	1,−1	−2,−2

FIGURE 4.11 Chicken

The table in Figure 4.11 is not dominance solvable (you should check this) and has no unique pure-strategy Nash equilibrium (you should check this too). Instead, this game illustrates the four essential features that define any game of chicken. First, each player has one strategy that is the "tough" strategy and one that is the "weak" strategy. Second, there are two pure-strategy Nash equilibria. These are the outcomes in which exactly one of the players is chicken, or weak. Third, each player strictly prefers that equilibrium in which the other player

---

[8]A slight variant was made famous by the 1955 James Dean movie, *Rebel Without a Cause*. There two players drive their cars in parallel, very fast, toward a cliff. The first to jump out of his car before it went over the cliff was the chicken. The other, if he left it too late, risked going over the cliff in his car to his death. The characters in the film referred to this as a "chicky-game." In the mid-1960s, the British philosopher Bertrand Russell and other peace activists used chicken as an analogy for the nuclear arms race between the United States and the U.S.S.R., and the game theorist Anatole Rappoport provided a formal game-theoretic analysis. Other game theorists have chosen to interpret the arms race as a prisoners' dilemma or as an assurance game, as mentioned earlier. For a review and interesting discussion, see Barry O'Neill, "Game Theory Models of Peace and War," in *The Handbook of Game Theory*, vol. 2, ed. Robert J. Aumann and Sergiu Hart (Amsterdam: North Holland, 1994), pp. 995–1053.

chooses chicken, or weak. Fourth, the payoffs when both players are tough are very bad for both players. In games like this, the real game becomes a test of how to achieve one's preferred equilibrium.

We are now back in a situation similar to that discussed in games of pure co-ordination. Players must use some method to arrive at one of the game's two equilibria. Again, rule making, strategic moves, or focal points can help attain an equilibrium. However, in chicken, players have strict preferences about which equilibrium they prefer, and each will want to try to influence the outcome. In addition, both players also want to try to avoid the bad (crash) outcome if at all possible. It may be the case that one player will try to create an aura of toughness that everyone recognizes in order to intimidate all rivals.[9] Another possibility is to come up with some other way to convince your rival that you will not be chicken, to make a *commitment* to going straight. We will discuss just how to make such commitment moves, a type of strategic move like the preemptive move mentioned above, in Chapter 9.

Finally, chicken can be "solved" by repetition. That is, if the teenagers played the game every Saturday night at midnight, they would have the benefit of knowing that the game had both a history and a future when deciding their equilibrium strategies. In such a situation, they might logically choose to alternate between the two equilibria, taking turns being the winner every other week. Repetition of games can be crucial to the potential for reaching an agreeable solution, not only in chicken but also in the prisoners' dilemma (as we will see in greater detail in Chapter 8) and in the battle-of-the-two-cultures game considered in the next section.

C. The Battle of the Two Cultures

Our third example of a game with multiple Nash equilibria has its own set of essential features that set it apart from the games we have already considered. It contains some similarities to chicken but is usually analyzed separately. The battle of the two cultures gets its name from C. P. Snow's famous essay entitled *The Two Cultures*, which pointed out and lamented the divide between scientists and humanists.[10] The game has two players, each of whom have the same two possible actions, one of which is preferred by each player. The game also has two pure-strategy Nash equilibria, one of which is strictly preferred by each player. The players have some common interests, though, in that they would both pre-

---

[9] Why would a potential rival play chicken against someone with a reputation for never giving in? The problem is that participation in chicken, as in lawsuits, is not really voluntary. Put another way, participation in chicken is, in itself, a game of chicken. As Schelling says, "If you are publicly invited to play chicken and say you would rather not, then you have just played [and lost]" (Thomas C. Schelling, *Arms and Influence*, New Haven, Conn.: Yale University Press, 1965, p. 118).

[10] C.P. Snow, *The Two Cultures and the Scientific Revolution* (New York: Cambridge University Press, 1959).

fer to choose the same strategy rather than doing different things. Again the need for coordination arises.[11]

As an example, consider a university that is contemplating renovation of an old, antiquated lecture hall. Two groups of faculty have suggested a use for the space. The scientists hope to have it made into a laboratory; the humanists would like to see it become a small theater. The university president has indicated her willingness to allocate funds for the renovation, but only in the presence of unified faculty opinion on the final product. Both groups of faculty would like to see *something* done with the space, although the scientists prefer the lab and the humanists prefer the theater (as fits in with their suggestions). Neither group wants to let the possibility of a renovation die, especially since this could jeopardize any hopes for future campus improvement projects.

The two faculty groups must decide which proposal to endorse, but their meetings on the subject are scheduled for exactly the same time. Thus they find themselves in a simultaneous-play game with a payoff table as illustrated in Figure 4.12. Notice that this game has no dominance and that cell-by-cell inspection yields two Nash equilibria along the diagonals of the table. The scientists prefer the (Lab, Lab) equilibrium and the humanists prefer the (Theater, Theater) equilibrium. Neither groups wants to end up in one of the off-diagonal cells.

		HUMANITIES FACULTY	
		Lab	Theater
SCIENCE FACULTY	Lab	2,1	0,0
	Theater	0,0	1,2

FIGURE 4.12 Battle of the Two Cultures

Notice that this game is similar to chicken in that the two players prefer different equilibria. A large difference between the two games, however, is that both players prefer both equilibria to either of the other possible outcomes. Furthermore, there is no exceptionally bad payoff associated with either out-of-equilibrium outcome; the penalty for not coordinating to reach an equilibrium here is not as high as it is in chicken.

The need for coordination, however, is still important. As before, rules (such

[11]This game appears in the literature as the *Battle of the Sexes* and is based on an interaction between a man and a woman, each of whom wants to spend an evening together, but the man prefers to go to a boxing match and the woman to a ballet. As our battle-of-the-two-cultures version shows, the properties of the game have no necessary connection to assumptions of such gender-stereotyped preferences.

as a fire code that restricts the capacity of the space in question), history (for example, if the scientists got new lab space in the last big renovation), or a focal point (say, the available space is in the art building) might help the faculty reach an equilibrium in this game. Players might also have some scope for bargaining with each other here (as indeed they might also in chicken) with the unpopular outcomes possible in the case of a disagreement. More specific analysis of the bargaining outcomes are addressed in Chapter 16.

D. Weak Dominance

We noted earlier that successive elimination of dominated strategies must be used with caution in games in which there might be multiple Nash equilibria. To make this point more clearly, we consider a specific example. Suppose a lottery has a prize of $10, and each ticket costs $5. There are just two contestants, A and B. Each can buy either zero, one, or two tickets, and they make these choices simultaneously and independently. Then, knowing the contestants' ticket-purchase decisions, the game organizer awards the prize. If neither player has bought any tickets, the prize is not awarded. If the two have bought equal numbers of tickets, the prize is split equally between them. If one has bought more tickets than the other, then the larger buyer gets the prize.

Each player's payoff equals the winnings *minus* the cost of tickets. Figure 4.13 shows the complete payoff table. Each player's strategies are the number of tickets bought; the payoffs are in dollars.

This game is dominance solvable. For player A, the strategy of buying two tickets is dominated by either of the other strategies, 0 or 1. Likewise for player B. Then for each player, 0 is dominated by 1, leaving (1, 1) as the outcome.

But suppose we use cell-by-cell inspection to solve the game. We will find that (0, 1) is also a Nash equilibrium. When player A is buying no tickets and B is buying 1, clearly B does worse to switch her strategies to buying 0 or 2. But when

		B		
		0	1	2
A	0	0, 0	0, 5	0, 0
	1	5, 0	0, 0	−5, 0
	2	0, 0	0, −5	−5, −5

FIGURE 4.13 Lottery

B is buying 1, A cannot benefit by switching to 1 or 2: her payoff stays at 0. Neither player has any clear incentive to change her own choice; thus (0, 1) complies with the definition of a Nash equilibrium. Similarly, (1, 0) is a Nash equilibrium too. Thus iterated elimination of dominated strategies has left us with a single Nash equilibrium, but it was not able to identify *all* of the Nash equilibria that existed in this game.

And the matter is even more complex. When we showed that this game was dominance solvable, we eliminated strategy 2 for both players first, and then we eliminated strategy 0 for both players. Suppose instead we had eliminated strategies for only one player at a time. Start with A, and eliminate 2. Look at B's strategies in the remaining two-by-three table: both 0 and 2 are dominated by 1. This leaves a two-by-one table where choices 0 and 1 remain for A and only 1 remains for B. Look at this from A's perspective: she gets the same payoff, namely 0, from both of her remaining choices. Therefore no elimination is possible, and both strategy combinations, (0, 1) and (1, 1), are seen as Nash equilibria. Of course, they are included in the full set of three that we found using cell-by-cell inspection, but one, namely (1, 0), is missing.

Similarly, if we had started the elimination with B, then we would have gotten a different selection of two of the three Nash equilibria, namely (1, 0) and (1, 1).

What goes wrong? The problem is that the dominance relations we are relying on for elimination in this game are weak, not strong. For example, after having eliminated A's strategy 2, we were left with a two-by-three table, in which we eliminated B's strategy of 0 because it was weakly dominated by 1. But B's payoff from 1 was actually better than that from 0 only if A was playing her 0; B's payoffs from the two strategies were equal when A was playing 1. Thus it remained possible that A playing 1 and B playing 0 could be an equilibrium, but our elimination forgot about this.

Which equilibria does weak elimination find if we choose a particular order of players? To determine the answer to this question, imagine for a moment that the rules of the game are switched from simultaneous to sequential play, where the two players take turns. The order of play is the *reverse* of the order in which we carry out the elimination of weakly dominated strategies. Thus if we begin by eliminating A's strategy 2, then in the hypothetical sequential-move game we should have A moving last. With this process, the Nash equilibria revealed by iterated elimination of weakly dominated strategies will be the same as the rollback equilibria of the sequential-move game. We reexamine and illustrate this idea when we consider sequential-move games in strategic form in Chapter 6.

More generally, if you use weak dominance solvability, be aware that you may have missed some Nash equilibria. If it is important to find all, then you should do a cell-by-cell inspection too.

12 NO EQUILIBRIUM IN PURE STRATEGIES

Each of the games that we have considered so far has had at least one Nash equilibrium in pure strategies. Some of these games, such as those in Section 11, had more than one equilibrium, while games in earlier sections had exactly one. Unfortunately, not all games that we come across in the study of strategy and game theory will have such easily definable outcomes in which players always choose one particular action as their equilibrium strategy. In this section, we look at games in which there is not even one pure-strategy Nash equilibrium—games in which none of the players would consistently choose one strategy as their equilibrium action.

A simple example of a game with no equilibrium in pure strategies is that of a single point in a tennis match. Imagine a match between two of the top women's players like Martina Hingis and Monica Seles. Hingis at the net has just volleyed a ball to Seles on the baseline, and Seles is about to attempt a passing shot. She can try to send the ball either down the line (DL; a hard, straight shot) or crosscourt (CC; a softer, diagonal shot). Hingis must likewise prepare to cover one side or the other. Each player is aware that she must not give any indication of her planned action to her opponent, knowing that such information would be used against her. Hingis would move to cover the side to which Seles is planning to hit, or Seles would hit to the side Hingis is not planning to cover. If we suppose that both are equally good at concealing their intentions until the last possible moment, then their actions are effectively simultaneous and we can analyze the point as a two-player simultaneous-move game.

Payoffs in this tennis-point game will be the fraction of times a player wins the point in any particular combination of passing shot and covering play. Given that a down-the-line passing shot is stronger than a crosscourt shot, and that Seles is more likely to win the point when Hingis moves to cover the wrong side of the court, we can work out a reasonable set of payoffs. Seles will be successful with a down-the-line passing shot 80% of the time if Hingis covers crosscourt; she will be successful with the down-the-line shot only 50% of the time if Hingis covers down the line. Similarly, Seles will be successful with her crosscourt passing shot 90% of the time if Hingis covers down the line. This is more often than when Hingis covers crosscourt, in which case Seles wins only 20% of the time.

Clearly, the fraction of times that Hingis wins this tennis point is just the difference between 100% and the fraction of time that Seles wins. Thus the game is zero-sum (more precisely, *constant-sum*, because the two payoffs sum to 100), and we can represent all the necessary information in the payoff table with just the payoff to Seles in each cell. Figure 4.14 shows the payoff table and the fraction of time Seles wins the point against Hingis in each of the four possible combinations of their strategy choices.

		HINGIS	
		DL	CC
SELES	DL	50	80
	CC	90	20

FIGURE 4.14   No Equilibrium in Pure Strategies

The rules for solving simultaneous-move games tell us to look first for dominant or dominated strategies, then to try minimax or cell-by-cell inspection to find a Nash equilibrium. It is a useful exercise to verify that no dominant strategies exist here. Going on to cell-by-cell inspection, we start with the choice of DL for both players. From that outcome, Seles can improve her success from 50 to 90% by choosing CC instead. But then Hingis can hold Seles down to 20% by choosing CC. Following this, Seles can raise her success again to 80% by making her shot DL, and Hingis in turn can do better with DL. In every cell, one player always wants to change her play, and we cycle through the table endlessly without finding an equilibrium.

There is an important message contained in the absence of a Nash equilibrium in this and similar games: what is important in games of this type is not what players *should* do, but what players should *not* do. In particular, neither player should always or systematically pick the same shot when faced with this situation. If either player engages in any determinate behavior of that type, the other can take advantage of it. (So, if Seles consistently went crosscourt with her passing shot, Hingis would learn to cover crosscourt every time and would thereby reduce Seles's chances of success with her crosscourt shot.) The most reasonable thing for players to do here is to act somewhat unsystematically, hoping for the element of surprise in defeating their opponents. An unsystematic approach would entail choosing each strategy part of the time. (Seles should be using her weaker shot with enough frequency to guarantee that Hingis cannot predict which shot will come her way. She should not, however, use the two shots in any set pattern because that too would lose her the element of surprise.) This approach, in which players randomize their actions, is known as *mixing strategies* and is the focus of Chapter 5. The game illustrated in Figure 4.14 may not have an equilibrium in pure strategies, but it can still be solved by looking for an equilibrium in mixed strategies, as we do in Chapter 5, Section 3.

SUMMARY

Simultaneous-move games differ from sequential-move games in that players make decisions without knowing their rivals' actions. Such games are illustrated using *game tables* where cells show payoffs to each player and the dimensionality of the table equals the number of players. Two-person *zero-sum games*, in which payoffs sum to the same value in each possible outcome, may be illustrated in shorthand with only one player's payoff in the cells of the game table.

Nash equilibrium is the solution concept used to solve simultaneous-move games; such an equilibrium exists when each player chooses the strategy that is best for her, given that all other players are using their equilibrium strategies. Nash equilibria may entail *pure* or *mixed strategies*, although the focus of this chapter is on pure strategies. Nash equilibria can be found by using a hierarchy of methods, beginning with the *search for dominant strategies* and *successive elimination of dominated strategies* from consideration as equilibrium moves. Other methods in the hierarchy include *minimax* and *cell-by-cell inspection*. When players have pure strategies that are continuous variables, the method of *best-response analysis* is used to find the equilibrium. In games with no single pure-strategy Nash equilibrium, there may be multiple equilibria (that may include a mixture) or no equilibrium (unless mixtures are considered).

Several specific games, including the *prisoners' dilemma, assurance, chicken,* and the *battle of the two cultures,* are analyzed in this chapter. Such games are important not only for their particular solutions but because their payoff patterns arise in a wide range of interactions in a variety of fields.

KEY TERMS

assurance game (107)

battle of the two cultures (107)

best-response analysis (100)

best-response curve (100)

cell-by-cell inspection (97)

chicken game (107)

constant-sum game (80)

convergence of expectations (109)

dominance-solvable game (90)

dominant strategy (83)

dominated strategies (84)

focal point (109)

game table (80)

game matrix (80)

maximin (94)

minimax (94)

minimax method (94)

mixed strategies (83)

Nash equilibrium (82)

non-zero-sum game (80)

normal form (80)

payoff table (80)

prisoners' dilemma (85)

pure strategies (82)

reaction curve (100)

strategic form (80)

strategic moves (108)
strictly dominant strategy (83)
successive or iterated elimination
   of dominated strategies (90)
symmetric game (92)

variable-sum game (80)
weakly dominant strategy
   (84)
zero-sum game (80)

EXERCISES

1. Find all Nash equilibria in pure strategies for the games in the tables below by checking for dominant strategies and using iterated dominance. Verify your answers using the minimax method.

(a)

		COLUMN	
		Left	Right
ROW	Up	1	4
	Down	2	3

(b)

		COLUMN	
		Left	Right
ROW	Up	1	2
	Down	4	3

(c)

		COLUMN		
		Left	Middle	Right
ROW	Up	5	3	1
	Straight	6	2	1
	Down	1	0	0

(d)

		COLUMN		
		Left	Middle	Right
ROW	Up	5	3	2
	Straight	6	4	3
	Down	1	6	0

2. Find all Nash equilibria in pure strategies in the following non-zero-sum games. Describe the steps you used in finding the equilibria.

(a)

		COLUMN	
		Left	Right
ROW	Up	2,4	1,0
	Down	6,5	4,2

(b)

		COLUMN	
		Left	Right
ROW	Up	1,1	0,1
	Down	1,0	1,1

(c)

		COLUMN		
		Left	Middle	Right
ROW	Up	0,1	9,0	2,3
	Straight	5,9	7,3	1,7
	Down	7,5	10,10	3,5

3. "If a player has a dominant strategy in a simultaneous-move game, then she is sure to get her best outcome." True or false? Explain and give an example of a game that illustrates your answer.

4. Find pure-strategy Nash equilibria for the non-zero-sum game illustrated below. Describe the process you use to find the equilibria. Use this game to

explain why it is important to describe an equilibrium using the strategies of the players, not merely their payoffs.

		COLUMN		
		Left	Center	Right
ROW	Up	1,2	2,1	1,0
	Level	0,5	1,2	7,4
	Down	−1,1	3,0	5,2

5. Check the following game for dominance solvability. Find a Nash equilibrium.

		COLUMN		
		Left	Middle	Right
ROW	Up	4,3	2,7	0,4
	Down	5,5	5,−1	−4,−2

6. An old lady is looking for help crossing the street. Only one person is needed to help her; more are okay but no better than one. You and I are the two people in the vicinity who can help; each has to choose simultaneously whether to do so. Each of us will get pleasure worth a 3 from her success (no matter who helps her). But each one who goes to help will bear a cost of 1, this being the value of our time taken up in helping. Set this up as a game. Write the payoff table, and find all pure-strategy Nash equilibria.

7. In the following game table, in each cell the first entry is the Row player's payoff. Find all Nash equilibria. Give the steps of your reasoning in four or five short sentences.

		COLUMN		
		Left	Middle	Right
ROW	Top	3,1	2,3	10,2
	High	4,5	3,0	6,4
	Low	2,2	5,4	12,3
	Bottom	5,6	4,5	9,7

8. Consider the following game of "pure" coordination:

		COLUMN	
		A	B
ROW	A	1, 1	0, 0
	B	0, 0	1, 1

Find all of the Nash equilibria of this game. If there are multiple equilibria, is one of them obviously focal? If not, what types of considerations might help players to coordinate?

9. Suppose two players, A and B, select from three different numbers, 1, 2, and 3. Both players get dollar prizes if their choices match, as indicated in the following table.

		B		
		1	2	3
A	1	10, 10	0, 0	0, 0
	2	0, 0	15, 15	0, 0
	3	0, 0	0, 0	15, 15

   (a) What are the Nash equilibria of this game? Which, if any, is likely to emerge as the (focal) outcome? Explain.
   (b) Consider a slightly changed game in which the choices are again just numbers but the two cells with (15, 15) in the above table become (25, 25). What is the expected (average) payoff to each player if each flips a coin to decide whether to play 2 or 3? Is this better than focusing on (1; 1) as a focal equilibrium? How should you account for the risk that A might do one thing while B does the other?

10. Two players, Jack and Jill, are put in separate rooms. Then each is told the rules of the game. Each is to pick one of six letters; G, K, L, Q, R, or W. If the two happen to choose the same letter, both get prizes as follows:

Letter	G	K	L	Q	R	W
Jack's prize	3	2	6	3	4	5
Jill's prize	6	5	4	3	2	1

If they choose different letters, each gets 0. This whole schedule is revealed

to both players, and both are told that both know the schedules, and so on.
  (a) Draw the table for this game. What are the Nash equilibria in pure strate-
      gies?
  (b) Can one of the equilibria be a focal point? Which one? Why?

11. Consider a game in which there is a prize worth $30. There are three contes-
    tants, A, B, and C. Each can buy a ticket worth $15 or $30, or not buy a ticket
    at all. They make these choices simultaneously and independently. Then,
    knowing the ticket-purchase decisions, the game organizer awards the prize.
    If no one has bought a ticket, the prize is not awarded. Otherwise, the prize is
    awarded to the buyer of the highest-cost ticket if there is only one such
    player, or split equally between two or three if there are ties among the high-
    est-cost ticket buyers. Show this game in strategic form. Find all pure-strategy
    Nash equilibria.

12. Yuppietown has two food stores, La Boulangerie, which sells bread, and La
    Fromagerie, which sells cheese. It costs $1 to make a loaf of bread and $2 to
    make a pound of cheese. If La Boulangerie's price is P_1 dollars per loaf of
    bread, and La Fromagerie's price is P_2 dollars per pound of cheese, their re-
    spective weekly sales, Q_1 thousand loaves of bread and Q_2 thousand pounds
    of cheese, are given by the following equations:

    $$Q_1 = 10 - P_1 - 0.5\,P_2, \qquad Q_2 = 12 - 0.5\,P_1 - P_2.$$

  (a) Following the same steps of reasoning and calculation as in the pizza-
      store example of the text, find the two stores' best-response curves and
      show that in the Nash equilibrium the prices are $P_1 = 4$ and $P_2 = 6$.
  (b) If the two stores collude and set prices jointly to maximize the sum of
      their profits, show that the prices will be lower, $P_1 = 3.17$ and $P_2 = 5.67$.
  (c) Bread and cheese are mutual *complements*. They are often consumed to-
      gether; that is why a drop in the price of one increases the sales of the
      other. The products of the two pizza stores were *substitutes* for each
      other; that was why a drop in the price of one hurt the sales of the other.
      How does this difference explain the differences between your findings
      for the best-response curves and the Nash equilibrium and joint profit-
      maximizing prices in this question, and the corresponding entities in the
      pizza-store example in the text?

13. [Optional] Remember the flagpole game in the text. The Ford dealer and the
    Subaru dealer can choose flagpoles of any height from 20 to 200 feet. Each
    wants his flagpole to be taller than that of the other but does not want to
    spend money beyond what is needed to get the taller pole. Draw the best re-
    sponse functions for this game, and show that it has a unique Nash equilib-
    rium in which both dealers get 200-foot flagpoles.

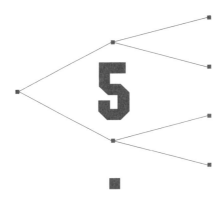

Simultaneous-Move Games
with Mixed Strategies

EARLIER CHAPTERS OUTLINED METHODS for determining equilibria in both sequential-move and simultaneous-move games. In most cases, the equilibrium strategies adopted by each player in the game could be described relatively simply. For sequential-move games, we used rollback to trace an equilibrium path through the game tree, then described the strategies that led to the equilibrium outcome. For simultaneous-move games, we looked for pure-strategy Nash equilibria in which players chose single actions in each play of the game. However, some of the purely simultaneous-move games that we considered did not have unique pure-strategy outcomes. We saw some games with multiple equilibria (assurance, chicken, battle of the two cultures). Perhaps worse, some games had no equilibria at all (tennis point). In order to predict outcomes for these games, we need an extension of our concepts of strategies and equilibria. This is to be found in the randomization of moves, which is the focus of this chapter.

The need for randomized moves in the play of a game usually arises when one player prefers a coincidence of actions, while her rival prefers to avoid it. The tennis-point example from the end of Chapter 4 is a good example. Martina Hingis wants to move to cover the direction of Monica Seles's shot, but Seles would like to surprise Hingis with a shot that goes to the side that she is not covering. Many similar situations can be found, often in games in which the players can be categorized as the attack and the defense. In the tennis case, Seles is on the attack, and Hingis on the defense. Other examples include basic war games and military conflicts, audit games between the IRS (attack) and taxpayers (defense), or games of timing in business (when or how to market a new product). In all of these games, players want to take advantage of the element of surprise; they want to be unpredictable. There is potentially much to be gained by using an unexpected strategy against an opponent, as when sports teams switch to new offensive or defensive configurations in the final moments of critical games. There is a skill to being unpredictable, though, and that skill requires understanding and being able to find the mixed-strategy equilibria of these games.

1 WHAT IS A MIXED STRATEGY?

We mentioned mixed strategies briefly in Chapter 4 to show their relationship to pure strategies in simultaneous-play games. To review, a pure strategy specifies a nonrandom plan of action for a game player. A mixed strategy specifies that an actual move be chosen randomly from the set of pure strategies with some specific probabilities. That is, players in a business pricing game may have the three pure strategies, High, Medium, and Low, which can be combined in numerous ways to create mixed strategies such as "high 25% of the time, medium 25% of the time, and low 50% of the time" or "high 80% of the time, medium 5% of the time, and low 15% of the time." Mixed strategies are just rules telling players to use each of their pure strategies a certain percentage of the time. They are a specific method of randomization.

   The most important aspect of mixed strategies from our perspective is that under very general conditions, every simultaneous-move game has a Nash equilibrium in mixed strategies. That is, there is a pair of mixed strategies, one for each player, such that each mixture is that player's best response to the other's mixture.[1] Such equilibria can often be the outcomes of a game that has no equilibrium in pure strategies, and sometimes when a game has many equilibria in pure strategies. The following sections consider how to find and describe these mixed-strategy equilibria.

2 WHAT DOES A MIXED STRATEGY ACCOMPLISH?

Recall the tennis-point example from Chapter 4, Section 12. In that game, Seles and Hingis consider their options during a particular point in which Seles must

---

[1] Mixed strategies can incorporate the playing of a particular pure strategy with zero probability. Thus, games with a unique pure-strategy Nash equilibrium can be said to have a mixed-strategy equilibrium in which the equilibrium pure strategy is played 100% of the time and other strategies are unused.

put her return down the line (DL) or crosscourt (CC) and Hingis simultaneously must decide which direction to cover. Hingis does better if her choice coincides with that made by Seles, but Seles does better if the two choices differ. Under the circumstances, there is no pair of pure strategies, one for each player, that represent the best responses to each other.

We can see this in a more precise way by using the minimax method developed in Chapter 4. Because this is a zero-sum game, Seles can assume that for each of her strategies Hingis will respond so as to keep Seles's success percentage as low as possible. In Figure 5.1 we show the payoff table, and write these minimum success values for each of Seles's strategies at the far right of the table: 50 for the DL row and 20 for the CC row. Then Seles would want to choose her strategy to get the highest of these lowest payoffs, or the maximin of 50. To achieve this, Seles would play DL.

		HINGIS		
		DL	CC	
SELES	DL	50	80	min = 50
	CC	90	20	min = 20
		max = 90	max = 80	

FIGURE 5.1 Seles's Success Percentages in the Tennis Point

Conversely, Hingis can assume that for each of her strategies, Seles will respond so as to achieve her highest-possible success percentage. We write the maxima of Seles's success values for each of Hingis's choices at the far bottom of the columns. These are 90 for the DL column and 80 for the CC column. Hingis, wishing to keep Seles's success as low as possible, would want to choose CC. This is her minimax strategy, yielding a minimax value of 80.

The maximin and minimax values for the two players are not the same: Seles's maximin success percentage (50) is less than Hingis's minimax (80). As we explained in Chapter 4, this shows that the game has no equilibrium in the (pure) strategies available to the players in this analysis. Let us see how equilibrium can be obtained by expanding the sets of strategies available to the players to include randomization of moves or mixed strategies.

A simple intuition underlies this analysis. Each player does relatively badly if her action is revealed to the other, because then the other can respond in a way that suits her best interest and, this being a zero-sum game, that hurts the first player. Therefore each player wants to *keep the other guessing*. That is exactly what acting at random achieves. Each player can achieve a better outcome—a higher maximin value for Seles and a lower minimax value for Hingis—by choosing a suitable random mixture of DL and CC.

A. The Row Player's Perspective

We begin by seeing this from Seles's perspective. In Figure 5.2 we expand the payoff table by adding a row that represents a mixed strategy for Seles: playing DL with probability p, and CC with probability $(1 - p)$.[2] We call this Seles's *p*-**mix** for short. Even when Seles has just two *pure* strategies, *mixing* between them generates a continuous range of choices, and in choosing a mixed strategy, what she actually chooses is the probability of playing one of the pure strategies. This probability is a variable that can range continuously from 0 to 1. We have already seen examples of players choosing from a continuous range of strategies in Chapter 4; now you will see that a mixed strategy is just a special case of a continuous strategy.

		HINGIS		
		DL	CC	
	DL	50	80	min = 50
SELES	CC	90	20	min = 20
	p-Mix	$50p + 90(1 - p)$	$80p + 20(1 - p)$	min = ?

FIGURE 5.2 Payoff Table with Seles's Mixed Strategy

To find Seles's best choice of p, we have to examine the consequences of all possible choices. We could do this by expanding the game table to have a row for every possible value of p. Because this would be impractical, we keep p as a general symbol instead and carry out all calculations algebraically. The process we use is essentially equivalent to finding Seles's maximin strategy; we look for the value of p that maximizes the minima of her payoffs among all of the possible values of p.

Payoffs associated with Seles's new strategy are **expected values.** That is, they show the success percentages that Seles can expect to attain on average when using her p-mix against each of Hingis's pure strategies. In Chapter 2 we saw how

---

[2] Placing a probability of 0.7 on playing DL is equivalent to saying that the player will play DL 70% of the time. The sum of the probabilities placed on all of her pure strategies must equal 1. That is, there must be a 100% chance (probability of 1) that she will do *something*. In the tennis-point example, if x is the probability that someone plays DL, then $1 - x$ must be the probability with which she plays CC.

to calculate such expected payoffs, or probability-weighted averages of payoffs; the Appendix to this chapter has more on this topic, and on the manipulation of probabilities more generally.

To calculate the expected-value payoffs in the present situation, remember that the p-mix involves Seles playing DL with a probability of p and CC with a probability of $(1 - p)$. Then, if Hingis plays her pure strategy of DL against Seles's mix, the outcome associated with (DL, DL) occurs a fraction p of the time (or $100p$ *percent* of the time), and the outcome associated with (CC, DL) occurs a fraction $(1 - p)$ of the time. Seles's payoff from using her p-mix against DL for Hingis is 50 times the fraction p of the time and 90 times the fraction $(1 - p)$ of the time. Written out mathematically, this payoff is $50p + 90(1 - p)$. Similarly, when Seles uses her p-mix and Hingis uses her pure strategy of CC, Seles's payoff is $80p + 20(1 - p)$.

Note also that the payoffs to both Seles and Hingis in this game are their actual success percentages; the payoffs are not simply rankings of the different possible outcomes for the two players. It is not enough to know that one payoff is bigger than another (as a ranking would indicate); it is important to know *how* much bigger. Remember that in arithmetic, counting numbers, which merely indicate order, are called *ordinal* while numbers that measure size or magnitude are called *cardinal*. Thus we now require payoffs to be measured in cardinal rather than ordinal units in order to calculate the appropriate probabilities in a player's mixed strategy. In games with payoffs based only on ordinal rankings, the differences between pairs of payoffs from ranks 1, 2, and 3 are not quantifiable; all we know is that 3 is "better" than 2, which in turn is "better" than 1. With cardinal payoffs, we know that 3 is bigger than 2 by exactly as much as 2 is bigger than 1, and we need this knowledge to calculate precise percentages in mixed strategies.

For each of her three strategies, Seles expects Hingis to choose that strategy which minimizes Seles's success percentage. As before, we want to show these minima at the extreme right of each row. The numbers for the two pure strategies are as before. But the minimum for the mixed strategy depends on the variable p. Therefore we have to examine what the minimum will be for each possible value of p, in the whole range from 0 to 1. Again, if we had a row of the table for every possible value of p, we would put minima in the third column of each row. Instead, we will do this graphically, using Figure 5.3.

When Hingis plays DL, Seles's p-mix yields her an expected payoff of $50p + 90(1 - p)$, as noted in the payoff table. We illustrate the possible values of this payoff for different values of p as the DL line on our graph. When $p = 0$, Seles's payoff against DL is 90; thus 90 is the vertical intercept of the DL line on the left side of the graph, at $p = 0$. When $p = 1$, Seles's payoff against DL is 50; 50 is the vertical intercept of the DL line on the right side of the graph, at $p = 1$. The height of the DL

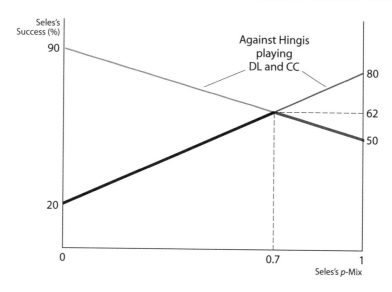

FIGURE 5.3 Diagrammatic Solution for Seles's *p*-Mix

line in between these two points shows Seles's payoff against DL for the values of *p* between 0 and 1. The CC line illustrates the same information for the success of Seles's *p*-mix against Hingis playing CC. Seles's payoff from her *p*-mix against CC is $80p + 20(1 - p)$, so Seles wins 20% of the time at $p = 0$ and 80% of the time at $p = 1$. Again, 20 and 80 are the vertical intercepts of the CC line at the left and right sides of the graph. In between, we see how Seles's payoff against CC changes as she alters the percentage of the time she chooses DL, or as she alters the value of *p* in her mix.

Notice that the two lines intersect each other at a unique value of *p*. That value is important to our determination of the mixed-strategy equilibrium, so we need to find its exact value. To do so, we set the expressions for the two payoff lines equal to each other and solve for *p:*

$$50p + 90(1 - p) = 80p + 20(1 - p)$$
$$90 - 40p = 60p + 20$$
$$70 = 100p$$
$$p = 7/10 = 0.7$$

The *p* value at the intersection of the DL and CC lines is 7/10, or 0.7, or 70%; for smaller values of *p* (those to the left of the intersection) the DL line is higher, and for larger values of *p* the CC line is higher. Notice further that at the point of intersection Seles's success percentage is 62. This value can be found by plugging the value 0.7 for *p* into the expression for either the DL line or the CC line. Using the former we get

$$50 \times 0.7 + 90 \times (1 - 0.7) = 50 \times 0.7 + 90 \times 0.3 = 35 + 27 = 62.$$

We could also calculate Seles's success percentage against either of Hingis's pure strategies for every different value of p that she could choose; we would simply plug a specific p value into the DL or CC expression to find the corresponding success percentage.

For each possible p-mix, Seles should expect Hingis to respond, in equilibrium, with the action that is best for her. Because the game is zero-sum, the action that is best for Hingis is the action that is worst for Seles. Thus, Seles should expect Hingis always to make the choice that corresponds to the lower of the two lines in the graph for any value of p. Given the graph we have drawn, Seles should expect Hingis to cover CC when Seles's DL percentage in her mix is less than 70 (when $p < 0.7$) and to cover DL when Seles's DL percentage is more than 70 (when $p > 0.7$). If Seles is mixing exactly 70% down the line and 30% crosscourt, then Hingis does equally well with either of her pure-strategy choices.

The lower segments of the DL and CC lines in Figure 5.3 have been made darker and thicker to highlight the lowest success percentage that Hingis can hold Seles to under each possible version of Seles's p-mix. This thick, inverted-V-shaped graph thus shows the minimum values of Seles's success percentage in relation to her choice of p in her mixed strategy. The whole inverted-V graph is what should be given as the minimum at the end of the p-mix row in Figure 5.2.

Indeed, having done this, we could even omit the first two rows that correspond to Seles's two pure strategies because they are included in the graph. The point at the extreme left, where $p = 0$, corresponds to Seles's pure choice of CC, and yields 20% success. The point at the extreme right, where $p = 1$, corresponds to the pure choice of DL and yields 50% success.

Most interestingly, there are points in between where Seles has success percentages above 50. This outcome illustrates the advantage of keeping Hingis guessing. If $p = 0.6$, for example, Hingis does best by playing her CC. This choice keeps Seles down to 20% success only for the 40% of the time that Seles plays CC; the rest of the time, 60%, Seles is playing DL and gets 80%. The expected value for Seles is $80 \times 0.6 + 20 \times 0.4 = 48 + 8 = 56$, which is more than the maximin of 50 that Seles could achieve by choosing one of her two pure strategies. Thus we see that a mix of 60% DL and 40% CC is better for Seles than either of her two pure strategies. But is it her *best* mix?

The figure immediately reveals to us Seles's *best* choice of mix: it is at the peak formed by the inverted-V-shaped thicker lines. That peak represents the maximum of her minima from all values of p and is at the intersection of the DL and CC lines. We already calculated that p is 70% at this intersection. Thus, when Seles chooses p to be 70%, she has a success percentage of 62 no matter what Hingis does. For any other p-mix Seles might choose, Hingis would respond with whichever one of her pure strategies would hold Seles down to a success rate of less than 62%, along either of the darker blue or black lines that slope away from the peak. Thus Seles's mix of 70% DL and 30% CC is the only one that cannot be exploited by Hingis to her own advantage (and therefore to Seles's disadvantage).

B. The Column Player's Perspective

In exactly the same way, Hingis can use mixed strategies to improve her minimax payoff. Suppose she plays DL with a probability of q and CC with a probability of $(1 - q)$; call this her **q-mix.** Figure 5.4 expands the payoff table of Figure 5.1 by introducing a third column for the q-mix. Again, q can take any value between 0 and 1, and we will examine the consequences of all choices of q. And as before, the payoffs to Hingis from using her q-mix against each of Seles's two pure strategies are expected values. You should verify the expressions with the same logic we just used for Seles's mixture.

		HINGIS		
		DL	CC	q-Mix
SELES	DL	50	80	$50q + 80(1 - q)$
	CC	90	20	$90q + 20(1 - q)$
		max = 90	max = 80	max = ?

FIGURE 5.4 Payoff Table with Hingis's Mixed Strategy

For each of her strategies, Hingis expects Seles to choose her best response. Therefore for each column, we show at the bottom the largest of Seles's payoffs. In the columns for Hingis's pure strategies DL and CC, these are the same numbers as in Figure 5.1. At the foot of the third column, corresponding to the q-mix, we should show the entire graph of the maxima corresponding to different values of q. Instead, we show this graph in Figure 5.5; its relation to the table of Figure 5.4 is the same as that between the graph in Figure 5.3 and the table of Figure 5.2. The horizontal axis shows the values of q ranging from 0 to 1; the vertical axis shows Seles's success percentages; the figure has two straight lines. The rising line shows Seles's success from playing CC against Hingis's q-mix. Higher values of q correspond to Hingis using her DL shot more often; therefore Seles succeeds better by using CC. Similarly, the downward-sloping line shows Seles's success from using DL against Hingis's q-mix.

Hingis should expect Seles to use the action that is best for her (and consequently worse for Hingis) at any given value of q. We see from the figure that Seles would choose DL for q values to the left of the intersection of the two lines and CC for q values to the right of the intersection. The dark blue and black V-shaped line segments in the diagram show the best that Seles can do (or the worst that can happen to Hingis) at each value of q. This is the graph of the maxima that should go at the foot of the third column in Figure 5.4. Of these maxima, Hingis should choose the minimum, so as to keep Seles's best (or her own worst) as small as possible. She would do so by choosing the value of q at the intersection of the two lines.

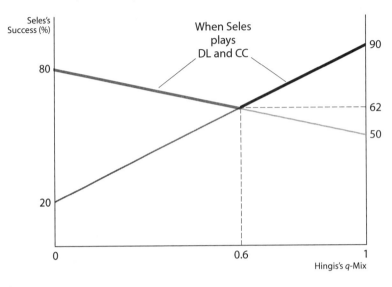

FIGURE 5.5 Diagrammatic Solution for Hingis's q-Mix

This q can be found algebraically by equating the expressions for the DL and CC lines from Figure 5.5; you should check that this process yields $q = 6/10 = 0.6$. Thus Hingis's best q-mix requires that she cover down the line 6/10 or (60%) of the time and that she cover crosscourt 4/10 or (40%) of the time. Plugging in this best q value into either payoff expression yields the average payoff from this mixture: $(50 \times 0.6) + (80 \times 0.4) = 30 + 32 = 62$. Recall that the zero-sum nature of this game means that large numbers are good for Seles but bad for Hingis. The payoff of 62 here is Seles's success percentage; Hingis's success percentage is, of course, just 100 *minus* this value, or 38. The 62, however, represents Hingis's minimax, and that is better for her than the 80 she could achieve by choosing from her two pure strategies. The intuition behind this advantage of mixing is the same as before—by keeping Seles guessing, Hingis prevents Seles from exploiting Hingis's choice to her own advantage and (this being a zero-sum game) to Hingis's detriment.

3 EQUILIBRIUM IN MIXED STRATEGIES

We examined the best choice of mixed strategies for Seles and Hingis separately; it only remains to put the two together and show that they constitute an equilibrium for the game.

We saw that when Seles chooses her best p-mix ($p = 0.7$), she gets the same expected payoff, namely 62, whether Hingis chooses DL or CC. In fact, that was the whole basis of the logic declaring 0.7 to be the best choice of p for Seles. But then Seles's expected payoff must be the same, namely 62, even when Hingis plays her q-mix against Seles's best p-mix: $62q + 62(1 - q) = 62$ for all values of q. Correspondingly, Hingis's expected payoff must be $100 - 62 = 38$ for all these situations.

Thus when Seles chooses her best p-mix, Hingis cannot gain any additional advantage by mixing, but neither does she lose by doing so. In other words, when Seles plays her best p-mix, all choices of q in the range from 0 to 1 are equally good from Hingis's point of view. In particular, she is perfectly happy to choose $q = 0.6$. Conversely, when Hingis chooses her best q-mix with $q = 0.6$, Seles gets the expected payoff of 62 from her two pure strategies or from any mixture. Therefore she is willing to go on choosing $p = 0.7$. Thus Hingis's $q = 0.6$ is a best response to Seles's $p = 0.7$, and vice versa. The two mixed strategies are mutually best responses, and therefore constitute an equilibrium of the game. The payoff to each player in this equilibrium, 62 to Seles and 38 to Hingis, represents the **value of the game** to that player.

Notice that if Hingis chooses a value of q other than 0.6, then Seles can choose a pure strategy that yields her a higher average payoff than is obtained against Hingis's optimal mix. Similarly, if Seles chooses a value of p other than 0.7, then Hingis can choose a pure strategy that yields her a higher average payoff than is obtained against Seles's optimal mix. Thus another way to think about an equilibrium mixture is that it leaves your rival no reason to choose a specific pure strategy to use against you.

When the players were restricted to using pure strategies, Seles's maximin was 50 and Hingis's minimax was 80. We noted that the use of mixed strategies would increase Seles's maximin and lower Hingis's minimax. Most remarkably, this happens to such an extent that when each uses the best mixture, the maximin equals the minimax; our solution method effectively extends the minimax method of Chapter 4 to games in which players consider mixed strategies. This gives yet another confirmation that the two mixed strategies constitute an equilibrium. It is, in fact, a consequence of a general result called the *minimax theorem for zero-sum games*; we provide a brief discussion of this theorem in Chapter 7. The first few times you solve zero-sum games with mixed strategies, you will find it a useful check on your calculations to verify that the row player's maximin and the column player's minimax are equal.

The mixed-strategy equilibrium can be shown more explicitly using the concept of **best-response curves,** or **reaction curves,** which we developed in Chapter 4. When each player can mix between two strategies, the strategy choice of each can be described by a single number, namely, the probability of choosing one of the strategies; the probability of choosing the other strategy is then 1 minus this number. In our example, Seles's strategy is described by p and Hingis's by q. (The two pure strategies are included by allowing p and q to take on the extreme values 0 or 1; for example, Seles's pure DL strategy corresponds to $p = 1$.) In equilibrium, each player's strategy is the best response to that of the other. To derive the reaction curves, then, we should find Hingis's best value of q for each of Seles's possible choices of p, and Seles's best value of p for each of Hingis's possible choices of q. The equilibrium occurs where both players' best responses occur simultaneously.

Most of the groundwork has already been laid, and Figure 5.6 uses it to construct the best-response curves. The left-hand panel shows Hingis's best response in her choice of q to Seles's choices of p. Therefore p is on the horizontal axis and q on the vertical axis. We know that for $p < 0.7$, Hingis does better by choosing pure CC ($q = 0$); this segment of her best response is the horizontal solid line along the bottom edge of the graph. For $p > 0.7$, Hingis does better by choosing pure DL; this segment of her best response is the horizontal solid line along the top edge of the graph. For $p = 0.7$, Hingis does equally well with all of her choices, pure and mixed, so the vertical solid line in the graph at $p = 0.7$ shows her best responses for this choice of Seles's. The whole solid line, consisting of three separate segments joined end to end, comprises Hingis's best-response curve.

The middle panel of Figure 5.6 shows Seles's best-response curve. Here, Seles's best p values are determined based on Hingis's various possible choices for q so q is on the horizontal axis and p on the vertical axis. For $q < 0.6$, Seles does better playing pure DL ($p = 1$); for $q > 0.6$ she does better with pure CC ($p = 0$); for $q = 0.6$ she does equally well with all choices, pure or mixed. The blue curve, consisting of three line segments joined end to end, is Seles's best-response curve.

The right-hand panel in Figure 5.6 combines the two previous panels, by reflecting the middle graph across the diagonal so that p is on the horizontal axis and q on the vertical axis, and then superimposing this on the left-hand graph.

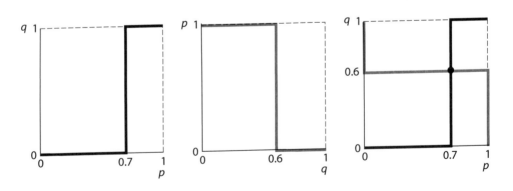

FIGURE 5.6 Best-Response Curves and Equilibrium

Now the black and blue curves meet at the point $p = 0.7$ and $q = 0.6$. Here each player's mixture choice is a best response to the other's choice, so we see clearly the derivation of our Nash equilibrium in mixed strategies.

This picture also shows that the best-response curves do not have any other common points. Thus the mixed-strategy equilibrium in this example is unique. What is more, this representation includes pure strategies as special cases corresponding to extreme values of p and q. We see that the best-response curves do not have any points in common at any of the corners of the square where each value of p and q equals either 0 or 1; thus we have another verification that the game does not have any pure-strategy equilibria.

Observe that each player's best response is a pure strategy for almost all values of her opponent's mixture. Thus Hingis's best response is pure CC for all of Seles's choices of $p < 0.7$, and it is pure DL for all of Seles's choices of $p > 0.7$. Only for the one crucial value $p = 0.7$ is Hingis's best response a mixed strategy, as is represented by the vertical portion of her three-segment best-response curve in Figure 5.6. Similarly, only for the one crucial value $q = 0.6$ of Hingis's mixture is Seles's best response a mixed strategy, namely the horizontal segment of her best-response curve in the figure. But these seemingly exceptional or rare strategies are just the ones that emerge in the equilibrium.

The best-response-curve method thus provides a very complete analysis of the game. Like the cell-by-cell inspection method, which examines all the cells of a pure strategy game, this is the method to use when one wants to locate *all* of the equilibria, whether in pure or mixed strategies, that a game might have. The best-response-curve diagram can show both types of equilibria in the same place. (It could also be used to show equilibria in which one player uses a mixed strategy, and the other player uses a pure strategy, although the diagram shows that such hybrids can be ruled out except in some very exceptional cases.)

If you had reasons to believe that a game would have just one mixed-strategy equilibrium, you could still find it directly without bothering calculating and graphing the whole best-response curves and finding their intersection. The equilibrium value, $p = 0.7$, comes from equating Seles's expected payoffs against Hingis's DL and CC strategies; in other words, p solves the algebraic equation $50p + 90(1 - p) = 80p + 20(1 - p)$. Similarly $q = 0.6$ comes from equating Hingis's expected payoffs against Seles's DL or CC strategies; it is the solution to $50q + 80(1 - q) = 90q + 20(1 - q)$. The substantive reason for this is the intuition we stated at the outset, namely that each player's mixture has to prevent her opponent from recognizing and exploiting any systematic pattern in her choice. For this reason, we call this the **prevent-exploitation method**.[3] It was developed in

---

[3] In recognition of this method's similarity to the minimax method of Chapter 4, we could also call it *minimax for mixes*.

the context of a zero-sum game where each player has two pure strategies and where players' interests are in direct conflict, and that is its domain of applicability. In Section 4, we will develop a related but different method that enables us to find mixed-strategy equilibria in more general non-zero-sum games when each player has two pure strategies. Later in this chapter, and again in Chapter 7, we will see how these methods must be modified when a player has three or more pure strategies.

4 MIXING IN NON-ZERO-SUM GAMES

The basic concepts of mixed strategies, and of equilibria involving them, are the same whether the game is zero-sum or not. Mixing two or more pure strategies means using a randomized procedure with certain probabilities of taking one action or another. A (Nash) equilibrium is a collection of mixed strategies, one for each player, such that the strategy of each is best for her, given those of the other players. But in non-zero-sum games the players' interests are not strictly opposed. Therefore one player acting in her own interest does not do so automatically to the detriment of the others. Each player is not necessarily out to exploit others, and interpretation of mixing as a way to prevent being exploited is not necessarily correct either. In this section we will explicate these matters using some simple examples.

A. Chicken

For our first example, we will use the chicken game from Chapter 4. Recall the two teenagers facing off in their cars, each determined to drive straight toward the other without swerving. Each player thus has two pure strategies, Straight and Swerve. In the two-by-two payoff table of Figure 4.11, we found two Nash equilibria in pure strategies in which one player swerves while the other continues straight. Now we add the possibility of mixed strategies for each player, and show the resulting three-by-three table in Figure 5.7. James's p-mix puts probability p on his strategy Swerve and probability $(1 - p)$ on his strategy Straight; similarly, Dean's q-mix puts probability q on Swerve and probability $(1 - q)$ on Straight. Note that we must show the separate payoffs for the two players in each cell due to the non-zero-sum nature of the game.

The table in Figure 5.7 is not as clean-looking as the game tables we have been working with so far in this chapter. The payoffs in the p-mix row and the q-mix column are, however, calculated in exactly the same way as was done earlier. For example, if James uses his pure strategy of Swerve against Dean's q-mix, for example, a fraction q of the time he gets a payoff of 0 and a fraction $(1 - q)$ of the time he gets -1. His expected payoff is $0q + (-1)(1 - q)$ or $q - 1$, as shown in the top right corner cell of the table. Dean's payoff in that cell shows his payoff from using his q-mix against James's pure strategy Swerve. Dean also gets 0 for a proportion q of the time in that situation but for a proportion $(1 - q)$ of the time he gets 1; therefore his expected payoff is $0q + 1(1 - q) = 1 - q$. The other mixture payoffs are found in a similar fashion. Where some payoff expressions can be simplified algebraically, we do so. For example, when James's p-mix encounters Dean going Straight, the resulting combination is (Swerve, Straight) for a fraction p of the times, with payoff 1 to Dean, and (Straight, Straight) for a fraction $(1 - p)$ of the times, with payoff -2 to Dean. Thus Dean's expected payoff is $1p + (-2)(1 - p) = p - 2(1 - p) = 3p - 2$.

		DEAN		
		Swerve	Straight	q-Mix
JAMES	Swerve	0,0	$-1,1$	$-(1 - q) = q - 1,$ $(1 - q)$
	Straight	$1,-1$	$-2,-2$	$q - 2(1 - q) = 3q - 2,$ $-q - 2(1 - q) = q - 2$
	p-Mix	$1 - p,$ $-(1 - p) = p - 1$	$-p - 2(1 - p) = p - 2,$ $p - 2(1 - p) = 3p - 2$	

FIGURE 5.7   Mixing Strategies in Chicken

Now we can characterize the optimal mixed strategies just as we did in the tennis game earlier, although we must recognize that the players' interests are not strictly opposed. Figure 5.8 shows the calculations graphically. The left panel shows Dean's payoffs from each of his pure strategies when James's p-mix is allowed to vary over its full range from 0 to 1.[4] The flatter straight line corresponds to Dean's choice of Swerve, and the steeper one to his choice of Straight. The two lines intersect when $p - 1 = 3p - 2$, or $p = 1/2 = 0.5$. For smaller values of p (when James favors Straight), Dean does better by Swerving (to reduce the risk of a crash), and for larger values (when James favors Swerve), Dean does better by going Straight (being relatively more likely to emerge as the victor). Dean's best choices, and the resulting payoffs, are shown by the kinked line that represents the higher of the two lines at each point; this is often called the **upper envelope**

[4] In the tennis example, we showed Seles's p-mix on the horizontal axis and Seles's payoffs on the vertical axis, and let Hingis choose the response that minimized Seles's payoff, because this also maximized Hingis's payoffs in that zero-sum game. This non-zero-sum game lacks that automatic connection between the two players' payoffs; therefore to find Dean's best choice we must show Dean's own payoffs.

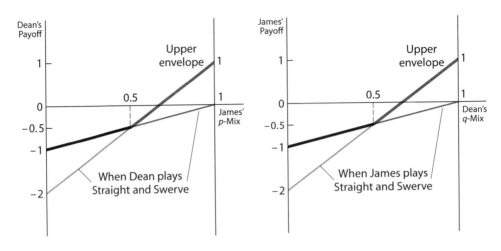

FIGURE 5.8 Optimal Responses with Mixed Strategies in Chicken

of the two separate lines. At the intersection where $p = 0.5$, Dean does equally well with either pure strategy, or indeed with any mixture of the two; his expected payoff from any of these choices is -0.5.

The right-hand panel of Figure 5.8 shows James's best choice of action for all values of Dean's q-mix; note that James's payoffs are being measured on the vertical axis. The reasoning and the results are identical to those for Dean. For $q < 0.5$, James does best with Swerve, which in mixed-strategy terms corresponds to choosing $p = 1$. For $q > 0.5$, James does best with Straight ($p = 0$). And for $q = 0.5$, James does equally well with either pure strategy or with any mixture; that is, all values of p between 0 and 1 are equally good.

Figure 5.9 translates the information in Figure 5.8 into best-response curves and shows the resulting equilibria of the game. The method of derivation is the same used to create Figure 5.6. In the left panel we show Dean's best response. This is pure Swerve ($q = 1$) for $p < 0.5$, pure Straight ($q = 0$) for $p > 0.5$, and any mix-

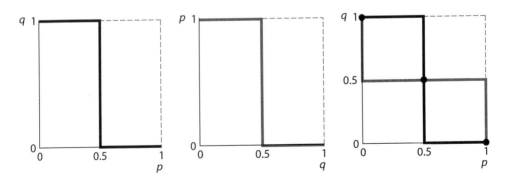

FIGURE 5.9 Best-Response Curves and Mixed Strategy Equilibria in Chicken

ture (the whole vertical line from $q = 0$ to $q = 1$) when $p = 0.5$. Thus Dean's best-response curve, shown solid, is composed of three straight-line segments joined end to end. The middle panel shows James's best response in a similar way; it is shown in blue. The right-hand panel superimposes the two by reversing the axes.

The two best-response curves have *three* points in common here. Two occur in the corners, one is where $p = 0$ and $q = 1$ (James going straight and Dean swerving), the other where $p = 1$ and $q = 0$ (the other way around). These are just the pure-strategy Nash equilibria we identified in Chapter 4. But now we see another equilibrium, namely $p = 0.5$ and $q = 0.5$. Here the two players' mixed strategies are mutual best responses, so we have a Nash equilibrium in mixed strategies. Once again, calculating and juxtaposing best-response functions and locating all their common points is the general infallible way to find all the equilibria that a game might have.

To find just the mixed-strategy equilibrium directly, we would use the equations that define the critical values of p and q in the above argument. These are the intersection points of the straight lines in Figure 5.8. The value of p is found by equating Dean's payoffs from his two pure choices when confronted with James's p-mix. In other words, the equilibrium value of p in James's mixture is such as to keep Dean indifferent between his two pure strategies. Conversely, the equilibrium value of q in Dean's mixture is such as to keep James indifferent between his two pure strategies. Notice that it is crucial here to equate *Dean's* payoffs from using each of his two pure strategies against the p-mix when finding *James's* optimal value of p, and to equate *James's* payoffs from using each of his two pure strategies against the q-mix when finding *Dean's* optimal value of q. *Each* player's mixture probability is such as to keep the *other* player indifferent between his pure strategies. This process differs from the prevent-exploitation method of Section 3 because the players' interests are not totally opposed. We call this method the **keep-the-opponent-indifferent** method.

In a zero-sum game, preserving your opponent's indifference would be tantamount to preserving your own indifference. And when your mixture is such that you are indifferent as to whether the opponent is playing one or another of her pure strategies, she cannot exploit you. Thus the prevent-exploitation method, which we stated earlier for zero-sum games, is a special case of the keep-the-opponent-indifferent method which applies to more general non-zero-sum games.

However, the prevent-exploitation method by itself does not work for non-zero-sum games. Thus for example, the table in Figure 5.7 shows that when James chooses his equilibrium mixture with $p = 0.5$, he gets $1 - p = 0.5$ against Dean's Swerve, and $p - 2 = -1.5$ against Dean's Straight. This is obvious: no matter what James does, it is better for him that Dean should Swerve. But Dean is pursuing his own interest, and is out to neither help nor hurt James as an end in itself.

In Chapter 7 we will study the relationship between these two methods in greater depth. Here we merely emphasize an important general point about mixed-strategy equilibria in any game. Do not attach any causal or purposive significance to the fact that each player's equilibrium mixture is such that the other player is indifferent about playing each of her pure strategies. Neither player deliberately *wants* to keep the other indifferent. Each is merely out to maximize her own expected payoff. The other's indifference is an accidental side effect of this behavior in a mixed-strategy equilibrium.

Look back now at the solution that we have found for the chicken game. The mixed-strategy equilibrium yields each player a payoff of –0.5. That payoff is worse than the 0 each player could get by agreeing to play (Swerve, Swerve).[5] Mixed-strategy equilibria in non-zero-sum games often yield such low values of the game. The problem is that the separate mixing on the part of the two players generates a positive chance of each outcome in the table. In this case, there is a probability $(1 - p)(1 - q)$, or 25% (!), of the (Straight, Straight) outcome occurring. This is a relatively large chance of the players making an expensive "mistake" and facing the bad outcome associated with a crash.

What might help here is coordinated randomization. In the absence of any coordination, each player might flip a coin to determine which strategy to use in a single play of the game.[6] With two coins, you would expect each of the four possible outcomes to be equally likely, occurring 25% of the time for each coin. Coordination could be attained by flipping just one coin, with the prearranged agreement that if it comes up heads, James will play Swerve (while Dean plays Straight), and if it comes up tails, Dean will Swerve (while James plays Straight). The players thus agree to play one of the game's Nash equilibria with heads and the other with tails. Under such an agreement, each player has a 50% chance of getting 1 (when he plays Straight) and a 50% chance of getting –1 (when he plays Swerve). On average, then, each player would get 0 instead of the –1/2 obtained with uncoordinated mixing. The coordinated randomization gives them an outcome just as good as both swerving, but at least not any worse. However, if the value of being chicken (playing Swerve against Straight) were still –1 but that of being Tough (playing Straight against Swerve) were 2 rather than 1, then the expected payoff from coordination would change. Players would get –1 half the time and 2 the other half of the time, leaving them with 1/2 on average, which would be better than the 0 associated with (Swerve, Swerve).

B. The Battle of the Two Cultures

We see the same type of results as in Section 4.A when we look for a mixed-strategy equilibrium in the battle-of-the-two-cultures game, also considered

---

[5] Of course, (Swerve, Swerve) is not a Nash equilibrium of the game, so it might be difficult to reach such an agreement.

[6] A flipped coin will land heads 50% of the time and tails 50% of the time, so it is a perfect randomization device for James or Dean, because the equilibrium mixtures entail a 50-50 split of Swerve and Straight.

originally in Chapter 4. Recall that our scenario entailed two groups of faculty voting simultaneously on a recommendation for the use of a particular campus space. The game in its original form has two Nash equilibria in pure strategies, both preferred by both players to either out-of-equilibrium outcome, but no obvious focal point. Thus there is a potential role for mixing strategies.

The original payoff table for the battle of the two cultures was shown in Figure 4.12. Now we augment it with mixed strategies and show the result in Figure 5.10. If the two groups of faculty were able to communicate with each other before their meetings, they could bargain and perhaps compromise by tossing a single coin and letting its outcome govern which way they would vote. If they are unwilling or unable to reach such a cooperative agreement, then the mixing of strategies comes into play. The science faculty will consider a mixture with a probability of p on Lab and a probability of $(1 - p)$ on Theater; the humanities faculty will consider a mixture with a probability of q on Lab and a probability of $(1 - q)$ on Theater.

		HUMANITIES FACULTY		
		Lab	Theater	q-Mix
SCIENCE FACULTY	Lab	2, 1	0, 0	$2q, q$
	Theater	0, 0	1, 2	$1 - q, 2(1 - q)$
	p-Mix	$2p, p$	$1 - p, 2(1 - p)$	

FIGURE 5.10   Mixing in the Battle-of-the-Two-Cultures Game

The calculation of best responses and equilibria proceeds along a route that should now be familiar. Figure 5.11 shows the payoffs for each faculty for each of its pure-strategy choices when confronted with the other faculty's mixture. The left-hand panel shows that the humanities faculty does better by choosing Lab ($q = 1$) when the science faculty's mixture has a high value of p, and by choosing Theater ($q = 0$) when the science faculty's mixture has a low value of p. The breakeven point is $p = 2/3$. The right-hand panel shows that the science faculty does better with Lab ($p = 1$) when the humanities faculty's mixture has a high value of q, and with Theater ($p = 0$) when the Humanities faculty's mixture has a low value of q. The breakeven point is $q = 1/3$.

The resulting best-response curves are juxtaposed in Figure 5.12. Now that you are familiar with the method (from Figures 5.6 and 5.9), we do not need to show all three panels, only the final one that superimposes the two best-response curves. We see two common points at the corners that correspond to the two pure-strategy equilibria we found in Section 11.3 of Chapter 4. One of these, at $p = q = 0$, corresponds to the pure-strategy equilibrium (Theater, Theater) while the other, at $p = q = 1$, corresponds to the (Lab, Lab) equilibrium. The new finding is the intersection at $p = 2/3$ and $q = 1/3$, which is a mixed-strategy equilibrium.

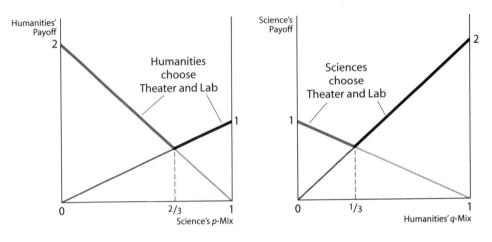

FIGURE 5.11   Best Responses with Mixed Strategies in the Battle of the Two Cultures

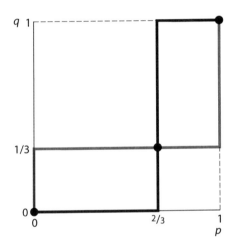

FIGURE 5.12   Mixed-Strategy Equilibria in the Battle of the Two Cultures

The mixed-strategy equilibrium value of p equates p and $(2 - 2p)$ and keeps the humanities faculty indifferent between voting for the lab or the theater. Solving for p yields a value of $2/3$, or 0.667, so the scientists would mix by voting for the lab $2/3$ of the time and the theater $1/3$ of the time. Given the symmetry of the game, it should not be surprising that the equilibrium q, which must equate the $2q$ and $(1 - q)$ payoffs for the scientists, is $1/3$, or 0.333. The humanities faculty will mix by voting for the theater $2/3$ of the time and the lab $1/3$ of the time. The value of the game to each player is $2/3$; the Humanists get $p = 2 - 2p = 2/3$ on average from either pure strategy, while the Scientists get $2q = 1 - q = 2/3$ from either of their pure strategies.

When each group mixes according to these probabilities, the two mixes constitute a Nash equilibrium. As in the chicken example, however, the mixed-strategy equilibrium yields an average payoff that is less than the average payoff the players could have achieved with coordinated randomization. If they played each of the two pure-strategy Nash equilibria with a probability of 0.5 (as would occur with the flipping of a coin to determine which to play), they would each receive 1.5 on average, as opposed to the 0.667 achieved with mixing. If the two groups valued the various possible outcomes slightly differently, we would obtain a slightly different (but relatively common) version of the payoffs for this game in which each player's payoff is –1 when they each vote for their least-preferred outcome. In that version of the game, the equilibrium mixed strategies entail a 75-25 split for each player. The Scientists would vote Lab 75% of the time and Theater 25% of the time, while the Humanists did the opposite; these mixtures would lead to an average payoff of 0.5 for each. This mixing outcome is even worse than in our original example, especially when you realize that the coordinated outcome still gives an average of 1.5.

Many situations, like chicken and battle of the two cultures, arise in real games. If players cannot coordinate or do not know how, mixing may be the only way to reach a Nash equilibrium. But if mixed strategies can lead to such bad outcomes, why would players use them? The answer can be complicated, but in general an appeal is made to the fact that not mixing leaves you open to exploitation and an even lower average payoff. We address this and other concerns about mixing in the next section.

5 FURTHER DISCUSSION OF MIXED STRATEGIES

A. The Odds Method

We begin with a simple formula for calculating and interpreting mixture probabilities, which applies when each of the players has only two pure strategies. People who are more used to thinking of the odds of an uncertain event than of its probability may find this way of expressing mixed strategies easier to follow.

The **odds** of a particular event represent the chances of that particular event occurring, as opposed to something else occurring. Odds are usually written as a pair of numbers, known as **oddments**, separated by a colon, as in 15:1. These odds would be read as "fifteen to one." You interpret these odds by summing the two oddments and taking ratios of each oddment to the sum. Thus if you have calculated that your odds of getting an A⁺ in a course are 15 : 1, you expect to get an A⁺ 15/16 (or 93.75%) of the time. The other 1/16 (6.25%)of the time, you expect to get something lower than an A⁺. (In horse racing, odds are often expressed as *odds against,* that is, the probability of the horse *losing* versus winning. Therefore odds of 15 : 1 indicate a long shot, a horse with a low probability of winning.)

Odds notation can be interpreted more generally as follows. Suppose the odds of winning versus losing are $W : L$. These odds indicate that a proportion $W/(W+L)$ of the time, you win; a proportion $L/(W+L)$ of the time you lose. Notice that the oddment associated with winning is listed first. In general, if you are describing the odds of any particular outcome, the oddment associated with that outcome is listed first. For example, if the odds of a player moving in a certain direction (say, right) versus moving in another direction (say, left) are 6 : 4, that would indicate a 60% chance of seeing a move to the right and a 40% chance of seeing a move to the left. Similarly, the odds against a move to the right are 4 : 6.

So if p is the probability that an event happens, the odds of it happening are $p : (1 - p)$, and the odds against it are $(1 - p) : p$. Since odds are such a binary concept (something happening versus not happening), they are a useful way of expressing probabilities of mixtures in a game only when there are only two pure strategies. In our tennis example, in Seles's equilibrium mixture the probability of her playing DL are 70% and that of playing CC are 30%. Therefore the odds in favor of DL are 7 : 3, or the odds against CC are 7 : 3.

We now develop a formula for calculating the equilibrium-mixture odds in a game where each player has two pure strategies. We do this in quite a general and abstract way. This will reinforce what you learned from the specific numerical examples above.

Let us call the two players Row and Column. The two pure strategies of Row are labeled Up and Down, and those of Column are labeled Left and Right. The game could be zero-sum or non-zero-sum; the odds method will work for either type of game. We show the payoff table in Figure 5.13, writing the payoffs of Row and Column resulting from the strategy choices (Up, Left) as (a, A), and similarly for the other three cells. The mixes of the two players and the resulting payoffs are shown as usual in the third row and column.[7]

This being a generic game that could be non-zero-sum, we must use the keep-the-opponent-indifferent method to find p and q. Thus the equation defining Row's equilibrium p is found by equating Column's expected payoffs from playing each of Left and Right against Row's p-mix: $pA + (1 - p)C = pB + (1 - p)D$. To cast this in the language of odds, we collect together the terms involving p on

---

[7] To save space we do not show the payoffs for the cell that pits mix against mix. But it is easy to calculate the probabilities of each of the four combinations of pure strategies, and verify that Row's expected payoff is $pqa + p(1 - q)b + (1 - p)qc + (1 - p)(1 - q)d$, and that of Column is $pqA + p(1 - q)B + (1 - p)qC + (1 - p)(1 - q)D$.

		COLUMN		
		Left	Right	q-Mix
ROW	Up	a, A	b, B	$qa + (1 - q)b,$ $qA + (1 - q)B$
	Down	c, C	d, D	$qc + (1 - q)d,$ $qC + (1 - q)D$
	p-Mix	$pa + (1 - p)c,$ $pA + (1 - p)C$	$pb + (1 - p)d,$ $pB + (1 - p)D$	

FIGURE 5.13 General Form of the Odds Method

one side and those involving $(1 - p)$ on the other side; thus $p(A - B) = (1 - p)(D - C)$. Then the ratio $p/(1 - p)$ in the optimal mix is equal to $(D - C)/(A - B)$, and the odds of Row playing Up are

$$p:(1 - p) = (D - C):(A - B).$$

In other words, the oddment on Up equals the difference between Column's payoffs when she meets Row's choice of Down with her choice of Right versus her choice of Left, and the oddment on Down equals the difference between Column's payoffs when she meets Row's choice of Up with her choice of Left versus her choice of Right. Note that there are two reversals in the calculation: the oddment on Up is a difference of Column's payoffs against Down and vice versa, and in one case we take the difference of Column's Right versus Left, and in the other the other way around.[8]

If the game happens to be zero-sum, that is, $A = -a$, $B = -b$, and so on, the previous formula can be expressed in terms of Row's payoffs, that is, $p : (1 - p) = (c - d) : (b - a)$. Working backward, this corresponds to $pa + (1 - p)c = pb + (1 - p)d$. This expression says that Row is indifferent as to whether Column plays Left or Right against Row's p-mix. Neither choice can do worse for Row, so Row cannot be exploited. Thus we see how the keep-the-opponent-indifferent method reduces to the prevent-exploitation method when the game is zero-sum.

If there is to be a genuine mixed-strategy equilibrium, then p should be a number between 0 and 1. Therefore both p and $(1 - p)$ should be positive, and so should their ratio. To ensure this, $(D - C)$ and $(A - B)$ should have the same sign: either $D > C$ and $A > B$, or $D < C$ and $A < B$. If you mechanically apply the odds method and find a negative ratio $p/(1 - p)$, then the game does not have a gen-

---

[8] This method is developed and applied in J. D. Williams, *The Compleat Strategyst* (McGraw-Hill, 1966).

uine mixed-strategy equilibrium, so you should go back and look for an equilibrium in pure strategies.

Column's odds of playing Left versus Right in this same generic game above are found from the condition that Row be indifferent between Up and Down when Column uses her q-mix: $qa + (1-q)b = qc + (1-q)d$, or $q(a-c) = (1-q)(d-b)$, or

$$q:(1-q) = (d-b) : (a-c).$$

For there to be a mixed-strategy equilibrium, $(d-b)$ and $(a-c)$ must have the same sign in the same way that $(D-C)$ and $(A-B)$ had to have the same signs.

Despite the seeming complexity of the formulas we have derived, the odds method is really a nice, simple alternative to the keep-the-opponent-indifferent method when you have a two-by-two game. We show this by using it to calculate the equilibrium mixed strategies for chicken. Consider the payoff table for chicken (Figure 4.11), repeated as Figure 5.14. Note that the outcome $(0, 0)$ in the upper left corner corresponds to the (a, A) outcome in Figure 5.13. Similarly, $(-1, 1)$ corresponds to (b, B), $(1, -1)$ to (c, C), and $(-2, -2)$ to (d, D). A quick check verifies that $(D-C)$ and $(A-B)$ have the same sign, and so do $(d-b)$ and $(a-c)$, as is required for a mixed-strategy equilibrium. Then, if we apply the odds-method formula verbatim, James's odds of choosing Swerve are $(D-C) : (A-B) = [-2-(-1)] : (0-1)$, or $-1:-1$, which is a positive ratio that we write as $1:1$, or $50:50$. Similarly for Dean, we find the odds of his choosing Swerve as $(d-b) : (a-c) = [-2-(-1)] : (0-1)$, or $-1:-1$ also. Again the ratio is positive, so we can express the odds as $1:1$, or $50:50$. Both players have equilibrium mixtures in which they choose Swerve 50% of the time and Straight 50% of the time.

		DEAN	
		Swerve	Straight
JAMES	Swerve	0, 0	−1, 1
	Straight	1, −1	−2, −2

FIGURE 5.14 Finding Odds in Chicken

For a zero-sum game like our tennis-point example, finding the odds is even easier. In that game, illustrated again in Figure 5.15, we have only 4 (rather than 8) payoff values to use in our calculations. Applying the formula from above, we get Seles's odds of playing DL as $(20-90):(50-80)$ or $70:30$ (dropping the two negative signs as before). Another way to find this is to use the differences in the row and column payoffs that we have included to the right and below the payoff table. Seles's odds of playing DL should equal the difference in the payoffs in the

CC row relative to the difference in payoffs in the DL row, or $70:30$. Similarly, Hingis's odds of playing DL equal the difference in payoffs in the CC column relative to the payoffs in the DL column, or $60:40$.

		HINGIS		
		DL	CC	
SELES	DL	50	80	$50 - 80 = -30$
	CC	90	20	$20 - 90 = -70$
		$50 - 90 = -40$	$20 - 80 = -60$	

FIGURE 5.15 Calculating Odds in the Tennis-Point Game

As a final point regarding the odds method, note that in sporting terminology, the play that is used most frequently (for example, in football, a run up the middle on third down and a yard to go), is called the **percentage play.** The odds method tells us *why* a particular play serves this role. Because the context of sports is often a zero-sum game, we can express Row's odds of playing Up as $p:(1-p) = (c-d):(b-a)$. Playing Up will be the percentage play if p is much bigger than $(1-p)$, which will happen if $(c-d)$ is much bigger than $(b-a)$. If $(b-a)$ is relatively small, then Up is a relatively safe play. It yields similar payoffs whether Column plays Left or Right; its success is not very sensitive to changes in the opponent's actions. Conversely, if $(c-d)$ is large, then playing Down is a *boom-or-bust play;* it succeeds spectacularly if Column plays Left but fails badly if Column plays Right. When viewed in these terms, it makes sense that a relatively safe play should be the percentage play, and a boom-or-bust play should have the function of one that is used occasionally in a mixture to keep the opponent guessing.

Note that the existence of a percentage play implies that mixed strategies do not necessarily entail the use of each pure strategy in the same proportion. Rather, pure strategies appear in the mix in proportion to their relative safety in the sense explained in the previous paragraph. As we will see later in this chapter, it is also possible that some pure strategies may not be used at all in a player's equilibrium mixture if there are more than two possible pure strategies available.

B. Counterintuitive Outcomes with Mixed Strategies

Games with mixed-strategy equilibria may exhibit some features that seem counterintuitive at first glance. The most interesting of these is the change in the equilibrium mixes that follows a change in the structure of a game's payoffs. To illustrate, we return again to Seles and Hingis and their tennis point.

Suppose that Hingis works on improving her skills covering down the line to the point where Seles's success using her DL strategy against Hingis covering DL drops from 50% to 30%. This improvement in Hingis's skill alters one of the payoffs in the simple two-by-two table of Figure 5.1. With a row and a column added to account for mixed strategies for each player, the new game table is as shown in Figure 5.16.

		HINGIS		
		DL	CC	q-Mix
SELES	DL	30	80	$30q + 80(1 - q)$
	CC	90	20	$90q + 20(1 - q)$
	p-Mix	$30p + 90(1 - p)$	$80p + 20(1 - p)$	

FIGURE 5.16   Counterintuitive Change in Mixture Probabilities

The new payoff value of 30 appears in the upper left-hand corner of the table where in Figure 5.1 this was a 50. This change in the payoff table does not lead to a game with a pure-strategy equilibrium because the players still have opposing interests; Hingis still wants their choices to coincide, and Seles still wants their choices to differ. We still have a game in which mixing will occur.

But how will the equilibrium mixes in this new game differ from those calculated in Section 5.3? At first glance, many people would argue that Hingis should cover DL more now that she has gotten so much better at doing so. Thus her equilibrium q-mix should be more heavily weighted toward DL and her equilibrium q should be higher than the 0.6 calculated earlier.

But when we calculate Hingis's q-mix by the usual condition of Seles's indifference between her two pure strategies, we get $30q + 80 (1 - q) = 90q + 20 (1 - q)$, or $q = 0.5$. The actual equilibrium value for q, 50%, is exactly the opposite of what many people's intuition predicts.

While the intuition seems reasonable, it misses an important aspect of the theory of strategy: the interaction between the two players. Seles will also be reassessing her equilibrium mix after the change in payoffs, and Hingis must take the new payoff structure *and* Seles's behavior into account when determining her new mix. Specifically, since Hingis is now so much better at covering DL, Seles will use CC more often in her mix. To counter that, Hingis will cover CC more often too.

We see this more explicitly by calculating Seles's new mixture. Her equilibrium value of p must equate $30p + 90(1 - p)$ and $80p + 20(1 - p)$; so p must be $7/12$, which is 0.583, or 58.3%. Comparing this new equilibrium value of p with the original 70% shows that Seles has significantly decreased the number of times she sends her shot down the line in response to Hingis's improved skills. Seles has taken into account the fact that she is now facing an opponent with better DL coverage, so she does better to play DL less frequently in her mixture. By virtue of this behavior, Seles makes it better for Hingis also to decrease the frequency of her DL play. Any other choice of mix by Hingis, in particular a mix heavily favoring DL, would now be exploited by Seles.

So is Hingis's skill improvement wasted? No, but we must judge it properly— not by how often one strategy or the other gets used but by the resulting payoffs. When Hingis uses her new equilibrium mix with $q = 0.5$, Seles's success percentage from either of her pure strategies is

$$30 \times 0.5 + 80 \times 0.5 = 90 \times 0.5 + 20 \times 0.5 = 55.$$

This is less than Seles's success percentage of 62 with the original numbers. Thus Hingis's average payoff also rises, from 38 to 45, and she does benefit by improving her DL coverage.

Unlike the counterintuitive result that we saw when we considered Hingis's strategic response to the change in payoffs, we see here that her expected payoff response is absolutely intuitive. In fact, players' expected payoff responses to changed payoffs can never be counterintuitive, although strategic responses, as we have seen, can be.[9] The most interesting aspect of this counterintuitive outcome in players' strategic responses is the message it sends to tennis players and to strategic game players more generally. The result here is equivalent to saying that Hingis should improve her DL coverage so she does not have to use it so often.

C. Using Mixed Strategies in Practice

Here we make note of some important things to remember when finding or using a mixed strategy. First, to use a mixed strategy effectively, a player needs to do more than calculate the equilibrium percentages with which to use each of her actions. Indeed, in our tennis game Seles cannot simply play DL five eighths of the time and CC three eighths of the time by mechanically rotating five shots down the line and three shots crosscourt. Why not? Because mixing your strategies is supposed to help you benefit from the element of surprise against your opponent. If you use a recognizable pattern of plays, your opponent is sure to discover it and exploit it to her advantage.

For a mixed strategy to work effectively, you need a truly random pattern of actions on each play of the game. You may want to rely on a computer's ability to generate random numbers for you, for example, from which you can determine

[9]For a general theory of the effect that changing the payoff in a particular cell has on the equilibrium mixture and the expected payoffs in equilibrium, see Vincent Crawford and Dennis Smallwood, "Comparative Statics of Mixed-Strategy Equilibria in Noncooperative Games," *Theory and Decision*, vol. 16 (1984), pp. 225–232.

your appropriate choice of action. If the computer generates numbers between 1 and 100, and you (like Hingis) want to mix pure strategies *A* and *B* in a 60:40 split, you may decide to play *A* for any random number between 1 and 60 and to play *B* for any random number between 61 and 100. Similarly, you could employ a device similar to the color-coded spinners provided in many children's games. For the same 60:40 mixture, you would color 60% of the circle on the spinner in blue, for example, and 40% in red; the first 216° of the circle would be blue, the remaining 144° in red. Then you would spin the spinner arrow and play *A* if it landed in blue but *B* if it landed in red. The second hand on a watch can provide the same type of device, but it is important that your watch not be so accurate and synchronized that your opponent can use an identical watch and figure out what you are going to do.

The importance of avoiding a predictable system of randomization is clearest in ongoing interactions of a zero-sum nature. Because of the diametrically opposed interests of the players in such games, your opponent will always benefit from exploiting your choice of action to the greatest degree possible. Thus, if you and your opponent play the same game against each other on a regular basis, she will always be on the lookout for ways to break the code you are using to randomize your moves. If she can do so, she will have a chance to improve her payoffs in future plays of the game. Of course, mixing is still justified in *single-meet* (sometimes called *one-shot*) zero-sum games because the benefit of tactical surprise remains important. In some non-zero-sum games, however, it may actually be crucial to develop a broadly understood system for mixing that can be used for mutual benefit. Coordination by flipping a single coin can help players avoid the particularly poor outcomes that sometimes occur when both players used mixed strategies.

Of course, players must understand and accept the fact that the use of mixed strategies can lead to such poor outcomes. If you make use of a mixed strategy in a situation in which you are responsible to a higher authority, however, you may need to plan ahead for the possibility of a low payoff. You may need to justify your use of such a strategy ahead of time to your coach or your boss, for example. They need to understand why you have adopted your mixture and why you expect it to yield you the best possible payoff, on average, even though it might yield an occasional low payoff as well. Even such advance planning may not work to protect your reputation, though, and you should prepare yourself for criticism in the face of a bad outcome.

Finally, notice that when your opponent plays her equilibrium mix, you have no direct, positive incentive to use yours. All of the pure strategies that appear in your equilibrium mix yield you the same expected payoff as the mix. Thus in theory, you could use one of your pure strategies against your opponent's mix and get the same payoff as you would get if you mixed. So why not do this? The reason for you to mix, in the end, is indirect. If you do not, your opponent will no longer play her equilibrium mix either. Figuring that you are using one of your pure strategies, she will start using the pure strategy of hers that gives her the greatest benefit. That change in her behavior will reduce the value of the game for you. Thus you mix when she mixes and the game attains a Nash equilibrium.

6 MIXING WHEN ONE PLAYER HAS THREE OR MORE PURE STRATEGIES

Our discussion of mixed strategies to this point has been confined to games in which each player has only two pure strategies, and mixes between the two. Obviously, in many strategic situations each player has available a larger number of pure strategies, and we should be ready to calculate equilibrium mixes for those cases as well. However, these calculations get complicated quite quickly. For truly complex games, we would turn to a computer to find the mixed-strategy equilibrium. But for some small games, it is possible to calculate equilibria by hand in a reasonable period of time. We will restrict our attention here to games of this type.

In this section we consider zero-sum games in which one of the players has only two pure strategies while the other has more. A very simple extension of the prevent-exploitation method works for such games. It also brings out a new point: the player who has three (or more) pure strategies will typically use only two of them in equilibrium. The others will not figure in her mix; they will get zero probabilities. Of course, we must determine which ones are used and which ones are not, and that requires more thought and more calculation.[10]

In Chapter 7 we go on to bigger games where both players have three or more pure strategies. We do the explicit calculation for some examples there, and again show how some pure strategies can go unused in equilibrium, and how to determine which ones. There we also state very briefly the general theory of mixed-strategy equilibria.

For our purposes here, let us alter the tennis-point game by giving Seles a third type of return. In addition to going down the line or crosscourt, she now can consider using a lob (a slower but higher and longer return). The equilibrium depends on the payoffs of the lob against each of Hingis's two defensive stances. We begin with the case that is most likely to arise, and then consider a coincidental or exceptional case.

---

[10] Even when a player has only two pure strategies, she may not use one of them in equilibrium. The other player then generally finds one of her strategies to be better against the one the first player does use. In other words, we have an equilibrium in pure strategies. A genuinely mixed-strategy equilibrium in which some pure strategies are unused requires at least three pure strategies for at least one of the players.

A. A General Case

Seles now has three pure strategies in her repertoire: DL, CC, and Lob. We leave Hingis with just two pure strategies, Cover DL or Cover CC. The payoff table for this new game can be obtained by adding a Lob row to the table in Figure 5.2. The result is shown in Figure 5.17.

		HINGIS		
		DL	CC	q-Mix
SELES	DL	50	80	$50q + 80(1 - q)$
	CC	90	20	$90q + 20(1 - q)$
	Lob	70	60	$70q + 60(1 - q)$
	p-Mix	$50p_1 + 90p_2 + 70(1 - p_1 - p_2)$	$80p_1 + 20p_2 + 60(1 - p_1 - p_2)$	

FIGURE 5.17 Payoff Table for Tennis Point with Lob

The payoffs in the first three rows of the table are straightforward. When Seles uses her pure strategies DL and CC, her payoffs against Hingis's pure strategies or the q-mix are exactly as in Figure 5.2. The third row is also analogous. When Seles uses Lob, we assume that her success percentages against Hingis's DL and CC are respectively 70% and 60%. When Hingis uses her q-mix, using DL a fraction q of the time and CC a fraction $(1 - q)$ of the time, Seles's expected payoff from Lob is $70q + 60(1 - q)$; therefore that is the entry in the cell where Seles's Lob meets Hingis's q-mix.

The really new feature here is the last row of the table. Seles now has three pure strategies, so she must now consider three different actions in her mix. The mix cannot be described by just one number p. Rather, we suppose that Seles plays DL with probability p_1, and CC with probability p_2, leaving Lob to get the remaining probability, $1 - p_1 - p_2$. Thus we need two numbers, p_1 and p_2, to define Seles's p-mix. Each of them, being a probability, must be between 0 and 1. Moreover, the two must add to something no more than 1; that is, they must satisfy the condition $p_1 + p_2 \leq 1$, because the probability $(1 - p_1 - p_2)$ of using Lob must be nonnegative.

Using this characterization of Seles's p-mix then, we see that her expected payoff, when Hingis plays her pure strategy DL, is given by $50p_1 + 90p_2 + 70(1 - p_1 - p_2)$. This is the entry in the first cell of the last row of the table in Figure 5.17. Seles's expected payoff from using her p-mix against Hingis's CC is similarly $80p_1 + 20p_2 + 60(1 - p_1 - p_2)$. We do not show the expression for the payoff of mix-against-mix because it is too long, and we do not need it for our calculations.

Even before we begin looking for a mixed-strategy equilibrium, we should verify that there is no pure-strategy equilibrium. This is easy to do, so we leave it to you, and turn to mixed strategies.

The key to solving for a mixed-strategy equilibrium when one player has two pure strategies and the other has three or more is to consider the situation from the perspective of the player who has just two pure strategies. Here that player is Hingis. What we need is a diagram analogous to Figure 5.5, as we have drawn in Figure 5.18.

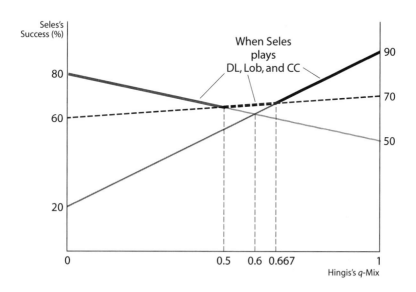

FIGURE 5.18 Diagrammatic Solution for Hingis's q-Mix

The two lines that represent Seles's success percentages from playing DL and CC respectively, as the q in Hingis's q-mix varies over its full range from 0 to 1, are exactly the same as before. The corresponding line for Lob is new. We can determine its intercepts as before; they are 60 when Hingis sets $q = 0$ and 70 when she sets $q = 1$.

If you draw three straight lines at random in a plane, they will generally not all meet in a single point. Similarly, given all the payoff numbers like 84 or 47 that could be assigned to the performance of one of Seles's three pure strategies against one of Hingis's two pure strategies, for only an exceptional or coincidental combination of these numbers will our three lines pass through a single point. For the particular numbers we have chosen here, they do not. Instead, they intersect in the more standard way, in pairs at three different points.

One value of q, at the left-most intersection in Figure 5.18, makes Seles indifferent between DL and Lob. That q must equate the two payoffs from DL and Lob when used against the q-mix. Setting those two expressions equal gives us $50q + 80(1 - q) = 70q + 60(1 - q)$ and $q = 20/40 = 1/2$. The first intersection thus occurs at $q = 0.5$, or 50%. The middle intersection occurs when the payoffs from DL and CC are the same. To solve for the q value there, we equate the DL and CC payoff expressions, $50q + 80(1 - q) = 90q + 20(1 - q)$, to find $q = 60/100$; the middle intersection is at $q = 0.6$, or 60%. Finally, the right-most intersection occurs when q equates the payoffs from CC and Lob. Solving for that value of q requires the CC and Lob expressions to be equal: $90q + 20(1 - q) = 70q + 60(1 - q)$. This yields $q = 40/60 = 2/3$ so the final intersection occurs at $q = 0.667$, or 66.7%.

Which of these intersections yields the equilibrium value of q? The keep-the-opponent-indifferent method can't tell us immediately. We must go behind the mechanics and use our understanding of what these lines actually mean.

The prevent-exploitation method gives the right insight for this zero-sum game. For any choice of q that Hingis could make, she should assume that Seles would respond with a strategy that is best for Seles (and therefore worst for Hingis). From Figure 5.18 we see that Seles's response is now of three different kinds. When $q < 0.5$ (the point of intersection of the DL and Lob lines), the DL line is higher than either of the other two, so DL is Seles's best response to Hingis's q-mix. When $q > 0.667$ (actually 2/3, the point of intersection of the CC and Lob lines), the CC line, being highest, is Seles's best response. When q lies between these two points of intersection ($0.5 < q < 0.667$), the Lob line is highest, and that is Seles's best response.

As in Figure 5.5, we show the best that Seles can do (or the worst that Hingis can do) at each value of q by means of thicker lines. Now this upper envelope consists of three rather than two segments. And Hingis wants to choose her equilibrium q to maximize her own success percentage, which is equivalent to keeping Seles's success percentage as low as possible. In other words, she should choose the value of q that gives the lowest possible point along the upper envelope in Figure 5.18. Visual inspection immediately shows that the correct q is the one at the leftmost pairwise intersection, $q = 0.5$. Seles's payoff is lower here than at the rightmost intersection at $q = 0.667$.

You might think that the correct value of q corresponds to the middle intersection ($q = 0.6$) because that value of q gives Seles an even lower success percentage. But our analysis has shown that intersection to be irrelevant. While Seles is indifferent there between DL and CC, she gets a higher payoff by using Lob, and therefore will use Lob when $q = 0.6$. Hingis must take this into account and realize that at $q = 0.6$ she cannot keep Seles's payoff down as far as she can at $q = 0.5$. Hingis should concentrate on the q values associated with intersections that actually lie on the upper envelope in the diagram.

Having established that Hingis uses the q-mix of the leftmost intersection, that is $q = 0.5$, we can calculate Seles's success percentage for each of her three pure strategies against Hingis's q-mix by plugging the value $q = 0.5$ into the three payoff expressions in the last column of the table in Figure 5.17. This process yields Seles an expected payoff of 65% when using pure DL against a 50-50 mix of DL and CC by Hingis. Her expected payoff from pure Lob is also 65%, but her payoff from pure CC is only 55%.

But how do we use this information to determine Seles's actual strategy in equilibrium? We have just calculated that when Hingis uses the q-mix corresponding to the left-most intersection in Figure 5.18, Seles is indifferent between DL and Lob, and that either of these gives her a better payoff than does CC. Therefore Seles will not use CC at all; CC will be an unused strategy in her equilibrium mix. We can now proceed with the equilibrium analysis as if this were a game with just two pure strategies for each player: DL and CC for Hingis, and DL and Lob for Seles. We are back in familiar territory.

Of course, we could not have started here, because we did not know in advance *which* of her three strategies Seles would not use. But we can be confident that in the general case there will be one such strategy. When the three lines are in the general positions, they intersect pairwise rather than at a single point, and the upper envelope has the valley shape described earlier. Its lowest point is defined by the intersection of the payoff lines associated with some two of the three strategies. The payoff from the third strategy lies below the intersection at this point, so the player choosing among the three strategies does not use that third one. (In Chapter 7, Section 5 we will discuss a general result of this kind when each player has several pure strategies.)

Once we know that Seles mixes between DL and the Lob, we can find her p-mix using the method of Figure 5.2. We begin by eliminating Seles's CC strategy from the payoff table of Figure 5.17, and show the resulting table in Figure 5.19. Of course, in the new p-mix we set p_2 for the CC strategy equal to zero, so Seles is mixing with a probability p_1 of DL and probability $(1-p_1)$ of Lob.

		HINGIS		
		DL	CC	q-Mix
SELES	DL	50	80	$50q + 80(1 - q)$
	Lob	70	60	$70q + 60(1 - q)$
	p-Mix	$50p_1 + 70(1 - p_1)$	$80p_1 + 60(1 - p_1)$	

FIGURE 5.19 Payoff Table after Eliminating Seles's CC Strategy

Figure 5.20 shows the solution for Seles's p-mix. The value of p_1 at the intersection of the DL and CC lines can be found by equating the expressions for Seles's payoffs from using her p-mix against Hingis using DL and against Hingis using CC. Thus $50p + 70(1 - p_1) = 80p + 60(1 - p_1)$ and $p_1 = 10/40 = 0.25$. Seles's op-

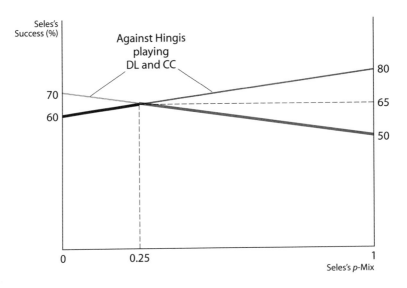

FIGURE 5.20 Diagrammatic Solution for Seles's *p*-Mix

timal mixture in this game entails her using a down-the-line return 25% of the time and a lob 75% of the time. She does not use her crosscourt return at all.

The value of the game to Seles can be calculated using her equilibrium value of *p*. For $p = 0.25$, Seles's success percentage against either Hingis's DL (or CC) is $50 \times 0.25 + 70 \times 0.75 = 65$ (or $80 \times 0.25 + 60 \times 0.75 = 65$). Seles's expected success percentage of 65% corresponds to her expected payoff from using either of the pure strategies that make up her mix against Hingis's equilibrium mix, exactly as we would expect. This outcome leaves Hingis with an expected success percentage of 35%.

B. A Coincidental Case

If Seles's success percentages take on just the right combination of values, the three lines in a graph like Figure 5.18 can all pass through the same point. For example, this happens if, in Figure 5.17, we lower Seles's success with the Lob against Hingis's CC from 60% to 50%, keeping all other payoffs unchanged. Figure 5.21 shows the new payoff table, and Figure 5.22 the corresponding diagram of Seles's success with each of her three pure strategies against Hingis's *q*-mix.

The DL and CC lines intersect when $50q + 80(1 - q) = 90q + 20(1 - q)$. We carried out this calculation before when the Lob strategy was not there; we know it yields $q = 0.6$ and a success rate of 62% for Seles. We follow exactly the same steps of calculation for the intersection of the DL and Lob lines to find $50q + 80(1 - q) = 70q + 50(1 - q)$ and $q = 3/5 = 0.6$. Then the common value of the success percentage of DL and Lob is also 62. We leave it to you to verify that the intersection of the CC and Lob lines also yields $q = 0.6$ and a success rate of 62%.

		HINGIS		
		DL	CC	q-Mix
SELES	DL	50	80	$50q + 80(1-q)$
	CC	90	20	$90q + 20(1-q)$
	Lob	70	50	$70q + 50(1-q)$
	p-Mix	$50p_1 + 90p_2$ $+ 70(1 - p_1 - p_2)$	$80p_1 + 20p_2$ $+ 50(1 - p_1 - p_2)$	

FIGURE 5.21   Payoff Table for Tennis Point with Lob: The Coincidence Case

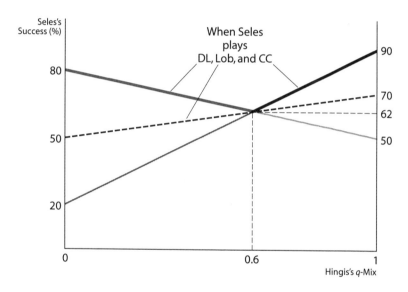

FIGURE 5.22   Diagrammatic Solution for Hingis's q-Mix: The Coincidence Case

To avoid being exploited, Hingis must now choose $q = 0.6$. Figure 5.22 shows that for any lower value of q, Seles would choose purely DL (the highest of the three lines for these values of q), while for any higher value of q, Seles would choose purely CC. The upper envelope of the three lines is again shown by the thickened segments. But now there are just two of them. The Lob strategy has only one point on the thickened lines, namely where all three lines intersect.

What does this imply about Seles's p-mix? In particular, does she mix only between DL and CC? As before, Seles's p-mix must be such that Hingis is indifferent between her two pure strategies. Using the payoffs on the last row of the table in Figure 5.21, this gives the equation

$$50p_1 + 90p_2 + 70(1 - p_1 - p_2) = 80p_1 + 20p_2 + 50(1 - p_1 - p_2)$$

which yields the relationship $p_2 = p_1 - 0.4$.

Since Hingis has only two pure strategies, there is only this one equation that equates their payoffs when each of them is matched against Seles's p-mix. But this one equation cannot be solved to give unique values for the two probabilities p_1 and p_2. All we know is that $p_2 = p_1 - 0.4$ or that $p_1 - p_2 = 0.4$. Therefore we say that Seles's equilibrium p-mix is **indeterminate.**

Seles's mix may be indeterminate, but it is not totally arbitrary. Figure 5.23 shows the possible combinations of p_1 and p_2 that satisfy the expression above. Each of p_1 and p_2 must be nonnegative, and they must not sum to more than 1, because the probability $(1 - p_1 - p_2)$ of the third pure strategy, Lob, cannot be negative. Therefore all logically permissible (p_1, p_2) combinations in Figure 5.23 must lie in the triangle OAB formed by the origin O, the point A where $p_1 = 1$ and $p_2 = 0$, and the point B where $p_1 = 0$ and $p_2 = 1$. Along side AB of this triangle, $p_1 + p_2 = 1$, so $1 - p_1 - p_2 = 0$.

In the same figure, we show the line of combinations of (p_1, p_2) that satisfy the indifference condition, namely $p_2 = p_1 - 0.4$. Only the portion of it within triangle OAB, shown by the blue line segment CD, is relevant. This line defines the total range of indeterminacy of Seles's p-mixes.

At point D, which lies on both the indifference line and the side AB, we know that $p_2 = p_1 - 0.4$ and $p_1 + p_2 = 1$. This yields a value of $p_1 = 0.7$ and then $p_2 = 0.7 - 0.4 = 0.3$, and $1 - p_1 - p_2 = 0$. Thus, at D, Seles does not use the Lob strategy at all but mixes only between DL and CC, and in the same proportions as she did in our earlier analysis when she did not have Lob available at all.

But that is not the only possibility. At point C, we have $p_1 = 0.4$ and $p_2 = 0$, so $1 - p_1 - p_2$ must be 0.6. Thus Seles could mix between DL and Lob alone with odds of $40 : 60$, leaving CC unused. Between points C and D there are several other possible

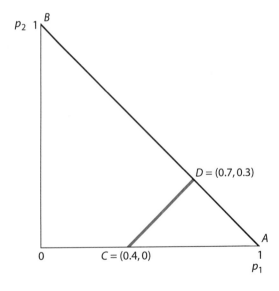

FIGURE 5.23 Seles's Indeterminate p-Mix in the Coincidence Case

mixtures including, for example, $p_1 = 0.5$, $p_2 = 0.1$, and $1 - p_1 - p_2 = 0.4$ (50% DL, 10% CC, 40% Lob); and $p_1 = 0.6$, $p_2 = 0.2$, and $1 - p_1 - p_2 = 0.2$ (60% DL, 20% CC, 20% Lob).

The techniques developed here and in the previous sections extend readily to the case where one player has two strategies and the other player has any number, say six. Choose the former to be the Column player and consider her q-mix. For each of the Row player's six pure strategies, draw the line showing the Row player's expected payoff when this strategy is played against Column's q-mix. Take the upper envelope of the six lines, and locate its lowest point.[11] That defines the equilibrium q-mix of the Column player. The intersection will generally occur at an intersection of just one pair of lines out of the six. Then Row must use only the pure strategies corresponding to those two lines in her equilibrium mixed strategy.

You can use the techniques developed here and in the previous sections to solve even larger games for equilibrium mixed strategies as well. In Chapter 7 we will return to these issues and show you some such calculations for three-by-three games. However, it becomes increasingly difficult to solve such games on paper. It is much simpler, for larger games, to write a computer program to solve the problem for you or to use one that someone else has already written. (Mathematically, such a problem is one in linear programming.) There are special algorithms that can be incorporated into computer programs to solve such problems, and several such items are now easily available in retail stores or on the Internet. Those of you who are interested in pursuing the study of game theory further might want to try out Gambit, which we have mentioned several times before, at *http://hss.Caltech.edu/~gambit/Gambit.html*.

SUMMARY

Many attack-and-defense-type games do not have equilibria in pure strategies because one player prefers a coincidence of actions while the other prefers the opposite. In these games, players want to be unpredictable and they do so by using a mixed strategy that specifies a probability distribution over their set of available pure strategies. A general result guarantees that every simultaneous-move game where each player has finitely many pure strategies has a Nash equilibrium in mixed strategies.

In zero-sum games, players choose their equilibrium mixtures to prevent exploitation by their rivals; players can improve their *maximin* payoff (or reduce their *minimax* payoff) with the correct equilibrium mix. Equilibrium mixtures for two players must be best responses to each other. Graphs of *best-response*

---

[11] Here is a source of another indeterminacy: one of the lines may be horizontal, and a segment of the upper envelope of the six lines may lie along this horizontal line. Then a whole range of values of q give an equal lowest rate of success for Row, and Column's best q-mix is indeterminate along this range.

curves for both players show all equilibria of the game (in pure or mixed strategies) at the intersection(s) of the curves. When games are non-zero-sum, the same basic concepts apply except that players choose their equilibrium mixtures to keep their rivals indifferent between their pure strategies.

If each player has only two pure strategies, equilibrium mixtures can be determined as the *odds* of using each of the two strategies. A player's equilibrium mixture also uses a pure strategy that has become stronger less often than originally because of the rival's decreased use of the strategy that does poorly against your newly strengthened action. Good use of a mixed strategy requires that there be no predictable system of randomization.

When one player has three pure strategies, two numbers are needed to describe the equilibrium mix and most often fewer than three strategies appear in that mix. The strategies appearing in the mix must all give the same *expected payoff* against the other's mix, and any unused strategy must give a lower expected payoff. Only by coincidence would all three strategies be used in equilibrium, in which case the specific probabilities in which they were to be used would be *indeterminate*.

KEY TERMS

best-response curves (133)
expected values (127)
indeterminate mix (158)
keep-the-opponent-indifferent
  method (139)
oddments (143)
odds method (143)
p-mix (127)

percentage play (147)
prevent-exploitation
  method (135)
q-mix (131)
reaction curves (133)
upper envelope (137)
value of the game (133)

EXERCISES

1. Find Nash equilibria in mixed strategies for the following games. In each case, verify your answer by using a second method to recalculate the equilibrium mixtures for each player.

   (a)

		COLUMN	
		Left	Right
ROW	Up	4	−1
	Down	1	2

(b)

		COLUMN	
		Left	Right
ROW	Up	3	2
	Down	1	4

(c)

		COLUMN	
		Left	Right
ROW	Up	4,0	−1,2
	Down	1,1	2,−1

2. "When a game has a mixed-strategy equilibrium, a player's equilibrium mixture yields her the same expected payoff against each of the other player's pure strategies." True or false? Explain and give an example of a game that illustrates your answer.

3. Recall Exercise 6 in Chapter 4 about an old lady looking for help crossing the street. Only one person is needed to help her; more are okay but no better than one. You and I are the two people in the vicinity who can help; each has to choose simultaneously whether to do so. Each of us will get pleasure worth a 3 from her success (no matter who helps her). But each one who goes to help will bear a cost of 1, this being the value of our time taken up in helping. In Exercise 6 you were asked to set this up as a game, write the payoff table, and find all of the pure-strategy Nash equilibria. Now find the mixed-strategy equilibrium of this game.

4. The following table illustrates the money payoffs associated with a two-person simultaneous-play game:

		COLUMN	
		Left	Right
ROW	Up	1,16	4,6
	Down	2,20	3,40

(a) Find the Nash equilibrium in mixed strategies for this game and the players' expected payoffs in this equilibrium.

(b) The two players jointly get most money when Row plays Down. However, in the equilibrium, Row does not always play Down. Why not? Can you think of ways in which a more cooperative outcome could be sustained?

5. "When a game has several Nash equilibria, then the outcome of playing the game will be a random mixture between all these equilibria." True or false? Explain and provide an example of a game that illustrates your answer.

6. Consider the following game:

		COLUMN			
		A	B	C	D
ROW	1	1,1	2,2	3,4	9,3
	2	2,5	3,3	1,2	7,1

(a) Find all the pure-strategy Nash equilibria.

(b) Now find a mixed-strategy equilibrium of the game.

7. Consider the following zero-sum game:

		COLUMN	
		Left	Right
ROW	Up	0	A
	Down	B	C

The entries are the Row player's payoffs, and the numbers A, B, and C are all positive. What other relations between these numbers (for example, $A < B < C$) must be valid for each of the following cases to arise?

(a) At least one of the players has a dominant strategy.

(b) Neither player has a dominant strategy, but there is a Nash equilibrium in pure strategies.

(c) There is no Nash equilibrium in pure strategies, but there is one in mixed strategies.

(d) Given that case (c) holds, write a formula for Row's probability of choosing Up. Call this probability p, and write it as a function of A, B, and C.

■

Appendix: Probability and Expected Utility

To calculate the expected payoffs and mixed-strategy equilibria of games in this chapter, we had to do some simple manipulation of probabilities. There are some simple rules that govern calculations involving probabilities. Many of you may be familiar with them, but we give a brief statement and explanation of the basics here by way of reminder or remediation, as appropriate. We also state how to calculate expected values of random numerical values.

We also discuss here the expected-utility approach to calculating expected payoffs. When the outcomes of your action in a particular game are not certain, either because your opponent is mixing strategies or because of some uncertainty in nature, you may not want to maximize your expected monetary payoff as we have generally assumed in our analysis up to this point; rather, you may want to give some attention to the *riskiness* of the payoffs. We mentioned in Chapter 2 that such situations can be handled by using the expected values (which are probability-weighted averages) of an appropriate nonlinear rescaling of the monetary payoffs. We offer here a brief discussion of how this can be done.

You should certainly read this material, but to get real knowledge and mastery of it, the best thing to do is to use it. The chapters to come, especially 7, 10, and 12, will give you plenty of opportunity for practice.

1 THE BASIC ALGEBRA OF PROBABILITIES

The basic intuition about the probability of an event comes from thinking about the frequency with which this event occurs by chance among a larger set of possibilities. Usually any one element of this larger set is just as likely to occur by chance as any other, so finding the probability of the event we are interested in is simply a matter of counting the elements corresponding to that event and dividing by the total number of elements in the whole large set.[1]

---

[1]When we say "by chance," we simply mean that a systematic order cannot be detected in the outcome, or that it cannot be determined using available scientific methods of prediction and calculation. Actually the motions of coins and dice are fully determined by the laws of physics, and highly skilled people can manipulate decks of cards, but for all practical purposes coin tosses, rolls of dice, or card shuffles are devices of chance that can be used to generate random outcomes. However, randomness can be harder to achieve than you think. For example, a perfect shuffle, where a deck of cards is divided exactly in half and then interleaved by dropping cards one at a time alternately from

In any standard deck of 52 playing cards, for instance, there are four suits (clubs, diamonds, hearts, and spades) and 13 cards in each suit (ace through 10 and the face cards—jack, queen, king). We can ask a variety of questions about the likelihood that a card of a particular suit, or value, or suit *and* value might be drawn from this deck of cards: How likely are we to draw a spade? How likely are we to draw a black card? How likely are we to draw a 10? How likely are we to draw the queen of spades? and so on. And we would need to know something about the calculation and manipulation of probabilities to answer such questions. If we had two decks of cards, one with blue backs and one with green backs, we could ask even more complex questions ("How likely are we to draw one card from each deck and have them both be the jack of diamonds?"), but we would still use the algebra of probabilities to answer them.

In general, a **probability** measures the likelihood of a particular event or set of events occurring. The likelihood that you draw a spade from a deck of cards is just the probability of the event "drawing a spade." Here the large set has 52 elements—the total number of equally likely possibilities—and the event "drawing a spade" corresponds to a subset of 13 particular elements. Thus you have 13 chances out of the 52 to get a spade, which makes the probability of getting a spade in a single draw equal to $13/52 = 1/4 = 25\%$. To see this another way, consider the fact that there are four suits of 13 cards each, so your chance of drawing a card from any particular suit is one out of four, or 25%. Of course, if you made a large number of such draws (each time from a complete deck), then out of 52 times you will not always draw exactly 13 spades; by chance you may draw a few more or a few less. But the chance averages out over different such occasions—over different sets of 52 draws. Then the probability of 25% is the average of the frequencies of spades drawn in a large number of observations.[2]

The algebra of probabilities simply develops such ideas in general terms and obtains formulas that you can then apply mechanically instead of having to do the thinking from scratch every time. We will organize our discussion of these probability formulas around the types of questions that one might ask when drawing cards from a standard deck (or two: blue backed and green backed).[3]

---

each, may seem a good way to destroy the initial order of the deck. But Cornell mathematician Persi Diaconis has shown that after eight of these shuffles the original order is fully restored. For slightly imperfect shuffles that people carry out in reality, he finds that some order persists through six, but randomness suddenly appears on the seventh! See "How to Win at Poker, and Other Science Lessons," *The Economist*, October 12, 1996. For an interesting discussion of such issues, see Deborah J. Bennett, *Randomness*, (Cambridge, Mass.: Harvard University Press, 1998), chaps. 6–9.

[2] Bennett, *Randomness*, chaps. 4 and 5, offers several examples of such calculations of probabilities.

[3] If you want a more detailed exposition of the following addition and multiplication rules, and more exercises to practice using these rules, we recommend David Freedman, Robert Pisani, and Roger Purves, *Statistics*, 3rd ed. (New York: Norton, 1998), chaps. 13 and 14.

This method will allow us to provide both specific and general formulas for you to use later. You can use the card-drawing analogy to help you reason out other questions about probabilities that you encounter in other contexts. One other point to note: In ordinary language it is customary to write probabilities as percentages, but the algebra requires that they be written as fractions or decimals; thus instead of 25% the mathematics works with 13/52 or 0.25. We will use one or the other depending on the occasion; be aware that they mean the same thing.

A. The Addition Rule

The first questions that we ask are: If we were to draw one card from the blue deck, how likely are we to draw a spade? And how likely are we to draw a card that is *not* a spade? We already know that the probability of drawing a spade is 25% because we determined that earlier. But what is the probability of drawing a card that is not a spade? It is the same likelihood of drawing a club *or* a diamond *or* a heart instead of a spade. It should be clear that the probability in question should be larger than any of the individual probabilities of which it is formed; in fact, the probability is 13/52 (clubs) + 13/52 (diamonds) + 13/52 (hearts) = 0.75. The *or* in our verbal interpretation of the question is the clue that the probabilities should be added together, because we want to know the chances of drawing a card from *any* of those three suits.

We could more easily have found our answer to the second question by noting that not getting a spade is what happens the other 75% of the time. Thus the probability of drawing "not a spade" is 75% (100% − 25%) or, more formally, $1 - 0.25 = 0.75$. As is often the case with probability calculations, the same result can be obtained here by two different routes, involving different ways of thinking about the event for which we are trying to find the probability. We will see other examples of this later in the Appendix, where it will become clear that the different methods of calculation can sometimes involve vastly different amounts of effort. As you develop experience, you will discover and remember the easy ways or shortcuts. In the meantime, be comforted that each of the different routes, when correctly followed, leads to the same final answer.

To generalize our calculation above, we note that if you divide the set of events, X, in which you are interested into some number of subsets, $Y, Z, \ldots$, none of which overlap (in mathematical terminology, such subsets are said to be **disjoint**), then the probabilities of each subset occurring must sum to the probability of the full set of events; if that full set of events includes all possible outcomes, then its probability is 1. In other words, if the occurrence of X requires the occurrence of *any one* of several disjoint $Y, Z, \ldots$, then the probability of X is the sum of the separate probabilities of $Y, Z, \ldots$. Using Prob(X) to denote the probability that X occurs and remembering the caveats on X (that it requires any one of

Y, Z, . . .) and on Y, Z, . . . (that they must be disjoint), we can write the **addition rule** in mathematical notation as $Prob(X) = Prob(Y) + Prob(Z) + \cdots$.

EXERCISE   Use the addition rule to find the probability of drawing two cards, one from each deck, such that the two cards have identical faces.

B. The Modified Addition Rule

Our analysis in the previous section covered only situations in which a set of events could be broken down into disjoint, nonoverlapping subsets. But suppose we ask, What is the likelihood, if we draw one card from the blue deck, that the card is either a spade *or* an ace? The *or* in the question suggests, as before, that we should be adding probabilities, but in this case the two categories "spade" and "ace" are not mutually exclusive because one card, the ace of spades, is in both subsets. Thus "spade" and "ace" are not disjoint subsets of the full deck. So if we were to sum only the probabilities of drawing a spade (13/52) and of drawing an ace (4/52), we would get 17/52. This would suggest that we had 17 different ways of finding either an ace or a spade when in fact we have only 16—there are 13 spades (including the ace) and three additional aces from the other suits. The incorrect answer, 17/52, comes from counting the ace of spades twice. In order to get the correct probability in the nondisjoint case, then, we must subtract the probability associated with the overlap of the two subsets. The probability of drawing an ace or a spade is the probability of drawing an ace plus the probability of drawing a spade minus the probability of drawing the overlap, the ace of spades; that is, $13/52 + 4/52 - 1/52 = 16/52 = 0.31$.

To make this more general, if you divide the set of events, X, in which you are interested into some number of subsets Y, Z, . . . , which may overlap, then the sum of the probabilities of each subset occurring minus the probability of the overlap yields the probability of the full set of events. More formally, the **modified addition rule** states that if the occurrence of X requires the occurrence of *any one* of the nondisjoint Y and Z, then the probability of X is the sum of the separate probabilities of Y and Z minus the probability that *both* Y and Z occur: $Prob(X) = Prob(Y) + Prob(Z) - Prob(Y \text{ and } Z)$.

EXERCISE   Use the modified addition rule to find the probability of drawing two cards, one from each deck, and getting at least one face card.

C. The Multiplication Rule

Now we ask, What is the likelihood that when we draw two cards, one from each deck, both of them will be spades? This event occurs if we draw a spade from the blue deck *and* a spade from the green deck. The switch from *or* to *and* in our interpretation of what we are looking for indicates a switch in mathematical operations from addition to multiplication. Thus the probability of two spades, one from each deck, is the product of the probabilities of drawing a spade from each deck, or $(13/52) \times (13/52) = 1/16 = 0.0625$, or 6.25%. Not surprisingly, we are much less likely to get two spades than we were in the previous section to get one spade. (Always check to make sure that your calculations accord in this way with your intuition regarding the outcome.)

In much the same way as the addition rule requires events to be disjoint, the multiplication rules requires them to be independent; if we break down a set of events, X, into some number of subsets $Y, Z, \ldots$, those subsets are independent if the probability of one occurring does not influence the probability of the other occurring. Our events—a spade from the blue deck and a spade from the green deck—satisfy this condition of independence; that is, drawing a spade from the blue deck does nothing to alter the probability of getting a spade from the green deck. If we were drawing both cards from the same deck, however, then once we had drawn a spade (with a probability of $13/52$), the probability of drawing another spade would no longer be $13/52$ (in fact, it would be $12/51$); drawing one spade and then a second spade from the *same* deck are not independent events.

The formal statement of the **multiplication rule** tells us that if the occurrence of X requires the simultaneous occurrence of *all* the several **independent** $Y, Z, \ldots$, then the probability of X is the *product* of the separate probabilities of Y, $Z, \ldots$: $\mathrm{Prob}(X) = \mathrm{Prob}(Y) \times \mathrm{Prob}(Z) \times \cdots$.

> EXERCISE   Use the multiplication rule to find the probability of drawing two cards, one from each deck, and getting a red card from the blue deck and a face card from the green deck.

D. The Modified Multiplication Rule

What if we are asking about the probability of an event that depends on two nonindependent occurrences? For instance, suppose that we ask, What is the likelihood that with one draw we get a card that is both a spade *and* an ace? If we think about this for a moment, we realize that the probability of this event is just the probability of drawing a spade *and* the probability that our card is an ace *given that* it is a spade. The probability of drawing a spade is $13/52 = 1/4$, and the probability of an ace given that we have a spade is $1/13$. The *and* in our question tells us to take the product of these two probabilities: $(13/52)(1/13) = 1/52$.

Of course, we could have gotten the same answer by realizing that our question was the same as asking, What is the likelihood of drawing the ace of spades? The calculation of that probability is straightforward; only 1 of the 52 cards is the ace of spades, so the probability of drawing it must be 1/52. As you see, how you word the question affects how you go about looking for an answer.

In the technical language of probabilities, the probability of a particular event occurring (like getting an ace), given that another event has already occurred (like getting a spade) is called the **conditional probability** of drawing an ace, for example, conditioned on having drawn a spade. Then, the formal statement of the **modified multiplication rule** is that if the occurrence of X requires the occurrence of both Y and Z, then the probability of X equals the product of two things: (1) the probability that Y alone occurs, and (2) the probability that Z occurs *given* that Y has already occurred, or the *conditional probability* of Z, conditioned on Y having already occurred: Prob(X) = Prob(Y alone) × Prob(Z given Y).

A third way would be to say that the probability of drawing an ace is 4/52, and the conditional probability of the suit being a spade given that the card is an ace is 1/4, so the overall probability of getting an ace of spades is (4/52) × 1/4. More generally, using the terminology just introduced, we have Prob(X) = Prob(Z) Prob(Y given Z).

> EXERCISE   Use the modified multiplication rule to find the probability that, when you draw two cards from a deck, the *second* card is the jack of hearts.

E. The Combination Rule

We could also ask questions of an even more complex nature than we have tried so far, in which it becomes necessary to use both the addition (or modified addition) and multiplication (or modified multiplication) rules simultaneously. We could ask, What is the likelihood, if we draw one card from each deck, that we draw *at least one* spade? As usual, we could approach the calculation of the necessary probability from several angles, but suppose that we come at it first by considering all of the different ways in which we could draw at least one spade when drawing one card from each deck. There are three possibilities: either we could get one spade from the blue deck and none from the green deck ("spade *and* none") *or* we could get no spade from the blue deck and a spade from the green deck ("none *and* spade") *or* we could get a spade from each deck ("spade *and* spade"); our event requires that one of these three possibilities occurs, each of which entails the occurrence of both of two independent events. It should be obvious now, using the *or*'s and *and*'s as guides, how to calculate the necessary probability. We find the probability of each of the three possible ways of getting at least one spade (which involves three products of two probabilities each) and sum these together: 1/4 × 3/4 + 3/4 × 1/4 + 1/4 × 1/4 = 7/16 = 43.75%.

The second approach entails recognizing that "at least one spade" and "not any spades" are disjoint events; together they constitute a sure thing. Therefore the probability of "at least one spade" is just 1 minus the probability of "not any spades." And the event "not any spades" occurs only if the blue card is not a spade 3/4 *and* the green card is not a spade 3/4, so its probability is $3/4 \times 3/4 = 9/16$. The probability of "at least one spade" is then $1 - 9/16 = 7/16$ as we found in the previous paragraph.

Finally, we can formally state the **combination rule** for probabilities: If the occurrence of X requires the occurrence of *exactly one* of a number of *disjoint* $Y, Z, \ldots$, while the occurrence of Y requires that of *all* of a number of *independent* $Y_1, Y_2, \ldots$ and the occurrence of Z requires that of *all* of a number of *independent* $Z_1, Z_2, \ldots$, and so on, then the probability of X is the sum of the probabilities of $Y, Z, \ldots$, which are the products of the probabilities $Y_1, Y_2, \ldots$, $Z_1, Z_2, \ldots$: or

$$
\begin{aligned}
\text{Prob}(X) &= \text{Prob}(Y) + \text{Prob}(Z) + \cdots \\
&= \text{Prob}(Y_1) \times \text{Prob}(Y_2) \times \ldots + \text{Prob}(Z_1) \times \text{Prob}(Z_2) \times \cdots + \cdots
\end{aligned}
$$

EXERCISE Suppose we now have a third (orange) deck of cards. Find the probability of drawing at least one spade when you draw one card from each of the three decks.

F. Expected Values

If a numerical magnitude (such as money winnings or rainfall) is subject to chance, and can take on any one of n possible values $X_1, X_2, \ldots, X_n$ with respective probabilities $p_1, p_2, \ldots, p_n$, then the *expected value* is defined as the weighted average of all its possible values using the probabilities as weights, that is, as

$$
p_1 X_1 + p_2 X_2 + \cdots + p_n X_n
$$

For example, suppose you bet on the toss of two fair coins. You win $5 if both coins come up heads, $1 if one shows heads and the other tails, and nothing if both come up tails. Using the rules for manipulating probabilities that we discussed earlier in this section, you can see that the probabilities of these events are, respectively, 0.25, 0.50, and 0.25. Therefore your expected winnings are

$$
0.25 \times \$5 + 0.50 \times \$1 + 0.25 \times \$0 = \$1.75.
$$

In game theory, the numerical magnitudes we need to average in this way are payoffs, measured in numerical ratings, or money, or as we will see later in this appendix, utilities. We will refer to the expected values in each context appropriately, for example, as *expected payoffs*, or *expected utilities*.

2 INFERRING EVENT PROBABILITIES FROM OBSERVING CONSEQUENCES

The rules given in Section 1 for manipulating and calculating the probability of events, particularly the combination rule, will soon prove useful in our calculations of payoffs when individuals use mixed strategies, and also when they face an uncertain environment for some other reason. The rule discussed in this section is somewhat harder, and will be used only in Chapter 12. There we discuss games in which the players are differently informed, and each is trying to find out the others' information by observing their actions. You can postpone your study of this rule until it is needed.

Sometimes we cannot actually observe some underlying condition about a player in a game, but we can observe the consequences or results of that underlying condition. Then we have to draw inferences about the likelihood of—estimate the probabilities of—the underlying conditions using the information that we have observed. The best way to understand this is by example. Suppose 1% of the population has a genetic defect that can cause a disease. A test that can identify this genetic defect has a 99% accuracy: when the defect is present, the test will fail to detect it 1% of the time, and the test will also falsely find a defect when none is present 1% of the time. We are interested in determining the probability that the defect is really there in a person with a positive test result. That is, we cannot directly observe the person's genetic defect (underlying condition), but we can observe the results of the test for that defect (consequences)—except that the test is not a perfect indicator of the defect; how certain can we be, given our observations, that the underlying condition does in fact exist?

We can do a simple numerical calculation to answer the question for our particular example. Consider a population of 10,000 persons in which 100 (1%) have the defect and 9,900 do not. Suppose they all take the test. Of the 100 persons with the defect, the test will be (correctly) positive for 99. Of the 9,900 without the defect, it will be (wrongly) positive for 99. That is 198 positive test results of which one half are right and one half are wrong. If a random person receives a positive test result, it is just as likely to be because the test is indeed right as because the test is wrong, so the risk that the defect is truly present for a person with a positive result is only 50%. (That is why tests for rare conditions must be designed to be have especially low error rates of generating "false positives.")

For general questions of this type, we use an algebraic formula called Bayes' theorem to help us set up the problem and do the calculations. Because information is often gradually revealed by observation in the course of making decisions, the method has great importance for games of uncertainty and information like those in Chapter 12. This theorem needs just a little more mathematical thinking than we have used thus far, however. You can omit the rest of this section for now, and return to it when you need to use its formula in Chapter 12.[4]

To generalize the example we used above, suppose there are two alternative underlying conditions, *A* and *B* (genetic defect or not, for example), and two observable consequences, *X* and *Y* (positive or negative test result, for example). Suppose that, in the absence of any information (over the whole population), the probability that *A* exists is *p*, so the probability that *B* exists is $(1 - p)$. When *A* exists, the chance of observing *X* is *a;* so the chance of observing *Y* is $(1 - a)$. (To use the language that we developed above, *a* is the probability of *X* conditional on *A*, and $(1 - a)$ is the probability of *Y* conditional on *A*.) Similarly, when *B* exists, the chance of observing *X* is *b*, so the chance of observing *Y* is $(1 - b)$.

This description shows us that there are four alternative combinations of events that could arise: (1) *A* exists and *X* is observed, (2) *A* exists and *Y* is observed, (3) *B* exists and *X* is observed, (4) *B* exists and *Y* is observed. Using the modified multiplication rule, we find the probabilities of the four combinations to be, respectively, pa, $p(1 - a)$, $(1 - p)b$, and $(1 - p)(1 - b)$.

Now suppose that *X* is observed; a person has the test for the genetic defect and gets a positive result. Then we restrict our attention to a subset of the four possibilities above, namely the first and third, both of which include the observation of *X*. These two possibilities have a total probability of $pa + (1 - p)b;$ this is the probability that *X* is observed. Within this subset of outcomes that involve observing *X*, the probability that *A also* exists is just pa, as we saw above. So we know how likely we are to observe *X* alone and how likely it is that both *X* and *A* exist.

But we are more interested in determining how likely it is that *A* exists, given that we have observed *X*, that is, the probability that a person has the genetic defect, given that the test is positive. This calculation is the trickiest one. Using the modified multiplication rule, we know that the probability of both *A and X* happening equals the product of the probability that *X* alone happens times the probability of *A* conditional on *X;* it is this last probability that we are after. Using the formulas for "*A* and *X*" and for "*X* alone," which we calculated just above, we get:

$$\text{Prob}(A \text{ and } X) = \text{Prob}(X \text{ alone}) \times \text{Prob}(A \text{ conditional on } X)$$

$$pa = [\, pa + (1 - p)b \,] \times \text{Prob}(A \text{ conditional on } X)$$

$$\text{Prob}(A \text{ conditional on } X) = \frac{pa}{pa + (1 - p)b}.$$

[4] However, we strongly recommend that you achieve an understanding of this formula at some point, because people often get it wrong (and get conditional probabilities wrong more generally), when thinking about many important practical problems involving probabilities. Bennett, *Randomness*, pp. 2–7, 175–181, gives several examples.

This formula gives us an assessment of the probability that A has occurred, given that we have observed X (and have therefore conditioned everything on this fact). The outcome is known as **Bayes' theorem** (or **rule** or **formula**).

In our example of testing for the genetic defect above, we had $\text{Prob}(A) = p = 0.01$, $\text{Prob}(X \text{ conditional on } A) = a = 0.99$ and $\text{Prob}(Y \text{ conditional on } A) = b = 0.01$. We can substitute these into Bayes' formula to get

Probability that the defect exists
given that the test is positive $= \text{Prob}(A \text{ conditional on } X)$

$$= \frac{(0.01)(0.99)}{(0.01)(0.99) + (1 - 0.01)(0.01)}$$

$$= \frac{0.0099}{0.0099 + 0.0099} = 0.5.$$

The probability algebra employing Bayes' rule confirms the arithmetical calculation we used earlier that was based on an enumeration of all of the possible cases. The advantage of the formula is that once we have it, we can apply it mechanically; this saves us the lengthy and error-prone task of enumerating every possibility and determining each of the necessary probabilities.

We show Bayes' rule in Figure 5A.1 in a tabular form, which may be easier to remember and to use than the formula above. The rows of the table show the alternative true conditions that might exist, for example, "genetic defect" and "no genetic defect." Here we have just two, A and B, but the method generalizes immediately to any number of possibilities. The columns show the observed events, for example, "test positive" and "test negative."

Each cell in the table shows the overall probability of that combination of the true condition and the observation; these are just the probabilities for the four alternative combinations we listed earlier. The last column on the right shows the sum across the first two columns for each of the top two rows. This sum is the total probability of each true condition (so for instance, A's probability is p, as we have seen). The last row shows the sum of the first two rows in each column. This sum gives the probability that each observation occurs. For example, the entry in

		OBSERVATION		Sum of Row
		X	Y	
TRUE CONDITION	A	pa	$p(1-a)$	p
	B	$(1-p)b$	$(1-p)(1-b)$	$1-p$
	Sum of Column	$pa + (1-p)b$	$p(1-a) + (1-p)(1-b)$	

FIGURE 5A.1 Bayes' Rule

the last row of the X column is the total probability that X is observed, either when A is the true condition (a true positive in our genetic test example) or when B is the true condition (a false positive).

To find the probability of a particular condition given a particular observation, then, Bayes' rule says that we should take the entry in the cell corresponding to the combination of that condition and that observation, and divide this by the column sum in the last row for that observation. As an example,

$$\text{Prob}(B \text{ given } X) = \frac{(1-p)b}{pa + (1-p)b}.$$

EXERCISE   In the genetic test example, suppose the test comes out negative (Y is observed). What is the probability that the individual does not have the defect (B exists)? Calculate this probability using Bayes' rule, and then check your answer by doing an enumeration of the 10,000 members of the population.

3 ATTITUDES TOWARD RISK AND EXPECTED UTILITY

In Chapter 2 we pointed out a difficulty about using probabilities to calculate the average or expected payoff for players in a game. Consider a game where players gain or lose money, and suppose we measure payoffs simply in money amounts. If a player has a 75% chance of getting nothing and a 25% chance of getting $100, then the expected payoff is calculated as a *probability-weighted average*; the expected payoff is the average of the different payoffs with the probabilities of each as weights. In this case, we have $0 with a probability of 75%, which yields $0.75 \times 0 = 0$ on average, added to $100 with a probability of 25%, which yields $0.25 \times 100 = 25$ on average. That is the same payoff as the player would get from a simple nonrandom outcome that guaranteed him $25 every time he played. People who are indifferent between two alternatives with the same average monetary value but different amounts of risk are said to be **risk-neutral.** In our example, one prospect is riskless ($25 for sure) while the other is risky, yielding either $0 with a probability of 0.75 or $100 with a probability of 0.25, for the same average of $25. In contrast are **risk-averse** individuals—those who, given a pair of alternatives each with the same average monetary value, would prefer the less risky option. In our example, they would rather get $25 for sure than face the risky $100-or-nothing prospect, and given the choice, would pick the safe prospect. Such risk-averse behavior is quite common; we should therefore have a theory of decision making under uncertainty that takes it into account.

We also said in Chapter 2 that a very simple modification of our payoff calculation can get us around this difficulty. We said we could measure payoffs not in money sums but using a nonlinear rescaling of the dollar amounts. Here we show explicitly how that rescaling can be done and why it solves our problem for us.

Suppose that when a person gets D dollars, we define the payoff to be something other than just D, perhaps $\sqrt{D}$. Then the payoff number associated with $0 is 0, and that for $100 is 10. This transformation does not change the way the person rates the two payoffs of $0 and $100; it simply rescales the payoff numbers in a particular way.

Now consider the risky prospect discussed above that involved getting $100 with probability 0.25 and nothing otherwise. After our rescaling, the expected payoff (which is the average of the two payoffs with the probabilities as weights) is $0.75 \times 0 + 0.25 \times 10 = 2.5$. This expected payoff is equivalent to the person's getting the dollar amount whose square root is 2.5; since $2.5 = \sqrt{6.25}$, a person getting $6.25 for sure would also receive a payoff of 2.5. In other words, the person with our square-root payoff scale would be just as happy getting $6.25 for sure as he would getting a 25% chance at $100. This indifference between a guaranteed $6.25 and a 1 in 4 chance of $100 reflects quite a strong aversion to risk; this person is willing to give up the difference between $25 and $6.25 in order to avoid facing the risk. Figure 5A.2 shows this nonlinear scale (the square root), the expected payoff, and the person's indifference between the sure prospect and the gamble.

What if the nonlinear scale we use to rescale dollar payoffs is the cube root instead of the square root? Then the payoff from $100 is 4.64, and the expected payoff from the gamble is $0.75 \times 0 + 0.25 \times 4.64 = 1.16$, which is the cube root of 1.56. Therefore a person with this payoff scale would accept only $1.56 for sure instead of a gamble that has a money dollar value of $25 on average; such a person is extremely risk-averse indeed. (Compare a graph of the cube root of x to a graph of the square root of x to see why this should be so.)

And what if the rescaling of payoffs from x dollars is done using the function x^2? Then the expected payoff from the gamble is $0.75 \times 0 + 0.25 \times 10,000 = 2,500$, which is the square of 50. Therefore a person with this payoff scale would be indifferent between getting $50 for sure and the gamble with an expected money value of only $25. This person must be a risk lover because he is not willing to give up any money to get a reduction in risk; on the contrary, he must be given an extra $25 in compensation for the loss of risk. Figure 5A.3 shows the nonlinear scale associated with a function like x^2.

So, by using different nonlinear scales instead of pure money payoffs, we can capture different degrees of risk-averse or risk-loving behavior. A concave scale like that of Figure 5A.2 corresponds to risk aversion, and a convex scale like that

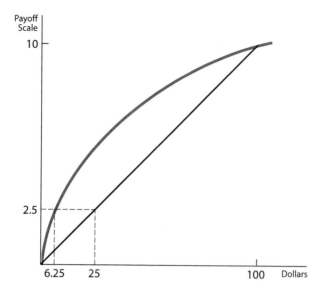

FIGURE 5A.2 Concave Scale: Risk Aversion

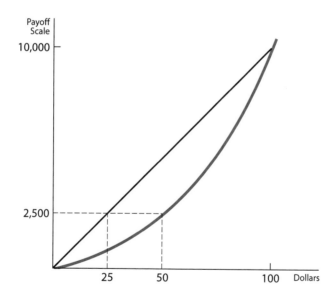

FIGURE 5A.3 Convex Scale: Risk-Loving

of Figure 5A.3 to risk-loving behavior. You can experiment with different simple nonlinear scales—for example, logarithms, exponentials, and other roots and powers—to see what they imply about attitudes to risk.[5]

[5]Additional information on the use of expected utility and risk attitudes of players can be found in many intermediate microeconomics texts, for example, Hal Varian, *Intermediate Microeconomics* (New York: Norton, 1987), pp. 216–220; Walter Nicholson, *Microeconomic Theory*, 7th ed. (New York: Dryden Press, 1998), pp. 211–226.

This method of evaluating risky prospects has a long tradition in decision theory; it is called the **expected utility approach.** The nonlinear scale that gives payoffs as functions of money values is called the **utility function;** the square root, cube root, and square functions referred to earlier are simple examples. Then the mathematical expectation, or probability-weighted average, of the utility values of the different money sums in a random prospect is called the expected utility of that prospect. And different random prospects are compared to each other in terms of their expected utilities; prospects with higher expected utility are judged to be better than those with lower expected utility.

Almost all of game theory is based on the expected utility approach, and it is indeed very useful, although is not without flaws. We will adopt it in this book, leaving more detailed discussions to advanced treatises.[6] However, we will indicate the difficulties it leaves unresolved by way of a simple example in Chapter 7.

SUMMARY

The *probability* of an event is the likelihood of its occurrence by chance from among a larger set of possibilities. Probabilities can be combined, using some rules. The *addition rule* says that the probability of any one of a number of disjoint events occurring is the sum of the probabilities of these events; the *modified addition rule* generalizes this to the case of overlapping events. According to the *multiplication rule,* the probability that all of a number of independent events will occur is the product of the probabilities of these events; the *modified multiplication rule* generalizes this to allow for lack of independence, using conditional probabilities. *Bayes' theorem* provides a formula for inferring the probabilities of hidden underlying conditions through observation of their consequences.

Judging consequences by taking expected monetary payoffs assumes *risk-neutral* behavior. *Risk aversion* can be allowed, using the *expected utility* approach. This involves using a *utility function,* which is a concave rescaling of monetary payoffs, and taking its probability-weighted average as the measure of expected payoff.

---

[6] See R. Duncan Luce and Howard Raiffa, *Games and Decisions* (New York: Wiley, 1957), chap. 2 and app. 1 for an exposition; and Mark Machina, "Choice Under Uncertainty: Problems Solved and Unsolved" *Journal of Economic Perspectives*, vol. 1, no. 1 (Summer 1987), pp. 121–154 for a critique and alternatives. While decision theory based on these alternatives has made considerable progress, it has not yet influenced game theory to any significant extent.

KEY TERMS

addition rule (166)

Bayes' theorem or rule or formula (172)

combination rule (169)

conditional probability (168)

disjoint (165)

expected utility (176)

independent events (167)

modified addition rule (166)

modified multiplication rule (168)

multiplication rule (167)

probability (164)

risk-averse (173)

risk-neutral (173)

utility function (176)

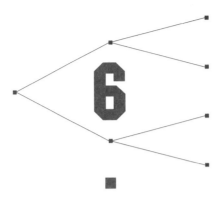

Combining Simultaneous and Sequential Moves

I N Chapters 3 and 4 we considered games of purely sequential and purely simultaneous natures. While games of this type lend themselves most easily to the tools of analysis covered in those chapters, many strategic situations contain elements of both types of interaction. Also, while we used game trees (extensive forms) as the sole method of illustrating sequential-move games and game tables (strategic forms) as the sole method of illustrating simultaneous-move games, it remains possible to use either form for any type of game. The first task of this chapter is to show how games can be illustrated in both extensive and strategic form and then to consider the sort of games that incorporate both kinds of moves. We describe how games can include both sequential and simultaneous moves, and show how to extend the solution methods discussed earlier to such situations.

In addition, this chapter considers the effects of changing the nature of the interaction in a particular game. Specifically, we look at the conversion of sequential-play games into simultaneous-play games. This topic gives us the opportunity to compare the equilibria found using the concept of rollback, in a sequential-move game, with those found using the Nash equilibrium concept, in the simultaneous version of the same game. From this comparison, we extend the concept of Nash equilibria to sequential-play games. It turns out that the rollback equilibrium is a special case, usually called a *refinement*, of these Nash equilibria. In addition, we investigate the change from simultaneous to sequential play. That investigation highlights and extends the discussion from Chapter 3 on the importance of the order of play.

1 ILLUSTRATING GAMES IN BOTH EXTENSIVE AND STRATEGIC FORM

Before we begin our analysis of games with multiple types of interaction and the effects of rule changes on strategic outcomes, we first show how to translate the information contained in extensive-form games to and from that contained in strategic form games. As illustrative examples, we will use the Senate race game from Chapter 3 and the battle of the Bismarck Sea from Chapter 4. Neither of these games is particularly large, so we will not see multiple-player or multiple-strategy translations. Translations for such games, however, require straightforward extensions of the techniques described later, and several of the chapter exercises provide practice with those techniques.

In the Senate race game, an incumbent senator (Gray) decides whether to use campaign ads and a potential challenger (Green) makes a decision about whether to enter the race. The sequential version of the game in which Gray first makes her advertising choice and Green then makes her entry decision is illustrated in its extensive form, as was done in Chapter 3, in Figure 6.1a. To show this game in strategic form requires a table of payoffs for all of the possible strategy combinations available to the two players. Thus we must identify the pure strategies available to each player in the game in order to determine the size of the table; we will then be able to use the information on payoffs from the extensive form to fill in each of its cells.

Gray has one decision node in the extensive form of the game at which she chooses between Ads and No Ads. If Gray is to be the row player in the payoff table for this game, the table will have two rows, one corresponding to her choice of Ads and the other corresponding to No Ads. Green's situation is slightly different. She has two decisions nodes and two choices, In or Out, at each, but her actions are contingent on the choice made earlier by Gray. As described in detail in Chapter 3, Section 3, Green's strategies are a set of rules that specify an action for her at each of her decision nodes, given the previous moves made by Gray. Green has four pure strategies. In shorthand—with the move at node b (where Gray uses ads) listed first, and that at node c (where Gray does not use ads) listed second—these pure strategies are: (1) In, In, (2) In, Out, (3) Out, In, and (4) Out, Out. The payoff table must then have four columns, one for each of these four pure strategies.

The strategic form for the Senate race game is illustrated in Figure 6.1b. Recall that the player who moves first is shown, as the row player and the player who moves second is shown as the column player. Note that it is a two-by-four table, so there are eight possible outcomes shown rather than the four shown in the extensive form. This result is due to the fact that more than one of Green's pure strategies can lead to the same outcome. Consider the outcome in which

Gray chooses No Ads and Green enters the race. This particular outcome will arise if Gray chooses No Ads and Green uses her first strategy in which she always enters. However, this outcome will also arise if Gray chooses No Ads and Green uses her third strategy of choosing to enter if Gray chooses No Ads but not if Gray chooses Ads. The same is true for all three of the other possible outcomes shown in the game's *extensive* form; given that Gray has made an advertising decision, two of Green's pure strategies will lead to each of these outcomes, as seen in the *strategic* form.

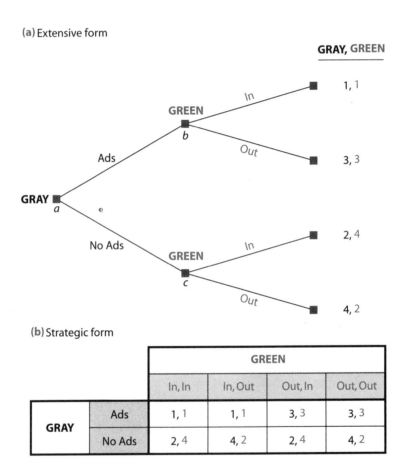

(a) Extensive form

GRAY, GREEN

(b) Strategic form

		GREEN			
		In, In	In, Out	Out, In	Out, Out
GRAY	Ads	1, 1	1, 1	3, 3	3, 3
	No Ads	2, 4	4, 2	2, 4	4, 2

FIGURE 6.1 Senate Race Game

Any sequential-move game can be illustrated with a game table using a similar technique. First determine the total number of pure strategies available to each player in order to determine the size of the table. Then simply fill in the payoffs associated with each combination of strategies. The more turns that are taken by each player, or the more players that are involved in the game, the larger the table will be. The payoff table for the sequential-move mall location game described in Chapter 3, Section 7, for example, is quite large, as we will see later in Section 5 of this chapter.

It is also relatively simple to translate the information contained in the strategic form of a simultaneous-play game into an extensive-form representation of the same game. The number of branches on the tree extending from the initial node will correspond to the number of pure strategies available to the row player, and the column player will have that number of decision nodes. Because the game is played simultaneously, however, the game tree must show that the column player cannot identify from which decision node she is moving. That is, we must indicate in some way that the column player makes her choice of action without knowing what the row player will choose. To do so, we surround all of the column player's decision nodes with a dotted line. The nodes within the "balloon" formed by the dotted line represent the player's **information set.** Such a set indicates the presence of imperfect information for the player; she cannot distinguish between the nodes in the set given her available information (because she cannot observe the row player's move before making her own). As such, her strategy choice from within a single information set must specify the same move at all the nodes contained in it, a stipulation that becomes clear when looking at an example.

Consider the battle-of-the-Bismarck-Sea game from Chapter 4. In that game the U.S. air forces and the Japanese Navy simultaneously choose whether to take a northern or southern route. The U.S. air forces hope to find and destroy the Japanese convoy, while the Japanese Navy hopes to avoid the U.S. planes. The strategic form of this game is illustrated in Figure 6.2a and its extensive form in Figure 6.2b.

The Japanese Navy's two decision nodes are contained within a single information set to indicate that the Japanese do not know from which node their decision is being made. That is, they must make their decision without knowing the choice of the United States. Moreover, Japan must choose either North at both nodes or South at both nodes. They cannot choose North at one and South at the other—much as they would like to have this ability and choose the opposite of what the United States chooses—because they cannot observe what the Americans have chosen.

Note that the ordering of players in the extensive form of this game does not affect the analysis of the game. The choice for the Japanese Navy could be shown at the initial node of the tree, and the United States could have two decision nodes within a single information set. Then the Americans would be restricted to making the same choice at both nodes in its information set. We leave the drawing of this alternative extensive form to you. Note that it leads to the same equilibrium as the version shown in Figure 6.2b.

Again, any simultaneous-play game can be illustrated in extensive form. You have only to remember to collect players' decision nodes within an information

(a) Strategic form

		JAPANESE NAVY	
		North	South
USAF	North	2	2
	South	1	3

(b) Extensive form

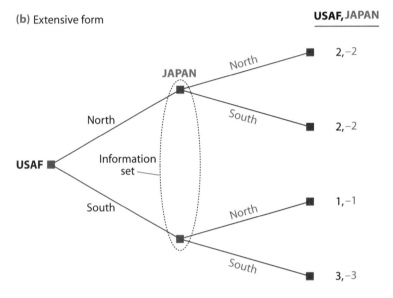

FIGURE 6.2 The Battle of the Bismarck Sea

set if they are all reached by branches from a decision another player is making simultaneously. Simultaneous-move games with a large number of players, particularly four or more, are probably easier to illustrate in extensive form than in strategic form. In a game table, a fourth player requires the addition of a fourth dimension (the first three being row, column, and page), which becomes complicated. A four-player simultaneous-play game in extensive form is simply represented as a larger tree in which three players have all of their decision nodes within single information sets.

2 GAMES WITH BOTH SIMULTANEOUS AND SEQUENTIAL MOVES

As mentioned several times thus far, most real games that you will encounter will be made up of numerous smaller components. Each of these components may entail simultaneous play or sequential play, so the full game requires you to be familiar with both. The most obvious examples of strategic interactions containing both sequential and simultaneous parts are those that occur between two (or more) players over an extended period of time. You may play a number of different simultaneous-play games against your roommate, for example, over the course of a week; your play, as well as hers, in previous situations will be important in determining how each of you decide to act in the next "round." Also, many sporting events, firm interactions, and political relationships are sequentially linked series of simultaneous-move games. Such games are analyzed by combining the tools presented in Chapters 3 and 4 and, for games in which mixed-strategy equilibria arise, with the tools from Chapter 5 as well. The only difference is that the actual analysis can become extremely complicated very quickly as the number of moves and interactions increases.

We consider two examples of combination games. In this section, we analyze a situation in which two players are involved in a sequence of simultaneous-move games, each of which has a unique *pure*-strategy equilibrium. In Section 3, we analyze a more complex example that consists of a sequence of simultaneous games with equilibria in *mixed* strategies.

Consider a game involving two electronics firms, Kumquat and Kiwifruit. The game takes place over the course of a year. During that year, the firms make private decisions about the amount of research and development to do in their quest to bring a new product to market. Both Kumquat and Kiwifruit are working on the same type of product, and each knows the other is doing so, but they do not publicize their R&D budgets. The only way to find out about a rival's R&D decision is to observe the quality, in terms of features provided, of the final product when it is announced and described at the industry's annual trade show. Once the two firms have seen each other's anticipated product at the trade show, they must decide on a price at which to offer their products.

To simplify matters, we assume that the firms can choose either a small R&D budget or a large one. A small budget yields a new product with a few special features; a large budget yields a new product with many special features. In addition, we assume that the only decision Kumquat and Kiwifruit make after seeing their rival's product announcement is a price decision on their own product. That price will be either High or Low.[1]

The extensive form for this combination game is shown in Figure 6.3. Firms first choose their R&D budgets, then observe the announced products, and finally choose a pricing strategy. The payoff numbers to the right of the 16 terminal nodes, rather than show specific profit values, indicate how Kumquat and

---

[1]A more realistic version of the game might allow firms to go back after the trade show and alter their R&D budget for a few months before releasing and pricing their product. Actual firms may do this type of back-and-forth observation and reaction numerous times before a product is priced and sent to stores for general consumption.

Kiwifruit rate each possible outcome. These ratings assume that full-featured products—the result of large R&D budgets—cost the firms more to produce and that consumers can identify, and are willing to pay more for, those full-featured goods. There is also assumed to be a group of consumers who prefer the cheaper product regardless of features.

Each firm has a number of strategies—complete plans of action—in this game determined by the number of decision nodes and branches from each node. In the game in Figure 6.3, both Kumquat and Kiwifruit have four decision nodes at which they choose a product price, each associated with a particular set of decisions for R&D budgets. This setup gives each firm 16 pure strategies of the following form: "Price high if both firms chose large R&D budgets; price high if our firm chose a large R&D budget but the other firm chose a small R&D budget; price high if we chose a small R&D budget but they chose a large R&D budget; and price high if both firms chose a small R&D budget."

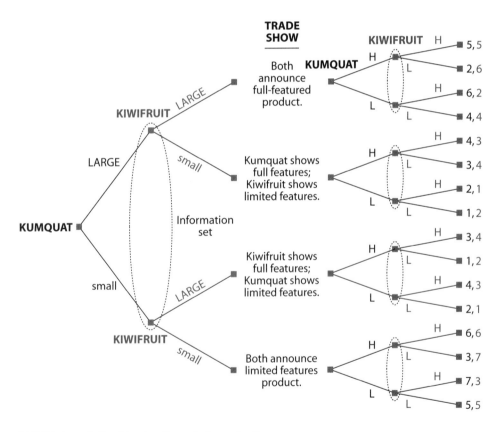

FIGURE 6.3 A Combination Game in Extensive Form

The firms actually end up facing four different pricing games. The two games in which both have chosen the same R&D budget (and so have the same number of features in their products), are prisoners' dilemmas; each firm has a dominant strategy to price low. The pricing game associated with both firms having large R&D budgets has outcomes that receive slightly lower ratings from the firms because of the higher costs involved in producing the products in that game. When the two firms choose different R&D budget levels, the outcomes of the two games that follow are the same for the small R&D budget firm in each and the same for the large R&D firm in each. These two pricing games are not prisoners' dilemmas, but each firm still has a dominant strategy; the small-budget firm prices low, and the large-budget firm prices high.

Given the outcomes in the pricing games, Kiwifruit and Kumquat can use rollback to decide their best R&D strategies. In the first stage of Figure 6.3, each firm has the weakly dominant strategy of choosing a small R&D budget. Thus the unique rollback equilibrium of the game has both firms choosing Small and then Low. The payoffs are rated a 5 by each firm.

Note that this game has a unique equilibrium by virtue of its particular payoff structure. If the firms put more emphasis on being the sole producer of a full-featured product, we might see higher ratings associated with the two middle pricing games on the game tree. If the ratings for the equilibrium outcomes in those games exceeded the 5 from the bottom prisoners' dilemma, the game's equilibrium would change. And, because the two middle games are simply mirror images of each other, we would end up with two equilibria, one of which was preferred by each firm. Under those circumstances, the firms would need to consider coordination, an option discussed in Chapter 4, or randomization, an option covered in Chapter 5. In either case, the fact that these large games can have a unique equilibrium, many equilibria, or none is noteworthy. Players must carefully determine the structure of payoffs in such games in order to be sure of their optimal course of play.

3 COMBINED SEQUENTIAL-PLAY AND MIXED-STRATEGY EQUILIBRIA

Now we illustrate another combination of the basic equilibrium types, where the game has some sequential moves, but at each step there are also simultaneous moves leading to an equilibrium in mixed strategies, rather than pure ones, as was the case in the Kiwifruit-Kumquat game above.

Our example comes from football. When the offense needs to gain 20 yards and has two plays to do it, commentators often discuss whether it is better to try for the whole lot in one play, or go for half the distance each time. We show you how to think this through, and how to startle your friends with your insight.

Suppose the offense has two more plays left, labeled the third down and the fourth down, to gain 20 yards, in which case it wins; otherwise it loses. The offense's coach has two plays for this situation, one that will cover 10 yards if successful, the other, 20. The opposing coach, knowing this, can set his defense to cover either. The offense's probabilities of success for the possible combinations of the two sides' choices in any one down are shown in Figure 6.4. The payoffs for the ultimate outcomes are as follows: if the offense wins, it gets 1; if it loses, it gets 0. This is a zero-sum game.

As usual, we solve this game backward, beginning with the fourth down. If the offense has not yet made its 20 yards and finds itself at fourth down, it must have either 10 or 20 yards to go, depending on what happened on the earlier third down. We have to consider these two cases separately.

		DEFENSE	
		10	20
OFFENSE	10	4/5	1
	20	1	1/2

FIGURE 6.4 Success Probability Table in Two-Play Football Example

First we look at the case where 20 yards are needed at the fourth down. Here the offense loses the game even if it successfully completes the 10-yard play. Therefore it is forced to use the 20-yard play. The defense, knowing this, uses the 20-yard defense. The payoff to the offense is 1/2. Technically, the payoff table for the offense is as shown in Figure 6.5. This game is dominance solvable, and the equilibrium is (20, 20) leading to an expected payoff of 1/2 for the offense, as we just argued.

Next we consider the case where only 10 yards are needed on the fourth down. Now either play, if successful, wins the game for the offense. Therefore its expected payoff equals its probability of success. The payoff table is shown in Figure 6.6.

		DEFENSE	
		10	20
OFFENSE	10	0	0
	20	1	1/2

FIGURE 6.5 Payoff Table for Fourth Down, 20 Yards to Go

		DEFENSE	
		10	20
OFFENSE	10	$4/5$	1
	20	1	$1/2$

FIGURE 6.6   Payoff Table for Fourth Down, 10 Yards to Go

In this payoff table there is no dominance for either pure strategy, nor a Nash equilibrium in pure strategies, so we look for a mixed-strategy Nash equilibrium. If the offense mixes strategies in proportions of p for the 10-yard play and $(1-p)$ for the 20-yard play, the condition for the defense to be indifferent between its two plays is

$$0.8p + (1-p) = p + 0.5(1-p)$$
$$1 - 0.2p = 0.5 + 0.5p,$$
$$0.7p = 0.5,$$

which yields $p = 5/7$. Similarly the defense's equilibrium mix has the probability $q = 5/7$ of choosing to cover the 10-yard play. The resulting expected payoff to the offense can be found by evaluating the result of the offense's p-mix when played against either of the defense's pure strategies, say its 10-yard play. This gives $1 - (5/7)/5 = 1 - 1/7 = 6/7$.

A more complicated way to get the same payoff is to calculate the average of all four possible strategy combinations using the right combined probabilities of their occurrence when both players mix:

$$5/7 \times 5/7 \times 4/5 + 5/7 \times 2/7 \times 1 + 2/7 \times 5/7 \times 1 + 2/7 \times 2/7 \times 1/2 = 42/49 = 6/7.$$

Now consider the third down. Here we need to trace the consequences of each pair of choices. For example, if each side uses the 10-yard move, then there are two possibilities: (1) with a probability of 4/5 the offense succeeds, gains 10 yards, and then on the fourth down with 10 yards to go has the expected payoff of 6/7 just calculated; (2) with a probability of 1/5 the offense fails, is left with fourth down and 20, and has the expected payoff of 1/2 calculated earlier. Therefore (10, 10) yields the expected payoff $4/5 \times 6/7 + 1/5 \times 1/2 = 55/70 = 11/14$ to the offense.

Similarly, (10, 20) yields $1 \times 6/7 = 6/7$, (20, 10) yields 1, and (20, 20) yields $1/2 \times 1 + 1/2 \times 1/2 = 3/4$. All of these calculations give us the payoff table for the third down, shown in Figure 6.7.

		DEFENSE	
		10	20
OFFENSE	10	11/14	6/7
	20	1	3/4

FIGURE 6.7 Payoff Table for Third Down

This table shows no pure strategy equilibrium, and so we look for a mixed-strategy equilibrium. We find the offense's p-mix by the usual condition,

$$11/14\,p + (1-p) = 6/7\,p + 3/4(1-p),$$

which yields

$$1 - 3p/14 = 3/4 + 3p/28$$
$$1/4 = 9p/28$$
$$p = 7/9.$$

Similarly, for the defense's mixture,

$$12 - q/14 = 3 + q/4$$
$$3/28 = 18q/56$$
$$q = 1/3.$$

The expected payoff to the offense is 5/7.

If the numbers in the initially specified table of success probabilities were different, these mixture probabilities and payoffs would also be different. But three general points stand out from the analysis. First, neither of the sports commentators' doctrines—take two stabs at 20 or "go for half the distance each time"—is appropriate as a pure strategy; mixing remains useful for the usual reason in zero-sum games, namely that any other consistent pattern would be exploited by the opponent.

Second, in the third-down mix, the offense's probability of choosing the 10-yard play is quite high. There are two different ways of looking at this: (1) The 10-yard play is safer—it is the percentage play in the sense that we saw in Chapter 5; its payoffs against the defense's choices are closer together than those of the 20-yard play: $6/7 - 11/14 = 1/14$ while $1 - 3/4 = 1/4$. (2) The defense covers the 20-yard play two-thirds of the time because losing it is very costly to the defense. Therefore the offense uses the less-covered 10-yard play more often.

And third, with two plays, an added point in favor of the 10-yard play on the third down is that if it succeeds, then the offense has the advantage of being able to mix strategies and keep the defense guessing on the fourth down. If the offense tries the 20-yard play on third down and fails, then it is forced to go for 20 yards on the fourth, and the defense can cover this. Hence the greater weight given to the 10-yard play on third down than to fourth and 10: 7/9 > 5/7.

4 RULES CHANGE I: CONVERTING FROM SEQUENTIAL TO SIMULTANEOUS PLAY

The games considered in previous chapters were presented as either sequential or simultaneous in nature. We used the appropriate tools of analysis to predict equilibria in each type of game. In Sections 2 and 3, we discussed games with elements of both sequential and simultaneous play. These games required both sets of tools to find solutions. But what about games that could be played either sequentially or simultaneously? How would changing the play of a particular game, and thus changing the appropriate tools of analysis, alter the expected outcomes?

The task of turning a sequential-play game into a simultaneous one requires changing only the timing or observability with which players make their choice of moves. Sequential-move games become simultaneous if the players involved cannot observe moves made by their rivals before making their own choices. In that case, we would analyze the game by searching for a Nash equilibrium in pure strategies rather than for a rollback equilibrium.

We considered a variety of sequential-play interactions in Chapter 3, including the Senate race game which had two players and a mall location game with three players. Let us look at the effects of simultaneous decisions rather than sequential ones in those two games. We begin first with the Senate race game, already described in Section 1. If that game is played simultaneously, Green (the challenger) cannot make her entry decision based on a previous move made by Gray (the incumbent); instead of the four pure strategies that she has in the sequential version of the game, in the simultaneous version she has only two: In and Out. Similarly, Gray has exactly two pure strategies, Ads and No Ads.

The game table for the simultaneous version of the Senate race game is shown in Figure 6.8. Using our hierarchy of solution methods from Chapter 4, we start by looking for dominant strategies in this game and we find that Gray's No Ads strategy dominates Ads. Then Green's equilibrium strategy is In, and the unique Nash equilibrium of the simultaneous-play version of the game is (No Ads, In) with payoffs of 2 to Gray and 4 to Green.

Recall from our analysis of the sequential-play game that the equilibrium in that version entailed Gray using Ads, Green using her strategy Out, In (so she stayed out of the race in equilibrium), and both players getting payoffs of 3. Here we find a completely different equilibrium outcome. Why? Notice that in the si-

| | | GREEN | |
		In	Out
GRAY	Ads	1,1	3,3
	No Ads	2,4	4,2

FIGURE 6.8   Simultaneous-Play Version of the Senate Race Game

multaneous-move game, Gray does worse than in the sequential-move game while Green does better. This is our clue to an explanation for the two different equilibria. When Gray moves first, she sets the stage for the game to follow and she can irreversibly commit to an advertising strategy that will deter Green's entry into the race; remember, this game has a first-mover advantage. In the simultaneous-play game, however, Gray cannot commit to what we see now is a dominated strategy; hence the different equilibrium outcome.

Obviously, Gray would prefer to be able to make the commitment to advertising, but she cannot do so in the simultaneous-move game. The only way she could achieve the better outcome—the one associated with the sequential-move game—would be if she made a strategic commitment move and effectively changed the simultaneous-play game back to a sequential one. We will analyze commitment moves in much greater detail in Chapter 9; here we simply note that they are equivalent to changing the rules of the game from simultaneous to sequential play.

We have now seen that the equilibria in the simultaneous and sequential versions of the same two-player game need not be the same. Does the same type of result hold for larger games as well? Recall now the mall location game from Chapter 3 in which Big Giant, Titan, and Frieda's are considering locations in two metropolitan area malls. There are only two large-enough locations available in the "choice" Urban Mall, and the stores would prefer these spaces; they would also prefer to be in a mall with at least one other department store. Each store has the option of trying to get into the Urban Mall or opting to try directly for a space in the alternative, Rural Mall. Any store that tries to get into Urban Mall but fails gets the lowest possible payoff. The sequential version of the game was illustrated in Figure 3.5, in a game tree with eight terminal nodes corresponding to the eight possible combinations of Urban Mall requests (U) and Rural Mall requests (R).

If this mall location game is played simultaneously, all three stores must decide whether to request a spot in Urban Mall or Rural Mall without knowing what their rivals will do. We then need a game table with two rows for Titan's two

FRIEDA'S

Urban

		BIG GIANT	
		Urban	Rural
TITAN	Urban	5, 5, 7	5, 2, 5
	Rural	2, 5, 5	4, 4, 3

Rural

		BIG GIANT	
		Urban	Rural
TITAN	Urban	5, 5, 2	3, 4, 4
	Rural	4, 3, 4	4, 4, 4

FIGURE 6.9   Simultaneous-Move Mall Location Game

strategies, two columns for Big Giant's two strategies, and two pages for Frieda's two strategies, as is done in Figure 6.9. As with all game tables, the payoffs are listed for the row player first, the column player second, and the page player third. Recall that when all three stores have requested spaces in Urban Mall simultaneously, the more prestigious Big Giant and Titan get the spaces and Frieda's request is rejected.

Checking this table first for dominance shows that no player has a dominant strategy; it is not always better to choose Urban over Rural for any store. Then we move on to try a cell-by-cell search. As in Chapter 4, a player's desire to change its strategy when arriving at a particular outcome is shown by a diagonal line through its payoff in that cell. This process leaves two outcomes that have no diagonal lines drawn through any payoffs and that therefore qualify as Nash equilibria of this simultaneous-play game.

Remember that we found only one equilibrium set of strategies in our analysis of this game in Chapter 3. Our rule change has led us from a sequential-move game with a single rollback equilibrium to a simultaneous-play game with two pure-strategy Nash equilibria. In the Senate race game, we found a new equilibrium in the simultaneous-move game, but only one. In Figure 6.9, we find not only a new equilibrium but also the equilibrium that we found for the sequential-play game. What happened to give the game multiple equilibria in its simultaneous-play version?

Look carefully at both equilibria found in Figure 6.9. One of them is (Rural, Rural, Rural); all three stores try directly for Rural Mall, even though it is the inferior of the two. This outcome leads to every store getting the same, relatively high payoff of 4. The other equilibrium is (Urban, Urban, Rural) in which Big Giant and Titan get their highest payoffs of 5 while Frieda's gets only 2. This is the equilibrium we found in Chapter 3.

The most important difference between the two versions of this game is that we give up a specific order of play when we move from sequential play to simultaneous play. In the sequential version, in which Frieda's chose first, then Big Giant, and then Titan, that ordering allowed later movers to observe the actions of their rivals. Because the stores prefer to be in Urban Mall but also prefer to be with at least one other store rather than alone, *even* if they are alone in Urban Mall, it is useful to know what earlier players have done. Titan could safely try for Urban Mall knowing that Big Giant had also; it was assured not only of getting into Urban Mall but also of having company there. In the simultaneous-play game, there is no similar guarantee that trying to locate in Urban Mall will get a company there with another department store. Thus the outcome in which all three pairs end up together in Rural Mall becomes a reasonable equilibrium outcome.

Both Big Giant and Titan strictly prefer the (Urban, Urban, Rural) equilibrium to the (Rural, Rural, Rural) equilibrium, although the reverse is true for Frieda's. Which equilibrium is likely to arise then in actual play of this game? We might argue that the (U, U, R) equilibrium is focal. Frieda's chooses Rural in both equilibria, so Big Giant and Titan can safely try for Urban with the knowledge that they will end up with their highest possible payoffs. Only minor coordination on the part of Big Giant and Titan would be required to guarantee this outcome. Of course, if the mechanism used to allocate the store locations in the event that all three stores requested Urban Mall were different, the game's outcomes could change also. That mall could, for instance, guarantee only a two-thirds probability, for each store, of getting a location there when all three stores want to be there. We pursue this example in Exercise 5.

5 SOLVING THE SEQUENTIAL-MOVE GAME FROM ITS STRATEGIC FORM

Note that changing the rules of the Senate race and mall location games led us, in Section 4, to analyzing strategic forms. Those strategic forms were for the simultaneous version of a game that had previously been played sequentially and analyzed in its extensive form. As we saw in Section 1 though, it is also possible to depict the sequential version of the game in strategic form. Would that make any difference to our analysis?

Recall that the strategic form of a sequential game must account for all the possible contingent strategies available to the players of the game. For our Senate race example, we showed the game table for the sequential-play game in Figure 6.1b, and we show it again here as Figure 6.10. Cell-by-cell inspection of this payoff table leaves two outcomes with no diagonal lines through either player's payoff. That is, when we analyze the sequential-play version of the Senate race game, we find *two* Nash equilibria.

		GREEN			
		In, In	In, Out	Out, In	Out, Out
GRAY	Ads	~~1~~,~~1~~	~~1~~,~~1~~	3, 3	~~3~~, 3
	No Ads	2, 4	4, ~~2~~	~~2~~, 4	4, ~~2~~

FIGURE 6.10 Senate Race Game in Strategic Form

This is a surprising outcome; we only found one Nash equilibrium of the sequential-play game in Chapter 3. Now we have found two, one of which is the equilibrium from Chapter 3 (Ads; Out, In) and the other of which looks like the one we found in our analysis of the simultaneous version of the game (No Ads; In, In). Does our analysis here mean that we get more equilibria when we use the strategic form of the game, even if we have not changed the rules?

The answer, thankfully, is no. The two Nash equilibria found in the strategic form of the sequential game represent all of the *possible* equilibria of the game. When we analyze a sequential-move game, however, we are looking for a specific equilibrium, the rollback equilibrium. It must be the case that only one of the Nash equilibria in Figure 6.10 satisfies the requirements of a rollback equilibrium. The other one must entail the use of a strategy that is not consistent with rollback—that is, a strategy in which a player does not always make the best possible choice available at every decision node.

And how do we determine which of the two equilibria is the rollback equilibrium? We simply use the tools we developed in Chapter 4 for predicting outcomes in games illustrated in strategic form. That is, we check Figure 6.10 for dominant or dominated strategies and use successive elimination if necessary. Recall that to find the rollback equilibrium of a sequential-move game when there might be weakly dominated strategies, we must eliminate strategies for players in the reverse of the order in which they move in the sequential-play game.

For the Senate race game, this process means that we begin by looking for dominated strategies for Green. This search shows that In, Out is strictly dominated by Out, In; that In, In is weakly dominated by Out, In; and that Out, Out is also weakly dominated by Out, In. Thus when we remove all of Green's dominated strategies from consideration, only one strategy remains—her rollback equilibrium strategy, Out, In. Gray's best choice is then Ads, and we are left with a unique equilibrium outcome that is exactly the rollback equilibrium we found in Chapter 3.

We can use the same approach in our mall location example, although this means constructing a payoff table that is 2 by 4 by 16; recall that Frieda's has two pure strategies, Big Giant has four, and Titan has 16 in the sequential play game. Using the same conventions as above for creating this table would require 16 "pages" of two-by-four payoff tables. We could portray the game that way, but we will opt instead for a reshuffling of the players so that Titan becomes the row player, Big Giant becomes the column player, and Freida's becomes the page player. Then "all" that is required is the 16 by 4 by 2 game table shown in Figure 6.11. Note that the order of payoffs still corresponds to our earlier convention in that they are listed row, column, page player; in our example, that means the payoffs are now listed in the order Titan, Big Giant, Freida's.

Cell-by-cell inspection of this payoff table would yield us eight possible Nash equilibria. (In Exercise 6 we ask you to find them all.) But we know that successive elimination should eliminate seven of these eight and leave only the roll-back equilibrium, (R, UU, UUUR), which we found in Chapter 3; recall that this notation indicates that Freida's plays the strategy "request Rural Mall," Big Giant plays "always request Urban Mall," and Titan plays "request Urban unless both previous players request Rural and then request Rural." Because we must do the elimination in reverse order of the moves made in the game, we consider first Titan, then Big Giant, and finally Freida's.

We begin thus with Titan's 16 strategies and make pairwise comparisons of the different rows of the table. Comparing strategies 1 and 2, we see that strategy 1 is weakly dominated by strategy 2 and can be removed from consideration as an equilibrium strategy. Continuing the comparison of strategy 2 with the rest of the available strategies shows that all the rest of Titan's strategies are weakly dominated by strategy 2. Amazingly, we have eliminated every possible strategy for Titan except for strategy 2, UUUR; this is Titan's rollback equilibrium strategy, which we found before.

We continue successive elimination by turning to Big Giant, which has four choices: always try for Urban Mall (UU), try for the same mall as Big Giant (UR), try for the opposite mall as Big Giant (RU), or always try for Rural Mall (RR). To compare Big Giant's payoffs from its different strategies, we look at its payoffs associated with UU, for example, in the eleventh row of *both pages* of the table and compare these to its payoffs associated with UR, for example. This pairwise comparison of Big Giant's payoffs across columns tells us that UU weakly dominates the other three strategies; only this strategy is consistent with the requirements of rollback.

Knowing that Titan uses the strategy UUUR and that Big Giant uses UU makes determining Freida's equilibrium strategy a breeze. There are only two possible cells of the table left to choose from. If Freida's chooses Urban Mall, it attains its worst possible payoff of 1, while if it chooses Rural Mall, it gets a payoff of 2. Freida's will choose Rural Mall.

Successive elimination of dominated strategies, performed in reverse of the order of moves in the sequential-play game, allows us to eliminate all but one of the possible Nash equilibria that we found in this game. It also allows us to elimi-

FRIEDA'S (Page player)

| | Urban (U) | | | | Rural (R) | | | |
| | GIANT (Column player) | | | | GIANT (Column player) | | | |
TITAN (Row player)	UU	UR	RU	RR	UU	UR	RU	RR
1: UUUU	5, 5, 1	5, 5, 1	5, 2, 5	5, 2, 5	5, 5, 2	3, 4, 4	5, 5, 2	3, 4, 4
2: UUUR	5, 5, 1	5, 5, 1	5, 2, 5	5, 2, 5	5, 5, 2	4, 4, 4	5, 5, 2	4, 4, 4
3: UURU	5, 5, 1	5, 5, 1	5, 2, 5	5, 2, 5	4, 3, 4	3, 4, 4	4, 3, 4	3, 4, 4
4: URUU	5, 5, 1	5, 5, 1	4, 4, 3	4, 4, 3	5, 5, 2	3, 4, 4	5, 5, 2	3, 4, 4
5: RUUU	2, 5, 5	2, 5, 5	5, 2, 5	5, 2, 5	5, 5, 2	3, 4, 4	5, 5, 2	3, 4, 4
6: UURR	5, 5, 1	5, 5, 1	5, 2, 5	5, 2, 5	4, 3, 4	4, 4, 4	4, 3, 4	4, 4, 4
7: URRU	5, 5, 1	5, 5, 1	4, 4, 3	4, 4, 3	4, 3, 4	3, 4, 4	4, 3, 4	3, 4, 4
8: RRUU	2, 5, 5	2, 5, 5	4, 4, 3	4, 4, 3	5, 5, 2	3, 4, 4	5, 5, 2	3, 4, 4
9: URUR	5, 5, 1	5, 5, 1	4, 4, 3	4, 4, 3	5, 5, 2	4, 4, 4	5, 5, 2	4, 4, 4
10: RURU	2, 5, 5	2, 5, 5	5, 2, 5	5, 2, 5	4, 3, 4	3, 4, 4	4, 3, 4	3, 4, 4
11: RUUR	2, 5, 5	2, 5, 5	5, 2, 5	5, 2, 5	5, 5, 2	4, 4, 4	5, 5, 2	4, 4, 4
12: URRR	5, 5, 1	5, 5, 1	4, 4, 3	4, 4, 3	4, 3, 4	4, 4, 4	4, 3, 4	4, 4, 4
13: RURR	2, 5, 5	2, 5, 5	5, 2, 5	5, 2, 5	4, 3, 4	4, 4, 4	4, 3, 4	4, 4, 4
14: RRUR	2, 5, 5	2, 5, 5	4, 4, 3	4, 4, 3	5, 5, 2	4, 4, 4	5, 5, 2	4, 4, 4
15: RRRU	2, 5, 5	2, 5, 5	4, 4, 3	4, 4, 3	4, 3, 4	3, 4, 4	4, 3, 4	3, 4, 4
16: RRRR	2, 5, 5	2, 5, 5	4, 4, 3	4, 4, 3	4, 3, 4	4, 4, 4	4, 3, 4	4, 4, 4

FIGURE 6.11   Mall Location Game in Strategic Form

nate all of the strategies, for each player, that are not consistent with rollback. Thus we arrive in the end at a unique outcome, which is the same outcome—(R, UU, UUUR)—we found using rollback in the extensive form of the game in Chapter 3.

6 SUBGAME-PERFECT EQUILIBRIA

As we saw in the previous section, cell-by-cell inspection of a game table may find several Nash equilibria when iterated dominance would find only one. Such a possibility arises because some of the several Nash equilibria involve a player's use of a strategy that does not satisfy the conditions required by rollback—that a player's equilibrium strategy be the one that entails using her best action at every one of her decision nodes in the game. Another interpretation of the rollback criterion is that it requires the players' equilibrium strategies not to rely on threats or promises that are not believable, or not **credible.**

Let us consider these ideas in greater detail. Remember that the action chosen by a player who moves early in the game—at an early node of the tree—depends on her expectations of what other players who move later will do when their turns come because she has looked forward and predicted their behavior. If a player who is to move later prefers a particular final outcome, she can try to bring it about by affecting the expectations of an earlier mover. With the right behavior, a later mover could induce an earlier mover to make a choice beneficial to the later mover.

In the Senate race game, for example, Green could make a false promise to stay out if Gray did not advertise (and, implicitly, to stay out when Gray advertised as in the equilibrium found earlier). If Gray believed this, she would foresee a payoff of 4 from No Ads and only 3 from Ads; Gray would then not advertise. Once Gray's decision was made, however, Green could then exercise her right to enter, and get her best outcome.

But Gray should anticipate that Green's promise would be broken. The incumbent should know that the promise was not credible, that Green would actually enter if given the right opportunity. Therefore Gray should not be fooled by the false promise. Such anticipation of others' behavior at later nodes in the game is just what the process of rollback embodies. More precisely, it postulates a particular form of an earlier mover's expectation about what later movers will do, namely that later movers will choose whatever is actually best for them when their turn comes. A later mover, say B, can assert that she intends to do something different than what is best for her when her turn comes, in order to influence an earlier player A into making a choice that works to B's own advantage. This assertion is typically made in the form of a threat or a promise. We will provide a more detailed analysis of the use of threats and promises, and the issues of credibility that arise when they are used, in Chapter 9. For now, we concentrate only on the fact that rollback says that A will not believe such bluffs. Instead, the principle of rollback says that A will assume that when B's turn comes, she will do what is best for her at that time; B will not carry out the threat, or not deliver on the promise.

To state this idea somewhat more precisely, we introduce the concept of subgame. At any one node of the full game tree, we can think of the part of the game that begins there as a game in its own right. This small portion of the game will be called a **subgame** of the full game. As successive players make their choices, the play of the game moves along a succession of nodes, and each move can be thought of as starting a subgame. The equilibrium derived using rollback corresponds to one particular succession of choices, and gives rise to one particular path of play. Certainly, other paths of play are consistent with the rules of the game. We call these other paths **off-equilibrium paths,** and we call any subgames that arise along these paths **off-equilibrium subgames,** for short.

With this terminology we can now say that the equilibrium path of play is itself determined by the players' expectations of what would happen if they chose a different action, that is, if they moved the game to an off-equilibrium path and started an off-equilibrium subgame. Rollback requires that all players make their best choices in every subgame of the larger game, whether or not the subgame lies along the path to the ultimate equilibrium outcome.

Strategies are complete plans of action. Thus a player's strategy must specify what she will do in each eventuality, or each and every node of the game, whether on or off the equilibrium path, where it is her turn to act. When one such node arrives, only the plan of action starting there—namely the portion of the full strategy that pertains to the subgame starting at that node—is pertinent. This portion is called the **continuation** of the strategy for that subgame. Rollback requires that the equilibrium strategy be such that its continuation in every subgame is optimal for the player whose turn it is to act at that node, whether or not the node and the subgame lie on the equilibrium path of play.

Therefore we call the equilibrium found using rollback a **subgame-perfect equilibrium (SPE):** it is a set of strategies (complete plans of action), one for each player, such that at every node of the game tree, whether or not the node lies along the equilibrium path of play, the continuation of the same strategy (plan of action) in the subgame starting at that node is optimal for the player who takes the action there. More simply, an SPE requires players to use strategies that constitute a Nash equilibrium in every subgame of the larger game.

In fact, as a rule, in games with finite trees and perfect information, where players can observe every previous action taken by all other players, rollback finds the *unique* subgame-perfect equilibrium of the game. Consider: if you look at any subgame that begins at the last decision node for the last player who moves, the best choice for that player is the one that gives her the highest payoff. But that is precisely the action chosen using rollback. As players move backward through the game tree, rollback eliminates all unreasonable strategies, including those involving incredible threats or promises, so that the collection of actions ultimately selected is the SPE.

As we saw in our analyses of the sequential-play Senate race and mall location games in their strategic form, iterated elimination of dominated strategies achieves the same outcome, eliminating all but the subgame-perfect equilibrium from the set of possible Nash equilibria of the game. It can be proven that iterated elimination of dominated strategies leads to a SPE in any two-player game with perfect information. The same method also works for some larger and more general examples, as in our three-player mall location game.[2]

We can now see how to apply directly the concept of subgame perfection to the mall location game. Recall that Titan's strategy 2, UUUR, is its rollback equilibrium strategy. This means that it must also be Titan's unique subgame-perfect strategy. Conversely, all of Titan's other strategies must not be subgame perfect for some reason. You can see this by considering the different components of the store's alternative strategies. Strategy 16, for example, requires that Titan always choose Rural Mall (RRRR). But that means that the store would be requesting a space in Rural Mall at every decision node, even at nodes (e) and (f), where the previous stores had each requested different malls. Requesting a space in Rural Mall in such a situation would not be in Titan's best interests; it would receive a higher payoff (by virtue of assuring itself a space in Urban with another store) if it simply chose to ask for Urban Mall instead. Another way to say this is that choosing Rural Mall in the subgames in which one of the earlier players has chosen Urban and the other has chosen Rural is not a Nash-equilibrium action for Titan in those subgames. Thus the choice of Rural in either case cannot be part of a SPE strategy for Titan either; strategy 16, RRRR, is not subgame perfect. Strategies 1 and 3 through 15 can all be shown to contain actions that are also not subgame perfect. Thus none of them ends up being Titan's equilibrium strategy. Only strategy 2, UUUR, is made up of a set of rules for action that are Nash-equilibrium choices at every one of Titan's decision nodes, in every possible subgame that could arise.

We also said that subgame perfection rules out the use of incredible threats or promises on the part of players. As noted earlier in this section, a promise from Green to stay out if Gray did not advertise in the Senate race game, for example, is not credible. Thus the SPE of that game could not include any strategy in which Green chooses Out after No Ads; neither Out, Out nor In, Out could be Green's subgame-perfect strategy. Because rollback requires that choices at *all* future nodes be optimal for the player making them whether or not the nodes actually arise in the course of play, it imposes exactly this requirement of *credibility* and guarantees the subgame perfection of the rollback-equilibrium outcome. Thus we see that subgame perfection not only imposes sequential rationality on the players (as in the mall location example); it also imposes a re-

---

[2] In more complex games, including ones in which players have different information and each player tries to infer others' information from their actions and each tries to influence other players' beliefs by her own actions, rollback may not lead to a unique outcome, and subgame-perfect equilibrium is a more general concept. We will not need this complexity in this introductory book, but more advanced analyses of game theory do need it, and also have to introduce other special subclasses, called *refinements*, of Nash equilibria.

quirement of credibility. Of course, players like Green would still like to use promises like "out if Gray does not advertise" and will thus search for ways to make them credible. Such behavior is the subject of Chapter 9.

7 RULES CHANGE II: CONVERTING FROM SIMULTANEOUS TO SEQUENTIAL PLAY

The conversion of sequential- to simultaneous-play games showed us that changing the rules of play could lead to the appearance of different equilibria in the different versions of the game. Not surprisingly, the same is true when we change simultaneous into sequential ones. Although some simultaneous-play games have the same Nash equilibria as their sequential-play versions, others do not. We also see that the ordering of players in sequential-move games, a nonissue in simultaneous games, plays a significant role in determining the equilibrium of the game.

Certain games will have the same Nash equilibria in both simultaneous and sequential versions and regardless of the order of play in the sequential-play game. This outcome generally only arises when *both*, or all, players have dominant strategies. We show that it holds for the prisoners' dilemma.

Consider the prisoners' dilemma game from Chapter 4, in which a husband and wife are being questioned regarding their role in a crime. The simultaneous version of that game, illustrated in Figure 6.12a, can be redrawn as *either* of the sequential-play games shown in Figures 6.12b and c. As in Figure 4.2, the payoff numbers indicate years in jail, so low numbers are better than high ones. In Figure 6.12b, Husband chooses his strategy before Wife does, so she knows what he has chosen before making her own choice; in Figure 6.12c the roles are reversed.

We can use rollback to solve the sequential versions of the prisoners' dilemma. In both cases illustrated in Figures 6.12b and c, the second player does best to confess if the first has confessed (10 rather than 25 years in jail) and the second player also does best to confess if the first has denied (1 rather than 3 years of jail). Given these choices by the second player, the first player does best to confess (10 rather than 25 years in jail). The equilibrium entails 10 years of jail for both players, exactly as we found for the simultaneous version of the game.

It is worth asking whether other games with dominant strategies also work the same way. In general, simultaneous-move games with dominant-strategy equilibria do not always yield the same equilibria when played sequentially. While the ordering of plays in the sequential-play prisoners' dilemma proved to

(a) Simultaneous play

		WIFE	
		Confess	Deny
HUSBAND	Confess	10 yr, 10 yr	1 yr, 25 yr
	Deny	25 yr, 1 yr	3 yr, 3 yr

(b) Sequential play—Husband moves first

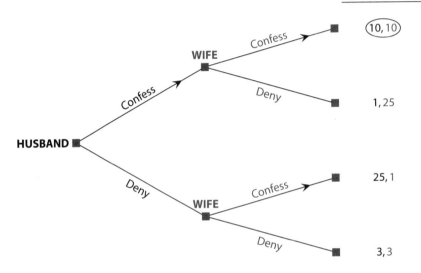

(c) Sequential play—Wife moves first

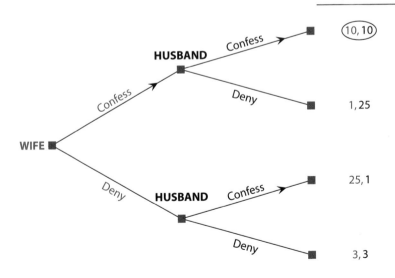

FIGURE 6.12 Three Versions of the Prisoners' Dilemma Game

be irrelevant, ordering is crucial to the outcome of the game when only one player has a (simultaneous-play) dominant strategy. We have already encountered this result, in our analysis of the Senate race game in Section 4.

Recall that when the game is played simultaneously, Gray has a dominant strategy not to advertise. Given Gray's dominant strategy, Green does best to enter, and the Nash equilibrium entails both players in the race with payoffs of 2 to Gray and 4 to Green. Also remember that No Ads (Gray's dominant strategy in the simultaneous-play game) was *not* her equilibrium strategy in the sequential version of the game in which she moved first. But if we were to look at the sequential version of the game in which *Green* moves first, as illustrated in Figure 6.13, we would see that Gray's equilibrium action in that game *is* No Ads; her full strategy is No Ads, Ads. Thus, one ordering of the sequential-play game leads to the outcome observed in the simultaneous-play game, but the other ordering does not.

Ordering, thus, does matter when we compare simultaneous and sequential versions of the same game, even when some players have dominant strategies in the simultaneous version. It also matters in games in which there are no dominant strategies, particularly if there is no unique pure-strategy Nash equilibrium with simultaneous play. Simultaneous-move games with multiple Nash equilibria in pure strategies, for example, will have only one subgame-perfect equilibrium when played sequentially; which equilibrium occurs will depend on the order of moves. Similarly, simultaneous-play games with no Nash equilibria in pure strategies generally have a unique subgame-perfect equilibrium in their

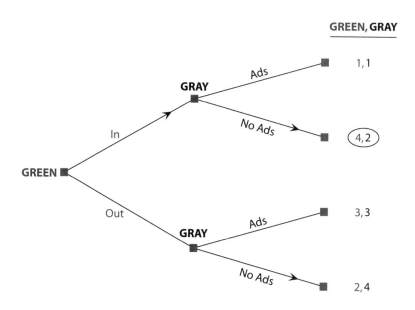

FIGURE 6.13 Senate Race Game with Green as First Mover

sequential versions; again, which outcome is the equilibrium will depend on which player moves first.

As an example of a game with multiple equilibria, let us return to chicken. The game in which two teenagers drive toward each other in their cars, both determined not to swerve, is depicted in its strategic and two extensive forms, one for each possible ordering of play, in Figure 6.14. Under simultaneous play, the two outcomes in which one player swerves (is "chicken") and the others goes straight (is "tough") are both pure-strategy Nash equilibria. Neither is focal. Our analysis in Chapter 4 suggested that coordinated play could help the players in this game, perhaps through an agreement to alternate between the two equilibria.

When we alter the rules of the game to allow one of the players the opportunity to move first, there are no longer two equilibria. Rather, we see that the second mover's equilibrium strategy is to choose the opposite action to that chosen by the first mover. Rollback then shows that the first mover's equilibrium strategy is Straight. We see in Figures 6.14b and 6.14c that allowing one person to move first and to be observed making the move results in a single SPE in which the first mover gets a payoff of 1 while the second mover gets a payoff of −1. Of course, the actual play of the game becomes almost irrelevant under such rules, which may make the sequential version uninteresting to many observers. While teenagers might not want to play such a game with the rule change, the strategic consequences of the change are significant.

Similarly significant changes arise in the tennis-point example from Chapters 4 and 5 when we make that game sequential as well. Recall that, in that game, Seles is planning the location of her return while Hingis considers where to cover. The version considered earlier assumed both players were skilled at disguising their intended moves until the very last moment so that they moved at essentially the same time. If Seles's movements as she goes to hit the ball belie her shot intentions, however, then Hingis can react and move second in the game. In the same way, if Hingis leans toward the side she intends to cover before Seles actually hits her return, then Seles is the second mover. Both orderings of the sequential-play game, along with the original simultaneous-play game, are illustrated in Figures 6.15 a–c.

Notice that in each ordering of the sequential version of this game, there is a unique subgame-perfect Nash-equilibrium outcome, but the equilibrium differs depending on who moves first. If the order is Seles, then Hingis, Hingis will choose to cover whichever direction Seles chooses and Seles will opt for a down-the-line shot. Each player is expected to win the point half the time in this equilibrium. If the order is reversed, Seles will choose to send her shot in the opposite direction from that which Hingis covers, so Hingis should move to cover down the line. In this case, Seles is expected to win the point 90% of the time.

(a) Simultaneous play

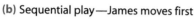

		DEAN	
		Swerve	Straight
JAMES	Swerve	0,0	−1,1
	Straight	1,−1	−2,−2

(b) Sequential play—James moves first

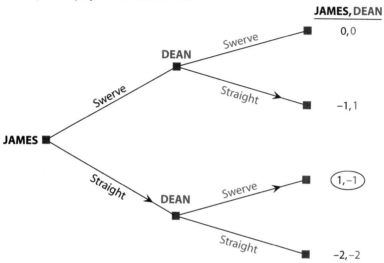

(c) Sequential play—Dean moves first

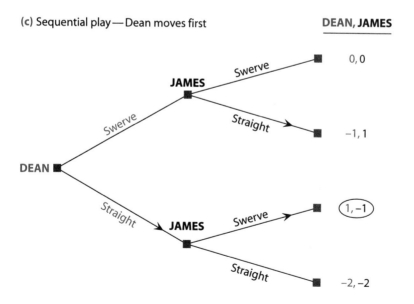

FIGURE 6.14   Chicken in Simultaneous- and Sequential-Play Versions

(a) Simultaneous play

		HINGIS	
		DL	CC
SELES	DL	50	80
	CC	70	20

(b) Sequential play — Seles moves first

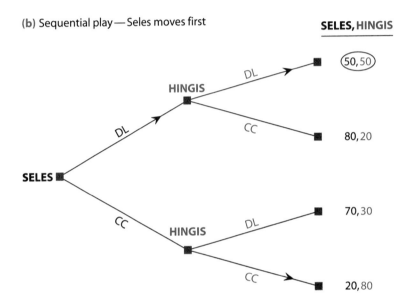

(c) Sequential play — Hingis moves first

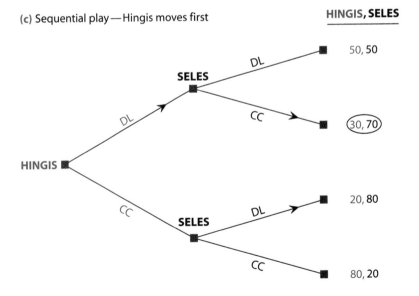

FIGURE 6.15 Tennis Game in Simultaneous and Sequential Versions

In both the chicken and tennis-point examples, the change from simultaneous to sequential play results in games with unique subgame-perfect Nash equilibria. The rule change works as a mechanism to allow a choice of exactly one of the possible equilibria in chicken. In tennis it reduces the need for mixing on the part of the players. In both cases the ordering of play also matters; one player is favored when the rules change. But the switch to sequential play from simultaneous play does not uniformly favor the same player. The player moving first in the sequential version of the chicken game gains an advantage; he basically is given the opportunity to choose which of the two pure-strategy equilibria he prefers. Chicken favors players who act over players who react. In the tennis-point example, however, the player moving second can do better than the player going first. That game incorporates an advantage to reaction rather than action; Seles prefers going second, so the game's equilibrium gives her a 70% success rate, while Hingis prefers holding that success rate down to 50% when she goes second.

SUMMARY

While earlier chapters analyzed sequential-play games with game trees and simultaneous-move games with game tables, either form can be used for any type of game. When illustrating a sequential-move game with a game table, you need to delineate every possible contingent strategy (provide a complete *plan* of action) for each player. When illustrating a simultaneous-play game with a game tree, you need to collect decision nodes in *information sets* when players make decisions without knowing at which specific node they find themselves. If a game has elements of both types of interactions, a combination of trees and tables may be used to illustrate the game. Such games are generally played over long time horizons, but their analysis is straightforward as long as precise payoff structures exist.

Changing the rules of a game to alter the timing of actions can alter the equilibrium outcome of the game. A sequential-move game generally has one rollback equilibrium, but the same game played simultaneously may have several Nash equilibria. Similarly, a sequential-play game analyzed from its strategic form may have many possible Nash equilibria. The number of potential equilibria may be reduced by using the criterion of *credibility* to eliminate some strategies as possible equilibrium strategies; this process leads to the *subgame-perfect equilibrium (SPE)* of the sequential-play game.

If a simultaneous-move game is changed to make moves sequential, it may or may not have the same equilibrium. When all players have dominant strategies in the simultaneous-play game, the sequentially played game has the same equilibrium. If only one player has a dominant strategy in the simultaneous game, the order of play in the sequential game determines the equilibrium outcome; the outcome may favor one player under one order of moves and the other player when the order is changed. The sequential version of a simultaneous-play game will always have an SPE outcome even if the simultaneous version has no or multiple equilibria; which is the outcome equilibrium depends on the order of play.

KEY TERMS

continuation (197)
credible (196)
information set (181)
off-equilibrium paths (197)

off-equilibrium subgames (197)
subgame (196)
subgame-perfect
  equilibrim (SPE) (197)

EXERCISES

1. Consider a game in which there are two players, A and B. A moves first and chooses either Up or Down. If Up, the game is over, and each player gets a payoff of 2. If A moves Down, then B gets a turn, and chooses between Left and Right. If Left, both players get 0; if Right, A gets 3 and B gets 1.

   (a) Draw the tree for this game, and find the subgame-perfect equilibrium.

   (b) Show this sequential-play game in strategic form, and find all the Nash equilibria. Which is/are subgame perfect and which is/are not? If any are not, explain why.

   (c) What method of solution could be used to find the subgame-perfect equilibrium from the strategic form of the game?

2. Return to the three sequential-move games illustrated in Exercise 2 in Chapter 3. Reexpress trees a and c in strategic (table) form. Find all of the pure-strategy Nash equilibria of each game. Indicate which of the equilibria are subgame perfect, and, for those equilibria that are not subgame-perfect, identify the reason (the source of the lack of credibility).

3. Consider the Airbus–Boeing example of Chapter 3, Exercise 4:

   (a) How many strategies does Airbus have, and how many strategies does Boeing have?

   (b) Show the game in the strategic form, and locate all Nash equilibria.

   (c) Which Nash equilibrium is subgame perfect (the rollback equilibrium in the extensive form?

   (d) For the Nash equilibrium or equilibria that are not subgame perfect, identify the problem of credibility that stops them from being subgame perfect.

4. Consider the cola industry, in which Coke and Pepsi are the two dominant firms. (To keep the analysis simple, just forget about all the others.) The market size is $8 billion. Each firm can choose whether to advertise. Advertising costs $1 billion for each firm that chooses to do so. If one firm advertises and the other doesn't, then the former captures the whole market. If both firms advertise, they split the market 50 : 50 and pay for the advertising. If neither advertises, they split the market 50 : 50 but without the expense of advertising.

   (a) Write down the payoff table for this game, and find the equilibrium when the two firms move simultaneously.

   (b) Write down the game tree for this game assuming it is played sequentially, with Coke moving first and Pepsi following.

   (c) Is either equilibrium in parts a and b best from the joint perspective of Coke and Pepsi? How could the two firms do better?

5. Recall the mall example, considered in this chapter in its simultaneous-play version. In the text we assumed that when all three stores requested space in Urban Mall, the two bigger stores were chosen. Suppose instead that the two are chosen by lottery in such a case, giving each store an equal chance of getting into Urban Mall. With such a system, if all three stores simultaneously requested a space in the Urban Mall, each would have a two-thirds probability (or a 66.67% chance) of getting in.

   (a) Draw the game table for this new version of the simultaneous-play mall location game. Find all of the Nash equilibria of the game.

   (b) Compare and contrast your answers in part (a) with the equilibria found in the text for the "big name" version of the game. Do you get the same Nash equilibria? Why? What equilibrium do you think is focal in this lottery version of the game?

6. Recall again the mall location example, this time in its sequential form. We saw in Figure 6.11 the strategic form for this game and claimed that there are eight Nash equilibria. What are those eight Nash equilibria? Why are there so many Nash equilibria even though there is only one reasonable equilibrium outcome for the game?

7. Along a stretch of a beach are 500 children in five clusters of 100 each. (Label the clusters A, B, C, D, and E in that order.) Two ice-cream vendors are deciding simultaneously where to locate. They must choose the exact location of one of the clusters.

   If there is a vendor in a cluster, all 100 children in that cluster will buy an ice cream. For clusters without a vendor, 50 of the 100 children are willing to walk to a vendor who is one cluster away, only 20 are willing to walk to a vendor two clusters away, and none are willing to walk the distance of three or more clusters. The ice cream melts quickly, so the walkers cannot buy for the nonwalkers.

If the two vendors choose the same cluster, each will get a 50% share of the total demand for ice cream. If they choose different clusters, then those children (locals or walkers) for whom one vendor is closer than the other will go to the closer one, and those for whom the two are equidistant will split 50% each.

Each vendor seeks to maximize her sales.

(a) Construct the five-by-five payoff table for their location game; the entries stated here will give you a start and a check on your calculations:

If both vendors choose to locate at A, each sells 85 units.

If the first vendor chooses B and the second chooses C, the first sells 150 and the second sells 170.

If the first vendor chooses E and the second chooses B, the first sells 150 and the second sells 200.

(b) Eliminate dominated strategies as far as possible.

(c) In the remaining table, locate all pure-strategy Nash equilibria.

(d) If the game is altered to one with sequential moves, where the first vendor chooses her location first and the second vendor follows, what are the locations and the sales that result from the subgame-perfect equilibrium? How and why is this answer different from the answer to part (c)?

8. Consider the following simplified version of baseball. The pitcher can throw either a fastball or a curveball; the batter can either swing at the pitch or take (not swing). These choices are simultaneous for each pitch. On the first pitch, if the batter swings at a curveball or takes a fastball, he strikes out and gets a 0. If the batter swings at a fastball, he has a probability of 0.75 of hitting a home run and getting a 1, and a probability of 0.25 of hitting a fly ball and getting a 0. If the batter takes a curveball, there is a second pitch.

On the second pitch, the first three combinations (swing at a curveball, take a fastball, and swing at a fastball) work as before; if the batter takes a curveball second pitch, he walks and gets 0.25.

This is a zero-sum game; the batter tries to maximize his expected score (his probability-weighted average payoff), and the pitcher tries to minimize the batter's expected score. Note that the two pitches constitute a sequential-move game while each individual pitch is a simultaneous-move game.

(a) Solve this game using rollback; construct a table of payoffs for the second pitch, and use these to determine the table of payoffs for the first pitch. Show that on the first pitch, the batter should take with a probability of 0.8.

(b) What is the pitcher's strategy in the subgame-perfect equilibrium?

(c) What is the batter's expected score in this equilibrium?

(d) Explain intuitively why the batter's probability of swinging is so small.

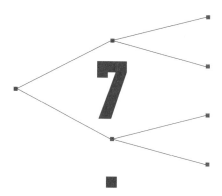

7

Consolidation, Extension, and Discussion

I<small>N</small> C<small>HAPTERS</small> 3 <small>THROUGH</small> 6 we introduced the basic concepts and techniques for analyzing games: rollback (or backward induction) and Nash equilibria in pure and mixed strategies. In this chapter we apply, extend, and develop all of these ideas and methods in several ways. We also discuss some basic questions about and criticisms of the concepts themselves.

This chapter is not light reading. Some of the arguments and criticisms of the Nash equilibrium concept get quite intricate. Some of the analysis is mathematically more demanding than in the rest of the book. We have tried to make the material as simple as possible, but much of it remains unavoidably challenging. Those of you who are not so quantitatively equipped or inclined, or who do not want to go into deeper conceptual issues but just want to get on with applications, may prefer to omit parts or even all of this chapter. They can do so without loss of continuity.

Nonetheless, we make no apologies for including this material. Our book aims to introduce game theory in all its facets. The theory has been found to have rich applications in economics, politics, and biology, and we develop and illustrate many such applications throughout the book. But the theory has also developed as a subject in its own right, and an introductory book should offer a glimpse of its higher reaches, hoping to attract some students to its further study. Instead of scattering such material throughout the book, where it might be harder for such students to identify and for the others to avoid, we have concentrated it in one chapter.

1 VALIDITY OF THE NASH EQUILIBRIUM CONCEPT

In Chapter 4 we introduced the pure-strategy Nash equilibrium for simultaneous-move games, and in Chapter 5 we extended the concept to the case of mixed strategies. The solution concept for sequential-move games in Chapter 3 was the rollback equilibrium. In Chapter 6 we looked at such games in their strategic form, and found that the rollback equilibrium of the game in extensive form is a special case, or refinement, which we called a subgame-perfect equilibrium, of the Nash equilibria of the same game in its strategic form. In other words, Nash equilibrium has been our basic solution concept for games, and we will continue to use it in the chapters that follow.

However, the Nash equilibrium concept is not without its critics, who come to their task from theoretical as well as practical perspectives. In this section we briefly review some such criticisms and some rebuttals, in each case using an example.[1] Some of the criticisms are mutually contradictory, and some can be countered by thinking of the games involved in a better way. We believe our discussion will leave you with renewed but cautious confidence in using the Nash equilibrium concept. But some serious doubts remain unresolved, indicating that game theory is not yet a settled science. Even this should give encouragement, not the opposite, to budding game theorists, for it shows that there is a lot of room for new thinking and new research in the subject. A totally settled science would be a dead science.

We begin by discussing the basic appeal of the Nash equilibrium concept. Most of the games in this book are noncooperative, in the sense that every player takes his action independently. Therefore it seems natural to suppose that if his action is not the best according to his own value system (payoff scale) given what everyone else does, then he will change it. In other words, it is appealing to suppose that every player's action will be the best response to the actions of all the others. Nash equilibrium has just this property of "simultaneous best responses"; indeed, that is its very definition. In any purported final outcome that is not a Nash equilibrium, at least one player could have done better by switching to a different action.

This is the consideration that leads eminent game theorist Roger Myerson to rebut those criticisms of the Nash equilibrium that are based on the intuitive appeal of playing a different strategy. His rebuttal simply shifts the burden of proof onto the critic. "When asked why players in a game should behave as in some Nash equilibrium," he says, "my favorite response is to ask 'Why not?' and to let the challenger specify what he thinks the players should do. If this specification

---

[1]David M. Kreps, *Game Theory and Economic Modelling* (Oxford: Clarendon Press, 1990) gives an excellent in-depth treatment.

is not a Nash equilibrium, then . . . we can show that it would destroy its own validity if the players believed it to be an accurate description of each other's behavior."[2]

We illustrate this rebuttal using an example provided by John Morgan, an economics professor at Princeton University. Figure 7.1 shows the game table. Cell-by-cell inspection quickly reveals that it has a unique Nash equilibrium, namely (A, A), yielding the payoffs (2, 2). But you may think, as did several participants in an experiment conducted by Morgan, that playing C has a lot of appeal, for the following reasons. It *guarantees* you the same payoff as you would get in the Nash equilibrium, namely 2, while if you play your Nash equilibrium strategy A, you will get a 2 only so long as the other player also plays A. Why take that chance? What is more, if you think the other player might use this rationale for playing C, then you would be making a serious mistake by playing A; you would get only a 0 when you could have gotten your guaranteed 2 by playing C.

		COLUMN		
		A	B	C
ROW	A	2, 2	3, 1	0, 2
	B	1, 3	2, 2	3, 2
	C	2, 0	2, 3	2, 2

FIGURE 7.1 A Game with a Questionable Nash Equilibrium

Myerson would respond, "Not so fast." If you really believe that the other player would think thus and would play C, then you should play B to get the payoff 3. And if you think the other person would think thus and so would play B, then your best response to B is A. And if you think the other would figure this out too, you should be playing your best response to A, namely A. Back to the Nash equilibrium!

As you can see, criticizing Nash equilibrium and rebutting the criticisms is itself something of an intellectual game, and quite a fascinating one. The following sections give a few other rounds in that game.

A. Is the Nash Equilibrium Concept Too Imprecise?

This criticism is based on the observation that many games have multiple Nash equilibria; therefore the concept fails to pin down outcomes of games suffi-

[2] Roger Myerson, *Game Theory* (Cambridge, Mass.: Harvard University Press, 1991), p. 106.

ciently precisely to give unique predictions. We have seen examples of games with multiple Nash equilibria, and in each case we described methods for selecting one that had better claims to be the outcome than the rest. In sequential-move games, the strategic form could have multiple Nash equilibria, but only the rollback equilibrium survived when we imposed the requirement of subgame perfectness, that is, we ruled out threats and promises that were not credible. In more complex games with information asymmetries or additional complications, other restrictions called *refinements* have been developed to identify and rule out Nash equilibria that are unreasonable in some way. In Chapter 12 we will meet one such refinement process that selects an outcome called a *Bayesian perfect equilibrium*. The motivation for each refinement is often specific to a particular type of game. A refinement stipulates how players update their information when they observe what moves other players made or failed to make. Each such stipulation is often perfectly reasonable in its context, and in many games it is not difficult to eliminate most of the Nash equilibria and therefore to narrow down the ambiguity in prediction.

In other games, one of the many Nash equilibria could emerge as a focal point if the players' expectations could converge on it. One outcome could be a focal point because of something intrinsic to the game and its payoffs; for example, if one Nash equilibrium gives uniquely higher payoffs to *all* players than the other Nash equilibria, then the players may reasonably expect that one to emerge. However, focal points are often governed by some associated historical, cultural, or linguistic features that assist the players' expectations to converge. And this is true not only in the assurance game of Chapter 4, in which both players' payoffs are increased by converging on the better of two equilibria, but also in games involving a lot of conflict.

Consider a very extreme case. Two players are asked to write down, simultaneously and independently, the shares they want out of a total prize of say $100. If the sums they write down add up to $100 or less, each player is given what he wrote. If the two add up to more than $100, neither gets anything. This game has numerous Nash equilibria: for any x, one player writing x and the other writing $(100 - x)$ is a Nash equilibrium. But in practice, 50:50 emerges as a focal point. This social norm of equality or fairness seems so deeply ingrained as to be almost an instinct; players who choose 50 say that it is the obvious answer. Of course, to be a true focal point it should not only be obvious to each, but everyone should know that it is obvious to each, and everyone should know that . . . ; in other words, its obviousness should be common knowledge. That need not always be the case, as we see when we consider a situation in which one player is a woman from an enlightened and egalitarian society who believes that 50:50 is obvious, and the other is a man from a patriarchal society, who believes it is obvious that in any matter of division a man should get three times as much as a woman. Then each will do what is obvious to her or him, and they will end up with nothing, because neither's obvious solution is obvious as common knowledge to both.

The existence of focal points is often a matter of coincidence, and creating them where none exist is basically an art that requires a lot of attention to the historical and cultural context of a game and not merely its mathematical description. This bothers many game theorists, who would prefer the outcome to depend only on an abstract specification of a game—players and their strategies should be identified by numbers without any external associations. We disagree. We think that historical and cultural contexts are just as important to a game as its purely mathematical description, and if such context helps in selecting a unique outcome from multiple Nash equilibria, that is all to the better.

The opposite of the criticism that some games may have too many Nash equilibria is that some games may have none at all—not just no equilibrium in pure strategies, but none in mixed strategies either. This difficulty can indeed arise, but it surfaces only in the higher reaches of game theory and not for the types of analysis and applications we deal with in this book, so we do not attempt to address it here.

B. Do Players in Actual Games Play Nash Equilibrium Strategies?

This criticism simply says that Nash equilibrium is unrealistic as a description of the outcomes of actual games. The criticism is often motivated by an abstract judgment: the critics argue that the concept of a Nash equilibrium is too subtle, and the calculation of Nash equilibrium strategy in an actual game too difficult, for players in real-life games. The rebuttal to the criticism is also partly based on equally abstract reasoning—that although players new to a game may not be able to figure out its structure all at once, experience will lead them to do so and to play their Nash equilibrium strategies. We will discuss these abstract arguments briefly, but first we consider some more direct practical evidence.

People have examined whether Nash equilibrium is observed in many real-life games like the prisoners' dilemma and chicken, and whether rollback calculations are done by participants in games like the ultimatum offer game. They often find violations of the predictions of the simple theory, but on closer examination it turns out that in many cases the theoretical analysis from which the predictions are derived is not so much simple as simplistic; it ignores aspects of payoffs in real life such as the benefits of reciprocity and reputation in repeated play. The fault lies not in the Nash equilibrium concept, but in its erroneous application. We commented on several such claims at various points in Chapters 3 through 6, and we will see in Chapters 8 and 11 that "fair" or "nice" behavior can be a Nash equilibrium in a repeated game.

Researchers have also conducted numerous laboratory experiments over the last three decades to test how people act when placed in certain interactive strategic situations. In particular, do they play Nash equilibrium strategies? Reviewing this work, Davis and Holt[3] conclude that in relatively simple single-move games with a unique Nash equilibrium, this outcome "has considerable drawing power . . . after some repetitions with different partners."[4] But in more complex or repeated situations, or when coordination is required because there are multiple Nash equilibria, the theory's success is more mixed. People do poorly at complicated rollback reasoning or calculations involving probabilities, particularly when they need to update probabilities using information revealed by actions at earlier stages of the game. They fail to take into account subtleties like the winner's curse, which we will meet when we study auctions in Chapter 15.

The behavior of players in some experimental situations does not conform to the experimenter's calculated Nash equilibrium for that game. For example, in bargaining games, people seem to "err" on the side of too much "fairness." But the reason may not be any failure to calculate or learn to play Nash equilibrium. It may instead be that the players' payoffs are different than those assumed by the experimenter. As with observations of naturally occurring games, participants in experimental situations also may know some complexities of the situation better than the experimenter knows. For example the *possibility* of repetition or a separate ongoing relationship with the other player may affect their choices in this game. Or the players' value systems may have internalized some social norms like fairness that have proven useful in the larger social context and that therefore carry over to their behavior in the experimental game.

Here we offer an encouraging tale, drawn from real life, of how players learn to play equilibrium strategies. Stephen Jay Gould discovered this beautiful example.[5] Through most of the 20th century, the best batting averages recorded in a baseball season have been declining. In particular, the number of instances of a player averaging .400 or better used to be much more frequent than they are now. Devotees of baseball history often explain this by invoking nostalgia: "There were giants in those days." A moment's thought should make one wonder why there were no corresponding pitching giants who would keep batting averages down. But Gould demolishes such arguments in a more systematic way. He points out that we should look at all batting averages, not just the top ones. The worst batting averages are not as bad as they used to be; there are also many

[3] Douglas D. Davis and Charles A. Holt, *Experimental Economics* (Princeton, N.J.: Princeton University Press, 1993).

[4] To focus on what players learn just from the physical (or cerebral) act of playing, it is important that experiments have each player's opponent in successive plays be anonymous or a different person drawn from a sufficiently large population. Otherwise players may be learning about how to play a repeated game against a known opponent; in such cases, different considerations arise, such as devising strategies to achieve tacit cooperation. However, the whole population of players needs to be stable. If novices without the same learning experience keep appearing and experimenting with different strategies, then the original group may unlearn what they had learned by playing against one another.

[5] "Losing the Edge," in *The Flamingo's Smile* (New York: Norton, 1985), pp. 215–229.

fewer .150 hitters in the major leagues than there used to be. He argues that this overall decrease in *variation* is a standardization or stabilization effect:

> When baseball was very young, styles of play had not become sufficiently regular to foil the antics of the very best. Wee Willie Keeler could 'hit 'em where they ain't' (and compile an average of .432 in 1897) because fielders didn't yet know where they should be. Slowly, players moved toward *optimal* methods of positioning, fielding, pitching, and batting—and variation inevitably declined. The best [players] now met an opposition too finely honed to its own perfection to permit the extremes of achievement that characterized a more casual age [emphasis added].

In other words, through a succession of adjustment of strategies to counter each other, the system settled down into its (Nash) equilibrium.

Gould marshals decades of hitting statistics to demonstrate that such a decrease in variation did indeed occur, except for occasional "blips." And indeed the blips confirm his thesis, because they occur soon after an equilibrium is disturbed by an externally imposed change. Whenever the rules of the game are altered (the strike zone is enlarged or reduced, the pitching mound is lowered, or new teams and many new players enter when an expansion takes place) or the technology changes (a livelier ball is used, or perhaps in the future, aluminum bats are allowed), the previous system of mutual best responses is thrown out of equilibrium. Variation increases for a while as players experiment, and some succeed while others fail. Finally a new equilibrium is attained, and variation goes down again. That is exactly what we should expect in the framework of learning and adjustment to a Nash equilibrium.

Observations of mixing in tennis provide similar evidence of players' abilities to learn to use Nash equilibrium strategies. Mark Walker and John Wooders of the University of Arizona examined the serve-and-return play of top-level players at Wimbledon. The server can serve to the receiver's forehand or backhand, and the receiver can guess to which side the serve will go and move that way. Since the receiver wants to guess correctly, and the server wants to wrong-foot the receiver, this interaction has a mixed-strategy equilibrium. If the players are using their equilibrium mixtures, the server should win the point with the same probability, whether he serves to the receiver's forehand or backhand. An actual tennis match contains a hundred or more points played by the same two players; thus there is enough data to test whether this implication holds. Walker and Wooders tabulated the results of serves in ten matches. Each match contains four kinds of serve-and-return combinations—A serving to B and vice versa, combined with service from the right or the left side of the court (Deuce or Ad). Thus they had data on forty serving situations, and found that in thirty-nine of them the server's success rates with forehand and backhand serves were equal to within acceptable limits of statistical error. The top-level players must have had enough general experience playing the game, and particular experience playing against the specific opponents, to have learned the general principle of mixing as well as the correct proportions to mix against the specific opponents.

Now we briefly mention some theoretical investigations of whether players will arrive at a Nash equilibrium through a process of playing the game repeatedly against different partners.[6] There are two related but different approaches. In the first, the group of players is unchanging. Each player tries out a strategy, observes how it performs, and experiments with something else if the performance of the original strategy is not satisfactory. The theorists examine the dynamics of this process mathematically but the answers are inconclusive; too much depends on exactly how players are assumed to judge satisfactory versus unsatisfactory performance, and what kinds of experiments with new strategies they are assumed to perform. For example, the players may continue using poor strategies if they hold wrong prior beliefs about the consequences of other strategies that prevent them from experimenting with these alternate strategies and discovering their error.

The second theoretical approach is to suppose that there is a population of players, each of whom follows one particular strategy. The strategies that perform well proliferate in the population; those that perform poorly die out. This is the basic idea of *evolutionary game theory*, which we will examine in Chapter 10. In the strict biological interpretation each player's strategy is fixed genetically and the proliferation or decay occurs by natural selection, but a more liberal interpretation allows strategies to be determined by culture or education, and allows proliferation to occur by imitation and decay by conscious avoidance. We will find that, in a specific sense which we will define in Chapter 10, if an evolutionary game has a stable outcome, it must be a Nash equilibrium. But evolutionary dynamics, like learning dynamics, can fail to converge, depending on the details of the selection mechanism.

Reviewing the empirical findings and the mathematical models, it seems the claim that we should expect experienced players to figure out and play Nash equilibrium finds better support from practical observations and experiments than from the purely theoretical analyses!

C. Does Rationality by Itself Imply Nash Equilibrium?

We argued earlier that the conceptual appeal of Nash equilibrium comes from its assumption that each player chooses his strategy to maximize his own payoff, given the strategies of the others. In other words, each player is assumed to behave rationally (as a calculating optimizer of his own values). While this assump-

---

[6] See Drew Fudenberg and David Levine, *The Theory of Learning in Games* (Cambridge, Mass.: MIT Press, 1998), and Jörgen Weibull, *Evolutionary Game Theory* (Cambridge, Mass.: MIT Press, 1995), for the complete analyses along these lines.

tion has been criticized by many psychologists and behavioral scientists, here we address a different problem—that the assumption of rationality alone is not enough to establish the case for Nash equilibrium.

First, every player may be rational, but if each does not know, or firmly believe, the others to be equally rational, then A may think that B will act in some particular irrational way and so will respond to that. Or A may think that B will think that A is irrational, which will affect B's choice and should therefore affect A's response . . . To rule out all such levels of thinking about others' rationality, it should be common belief among the players that everyone is rational.

But even that is not enough. A rational player will choose his best response to what he believes the others will do. Why should he believe that the others will play their Nash equilibrium strategy? Is it because of the common belief in rationality? But whether that implies Nash behavior is exactly the question we are investigating. We cannot proceed as if the answer is in the affirmative even before we have found the answer; that would be circular reasoning.

Such a common belief that everyone plays the Nash equilibrium strategy could arise because the players have all read the same book or employ consultants who have (and this fact is common knowledge among them). But if the common belief that everyone plays the Nash equilibrium strategy does not arise, what can we conclude on the basis of common knowledge of rationality alone?

Although a player who knows the others to be rational cannot predict exactly what they will choose unless he knows what they believe about what he will choose, he *can* predict what they will *not* choose—a strictly dominated strategy. And he can be sure they will believe the same of him. In other words, players with common knowledge of rationality can confidently carry out iterated elimination of strictly dominated strategies. We saw in Chapter 4 that if this procedure culminates in a unique outcome, then that is the (unique) Nash equilibrium of the game. But quite often the procedure fails to eliminate all but one outcome. Then any remaining outcomes can in principle be justified by players' beliefs about what others will do; we call this remaining set of outcomes **rationalizable.** Anything in this set can be justified, or rationalized, as a final outcome of the game, purely on the assumption that the players are rational (and this rationality is common knowledge).[7]

In the example of Figure 7.1, there are no dominated strategies, so every outcome is rationalizable. The outcome where Row plays A and Column plays B can be rationalized by the set of beliefs in which Column believes (incorrectly, as it turns out) that Row will play C, and Row believes (correctly, as it turns out) that Column will think thus. Of course, this situation will not last if the players are

---

[7]Our definition of rationalizability is a simplified one. For the more general formulation, first developed by Douglas Bernheim and David Pearce, see Drew Fudenberg and Jean Tirole, *Game Theory* (Cambridge, Mass.: MIT Press, 1991), pp. 48–53. Our definition coincides with the more general one in two-person games.

learning from the experience of playing the game. Observation will change their beliefs, unless they are playing a Nash equilibrium in which the beliefs are confirmed. Repetition thus strengthens the case for Nash equilibrium. But the dynamics of belief revision do not have to converge to a Nash equilibrium; the players may go around in cycles through the set of rationalizable outcomes forever.

D. Does the Nash Equilibrium Account for Risks Properly?

Given the strategies and payoffs of a game, a player can calculate its Nash equilibrium. But if he has specified the game incorrectly, his calculation may lead him to a strategy that is not the correct one for him to play in the true game. Might this lead to disaster? And should that possibility detract from the appeal of the Nash equilibrium?

We illustrate this possibility using the game of Figure 7.2. Before analyzing this game theoretically, you should pretend that you (as player A) are actually playing the game. Which of your two actions would you choose?

		B	
		Left	Right
A	Up	9,10	8,9.9
	Down	10,10	−1,000,9.9

FIGURE 7.2 Disastrous Nash Equilibrium?

Keep in mind your answer to this question as we proceed to analyze the game. If we start by looking for dominant strategies, we see that player A has no dominant strategy but player B does. Playing Left guarantees B a payoff of 10, no matter what A does, versus the 9.9 gained by playing Right (also no matter what A does). Thus player B should play Left. Given that player B is going to play Left, player A does better to play Down. The unique pure-strategy Nash equilibrium of this game is (Down, Left); each player achieves a payoff of 10 at this outcome. (It is also the unique rationalizable outcome.)

The problem that arises here is that many people (but not all, of course) in A's position would not choose to play Down. (What did you choose?) This is as true for those who have been students of game theory for years as for those who have never heard of the subject. If A has *any* doubts about *either* B's payoffs *or* B's rationality, then it is a lot safer for A to play Up than to play his Nash equilibrium strategy of Down. What if A thought the payoff table was as illustrated in Figure 7.2 but in reality B's payoffs were the reverse—the 9.9s went with Left and the 10s went with Right? What if the 9.9s were only an approximation and the exact payoffs were actually 10.1? What if B were a player with a substantially different value system, or not a truly rational player who might choose the "wrong" action just for fun? Obviously, our assumptions of perfect information and rationality can really be crucial to the analysis we use in the study of strategy. Doubts of players can alter equilibria from those we would normally predict and can call the reasonableness of the Nash equilibrium concept into question.

However, the real problem with many such examples is not that the Nash equilibrium concept is inappropriate, but that the examples choose to use it in an inappropriately simplistic way. In this example, any doubts about B's payoffs should be made an integral part of the analysis. If A does not know B's payoffs, this becomes a game of asymmetric information. Although we will not develop the general techniques for studying such games until Chapter 12, this particular example is a relatively simple game of that kind, which we can figure out using the tools of probability from the Appendix to Chapter 5.

Suppose A thinks there is a probability p that B's payoffs from Left and Right are the reverse of those shown in Figure 7.2, so $(1 - p)$ is the probability that B's payoffs are as stated in that figure. Since A must act without knowing which is the case, he must choose his strategy to be "best on the average." In this game the calculation is simple, because in each case B has a dominant strategy; the only problem for A is that in the two different cases, different strategies are dominant for B. With the probability $(1 - p)$, B's dominant strategy is Left (the case shown in the figure), and with probability p, it is Right (the opposite case). Therefore if A chooses Up, then with the probability $(1 - p)$ he will meet B playing Left and so will get a payoff of 9, and with the probability p he will meet B playing Right and so will get a payoff of 8. Thus A's expected payoff from playing Up is $9(1 - p) + 8p$. Similarly, A's expected payoff from playing Down is $10(1 - p) - 1,000p$. Therefore it is better for A to choose Up if

$$9(1 - p) + 8p > 10 (1 - p) - 1,000p$$
$$p > 1/1,009.$$

Thus even if there is only a very slight probability that B's payoffs are the opposite of those in Figure 7.2, it is optimal for A to play Up. Analysis based on rational behavior, when done correctly, contradicts neither the intuitive suspicion nor the experimental evidence after all.

E. Is the Expected-Payoff Maximization Assumption Reasonable?

The calculation we used in Section 1.D to justify the use of Up by A in Figure 7.2, and indeed the whole framework of game theory, has been based on the assumption that the players' objectives are their expected payoffs. As we explained in Chapter 2 and in the Appendix to Chapter 5, payoffs do not have to be measured in money units; a nonlinear scale of payoffs can capture a player's aversion to risk. Nevertheless, the expected-payoff objective is still sometimes responsible for producing equilibria that seem unreasonable. Here is an example.

When we studied the use of mixed strategies in Chapter 5, we saw which pure strategy is the percentage play—the strategy used with high probability in the equilibrium mixture in a zero-sum game. It is the pure strategy that gives you nearly equal payoffs against all the pure strategies used by the opponent. By contrast, the nonpercentage play is a risky strategy; it gives high payoffs against some of the opponent's pure strategies but very poor payoffs against others. However, people often say that if the occasion is really important, in the sense that winning versus losing is a big difference in payoffs, then one should use the percentage play more often. Thus they argue that the offense may throw a long pass on third down with a yard to go in an ordinary season game, but the risk of this play is so large that it should not be used in the Super Bowl. Let us try to fit this intuition into our formal theory of mixing strategies.

Most of our examples have used illustrative numbers for payoffs, but here we want to emphasize the generality of the problem. Therefore we let the payoffs be general algebraic symbols, subject only to some conditions that reflect the basic issue being considered.

Consider any zero-sum game in which you have two pure strategies. Let us call the relatively safe strategy (the percentage play) P, and the more risky strategy (the nonpercentage play) R. The opponent has two pure strategies, also P and R; his P is his best response to your P, and his R to your R. Figure 7.3 shows the table of probabilities that your play succeeds; these are not your payoffs. The sense of "safe" and "risky" is captured by requiring that $a > b > c > d$. The risky play does really well if the opponent is not prepared for it (your success probability is a), but really badly if he is (your success probability is d), while the percentage play does moderately well in either case (you succeed with a probability of b or c), but a little worse if the opponent expects it ($c < b$).

Let your payoff or utility be W if your play succeeds and L if it fails. A "really big occasion" is when W is much bigger than L. Note that W and L are utilities,

		OPPONENT EXPECTS	
		P	R
YOU PLAY	P	c	b
	R	a	d

FIGURE 7.3   Table of Success Probabilities of Risky and Percentage Plays

not necessarily money amounts, so they are already intended to capture any aversion to risk. Now we can fill in the table of expected payoffs from the various strategy combinations, as shown in Figure 7.4. Note how this table is constructed. For example, if you play P and your opponent expects R, then you get utility W with a probability of b and utility L with a probability of $(1 - b)$; your expected payoff is $bW + (1 - b)L$. This is a zero-sum game, so the opponent's payoffs in each cell are just the negative of yours.

		OPPONENT EXPECTS	
		P	R
YOU PLAY	P	$cW + (1 - c)L$	$bW + (1 - b)L$
	R	$aW + (1 - a)L$	$dW + (1 - d)L$

FIGURE 7.4 Payoff Table with Risky and Percentage Plays

In the mixed-strategy equilibrium your probability p of choosing P is defined by

$$p[cW + (1 - c)L] + (1 - p)[aW + (1 - a)L] = p[bW + (1 - b)L] + (1 - p)[dW + (1 - d)L].$$

This simplifies to

$$p = \frac{a - d}{a - d + b - c}.$$

You can similarly calculate the opponent's q–mix to find

$$q = \frac{a - c}{a - c + b - d}.$$

The surprise is that the expressions for p and q are completely independent of W and L. Therefore the theory does not bear out the usual intuition. The theory says that you should mix the percentage play and the risky play in exactly the same proportions on a big occasion as you would on a minor occasion.

So which is right: theory or intuition? We suspect you will be divided on this issue. Some of you will think that the sports commentators are wrong, and will be glad to have found a theoretical argument to refute their claims. Others will side with the commentators and argue that bigger occasions call for safer plays. Still others of you may think that bigger risks should be taken when the prizes are bigger, but even so will find no support in the theory, which says that the size of the prize or the loss should make no difference to the mixture probabilities.

On many previous occasions where discrepancies between theory and intuition arose, we argued that the discrepancies were only apparent, that they were the result of failing to make the theory sufficiently general or rich enough to capture all the features of the situation that created the intuition, and that improving the theory removed the discrepancy. This situation is different; the problem is fundamental to the use of expected utility in constructing the payoffs. And almost all of existing game theory has this starting point.

The real problem is that expected payoffs are linear in probabilities. For example, in the expression $cW + (1 - c)L = L + c(W - L)$, if c increases from 0% to 1%, the expected payoff will increase by $0.01(W - L)$, and if c increases from 90% to 91%, the expected payoff will increase by the same amount: $(0.91 - 0.90)(W - L) = 0.01(W - L)$. Many people will not regard the two changes as equivalent in their calculations, however. Some may prefer avoiding total disaster, others may prefer getting close to certainty, but exact equality of effects is a very special case. However, that is the case inherent in the expected utility framework.

Theories of decisions (one-person choices) under uncertainty with objectives other than expected utility have proliferated.[8] But these alternatives have not been much used to create an alternative theory of interactive decisions (games).[9] Here is an interesting new research opportunity.

WE HAVE PRESENTED SEVERAL ARGUMENTS criticizing the Nash equilibrium concept, and others supporting it or countering the criticisms. We can now sum up. On the whole, we believe you should have considerable confidence in using the Nash equilibrium concept when the game in question is played frequently by players from a reasonably stable population and under relatively unchanging rules and conditions. When the game is new, or played just once, and the players are inexperienced, you should use the equilibrium concept more cautiously, and should not be surprised if the outcome you observe is not the equilibrium you calculate. But even then, your first step in the analysis should be to look for a Nash equilibrium; then you can judge whether it seems a plausible outcome, and if not, proceed to the further step of asking why not. If the game is played repeatedly by the same known groups of players, then you should treat the repeated game as a game in its own right and look for its equilibria, which may allow for possibilities, such as tacit cooperation, that cannot arise in a single play.

Sometimes you may need a refinement to eliminate unreasonable possibili-

---

[8] Probably the best known of these alternative formulations is Daniel Kahneman and Amos Tversky, "Prospect Theory: An Analysis of Decision Under Risk," *Econometrica*, vol. 47 (1979), pp. 263–291. Mark Machina, "Choice Under Uncertainty: Problems Solved and Unsolved," *Journal of Economic Perspectives*, vol. 1 (1987), pp. 121–154 gives an excellent survey.

[9] Exceptions include Vincent P. Crawford, "Equilibrium Without Independence," *Journal of Economic Theory*, vol. 50 (1990), pp. 127–154, and James Dow and Sergio Werlang, "Nash Equilibrium Under Knightian Uncertainty: Breaking Down Backward Induction," *Journal of Economic Theory*, vol. 64 (1994), pp. 305–524.

ties among multiple Nash equilibria. At other times outcomes other than Nash equilibria may be deemed rationalizable. In still other situations you may be able to identify particular departures from rationality that make sense. But Nash equilibrium will remain the central point from which to consider these departures. Used with such caution and adaptability, you will find that the concept remains useful in practice far more often than you might expect.

2 GAMES OF DYNAMIC COMPETITION

Here we turn from the conceptual level to the analytical, and present an example whose main purpose is to improve your skills in using rollback reasoning and doing probability calculations. But the example also has a substantive purpose: it offers some general lessons about the dynamics of competitive strategies or races. In many such situations, there are two objectives: getting a certain absolute level of achievement, and reaching a certain level ahead of your rivals. For example, a pharmaceutical company developing a new drug wants it to be effective (absolute achievement), and to be sufficiently better than other similar drugs in order to be successful in the market (achievement relative to one's rivals). Research is a lengthy process, and as each company proceeds through the stages of trying various compounds, patenting them, taking the promising ones through animal trials and then human trials, getting approval from the regulating authorities, such as the Food and Drug Administration, and so on, it not only thinks about its own progress but also watches how its competitors are doing. It has to decide whether it should intensify its efforts or slacken them, or even whether it should simply give up.

Game-theoretic modeling of such processes gets quite complex, but fortunately there is a simple and familiar situation with broad similarities—tennis. To win a game in tennis, you have to win four or more points (absolute achievement) with a margin over the opponent of at least two (relative achievement). We analyze this well known game, find results that accord well with intuition, and then translate the findings back to the context of business competition.

Just in case you didn't know, *game* is a technical term in tennis. It consists of a sequence of *points*. In each point, one player, the server, puts the ball in play, and the other player, the receiver, hits it back. They continue hitting the ball back and forth until one player hits it out of the field of play or into the net; the other then wins the point.[10] To win a single *game* of tennis, you have to win four or more points and you have to win by a margin of at least two. The scoring is peculiar: the

---

[10] We are leaving out some complications; for example, a serve has to land in a limited part of the whole field for it to be legal, the server is allowed two attempts (serves) at putting the ball into play, and so on. These can be incorporated into the analysis, but the calculations are a little more involved.

first point in a game is called 15, the second 30, the third 40.[11] Thus if the server has won two points and the receiver one point, the score is 30–15. (By convention, the server's score is always given first.) If the score reaches 40–40, the situation is called *deuce*. Then the game continues until one player is ahead by two points. Being ahead by one point at this juncture is called having the *advantage*, or, simply, *ad*. When the server is one point ahead here, we call that "ad to the server," or AS for short, and when the receiver is one point ahead, "ad to the receiver," or AR. (The terms *ad in* and *ad out*, respectively, are also used.) In principle a game could go on forever, as the players switch back and forth between deuce and advantage. But in practice, all games played so far have ended in finite time.

In tennis, a game is part of a larger structure. In a *set*, the players take turns serving alternate games. A set is won by winning six or more games with a margin of at least two games. If a set goes to six games each, a further game called the *tie-breaker* is played under somewhat different rules, and its winner wins the set 7–6. A set in turn is part of a complete *match*. Women play matches that are best-of-three sets, while men play best-of-five sets in the major *grand slam* tournaments and best-of-three in most others. At Wimbledon and in the international Davis Cup matches, the last set does not have a tie-breaker, but must be played out until one player gets a two-game lead.

We consider just one game, although the method developed here can easily be extended to analyze whole sets or matches using a computer. We can show possible configurations that can arise in a tennis game using a two-dimensional grid, as in Figure 7.5. The server's score is shown along the horizontal dimension (the x-axis) moving from left to right; the receiver's score is along the vertical dimension (the y-axis) from the bottom to the top. Each crossing point of the grid represents a particular score; for example, the score at the bottom left corner is 0–0, the point three units to the right and two units up from this corner is 40–30, and so on. The thick lines on the right side of the diagram represent points where the server wins the game; those on the top side represent points where the receiver wins.

We can trace the progress of an actual game by a succession of horizontal and vertical moves along the grid lines starting at 0–0. For example, the server may win the first point, taking the score to 15–0, win the second (30–0), lose the next three (30–15, 30–30, and 30–40), and then win the next three—deuce (D in Figure 7.5), ad to the Server (AS), and game to the Server. One player wins if any such sequence reaches the thick line on his side of the diagram.[12]

---

[11] This is presumably because scores used to be kept on a clocklike dial by rotating a pointer like the minute hand through a quarter circle for each point. Then 40 would be a misnomer for 45.

[12] Figure 7.5 shows that deuce and 30–30 are equivalent in the sense that the subsequent paths that lead to the server or the receiver winning the game are the same from these two starting positions. For example, starting at deuce, the server can win the game by winning the next two points, or by losing the next point but winning the three points to follow, and so on; the same is true starting at 30–30. Similarly, AS is equivalent to 40–30, and AR is equivalent to 30–40. We maintain the name distinctions between these scores because it is conventional in tennis to do so.

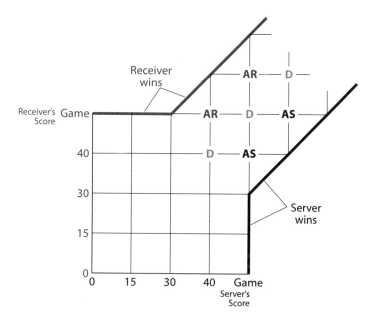

FIGURE 7.5   Possible Score Combinations in a Tennis Game

A. Calculating the Probabilities of Winning at Tennis

We begin with some simple calculations of the dynamics of a tennis game, without considering any choice of strategies at all. We want to know how it is likely to proceed—who is likely to win each point, and then who is likely to win the game. Generally, the server is more likely to win any given point. We will let x stand for the probability that the server wins any single point; the probability that the receiver wins the point is then $(1 - x)$. For now, we will assume that we know the value of x and that it is the same regardless of the current position in the game. If the server is more likely to win a point, then $x > 0.5$. In top-level men's singles, for instance, a typical value of x might be 0.65 (65%). For convenience (and comparisons), we will also look at the case where $x = 0.5$.

Given x, can we calculate the probability that the server wins the whole game? A hard way to calculate this probability is to figure out all of the possible different ways the server can win (corresponding to all the zigzag paths leading from 0–0 to the server's winning line on the right), find the probability of each, and add all these probabilities. The trouble is that there are infinitely many zigzags between deuces and ads, so the probabilities form a geometric series; summing it is a complicated and error-prone task. Rollback reasoning lets us cut through all that. We can begin by calculating the probabilities of the server winning from one of the later positions in the game (deuce, ad in, or ad out) first, and then use this information to calculate the probabilities of the server winning, starting from successively earlier positions, until we know the probability of winning from a 0–0 start.

To do this, we define a bit of notation and do a bit of algebra. Let us write $P(i, j)$ for the probability that the server wins the game, starting from the point (i, j), where i denotes the server's score and j the receiver's score within the game. For example, $P(30, 15)$ is the probability of the server winning, starting from the score 30–15.[13] Then the probability of the server winning from any particular point equals the sum of two products: the probability of the server *winning* that point times the probability of the server winning the game after winning the point, plus the probability of the server *losing* the point times the probability of the server winning the game after losing the point.[14] In mathematical notation, starting at $(0, 0)$,

$$P(0, 0) = x \, P(15, 0) + (1 - x) \, P(0, 15).$$

We emphasize two things about this formula: (1) All the probabilities $P(i, j)$ are for the *server* winning the game eventually. The different points (i, j) represent the different scores, starting from which the server is supposed to win. (2) All the values of $P(i, j)$ are unknown thus far; we have to solve for them using equations, like the one just given.

To solve for $P(0, 0)$, we would have to know $P(15, 0)$ and $P(0, 15)$. But to solve for those we would need to know $P(30, 0)$, $P(15, 15)$, and $P(0, 30)$. In turn, to get those we would need $P(40, 0)$, and so on. Thus we need to start at the end of the game and work our way back, as noted earlier; that is, we need to use rollback. To do so, we have to consider how a tennis game can end after it has gone to $(40, 40)$ or beyond. The end can occur in three possible ways. We use this fact and link three probabilities, all for the *server* to win the game, starting from each of the different end situations:

$P(AS)$ = probability of the server winning when initially he had the ad,
$P(D)$ = probability of the server winning when initially the game is at deuce,
$P(AR)$ = probability of the server winning when initially the receiver had the ad.

To win the game starting with having the ad, the server either wins the next point and thereby the game (probability x), or loses the first point (probability $1-x$) and then wins the game starting at deuce [probability P(D)]. Therefore

$$P(AS) = x + (1 - x) P(D).$$

Similar arguments starting at deuce and ad to receiver tell us that

$$P(D) = x \, P(AS) + (1 - x) \, P(AR) \quad \text{and} \quad P(AR) = x \, P(D).$$

---

[13] We will calculate only the probabilities of the *server* winning the game. Of course, $[1 - P(30, 15)]$ is the probability that the receiver wins the game starting from a score of $(30, 15)$, and similarly from any other starting point.

[14] The Appendix to Chapter 5 explains in more detail and more general form the rules used here for manipulating probabilities.

To recapitulate the reasoning: From point AS, if you (the server) win the point, you win the game, but if you lose the point, you are at deuce (D). From D, if you win the point, you are at AS, but if you lose the point, you are at AR. From AR, if you win the point, you are at D, but if you lose the point, you lose the game.

Now we have three linear equations in three unknowns, $P(AS)$, $P(D)$, and $P(AR)$. These three can be solved simultaneously. A simple way is to plug the expressions for $P(AR)$ and $P(AS)$, which only have $P(D)$ in them, into the expression for $P(D)$. This gives us just one equation in $P(D)$, which solve and substitute back to get the other two equations. The answers are

$$P(D) = \frac{x^2}{1 - 2x + 2x^2}$$

$$P(AR) = \frac{x^3}{1 - 2x + 2x^2}$$

$$P(AS) = \frac{x(1 - x + x^2)}{1 - 2x + 2x^2}.$$

Once we know these three probabilities, we can proceed to find the probabilities that server wins from earlier points in game. For example, starting at 40–30 gives us[15]

$$P(40, 30) = x + (1 - x)P(D).$$

Since we know $P(D)$, we can find $P(40, 30)$. Similarly, we can find $P(30, 40)$; then we can do a similar solution for each preceding point in turn.

The results of these probability calculations are shown schematically in Figure 7.6. Note that the grid arrangement is exactly like that of Figure 7.5 except that at each grid point we show the $P(i, j)$ corresponding to that score. The figure shows two cases: (a) for $x = 0.50$, and (b) for $x = 0.65$. The latter is typical for the top professional level in the men's game. You can experiment with other values by plugging different values of x into the probability expressions.

In Figure 7.6a, we see a symmetry about the 45° line. If the server is not favored to win any particular point, he is not favored to win the game, starting from any point on the 45° line, where the server and receiver have equal scores. And the probabilities of the server's winning the game from any two situations symmetrically on either side of the 45° line add to 1. For example, starting from 30–15 the server will win 69% of the time, whereas starting from 15–30 he will win 31% of the time.

In the Figure 7.6b, the server is quite heavily favored to win any one point. Therefore the probabilities of the server winning the game, starting from any symmetric initial score, are also greater than 50%. These probabilities get bigger

---

[15] Note that the equation for $P(40, 30)$ is the same as the equation for $P(AS)$ on the previous page. This reinforces the point, mentioned earlier in footnote 12, about the equivalence of these scoring situations.

as we move down along the 45° line to the start of the game: the probability of the server winning the game from deuce or 30–30 is 78%, from 15–15 it is 80%, and from 0–0 it is 83%. The reason is that the earlier the starting point, the more opportunities the server has for recovery from a setback.

The server has good chances to win the game even from many initially unfavorable situations. For example, he has an almost even chance (48%) of

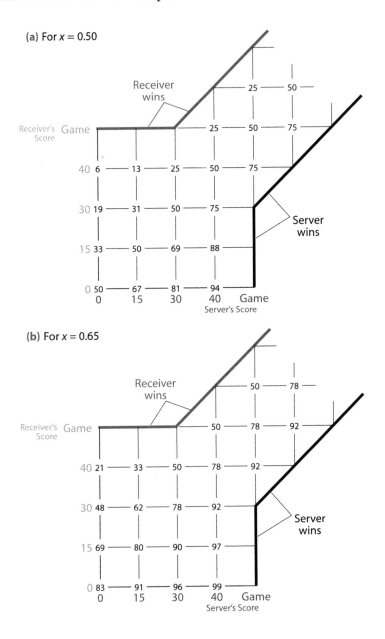

FIGURE 7.6 *P* Values in Tennis

retrieving a game from 0–30. If you are surprised, keep track of the scoring the next time you watch a top-level men's match.

B. Strategy in Tennis

All this information is of interest, but it is not based on any strategic game or decision situation. Neither player is actually choosing anything; the probability x is just a given, based on an assessment of the relative skills of the two players. We can easily graft strategic considerations onto the example, however. Suppose the server has a special serve that is much more effective than his ordinary serve. He cannot use the special serve all the time, perhaps because it takes a lot of effort, but more likely because if it is overused, the other player will learn how to counter it and its effectiveness will diminish.

Then the strategic question for the server entails determining when to use his special serve. The instinctive answer is of course that it should be used on the most "important" or "crucial" situations in the game. But that just raises another question—what makes one situation more important than another? This needs a little more thought, but the answer is simple. The ultimate objective here is to win the whole game. A particular point is crucial when winning versus losing *that* point makes the biggest difference to the chances of winning the full game.

We can construct quantitative measures of this importance from our knowledge of the probabilities $P(i, j)$, and show how to construct it for any one score i–j, say 30–15. If the server wins the point after 30–15, the probability of his winning the game becomes $P(40, 15)$. If the server loses the point after 30–15, the probability that he wins from then on is $P(30, 30)$. If the difference $P(40, 15) - P(30, 30)$ is large, then the server's probability of winning the game is greatly enhanced by winning the point after 30–15, compared to what the probability of winning would be if he lost this point. Therefore we use this difference to quantify the *importance* of this point to the server. Using the symbol $I(30, 15)$ for this measure, we therefore write

$$I(30, 15) = P(40, 15) - P(30, 30).$$

Consider the case of $x = 0.50$, shown in Figure 7.6a. Using the actual numbers there, $P(40, 15) = 88\%$ and $P(30, 30) = 50\%$; therefore $I(30, 15) = 88 - 50 = 38\%$.

Similar calculations can be done for all scores, and Figure 7.7 shows the resulting measures of importance of all positions for our two x values: (a) $x = 0.50$ and (b) $x = 0.65$. In Figure 7.7a, for example, the number at the 30–15 location is 38, as we calculated in the previous paragraph. You can verify all the other values similarly, using the numbers in Figure 7.6 and doing the appropriate subtractions.

What about the importance of a point to the receiver? The probability that he wins starting from any point is 1 minus the probability that the server wins starting from the same point. Consider the same starting score 30–15 as we did above.

If the receiver wins it, the score goes to 30–30; if he loses it, the score goes to 40–15 in favor of the server. Therefore the importance from the receiver's perspective is

$$[1 - (30, 30)] - [1 - P(40, 15)] = P(40, 15) - P(30, 30)$$
$$= I(30, 15).$$

This is exactly the same as the importance of 30–15 to the server! The two have exactly the opposite objectives, but they are in full agreement as to which situations are more important.

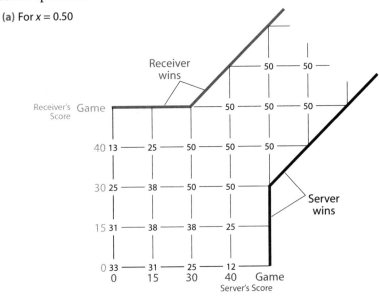

(a) For $x = 0.50$

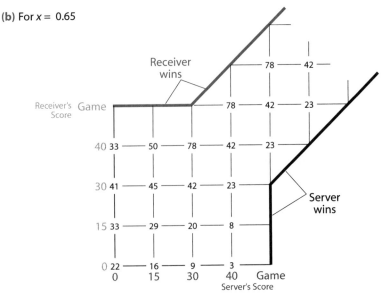

(b) For $x = 0.65$

FIGURE 7.7 I Values

These measures of importance have several noteworthy properties:

1. The importance of points with equal initial scores rises steadily as we move northeast along any 45° line. Why? As we get closer to the end of a game, there is less opportunity to retrieve the loss of a point. Therefore it is more vital to win than lose the point. The only slight exception arises when comparing 30–0 and 40–15. At the latter point the server has almost won the game, so winning that point is not so important, whereas losing the former point takes the score to 30–15, which creates some chances for the receiver to win.

2. In Figure 7.7a, where $x = 0.5$, the importance is highest along the central 45° line and falls off to each side (northwest and southeast) of it. Why? If you are very far ahead (bottom right of diagram), you are quite likely to win even if you lose this point. If you are very far behind (upper left of diagram), the game may be lost even if you win this point. In neither case does winning or losing this point make much of a difference. Only when the initial situation is reasonably balanced does winning or losing this point make a big difference.

3. In Figure 7.7b, where $x = 0.65$, the line of greatest importance runs from 0–15 to AR. The points along this diagonal represent scores at which the server has fallen behind by one point. Why is the line of greatest importance *not* that of equal scores, as it was for $x = 0.5$? If you are favored to win any single point, then you have good prospects of coming from behind to win the game. Losing one point when you are initially equal does not matter all that much, so the line of exact equality is less crucial than the line where you are a little behind. In the $x = 0.65$ case, the most crucial point in the game is when the score is 30–40 or ad out. This conforms to the accepted wisdom of tennis commentators, and also to the practice of the top men players. You see them use their best serves in these situations. Pete Sampras at his best goes a level beyond that. His probability of winning a single point is even higher than 65%. Therefore the most important points to him are when he is serving and the score is 15–40 or even 0–40; that is when he delivers in succession two or even three very special, unreturnable serves.

C. Applying Tennis Analysis to Business

We asked you to regard the tennis game as a metaphor for business competition, especially an R&D race for a new product, where the objective is to combine some absolute standard of achievement and some relative level of superiority over your rivals. Now we can offer valuable insights into some aspects of this competition. The most crucial stages are when the rivals are reasonably close to each other and both products are reasonably close to completion—that is when one misstep would be difficult to correct in time. That is also when both firms should exert maximum effort. If the two firms are unequal in their innate strengths, then the most crucial situation comes when, through some combination of luck or skill, the weaker firm has stolen a lead over the stronger one. If Microsoft is favored to dominate almost any new software product, and Netscape forges ahead with its Web browser, that is when Microsoft will work hardest to get back to its customary leading position, and Netscape will also work hardest to protect its lead. You should be able to think of several other examples of this principle.

3 NASH EQUILIBRIUM WITH CONTINUOUS STRATEGIES

In Chapter 4 we saw a simple example involving pizza pricing, where each player's strategy was a continuous variable, and a Nash equilibrium was found by combining the best-response curves of the two players. Here we reinforce that idea by developing another example, this one drawn from politics and requiring more mathematics background than we normally assume.

Consider an election contested by two parties or candidates. Each is trying to win votes away from the other by advertising—either positive ads that highlight the good things about oneself, or negative, attack ads that emphasize the bad things about the opponent. To keep matters simple, suppose the voters start entirely ignorant and unconcerned, and are moved solely by the ads. (Many people would claim this is a pretty accurate description of U.S. politics, but more advanced analyses in political science do recognize that there are informed and strategic voters. We address the behavior of such voters in detail in Chapter 14.) Even more simply, suppose the vote share of a party equals its share of the total campaign advertising that is done. Call the parties or candidates L and R; when L spends $\$x$ million on advertising and R spends $\$y$ million, L will get a share $x/(x+y)$ of the votes and R will get $y/(x+y)$. Once again, readers who get interested in this application can find more general treatments in specialized political science writings.

Of course, raising money to pay for these ads involves a cost: money to send letters and make phone calls; time and effort of the candidates, party leaders, and activists; the future political payoff to large contributors; and possible future political costs if these payoffs are exposed and lead to scandals. For simplicity of analysis, let us suppose all these costs are proportional to the direct campaign expenditures x and y. Specifically, let us suppose that party L's payoff is measured by its vote share *minus* its advertising expenditure, $x/(x+y) - x$. Similarly party R's payoff is $y/(x+y) - y$.

Now we can find the best responses. Because this cannot be done without calculus, we derive the formula mathematically and then explain its general meaning intuitively, in words. For a given strategy x of party L, party R chooses y to maximize its payoff. The calculus first-order condition—holding x fixed and setting the derivative of $y/(x+y) - y$ with respect to y equal to zero—is $x/(x+y)^2 - 1 = 0$, or

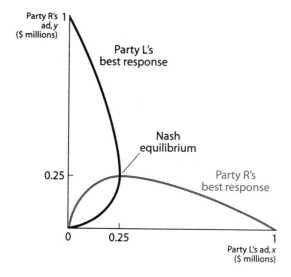

FIGURE 7.8   Best Responses and Equilibrium in the Campaign Advertising Game

$$y = \sqrt{x} - x.$$

Figure 7.8 shows the graph of this and of the analogous best-response function of party L, namely

$$x = \sqrt{y} - y.$$

Look at the best-response curve of party R. As the value of party L's x increases, party R's y increases for a while and then decreases. If the other party is advertising very little, then one's own ads have a high reward in the form of votes, and it pays to respond to a small increase in the other's expenditures by spending more oneself to compete harder. But if the other party already has a massive expenditure, then one's own ads get only a small return in relation to their cost, so it is better to respond to the other's increase by scaling back.

As it happens, the two parties' best-response curves intersect at their peak points. Again, some algebraic manipulation of the equations for the two curves yields us exact values for the equilibrium values of x and y. You should verify that here x and y are each equal to 1/4, or $250,000. (This is presumably a little local election.)

As in the pricing game, we have a prisoners' dilemma. If both parties cut back on their ads in equal proportions, their vote shares would be entirely unaffected, but both would save on their expenditures and so both would have a larger payoff. And unlike a producers' cartel that keeps prices high and hurts consumers, a politicians' cartel to advertise less would probably benefit voters and society. We could all benefit from finding ways to resolve this particular prisoners' dilemma. In fact, Congress has been trying to do just that for several years, and has imposed some partial curbs, but political competition seems too fierce to permit a full or lasting resolution.

What if the parties are not symmetrically situated? Suppose R's advertising dollars are twice as effective as L's; that is, L's vote share is $x/(x + 2y)$ and R's is $2y/(x + 2y)$. Does the more favorably situated party take advantage of its position and spend less? No; the outcome is just the opposite. It turns out that equilibrium is on the downward portion of L's best-response curve and on the upward portion of R's best-response curve. That is to say, the favored party R spends more than the amount that would bring forth the maximum response from party L, and underdog party L spends less than the amount that would bring forth the maximum response from party R. The favored party spends so much that the other gives up, while the underdog party spends sufficiently little that the other eases up.

4 MIXING AMONG MANY STRATEGIES: EXAMPLES

In Chapter 5 we showed how the mixed-strategy equilibrium can be computed when at least one of the players has just two pure strategies. When each player has at least three pure strategies, the calculation can quickly get quite complicated. For truly complicated games with many pure strategies for each player, one must rely on a computer to do the job. Many such computer programs exist. *Mathematica* has a routine of this kind, and Gambit, which we mentioned in Chapters 3 through 5 also, is probably the most widely available. However, it is useful to acquire some understanding of how such calculations are done in order to get a better feel for the results, even when a computer does the messy arithmetic. Therefore we now illustrate the general idea using a few relatively simple examples. We then set out the more general theory of mixing with many available strategies in the following section.

A. A Zero-Sum, Three-by-Three Game

Some larger games (and for now we concentrate on zero-sum games) can be solved easily for equilibrium mixes using the prevent-exploitation method of Section 3 in Chapter 5. One such game is rock–paper–scissors (RPS), a game we described in Chapter 4, Section 1. Other situations present more difficulties, which we discuss later in the chapter. We begin, however, with RPS.

As noted in Chapter 4, RPS is a simple game in which two players show Rock, Paper, or Scissors, and each player wins, loses, or ties. Recall that ties occur when both players show the same item, while wins occur when paper "covers" rock, scissors "cut" paper, or rock "breaks" scissors. The simple game table in Figure 7.9, which repeats Figure 4.1b, shows a symmetric game with no Nash equilibrium in pure strategies. To find an equilibrium requires mixed strategies.

		PLAYER 2		
		R	P	S
	R	T	L	W
PLAYER 1	P	W	T	L
	S	L	W	T

FIGURE 7.9   Outcomes in the Rock–Paper–Scissors Game

To simplify our calculation of the equilibrium mixes, we associate numerical values with the three possible outcomes (win, lose, and tie). Winning earns a player a 1, losing earns him a –1, and a tie is worth 0. We also extend the original table to allow for each player's mixed strategy, as shown in Figure 7.10. With three possible pure strategies we can no longer have a single strategy played with probability p or q and the other with probability $(1 - p)$ or $(1 - q)$. Rather, we say that Player 1 associates probability p_1 with the pure strategy Rock, and probability p_2 with the pure strategy Paper, leaving a probability of $(1 - p_1 - p_2)$ for the pure strategy Scissors. A similar notation is used for Player 2, who will assign probability q_1 to Rock, q_2 to Paper, and $(1 - q_1 - q_2)$ to Scissors.

		PLAYER 2			
		R	P	S	q-Mix
	R	0	–1	1	$-q_2 + (1 - q_1 - q_2)$
PLAYER 1	P	1	0	–1	$q_1 - (1 - q_1 - q_2)$
	S	–1	1	0	$-q_1 + q_2$
	p-Mix	$p_2 - (1 - p_1 - p_2)$	$-p_1 + (1 - p_1 - p_2)$	$p_1 - p_2$	

FIGURE 7.10   Payoffs in the RPS game with Mixed Strategies

Note that the payoffs in the last row and the last column in Figure 7.10 are determined just as they were in Figures 5.2 and 5.4. The cells in the last row show Player 1's payoffs from using his p-mix against each of Player 2's strategies. The cells in the last column show Player 2's payoffs from using his q-mix against each of Player 1's three pure strategies. As usual, we omit the payoff from the bottom right-hand corner cell because it is simply too long.

Recall that Player 1's equilibrium p-mix guarantees him the same expected payoff against Player 2 using each of his three pure strategies. That is, the equilibrium p-mix keeps $[p_2 - (1 - p_1 - p_2)]$, $[-p_1 + (1 - p_1 - p_2)]$, and $(p_1 - p_2)$ all equal.

To determine actual values for the different probabilities, we take two of the payoffs at a time, equate them, and consider what they tell us. For instance, consider the first two payoffs. When equated, they indicate

$$p_2 - (1 - p_1 - p_2) = -p_1 + (1 - p_1 - p_2)$$
$$p_1 + p_2 = 2/3.$$

This information does not tell us actual values for the probabilities p_1 and p_2, but it does tell us about the relationship between the two. We can move on to another pairing of the payoffs, the second and third, for example, to get a second relation. When equated, those two payoffs show that

$$-p_1 + (1 - p_1 - p_2) = p_1 - p_2$$
$$3p_1 = 1.$$

As it happens, this equation reduces to the specific solution $p_1 = 1/3$. Using this in the earlier relation linking p_1 and p_2, we get $p_2 = 2/3 - p_1 = 1/3$. Finally, $1 - p_1 - p_2 = 1 - 1/3 - 1/3$, which also reduces to $1/3$. The equilibrium mixture for Player 1 entails playing each of Rock, Paper, and Scissors one-third of the time. The symmetry of the game guarantees that $q_1 = q_2 = 1 - q_1 - q_2 = 1/3$ also. (You should verify this result.)

The derivation of the mixed-strategy equilibrium for this particular game was straightforward. We did not make use of a graph, as in Chapter 5, because of the complications arising from the addition of the third pure strategy for each player. While the graph in Figure 5.3 uses two lines to show Seles's success percentage from her mix against Hingis's two pure strategies, a similar construction here would need to be in three dimensions. Player 1's payoffs from his p-mix would depend on the values of both p_1 and p_2 and would need to be represented as planes rather than lines. Such a diagram is not impossible to draw; in fact, we present the three-dimensional diagram for this game in Figure 7.11. But free-hand drawing of such a complex figure is difficult, and graphs become impossible to use when more than three strategies are used in an equilibrium mix. We therefore rely on algebra to find solutions in larger mixing games.

Our analysis of the RPS game assumed that each player used all of his pure strategies in his equilibrium mixture. This assumption entered into the way we set up the prevent-exploitation conditions. We assumed that Player 1's mixture had to be equally effective against each of Player 2's three pure strategies. If Player 2 does not use one of his pure strategies in his equilibrium mix,

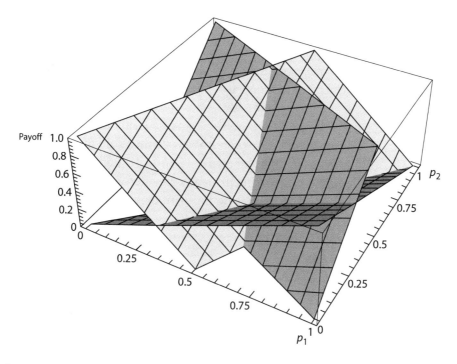

FIGURE 7.11   Graph for the Rock–Paper–Scissors Game

however, then Player 1's mix need not be equally effective against that strategy. In fact, it must be *more* effective: Player 1's mix must yield him a *higher* expected payoff against the other player's unused strategy. To see why, remember that this is a zero-sum game. If Player 2 confronts Player 1's equilibrium mixed strategy with a particular pure strategy that yields Player 1 a higher expected payoff, a lower expected payoff will automatically ensue for Player 2. In fact, that is exactly why Player 2 does not use that pure strategy in his own equilibrium mixture.

Therefore, when we recognize the possibility that some pure strategies can go unused in equilibrium, we must change the system of equations that defines the equilibrium mix into a system of equations and inequalities. Player 1's equilibrium *p*-mix should yield him the same payoff against each of Player 2's pure strategies that Player 2 actually uses in his equilibrium mix, and an even better payoff against those pure strategies that Player 2 does not use in his equilibrium mix. A corresponding system of equations and inequalities applies to Player 2's *q*-mix. Of course, we do not know in advance which strategies are used in equilibrium and which are not; therefore we do not know which payoff expressions go into equations and which ones into inequalities. We have to sort these out as a part of an internally consistent solution, which is what makes the calculation so difficult. We now illustrate all this in a relatively simple example.

B. A Three-by-Three Example with Unused Strategies

To show you how to calculate mixed-strategy equilibria when some strategies go unused, we consider a variant of the football example from Chapter 4. We simplify it a little by removing the Medium Pass strategy from the offense's playbook. Therefore the offense has three pure strategies—Run, Short Pass, and Long Pass—and the defense also has three pure strategies—counter the run, counter a pass, and Blitz. We also change the payoff numbers slightly from those shown in Figure 4.5. The resulting payoff table is shown in Figure 7.12. Remember that this is a zero-sum game, and the payoffs shown are for the offense.

It is easy to verify that the game has no pure-strategy Nash equilibrium. So we look for mixed strategies. We begin by supposing, as in the RPS game, that every strategy is used in the equilibrium mix. Accordingly, we stipulate a p-mix for the offense consisting of using the Run, the Short Pass, and the Long Pass in proportions p_1, p_2, and $(1 - p_1 - p_2)$ respectively, and a q-mix for the defense, where the Run defense, the Pass defense, and Blitz are used in proportions q_1, q_2, and $(1 - q_1 - q_2)$ respectively. The payoffs for each of these mixes against the opponent's pure strategies are as usual shown in the p-mix row and the q-mix column in Figure 7.12.

		DEFENSE			
		Run	Pass	Blitz	q-Mix
OFFENSE	Run	2	8	13	$2q_1 + 8q_2 + 13(1 - q_1 - q_2)$
	Short Pass	5	3	6	$5q_1 + 3q_2 + 6(1 - q_1 - q_2)$
	Long Pass	10	7	-2	$10q_1 + 7q_2 - 2(1 - q_1 - q_2)$
	p-Mix	$2p_1 + 5p_2 + 10(1 - p_1 - p_2)$	$8p_1 + 3p_2 + 7(1 - p_1 - p_2)$	$13p_1 + 6p_2 - 2(1 - p_1 - p_2)$	

FIGURE 7.12 Mixed Strategies in the Football Example

Now we proceed as we did in the RPS example. We state the offense's conditions to prevent exploitation, namely that the p-mix of the offense should yield the offense the same payoff, no matter which pure strategy the defense is playing. This gives us the two equations

$$2p_1 + 5p_2 + 10(1 - p_1 - p_2) = 8p_1 + 3p_2 + 7(1 - p_1 - p_2)$$

and

$$2p_1 + 5p_2 + 10(1 - p_1 - p_2) = 13p_1 + 6p_2 - 2(1 - p_1 - p_2),$$

which simplify to

$$9p_1 + p_2 = 3 \quad \text{and} \quad 23p_1 + 13p_2 = 12,$$

respectively. Then it is not hard to solve the two to find that $p_1 = 27/94$, $p_2 = 39/94$, and then $1 - p_1 - p_2 = 28/94$.

So far so good. Now look at the corresponding conditions for the defense's q-mix. They are

$$2q_1 + 8q_2 + 13(1 - q_1 - q_2) = 5q_1 + 3q_2 + 6(1 - q_1 - q_2)$$

and

$$2q_1 + 8q_2 + 13(1 - q_1 - q_2) = 10q_1 + 7q_2 - 2(1 - q_1 - q_2)$$

or

$$10q_1 + 2q_2 = 7 \quad \text{and} \quad 23q_1 + 14q_2 = 15.$$

These give a final solution of $q_1 = 68/94$, $q_2 = -11/94$, and so $1 - q_1 - q_2 = 37/94$.

Now we have a problem. The solution has a negative value of q_2, which is meaningless, since a probability must be nonnegative. This result implies that there cannot be a mixed-strategy equilibrium in which both players use all three pure strategies.

One could examine all possible combinations of used and unused strategies, and check which one satisfies all the conditions of equilibrium. But there is a better method specific to the structure of this example; in fact, the nature of the difficulty immediately suggests which combination we should first try. Since q_2 is "trying to go negative," let it go as far as it can within the constraints of the problem, namely to zero. That is, suppose the defense does not use its second pure strategy (the Pass defense) in its equilibrium mix. On this supposition, the game reduces to the case where the defense has only two pure strategies, the Run defense and Blitz. The offense in principle could still mix among its three pure strategies. But the game has been simplified to one exactly like the tennis-lob example of Chapter 5. It can be solved by the same method, and we leave this task to you.

You will find that in equilibrium the offense also uses only two strategies, namely Run and Long Pass. It mixes them with probabilities 12/23 and 11/23, respectively; this is its equilibrium p-mix. The defense mixes its Run defense and Blitz with probabilities of 15/23 and 8/23; this is its equilibrium q-mix.

We can then calculate the payoffs of these equilibrium mixtures against the opponent's pure strategies. The result is shown in Figure 7.13, which merely replaces the algebraic expressions for the general mixes of Figure 7.12 with the particular numbers that come from the equilibrium mixes.

		DEFENSE			
		Run	Pass	Blitz	Mix
OFFENSE	Run	2	8	13	5.83
	Short Pass	5	3	6	5.35
	Long Pass	10	7	–2	5.83
	Mix	5.83	7.52	5.83	5.83

FIGURE 7.13 Mixed-Strategy Equilibrium in the Football Example

All the pure strategies that are actually used in the equilibrium mix yield the payoff of 5.83; this is the expected number of yards the offense will gain in a single play. But we have also shown the expected payoffs that the unused strategies *would* yield if they *were* pitted against the opponent's equilibrium mix, and these numbers do not equal 5.83. The offense's unused strategy (Short Pass) would yield only 5.35; that is why the offense does not use it. Conversely, the defense's unused strategy (the Pass defense) would give up an expected 7.52 yards if it were used against the offense's equilibrium mix, which is why the defense does not use it.

These ideas are quite general. A player gets the same expected payoff from all the pure strategies he actually uses in his own mix, when each is pitted separately against the opponent's equilibrium mix. Each of his unused strategies should yield him a lower expected payoff if each were individually pitted against the opponent's equilibrium mix. In the next section, we explain these ideas in a more general algebraic or theoretical way.

5 MIXING AMONG MANY STRATEGIES: GENERAL THEORY

Here we generalize some of the ideas that were introduced in Section 4 as simple examples. Such general theory unavoidably requires some algebra and some abstract thinking. Readers unprepared for, or averse to, such mathematics can omit this section without loss of continuity.

Suppose the row player has available the pure strategies $R_1, R_2, \ldots, R_m$, and the column player has $C_1, C_2, \ldots, C_n$. Write Row's payoff from the strategy combination (i,j) as A_{ij} and Column's as B_{ij}, where the index i ranges from 1 to m, and the index j ranges from 1 to n. We allow each player to mix the available pure strategies. Suppose Row's p-mix has probabilities P_i and Column's q-mix has Q_j.

Of course, all these probabilities must be nonnegative, and each set must add to 1, so

$$P_1 + P_2 + \cdots + P_m = 1 = Q_1 + Q_2 + \cdots + Q_n.$$

We write V_i for Row's expected payoff from using his pure strategy i against Column's q-mix. Using the reasoning we have already employed in several examples, we have

$$V_i = A_{i1}Q_1 + A_{i2}Q_2 + \cdots + A_{in}Q_n = \sum_{j=1}^{n} A_{ij}Q_j,$$

where the last expression on the right is just the summation notation used in algebra. When Row plays his p-mix and it is matched against Column's q-mix, Row's expected payoff is

$$P_1V_1 + \cdots + P_mV_m = \sum_{i=1}^{m} P_iV_i = \sum_{i=1}^{m}\sum_{j=1}^{n} P_iA_{ij}Q_j.$$

The row player will choose his p-mix to maximize this expression.

Similarly, letting W_j be Column's expected payoff when his pure strategy j is pitted against Row's p-mix, we have

$$W_j = P_1B_{1j} + P_2B_{2j} + \cdots + P_mB_{mj} = \sum_{i=1}^{m} P_iB_{ij}.$$

Pitting mix against mix, Column's expected payoff is

$$Q_1W_1 + \cdots + Q_nW_n = \sum_{j=1}^{n} Q_jW_j = \sum_{i=1}^{m}\sum_{j=1}^{n} P_iB_{ij}Q_j,$$

and he will choose his q-mix to maximize this expression.

We will have a Nash equilibrium when simultaneously each player chooses his best mix, given that of the other. That is, Row's p-mix should be his best response to Column's q-mix, and vice versa. Let us begin by finding Row's best-response rule. That is, let us temporarily fix Column's q-mix, and consider Row's choice of his p-mix.

Suppose $V_1 > V_2$. Then Row can increase his payoff by shifting some probability from strategy R_2 to R_1, that is, reducing the probability P_2 of playing R_2 and increasing the probability P_1 of playing R_1 by the same amount. Since the expressions for V_1 and V_2 do not involve any of the probabilities P_i at all, this is true no matter what the original values of P_1 and P_2 were. Therefore Row should reduce the probability P_2 of playing R_2 to as low as possible, namely to zero.

The idea generalizes immediately. Row should rank the V_i in descending order. At the top there may be just one strategy, in which case it should be the only one used; that is, Row should then use a pure strategy. Or there may be a tie among two or more strategies at the top, in which case Row should mix solely among these strategies, and not use any of the others.[16] When there is such a tie,

---

[16] In technical mathematical terms, the expression $\Sigma_i P_i V_i$ is *linear* in P_i; therefore its maximum must occur at an extreme point of the set of permissible P_i.

all mixtures between these strategies gives Row the same expected payoff. There-fore this consideration alone does not serve to fix Row's equilibrium p-mix. We will see later how, in a way that may seem somewhat strange at first, Column's in-difference condition does that job.

In general, for most values of Q_1, Q_2, ... Q_n that we hold fixed in Column's q-mix, Row's V_1, V_2, ..., V_m will not have any ties for the highest value, and therefore Row's best response will be a pure strategy. We saw this in Chapter 5, for example in Figures 5.6, 5.9, and 5.12. For most values of p in Row's p-mix, the best value of q for Column was either 0 or 1, and vice versa. For only one critical value of Row's p-mix was it optimal for Column to mix (that is, choose any value of q between 0 and 1), and vice versa.

The same argument applies to Column. He should use only the pure strategy that gives him the highest W_j, or mix only among those of his pure strategies C_j whose payoffs W_j are tied for the top value. If there is such a tie, then all mixtures are equally good from Column's perspective; the probabilities of the mix are not fixed by this consideration alone.

All this constitutes the complicated set of equations and inequalities that, when simultaneously satisfied, defines the mixed-strategy Nash equilibrium of the game. To understand it better, suppose for the moment that we have done all the work and found which strategies are used in the equilibrium mix. We can al-ways relabel the strategies so that Row uses, say, the first g pure strategies, R_1, R_2, ..., R_g, and does not use the remaining $(m - g)$ pure strategies, R_{g+1}, R_{g+2}, ..., R_m, while Column uses his first h pure strategies, C_1, C_2, ..., C_h, and does not use the remaining $(n - h)$ pure strategies, C_{h+1}, C_{h+2}, ... C_n. We write V for the tied value of Row's top expected payoffs V_i, and similarly W for the tied value of Column's top expected payoffs W_j. Then the equations and inequalities can be written as follows. First, for each player we set the probabilities of the unused strategies equal to zero and require those of the used strategies to sum to 1:

$$P_1 + P_2 + \cdots + P_g = 1, \qquad P_{g+1} = P_{g+2} = \cdots = P_m = 0 \tag{7.1}$$

and

$$Q_1 + Q_2 + \cdots + Q_h = 1, \qquad Q_{h+1} = Q_{h+2} = \cdots = Q_n = 0. \tag{7.2}$$

Next we set Row's expected payoffs for the pure strategies he uses equal to the top tied value,

$$V = A_{i1} Q_1 + A_{i2} Q_2 + \cdots + A_{ih} Q_h \quad \text{for } i = 1, 2, \cdots, g, \tag{7.3}$$

and note that his expected payoffs from his unused strategies must be smaller (which is why they are unused):

$$V > A_{i1} Q_1 + A_{i2} Q_2 + \cdots + A_{ih} Q_h \quad \text{for } i = g + 1, g + 2, \cdots, n. \tag{7.4}$$

Next we do the same for Column, writing W for his top tied payoff value:

$$W = P_1 B_{1j} + P_2 B_{2j} + \cdots + P_g B_{gj} \quad \text{for } j = 1, 2, \cdots, h, \quad (7.5)$$

and

$$W > P_1 B_{1j} + P_2 B_{2j} + \cdots + P_g B_{gj} \quad \text{for } j = h + 1, h + 2, \cdots, n. \quad (7.6)$$

To find the equilibrium, we must take this whole system, regard the choice of g and h as well as the probabilities $P_1, P_2, \ldots, P_g$, and $Q_1, Q_2, \ldots, Q_h$ as unknowns, and attempt to solve for them.

There is always the exhaustive search method. Try a particular selection of g and h, that is, choose a particular set of pure strategies as candidates for use in equilibrium. Then take Eqs. (7.1) and (7.5) as a set of $(h + 1)$ simultaneous linear equations regarding $P_1, P_2, \ldots, P_g$ and W as $(g + 1)$ unknowns, solve for these unknowns and check to see whether the solution satisfies all the inequalities in Eqs. (7.6). Similarly, take Eqs. (7.2) and (7.3) as a set of $(g + 1)$ simultaneous linear equations in the $(h + 1)$ unknowns $Q_1, Q_2, \ldots, Q_h$ and V, solve for them, and check whether the solution satisfies all the inequalities in Eq. (7.4). If all these things check out, we have found an equilibrium. If not, take another selection of pure strategies as candidates, and try again. There is only a finite number of selections—there are $(2^m - 1)$ possible selections of pure strategies that can be used by Row in his mix, and $(2^n - 1)$ possible selections of pure strategies that can be used by Column in his mix. Therefore the process must end successfully after a finite number of attempts.

When m and n are reasonably small, exhaustive search is manageable. Even then, shortcuts suggest themselves in the course of calculating each specific problem. Thus in our earlier three-by-three football example, the *way* the attempted solution using all strategies failed told us which strategy to discard in the next attempt.

Even for moderately large problems, however, a solution based on exhaustive search or ad hoc methods becomes too complex. That is when one must resort to more systematic computer searches or algorithms. What these computer algorithms do is to search simultaneously for solutions to two linear maximization (or *linear programming*) problems: given a q-mix, and therefore all the V_i values, choose a p-mix to maximize Row's expected payoff $\Sigma_i P_i V_i$, and given a p-mix, and therefore all the W_j values, choose a q-mix to maximize Column's expected payoff $\Sigma_j Q_j W_j$. However, for a typical q-mix, all the V_i values will be unequal. If Column were to play this q-mix in an actual game, Row would not mix strategies, but would instead play just the one pure strategy that gave him the highest V_i. But in our numerical solution method we should not adjust Row's strategy in this drastic fashion. If we did, then at the next step of our algorithm, Column's best q-mix would also change drastically, and Row's chosen pure strategy would no longer look so good. Instead, the algorithm should take a more gradual step, adjusting the p-mix a little bit to improve Row's expected payoff.

Then, using this new p-mix for Row, the algorithim should adjust Column's q-mix a little bit to improve his expected payoff. Then back again to another adjustment in the p-mix. The method proceeds in this way until no improvements can be found; that is the equilibrium.

We will not need to know the details of such procedures, but the general ideas which we have developed above already tell us a lot about equilibrium. Here are some important lessons of this kind:

1. We solve for Row's equilibrium mix probabilities P_1, P_2, ..., P_g from Eqs. (7.1) and (7.5). The former is merely the adding-up requirement for probabilities. The more substantive equation is (7.5), which gives the conditions under which Column will get the same payoff from all the pure strategies he uses. This seems puzzling. Why should Row adjust his mix so as to keep Column indifferent? Row is concerned only about his own payoffs, not Column's. Actually the puzzle is only apparent. We derived those conditions (7.5) by thinking about Column's choice of his q-mix, motivated by concerns about his own payoffs. We argued that Column would use only those strategies that gave him the best (tied) payoffs against Row's p-mix. This is the requirement embodied in Eqs. (7.5). Even though it appears *as if* Row is deliberately choosing his p-mix so as to keep Column indifferent, the actual force that produces this outcome is Column's own purposive strategic choice.

The general idea that each player's indifference conditions constitute the equations that can be solved for the other player's equilibrium mix is the basis of the keep-the-opponent-indifferent approach of Chapter 5. We claimed then that it worked for general games—zero-sum and non-zero-sum—and now we see how.

2. However, in the zero-sum case the idea that each player chooses his mixture to keep the other indifferent is not just an *as if* matter; there is a genuine reason why a player should behave in this way. When the game is zero-sum, we have a natural link between the two players' payoffs: $B_{ij} = -A_{ij}$ for all i and j, so similar relations hold among all the combinations and expected payoffs too. Therefore we can multiply Eqs.(7.5) and (7.6) by -1 to express them in terms of Row's payoffs rather than Column's. We write these "zero-sum versions" of the conditions as Eqs. (7.5z) and (7.6z):

$$V = P_1 A_{1j} + P_2 A_{2j} + \cdots + P_g A_{gj} \quad \text{for } j = 1, 2, \cdots, h \tag{7.5z}$$

and

$$V < P_1 A_{1j} + P_2 A_{2j} + \cdots + P_g A_{gj} \quad \text{for } j = h+1, h+2, \cdots, n. \tag{7.6z}$$

Note that multiplying by -1 to go from (7.6) to (7.6z) reverses the direction of the inequality.

Of these, Eqs. (7.5) and (7.5z) tell us that so long as Row is using his equilibrium mix, Column (and therefore Row, too, in this zero-sum game) gets the same payoff from any of the pure strategies that he actually uses in equilibrium. Column cannot do any better for himself—and therefore in this zero-sum game cannot cause any harm to Row—by choosing one of those strategies rather than another. What is more, (7.6z) tells us that were Column to use any of the other strategies, Row would do even better. In other words, these conditions tell us that Row's equilibrium mix cannot be exploited by Column. Thus we see how the prevent-exploitation method of Chapter 5 arises, and why it works only for zero-sum games.

3. Now we return to the general, non-zero-sum, case. Note that the system comprised of (7.1) and (7.5) has $(h + 1)$ linear equations and $(g + 1)$ unknowns. In general, such a system has no solution if $h > g$, has exactly one solution if $h = g$, and has many solutions if $h < g$. Conversely, the system (7.2) and (7.3) has $(g + 1)$ linear equations and $(h + 1)$ unknowns. In general, such a system has no solution if $g > h$, has exactly one solution if $g = h$, and has multiple solutions if $g < h$. Since in equilibrium we want both systems to be satisfied, in general we need $g = h$. Thus in a mixed-strategy equilibrium the two players use equal numbers of pure strategies.

We keep on saying "in general" because exceptions can arise for fortuitous combinations of coefficients and right-hand sides of equations. In particular, it is possible for too many equations in too few unknowns to have solutions. This happened in one case of the Seles–Hingis tennis-point-with-lob game in Chapter 5. In Figure 5.20, the three lines representing Monica Seles's success probabilities as a function of Martina Hingis's mix happened to intersect in a common point. Then it was possible that Hingis would use all three pure strategies. That is, the two probabilities in Hingis's mix could simultaneously satisfy three equations. Moreover, in that case Seles's mix was indeterminate. Now we see how that fits with the general theory. Suppose the equilibrium has $h > g$; it is possible for the too few strategies in Row's mix to satisfy the too many equations of Column's indifference. But then the other system has multiple solutions. Column's mix has too many variables in relation to the number of equations required for Row's indifference. There are degrees of freedom, and Column's mix is indeterminate.

4. We observe a very particular relation between the use of strategies and their payoffs. Row uses strategies P_1 to P_g with positive probabilities, and Eq. (7.3) shows that he gets exactly the payoff V when any one of these pure strategies is played against Column's equilibrium mix. For the remaining pure strategies in Row's armory, Eq. (7.4) shows that they yield a payoff less than V when played against Column's equilibrium mix, and then they are not used; that is, P_{h+1} to P_m are all zero. In other words, for any i, it is *impossible* to have

$$\text{both} \quad V > A_{i1} Q_1 + A_{i2} Q_2 + \cdots + A_{ih} Q_h \quad \text{and} \quad P_i > 0.$$

At least one of these inequalities must collapse into equality. This is known as the principle of **complementary slackness,** and it is of great importance in the general theory of games and equilibria, as well as in mathematical optimization (programming).

5. Back to the zero-sum case. When both players choose their equilibrium mix, Row's expected payoff is

$$V = \sum_{i=1}^{m} \sum_{j=1}^{n} P_i A_{ij} Q_j,$$

and Column's expected payoff is just the negative of this. Moreover, the equilibrium comes about when Row, for the given q-mix, chooses his p-mix to maximize this expression, and simultaneously Column, for the given p-mix, chooses his q-mix to maximize the negative of the same expression, or to minimize the same expression. If we regard the expression as a function of all the P_i and the Q_j, therefore, and graph it in a sufficiently high dimensional space, it will look like a saddle. The front-to-back cross section of a saddle is shaped like a valley or a U, with its minimum at the middle, while the side-to-side cross section looks like a peak or an inverted U, with its maximum at the middle. If each player has just two pure strategies, the p-mix and q-mix can each be described by a single number, say the probability of choosing the first strategy. (Those of the second player are then just 1 minus that of the first.) We can then draw a graph in three dimensions, where the x- and y-axes lie in a horizontal plane and the z-axis points vertically upward. The p-mix is shown along the x-axis, the q-mix along the y-axis, and the value V along the z-axis. The cross section of this graph along the x direction will show the maximization of V with respect to p, therefore a peak. And the cross section along the y direction will show the minimization of V with respect to q, therefore a valley. Thus the graph will look like a saddle, as illustrated in Figure 7.14. Such an equilibrium is called a **saddle point.**

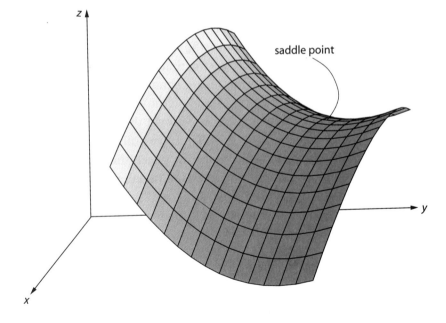

FIGURE 7.14 Saddle Point

The value of V in equilibrium—that is, the simultaneous maximum with respect to the P_i and the minimum with respect to the Q_j—is called the *minimax* value of the zero-sum game. The idea of such an equilibrium and the formulation of the conditions like (7.5z) and (7.6z) that define it were the first important achievement of game theory, and appeared in the work of John von Neumann and Oscar Morgenstern in the 1940s. It is called their minimax theorem.

SUMMARY

Despite the considerable appeal of the Nash equilibrium concept, numerous criticisms of the concept have come from both experimentalists and theorists. The debate has asked whether the Nash concept is too imprecise, whether it requires too much calculation, whether it can be based solely on rationality, whether it accounts properly for risks, and whether the expected payoff-maximization assumption is reasonable. You should have considerable confidence in using the Nash concept, however, especially in games played frequently by a stable set of players under a relatively unchanging set of rules. Even in other circumstances, you should use it as the central starting point for your analysis.

Tennis can be used as a metaphor for many games in which players must attain both an absolute level of achievement and a certain margin of superiority over rivals. The structure of a tennis game permits calculation of the probabilities of a win, for either player, from any point in the game. These probabilities can be used to determine when it is crucial for a player to win the next point—when rivals are closely matched and the game is almost over or when the rivals are unevenly matched but the weaker one has forged ahead.

The analysis of games with continuous strategies, from Chapter 4, can be extended to more complex situations if we use calculus. Also, the prevent-exploitation method of Chapter 5 can be extended to find equilibrium mixtures in games where players have more than two pure strategies. When the conditions for preventing exploitation lead to outcomes for probabilities that are negative or larger than 1, some pure strategies must remain unused in the equilibrium mixed strategy. Players must get the same expected payoff from all of the pure strategies used in their mixture when using those strategies against the opponent's equilibrium mix; strategies unused in a player's equilibrium mix must yield less when used against the opponent's equilibrium mix. This theory can be generalized for games of any size, zero-sum or non-zero-sum.

complementary slackness (245)   saddle point (246)
rationalizable (217)

EXERCISES

1. Recall the game from Exercise 11 in Chapter 4: Three contestants (A, B, and C) are participating in a game with a prize worth $30. Each can buy a ticket worth $15 or $30, or buy no ticket at all. They make these choices simultaneously and independently. Then, knowing the ticket-purchase decisions, the game organizer awards the prize. If none of the contestants has bought a ticket, the prize is not awarded. If only one has bought a ticket, that player gets the prize. If two or more have bought tickets, then the one with the highest-cost ticket gets the prize if there is only one such contestant. If there are two or three contestants, all having bought highest-cost tickets, the prize is split equally among them.

   Find a symmetric equilibrium in mixed strategies for this game, where each contestant mixes his ticket-purchase strategies and the mixture probabilities are the same for all three contestants.

2. Recall the game from Exercise 7 in Chapter 6 involving ice-cream vendors on the beach. Construct the five-by-five table for the game, and find all Nash equilibria in mixed strategies. Explain why some pure strategies are unused in the equilibrium mixtures, and verify (using the opponent's equilibrium mixture) that they should not be included.

3. Consider a variant of Chicken in which each driver can go straight, swerve to his left, or swerve to his right. If one driver swerves to his left and the other swerves to his (own) right, then the cars will collide, just as they will if both drivers go straight. Thus the payoff table becomes:

		DEAN		
		Left	Straight	Right
JAMES	Left	0,0	−1,1	−2,−2
	Straight	1,−1	−2,−2	1,−1
	Right	−2,−2	−1,1	0,0

Show that the game has a mixed-strategy equilibrium in which each player chooses Left with a probability of 1/6, Straight with a probability of 2/3, and Right with a probability of 1/6.

Remember that in the standard (Straight, Swerve) version of Chicken (whose payoff table can be found by omitting the last row and the last column of the above three-by-three table and replacing Left by Swerve), the probability that a player chooses Straight is 1/2. What is the intuitive reason for the probability of Straight being larger in this three-by-three version?

4. Two French aristocrats, Chevalier Chagrin and Marquis de Renard, fight a duel. Each has a pistol loaded with one bullet. They start 10 steps apart and walk toward each other at the same pace, one step at a time. After each step, either may fire his gun. When one shoots, the probability of scoring a hit depends on the distance; after k steps it is $k/5$, so it rises from 0.2 after the first step, to 1 (certainty) after 5 steps, when they are right up against each other. If one player fires and misses while the other has yet to fire, the walk must continue even though the bulletless one now faces certain death; this is dictated by the code of the aristocracy. Each gets a payoff of −1 if he himself is killed and 1 if the other is killed. If neither or both are killed, each gets 0.

   This is a game with five sequential steps, and simultaneous moves (shoot or not shoot) at each step. Find the rollback (subgame-perfect) equilibrium. Begin at step 5, where they are right up against each other. Set up the two-by-two table of the simultaneous-move game at this step, and find its Nash equilibrium. Now move back to step 4, where the probability of scoring a hit is 4/5, or 0.8, for each. Set up the two-by-two table of the simultaneous-move game at this step, correctly specifying what happens in the future. For example, if one shoots and misses while the other does not shoot, then the other will wait until step 5 and score a sure hit; if neither shoots, then the game will go to the next step, for which you have already found the equilibrium. Using all this information, find the payoffs in the two-by-two table of step 4, and find the Nash equilibrium at this step. Work backward in the same way. Show that in the outcome of the subgame-perfect equilibrium, both will shoot at step 3.

5. Think of an example of business competition that is similar in structure to the duel in Exercise 4.

6. You can think of a penalty kick in soccer as a two-person zero-sum game between the shooter and the goalkeeper. If the shooter can conceal his intentions of where he will shoot, and the goalie can similarly hide which way he will move to block the shot until the last fraction of a second, then the game effectively has simultaneous moves. Suppose the shooter has six distinct strategies: to shoot high and to the left (HL), low and to the left (LL), high and in the center (HC), low and in the center (LC), high right (HR), and low right (LR). The goalkeeper has three distinct strategies: to move to the shooter's left (L), to move to the shooter's right (R), or to stand facing straight (C). The probabilities of the shooter's success are shown in the following table. If we give

the shooter the payoff 1 when he succeeds and 0 when he does not, the table also shows the shooter's expected payoffs.

		GOALKEEPER		
		L	C	R
SHOOTER	HL	0.50	0.85	0.85
	LL	0.40	0.95	0.95
	HC	0.85	0.00	0.85
	LC	0.70	0.00	0.70
	HR	0.85	0.85	0.50
	LR	0.95	0.95	0.40

The explanation of these probabilities is as follows: Shooting high runs some risk of missing the goal even if the goalkeeper goes the wrong way; that is why $0.85 < 0.95$. If the goalkeeper guesses correctly, he has a better chance of collecting or deflecting a low shot than a high one; that is why $0.40 < 0.50$. And if the shot is to the center while the goalkeeper goes to one side, he has a better chance of using his feet to deflect a low shot than a high one; that is why $0.70 < 0.85$.

Verify that in the mixed-strategy equilibrium of this game, the goalkeeper uses L and R 42.2% of the time each, and C 15.6% of the time, while the shooter uses LL and LR 37.8% of the time each, and HC 24.4% of the time. Also verify that the value of the game (the probability of the shooter's success) is 71.8%. (This conforms well with the observation of three to four successes out of every five attempts in recent Soccer World Cup penalty shoot-outs.)

Note that to verify the claim, all you have to do is to take the given numbers and show that they satisfy the conditions of equilibrium, namely Eqs. (7.3) to (7.6). But if you have access to a computer program that calculates mixed-strategy equilibria, use it to obtain the equilibrium mixture from scratch. If you can, experiment by changing some of the probability numbers in the table, and see how these changes alter the equilibrium mixtures and the value of the game.

7. [Optional] Now suppose the duelists from Exercise 4 have guns with silencers. If one player fires and misses, the other does not know this has happened, and cannot follow the strategy of then holding his fire until the final step to get a sure shot. Each must formulate a strategy at the outset that is not conditional on the other's intermediate actions. Thus we have a simultaneous-move game, with strategies of the form "Shoot after n steps if still alive."

Each player has five such strategies corresponding to the five steps. Show that the five-by-five payoff table of this game in strategic form is as follows:

		CHAGRIN				
		1	2	3	4	5
RENARD	1	0.00	−0.12	−0.28	−0.44	−0.60
	2	0.12	0.00	0.04	−0.08	−0.20
	3	0.28	−0.04	0.00	0.28	0.20
	4	0.44	0.08	−0.28	0.00	0.60
	5	0.60	0.20	−0.20	−0.60	0.00

Verify that there is a mixed-strategy Nash equilibrium with the following strategy for each player: strategies 1 and 4 are unused, while strategies 2, 3, and 5 are used in the proportions 5/11, 5/11, and 1/11, respectively. That is, check that these numbers satisfy the conditions for Nash equilibrium in mixed strategies of a zero-sum game, namely Eqs. (7.3) through (7.6).

If you have access to a computer program that calculates mixed-strategy equilibria, do not merely check the given answer, but actually obtain the answer using the program. What is the expected payoff for each player in equilibrium?

8. [Optional] Suppose the probability that the server wins a single point in tennis is x. Find the probability that the player who serves to start a tie-breaker game wins it. Remember that to win a tie-breaker a player has to have won seven or more points with a margin of two or more. The one who opens serves one point, then the other serves two, and they continue to alternate serving two points each.

PART THREE

■

Some Broad
Classes of Games
and Strategies

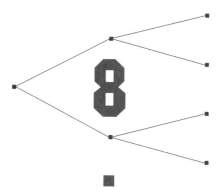

The Prisoners' Dilemma Game

WE HAVE NOW COMPLETED the task of equipping you with a working knowledge of the basic concepts and techniques used in analyzing games of strategy. These tools will be invaluable to you as you proceed through Chapters 8 through 12. In these chapters, we present some broad classes, or categories, of games that arise frequently in actual strategic interactions. We will draw heavily on the material presented earlier and also develop new techniques of analysis as needed.

The first broad class of games we consider is the prisoners' dilemma. It is probably *the* classic example of the theory of strategy and its implications for predicting the behavior of game players, and most people who learn only a little bit of game theory learn about it. Even people who know *no* game theory may know the basic story behind this game or they may have at least heard that it exists. The prisoners' dilemma is a game in which each player has a dominant strategy, but the equilibrium that arises when all players use their dominant strategy provides a worse outcome for every player than would arise if they all used their dominated strategy instead. The paradoxical nature of this equilibrium outcome leads to several more complex questions about the nature of the interactions involved that only a more thorough analysis can hope to answer. The purpose of this chapter is to provide that additional thoroughness.

We already discussed the prisoners' dilemma in Section 4 of Chapter 4. There we took note of the curious nature of the equilibrium that is actually a "bad" outcome for the players. The "prisoners" can find another outcome they both prefer to the equilibrium outcome, but they find it difficult to bring about. The focus of this chapter is the potential for achieving that better outcome. That is, we consider whether and how the players in a prisoners' dilemma can attain and sustain their mutually beneficial cooperative outcome, overcoming their separate incentives to cheat for individual gain. We first review the standard prisoners' dilemma game and then develop three categories of solutions. The first and most important method of solution involves repetition of the standard one-shot game. We then discuss solutions that involve punishment schemes and the role of leadership. As we consider each potential solution, the importance of the costs of cheating and the benefits of cooperation will become clear.

This chapter concludes with a discussion of some of the experimental evidence regarding the prisoners' dilemma as well as several examples of actual dilemmas in action. Experiments generally put live players in a variety of prisoners' dilemma–type games and show some perplexing as well as some more predictable behavior; experiments conducted using computer simulations yield additional interesting outcomes. Our examples of real-world dilemmas that end the chapter are provided to give a sense of the diversity of situations in which prisoners' dilemmas arise and to show how, in at least one case, players may be able to create their own solution to the dilemma.

1 THE BASIC GAME (REVIEW)

Before we dive into discussing methods for avoiding the "bad" outcome in the prisoners' dilemma, we briefly review the basics of the game. Recall our example from Chapter 4 involving the husband and wife suspected of murder. Each is interrogated separately and can choose to confess to the crime or to deny any involvement. The payoff matrix that they face is shown in Figure 8.1. The numbers shown indicate years in jail, so low numbers are better for both players.

Both players here have a dominant strategy. Each does better to confess, regardless of what the other player does. The equilibrium outcome entails both players deciding to confess and each getting 10 years in jail. If they both chose to deny any involvement, however, they would have been better off, with only three years of jail time to serve.

		WIFE	
		Confess	Deny
HUSBAND	Confess	10 yr, 10 yr	1 yr, 25 yr
	Deny	25 yr, 1 yr	3 yr, 3 yr

FIGURE 8.1 Payoffs for the Standard Prisoners' Dilemma

In any prisoners' dilemma game, there is always a *cooperative strategy* and a *cheating* or *defecting strategy*. In Figure 8.1, Deny is the cooperative strategy; both players using that strategy yields the best outcome. Confess is the cheating or defecting strategy; when the players do not cooperate with each other, they choose to Confess in the hopes of attaining individual gain at the rival's expense. Thus, players in a prisoners' dilemma can always be labeled, according to their choice of strategy, as either *cheaters* or *cooperators*. We will use this labeling system throughout our discussion of potential solutions to the dilemma.

2 SOLUTIONS I: REPETITION

Of all the mechanisms that can sustain cooperation in the prisoners' dilemma, the most well known and the most natural is **repeated play** of the game. Repeated or ongoing relationships between players imply special characteristics for the games they play against each other. In the prisoners' dilemma, this result plays out in the fact that each player fears that one instance of cheating will lead to a collapse of cooperation for the future. If the value of future cooperation is large and exceeds what can be gained in the short term by cheating, then the long-term individual interest of the players can automatically and tacitly keep them from cheating, without the need for any additional punishments or enforcement by third parties.

We consider here the dilemma faced by the two pizza stores originally introduced in Chapter 4. In that game, Pierce's Pizza Pies and Donna's Deep Dish set the price of a pizza at one of three levels: High, which yields a profit margin of $12 per pie; Medium, which yields a profit of $10 per pie; and Low, which yields only $5 per pie. Our focus here is on the new issue of cooperation, and to explain that in the simplest possible way, we omit the Low strategy. This leaves us the two-by-two table shown in Figure 8.2. Recall that the payoff numbers (in thousands of dollars) are based on the assumption that each store has a loyal customer base that purchases 3,000 pies per week, regardless of price, and that the stores compete for a remaining 4,000 pies' worth of floating demand. The floating demand goes to the lower-priced store, or is split equally between the two stores if they charge the same price. The dominant strategy for each store is to price medium, but both get a lower profit in that case than if they had been able to cooperate and both charge high prices.

Let us start our analysis by supposing that the two stores are initially in the cooperative mode, each charging a high price. If one store, say Pierce's, deviates from this pricing strategy, it can increase its profit from 60 to 70 ($60,000 to $70,000) for one week. But then cooperation has dissolved and Pierce's rival, Donna's, will see no reason to cooperate from then on. Once cooperation has

		PIERCE'S PIZZA PIES	
		High	Medium
DONNA'S DEEP DISH	High	60, 60	36, 70
	Medium	70, 36	50, 50

FIGURE 8.2 Pizza Stores in a Prisoners' Dilemma ($000)

broken down, the profit for Pierce's is 50 each week instead of the 60 it would have been if Pierce's had never cheated in the first place. By gaining 10 in one week of cheating, Pierce's gives up 10 each week thereafter by destroying cooperation. Even if the relationship lasts as little as three weeks, it seems that cheating is not in Pierce's best interests. A similar argument can be made for Donna's. Thus if the two stores competed on a regular basis for at least three weeks, it seems we might see cooperative behavior and high prices rather than the cheating behavior and medium prices predicted by theory for the one-shot game.

A. Finite Repetition

But the solution of the dilemma is not actually that simple. What if the relationship did last exactly three weeks? Then strategic pizza stores would want to analyze the full three-week game and choose their optimal pricing strategies. Each would use rollback to determine what price to charge each week. Starting their analysis with the third week, they would realize that at that point there was no future relationship to consider. Each store would find that it had a dominant strategy to cheat. Given that, there is effectively no future to consider in the second week either. Each player knows that there will be mutual cheating in the third week, and therefore both will cheat in the second week; cheating is the dominant strategy in week 2 also. Then the same argument applies to the first week as well. Knowing that both will cheat in weeks 2 and 3 anyway, there is no future value of cooperation in the first week. Both players cheat right from the start, and the dilemma is alive and well.

   This result is very general. As long as the relationship between the two players in a prisoners' dilemma game lasts a fixed and known length of time, the dominant-strategy equilibrium with cheating should prevail in the last week. Once the players arrive at the end of the game, there is never any value to continued cooperation, so cheating occurs. Then rollback predicts mutual cheating all the way back to the very first play. However, in practice, finitely repeated prisoners' dilemma games show a lot of cooperation; more on this to come.

B. Infinite Repetition

Analysis of the finitely repeated prisoners' dilemma shows that even repetition of the game cannot guarantee the players a solution to their dilemma. But what would happen if the relationship did not have a predetermined length? What if the two stores expected to continue competing with each other indefinitely? Then our analysis must change to incorporate this new aspect of their interaction, and we will see that the incentives of the players change also.

In repeated games of any kind, the sequential nature of the relationship means that players can adopt strategies that depend on behavior in previous plays of the games. Such strategies are known as **contingent strategies** and several specific examples are used frequently in the theory of repeated game. Most contingent strategies are **trigger strategies.** A player using a trigger strategy plays cooperatively as long as her rival(s) do so, but any defection on their part "triggers" a period of **punishment,** of specified length, in which she plays noncooperatively in response. Two of the best-known trigger strategies are the grim strategy and tit-for-tat. The **grim strategy** entails cooperating with your rival until such time as she defects from cooperation; once a defection has occurred, you punish your rival (by choosing the Cheat, or Defect, strategy) on every play for the rest of the game.[1] **Tit-for-tat (TFT)** is not as harshly unforgiving as the grim strategy and is famous (or infamous) for its ability to solve the prisoners' dilemma without requiring permanent punishment. Playing TFT means cooperating with your rival if she cooperated during the most recent play of the game and cheating if your rival cheated during the most recent play of the game; thus the punishment phase lasts only as long as your rival continues to cheat. The behavior described earlier, in which cooperation between the pizza stores broke down after an incident of cheating, is an example of how play might proceed when players use trigger strategies like TFT.

Let us consider how play might proceed in the repeated pizza store game if one of the players uses the contingent strategy tit-for-tat. We have already seen that if Pierce's Pizza Pies cheats one week, it could add 10 to its profits (70 instead of 60). But if Pierce's rival is playing TFT, then such cheating would induce Donna's Deep Dish to punish the next week in retaliation. At that point, Pierce's has two choices. It can continue to cheat by pricing medium, and Donna's will continue to punish according to TFT; in this case, Pierce's loses 10 (50 rather than 60) for every week thereafter in the foreseeable future. Pierce's *could* get back to cooperation too, if it so desired. By cooperating the week after it cheated, Pierce's would suffer a loss in profit of 24 while it was being punished (36 rather than the 60 that would have been earned without any cheating). But Donna's, committed

---

[1] Cheating as retaliation under the requirements of a trigger strategy is often termed *punishing,* to distinguish it from the original decision to deviate from cooperation.

to TFT, would return to cooperating the following week, and both stores would return to earning 60 each week. Given that Pierce's has cheated once already and must be punished by Donna's, neither continued cheating nor an immediate return to cooperation seems to make the original gain worthwhile.

It is important to realize here, however, that Pierce's extra $10 from cheating is gained in the first week. Its losses are ceded in the future. Therefore the relative importance of the two depends on the relative importance of the present versus the future. Here, because payoffs are calculated in dollar terms, an objective comparison can be made. Generally, money (or profit) that is earned today is better than money that is earned later because, even if you do not need (or want) the money until later, you can invest it now and earn a return on it until you need it. So Pierce's should be able to calculate whether it is worthwhile to cheat, based on the total rate of return on its investment, r (including capital gains and/or dividends and/or interest depending on the type of investment).

Note that we can calculate whether it is in Pierce's interest to cheat because the firms' payoffs are given in dollar terms, and not as simple ratings of outcomes, as with some of the games in earlier chapters (the mall location game in Chapters 3 and 6, for example). This means that payoff values in different cells are directly comparable; a payoff of 4 (dollars) is twice as good as a payoff of 2 (dollars) here, whereas a payoff of 4 is not necessarily exactly twice as good as a payoff of 2 in any two-by-two game in which the four possible outcomes are ranked from 1 (worst) to 4 (best). As long as the payoffs to the players are given in measurable units, the calculations necessary to determine whether it is worthwhile to cheat in a prisoners' dilemma game can be completed.

I. IS IT WORTHWHILE TO CHEAT ONLY ONCE AGAINST A RIVAL PLAYING TFT?   One of Pierce's options when playing against a rival using TFT is to cheat just once after a cooperative start and then to return to cooperating. This particular strategy gains the store 10 in the first week (the week during which it cheats) but loses it 24 in the second week. By the third week, cooperation is restored. Is cheating for only one week worth it?

We cannot directly compare the 10 gained in the first week with the 24 lost in the second week because the additional money value of *time* must be incorporated into the calculation. That is, we need a way to determine how much the 24 lost in the second week is worth during the first week. Then we can compare that number to 10 to see whether cheating once is worthwhile. What we are looking for is the **present value (PV)** of 24 or, how much in profit earned this week (in the *present*) is equivalent to (has the same *value* as) 24 earned next week. That is, we need to determine the number of dollars earned today that, with interest, would give us 24 next week; we call that number PV, the present value of 24.

This value solves the expression

$$PV + r\,PV = 24.$$

Given that the (weekly) total rate of return is r, getting PV today and investing it until next week yields a total next week of $PV + r\,PV$. When that total is exactly 24, then PV equals the present value of 24. The expression above simplifies to $PV(1 + r) = 24$, or:

$$PV = \frac{24}{1 + r}.$$

For any value of r, we can now determine the exact number of dollars that, earned today, would be worth 24 next week.

From the perspective of Pierce's Pizza Pies, the question remains whether the gain of 10 this week is offset by the loss of 24 next week. The answer depends on the value of PV. Pierce's must compare the gain of 10 with the PV of the loss of 24. It is worthwhile to cheat once (and then return to cooperation) only if $10 > 24/(1 + r)$. This is the same as saying that cheating once is beneficial only if $10(1 + r) > 24$, which reduces to $r > 1.4$. Thus Pierce's Pizza Pies should only choose to cheat once against a rival playing TFT if the weekly total rate of return exceeds 140%. Obviously this outcome is very unlikely; given the payoff structure in this particular game, it is in Pierce's better interest to continue cooperating than to try a single instance of cheating when Donna's is playing TFT.

II. IS IT WORTHWHILE TO CHEAT FOREVER AGAINST A RIVAL PLAYING TFT? What about the possibility of cheating once and then continuing to cheat forever? This second option of Pierce's gains the store 10 in the first week but loses it 10 in every week thereafter into the future if the rival store plays TFT. To determine whether such a strategy is in Pierce's best interest again depends on the present value of the losses incurred. But this time the losses are incurred over an **infinite horizon** of future weeks of competition.

We need to figure out the present value of all of the 10s that are lost in future weeks, add them all up, and compare them with the 10 gained during the week of cheating. The PV of the 10 lost during the first week of continued cheating is just $10/(1 + r)$; the calculation is identical to that used in Section 2.B.I. to find that the PV of 24 was $24/(1 + r)$. For the following week, the PV must be the dollar amount needed today that, with two weeks of **compound interest,** would yield 10 in two weeks. If the PV is invested today, then in one week the investor would have that principal amount plus a return of $r\,PV$, for a total of $PV + r\,PV$, as before; leaving this total amount invested for the second week means that the investor, at the end of two weeks, has the amount invested at the beginning of the second week $(PV + r\,PV)$ plus the return on that amount, which would be $r(PV + r\,PV)$. The PV of the 10 lost two weeks from now must then solve the equation: $PV + r\,PV + r(PV + r\,PV) = 10$. Working out the value of PV here yields $PV(1 + r)^2 = 10$, or $PV = 10/(1 + r)^2$. You should see a pattern developing. The PV of the 10 lost in the third week of continued cheating is $10/(1 + r)^3$, and the PV of the 10 lost in the fourth week is $10/(1 + r)^4$. In fact, the PV of the 10 lost in the nth week of continued cheating is just $10/(1 + r)^n$. Pierce's loses an infinite sum of 10s, and the PV of each of those gets smaller each week.

More precisely, Pierce's loses the sum, from $n = 1$ to $n = \infty$ (where n labels the weeks of continued cheating *after* the initial week), of $10/(1 + r)^n$. Mathematically, this is written as the sum of an infinite number of terms:[2]

$$\frac{10}{1 + r} + \frac{10}{(1 + r)^2} + \frac{10}{(1 + r)^3} + \frac{10}{(1 + r)^4} + \cdots \cdot$$

Because r is a rate of return and presumably a positive number, the ratio of $1/(1 + r)$ will be less than 1; this ratio is generally called the **discount factor** and referred to as d (or the Greek letter δ). With $d = 1/(1 + r) < 1$, the mathematical rule for infinite sums tells us this sum converges to a specific value; in this case $10/r$.

It is now possible to determine whether Pierce's Pizza Pies will choose to cheat forever. The store compares its gain of 10 to the PV of all the lost 10s, or $10/r$. Then it cheats forever only if $10 > 10/r$, or $r > 1$; cheating forever is only beneficial in this particular game if the weekly rate of return exceeds 100%, an unlikely event. Thus we would not expect Pierce's to cheat against a cooperative rival when both are playing tit-for-tat. When both Donna's Deep Dish and Pierce's Pizza Pies play TFT, the cooperative outcome in which both price high is a Nash equilibrium of the game. The TFT strategy has solved the prisoners' dilemma for them.

Remember that tit-for-tat is only one of many trigger strategies that could be used by players in a repeated prisoners' dilemma. And it is one of the "nicer" ones. Thus if TFT can be used to solve the dilemma for the pizza stores, other, harsher trigger strategies should be able to do the same. The grim strategy, for instance, can also be used to sustain cooperation in infinitely repeated games.

C. Games of Unknown Length

In addition to considering games of finite or infinite length, we can also incorporate a more sophisticated tool to deal with games of unknown length. It is possible that in some repeated games, players might not know for certain exactly how long their interaction will continue. They may, however, have some idea of the *probability* that the game will continue for another period. For example, our pizza stores might believe that their repeated competition will continue only as long as their customers find pizza to be the takeout dinner of choice; if there were some probability each week that Thai food would replace pizza in that role, then the nature of the game is altered.

---

[2] The Appendix to this chapter contains a detailed discussion of the solution of infinite sums.

Recall that the present value of a loss next week is already worth only $d = 1/(1 + r)$ times the amount earned. If in addition there is only a probability p (less than 1) that the relationship will actually continue to the next week, then next week's loss is worth only p times d times the amount lost. For Pierce's Pizza Pies, this means that the PV of the 10 lost with continued cheating is worth $10 \times d$ [the same as $10/(1 + r)$] when the game is assumed to be continuing with certainty but only $10 \times d \times p$ when the game is assumed to be continuing with probability p. Incorporating the probability that the game may end next period means that the present value of the lost 10 is smaller, because $p < 1$, than it is when the game is definitely expected to continue (when p is assumed to equal 1).

The effect of incorporating p is that we now effectively discount future payoffs by the factor $d \times p$ instead of simply by d. We call this **effective rate of return** R, where $1/(1 + R) = d \times p$, and R depends on d and p as shown:[3]

$$\frac{1}{1 + R} = dp$$

$$1 = dp\,(1 + R)$$

$$R = \frac{1 - dp}{dp}$$

With a 5% **actual rate of return** on investments ($r = 0.05$, so $d = 1/1.05 = 0.95$) and a 50% chance that the game continues for an additional week ($p = 0.5$), then $R = [1 - (0.5)(0.95)] / (0.5)(0.95) = 1.1$, or 110%.

Now the high rates of return needed to sustain cooperation in these examples seem more realistic, if we interpret them as effective rather than actual rates of return. It becomes conceivable that cheating forever, or even once, might actually be to one's benefit if there is a large enough probability that the game will end in the near future. Consider Pierce's decision whether to cheat forever against a TFT-playing rival. Our earlier calculations showed that permanent cheating is beneficial only when r exceeded 1 or 100%. If Pierce faces the 5% actual rate of return and the 50% chance that the game will continue for an additional week, as we assumed in the preceding paragraph, then the effective rate of return of 110% will exceed the critical value needed for it to continue cheating. Thus the cooperative behavior sustained by the TFT strategy can break down if there is a sufficiently large chance that the repeated game might be over by the end of the next period of play—that is, by a sufficiently small value of p.

D. General Theory

We can easily generalize the ideas about when it is worthwhile to cheat against TFT-playing rivals so that you can apply them to any prisoners' dilemma game you encounter. To do so, we use a table with general payoffs (delineated in ap-

---

[3] We could also express R in terms of r and p, in which case $R = (1 + r)/p - 1$.

propriately measurable units) that satisfy the standard structure of payoffs in the dilemma as in Figure 8.3. The payoffs in the table must satisfy the relation $H > C > D > L$ for the game to be a prisoners' dilemma, where C is the *cooperative* outcome, D is the payoff when both players *defect* from cooperation, H is the *high* payoff that goes to the *cheater* when one player cheats while the other cooperates, and L is the *low* payoff that goes to the *loser* (the cooperator) in the same situation.

In this general version of the prisoners' dilemma, a player's one-time gain from cheating is $(H - C)$. The single-period loss for being punished while you return to cooperation is $(C - L)$, and the per-period loss for perpetual cheating is $(C - D)$. To be as general as possible, we will allow for situations in which there is a probability $p < 1$ that the game continues beyond the next period, and discount payoffs using an effective rate of return of R per period. If $p = 1$, as would be the case when the game is guaranteed to continue, then $R = r$, the simple interest rate used in our previous calculations. Replacing r with R, we find that the results we attained earlier generalize almost immediately.

		COLUMN	
		Defect	Cooperate
ROW	Defect	D, D	H, L
	Cooperate	L, H	C, C

FIGURE 8.3 General Version of the Prisoners' Dilemma

We found earlier that a player cheats exactly once against a rival playing TFT if the one-time gain from cheating exceeds the present value of the single-period loss from being punished. In this general game, that means that a player cheats once against a TFT-playing opponent only if $(H - C) > (C - L) / (1 + R)$, or $(1 + R)(H - C) > C - L$, or

$$R > \frac{C - L}{H - C} - 1.$$

Similarly, we found that a player cheats forever against a rival playing TFT only if the one-time gain from cheating exceeds the present value of the infinite sum of the per-period losses from perpetual cheating. For the general game, then, a player cheats forever against a TFT-playing opponent only if $(H - C) > (C - D)/R$, or

$$R > \frac{C - D}{H - C}.$$

These results show which components of the calculation are the most crucial when a player considers whether to cheat. The most obvious elements are the immediate gain from cheating and its future costs. The other two crucial elements in the calculation are the discount factor and the probability of the game's continuation, the two of which jointly determine the value of R; these values provide a measure of "the importance of the future relative to the present." This measure is inversely related to the total effective rate of return, so the more important the future is, the less important the present is, and the smaller would be the equilibrium effective rate of return. Similarly, a small effective rate of return is associated with either a large discount factor (for a fixed value of p) or a large probability of continuation (for a fixed value of d). We use the discount factor $[1/(1 + r)]$ to quantify the importance of the future relative to the present, noting that it is large (close to 1) when r is small and small (close to 0) when r is large.

Now we can consider how the four crucial elements of the calculation interact. For fixed levels of the first two (gains and losses) and the probability of continuation, players are more likely to cheat when the future is relatively unimportant, so the discount factor is small (and the effective rate of return is large). For a given level of gain and a fixed effective rate of return (fixed d and n), players are more likely to cheat when the future costs are small. For a given cost of cheating and a given effective rate of return, players cheat more when the gains are high. And finally, with fixed gains and losses as well as a fixed discount factor, players are more likely to cheat when the probability of continuation is small, implying that there is little future to consider.

It is important to consider the types of circumstances under which these elements take on different values. The immediate gain from cheating, for instance, will be large if cheating is very profitable and also if cheating is not detected very quickly. Similarly, the future costs of cheating will be large if the gains from cooperation are large or if punishment is quick, severe, or sure. Finally, the discount factor and probability of continuation will be large, as will the importance of the future relative to the present (and a small effective rate of return), if the relationship between the players is likely to continue for a long time and if the players are very patient.

You can use these ideas to guide you in when to expect more cooperative behavior between rivals and when to expect more cheating and cutthroat actions. If times are bad and an entire industry is on the verge of collapse, for example, so that businesses feel that there is no future, competition may become more fierce (less cooperative behavior may be observed) than in normal times. Even if times are temporarily good but this is not expected to last, firms may want to make a quick profit while they can, so cooperative behavior might again break down. Similarly, in an industry that emerges temporarily because of a quirk of fashion and is expected to collapse when fashion changes, we should expect less cooperation. Thus, a particular beach resort might become *the* place to go, but all the hotels there will know that such a situation can't last, so they cannot afford to collude on pricing. If, on the other hand, the shifts in fashion occur among products made by an unchanging group of companies in long-term relationships with one another, cooperation might persist. For example, even if all the children want cuddly bears one year and Power Ranger–type toys the next, collusion in pricing may occur if the same small group of manufacturers makes both items.

This discussion also highlights the importance of the detection of cheating. Decisions about whether to continue along a cooperative path depend on how long cheating might be able to go on before it is detected, on how accurately it is detected, and on how long any punishment can be made to last before an attempt is made to revert back to cooperation. If cheating can be detected accurately and quickly, its benefit will not last long, and the subsequent cost will have to be paid more surely. Therefore the success of any method of resolving a prisoners' dilemma depends on how well (both in speed and accuracy) it can detect cheating. This is one reason why the TFT strategy is often considered dangerous; it is too unforgiving. Slight errors in the execution of actions or in the perception of those actions can send players into continuous rounds of punishment from which they may not be able to escape for a long time, until a slight error of the *opposite kind* occurs.

In Chapter 11 we will look at prisoners' dilemmas that arise with many players, and examine when and how players can overcome these dilemmas and achieve outcomes better for them all.

3 SOLUTIONS II: PENALTIES AND REWARDS

Although repetition is the major vehicle for solution of the prisoners' dilemma, there are also several others that can be used to achieve this purpose. One of the simplest ways to avert the prisoners' dilemma in the one-shot version of the game is to inflict some direct **penalty** on the players when they cheat. Once the payoffs have been altered to incorporate the cost of the penalty, players may find that the dilemma has been resolved.

Consider the husband-wife dilemma from Section 1. If only one player cheats, the game's outcome entails one year in jail for the cheater and 25 years for the cooperator. The cheater, though, getting out of jail early, might find the cooperator's friends waiting outside the jail. The physical harm caused by those friends might be equivalent to an additional 20 years in jail. If so, and if the players account for the possibility of this harm, then the payoff structure of the original game has changed.

The "new" game, with the physical penalty included in the payoffs, is illustrated in Figure 8.4. With the additional 20 years in jail added to each player's sentence when they confess while the other denies, the game is completely different.

		WIFE	
		Confess	Deny
HUSBAND	Confess	10 yr, 10 yr	21 yr, 25 yr
	Deny	25 yr, 21 yr	3 yr, 3 yr

FIGURE 8.4 Prisoners' Dilemma with Penalty for the Lone Cheater

A search for dominant strategies in Figure 8.4 shows there are none. A cell-by-cell check then shows there are now two pure-strategy Nash equilibria. One of these is the (Confess, Confess) outcome; the other is the (Deny, Deny) outcome. Now each player finds that it is in his or her best interest to cooperate if the other is going to do so. The game has changed from being a prisoners' dilemma to an assurance game, which we studied in Chapter 4. Solving the new game requires selecting an equilibrium from the two that exist. Of course, one of them—the cooperative outcome—is clearly better than the other from the perspective of both players. Therefore it may be easy to sustain it as a focal point if some convergence of expectations can be achieved.

Notice that the penalty in this scenario is inflicted on a cheater only when his or her rival does *not* cheat. However, stricter penalties can be incorporated into the prisoners' dilemma, such as penalties for *any* confession. Such discipline typically must be imposed by a third party with some power over the two players, rather than by the other player's friends, because the friends would have little standing to penalize the first player when their associate also cheats. If both prisoners are members of a special organization (like a gang or crime mafia), and the organization has a standing rule of never confessing to the police under penalty of extreme physical harm, the game changes again to the one illustrated in Figure 8.5.

		WIFE	
		Confess	Deny
HUSBAND	Confess	30 yr, 30 yr	21 yr, 25 yr
	Deny	25 yr, 21 yr	3 yr, 3 yr

FIGURE 8.5 Prisoners' Dilemma with Penalty for Any Cheating

Now the equivalent of an additional 20 years in jail is added to *all* payoffs associated with the Confess strategy. (Compare Figures 8.5 and 8.1.) In the new game, each player has a dominant strategy, as in the original game. The difference is that the change in the payoffs makes Deny the dominant strategy for each player. And (Deny, Deny) becomes the unique pure-strategy Nash equilibrium. The stricter penalty scheme achieved with third-party enforcement makes cheating so unattractive to players that the cooperative outcome becomes the new equilibrium of the game.

In larger prisoners' dilemma games, difficulties arise with the use of penalties. In particular, if there are many players and some uncertainty exists, penalty schemes may be more difficult to maintain. It becomes harder to decide whether actual cheating is occurring, or it's just bad luck or a mistaken move. In addition, if there really is cheating, it is often difficult to determine the identity of the cheater from among the larger group. And if the game is one-shot, there is no opportunity in the future to correct a penalty that is too severe or to inflict a penalty once a cheater has been identified. Thus penalties may be less successful in large one-shot games than in the two-person game we consider here. We study the issues involved in prisoners' dilemmas with a large number of players in greater detail in Chapter 11.

A further interesting possibility arises when a prisoners' dilemma that has been solved with a penalty scheme is considered in the context of the larger society in which the game is played. It might be the case that, while the dilemma equilibrium outcome is bad for the players, it is actually good for the rest of society or for some subset of individuals within the rest of society. If so, social or political pressures might arise to try to minimize the ability of players to break out of the dilemma. When third-party penalties are the solution to a prisoners' dilemma, as is the case with crime mafias who enforce a no-confession rule, for instance, society can come up with its own strategy to reduce the effectiveness of the penalty mechanism. The Federal Witness Protection Program is an example of a system that has been set up for just this purpose. The U.S. government removes the threat of penalty in return for confessions and testimonies in court.

Similar situations can be seen in other prisoners' dilemmas, such as the pricing game between our two pizza stores. The equilibrium there entailed both stores charging a medium price even though they enjoy higher profits when charging high prices. Although the stores want to break out of this "bad" equilibrium—and we have already seen how the use of trigger strategies can help them do so—their customers are happier with the medium prices offered in the Nash equilibrium of the one-shot game. The customers then have an incentive to try to destroy the efficacy of any enforcement mechanism or solution process used by the stores. For example, because some firms facing prisoners' dilemma pricing games attempt to solve the dilemma through the use of a "meet the competition" or "price matching" campaign, customers might want to press for legislation banning such policies. We provide a more complete analysis of the effect of such price-matching strategies in Section 6.

Just as a prisoners' dilemma can be resolved by penalizing cheaters, it can also be resolved by rewarding cooperators. Because this solution is more difficult to implement in practice, we mention it only briefly.

The most important question is who is to pay the rewards. If it is a third party, that person or group must have sufficient interest of its own in the cooperation achieved by the prisoners to make it worth its while to pay out the rewards. A rare example of this occurred when the United States brokered the Camp David accords between Israel and Egypt by offering large promises of aid to both.

If the rewards are to be paid by the players themselves to each other, the trick is making the rewards contingent (paid out only if the other player cooperates) and credible (guaranteed to be paid if the other player cooperates). Meeting these criteria requires an unusual arrangement; for example, the player making the promise should deposit the sum in advance in an escrow account held by an honorable and neutral third party, who will hand the sum over to the other player if she cooperates, and return it to the promisor if the other cheats. Exercise 8 in this chapter shows you how this can work, but we admit its artificiality.

4 SOLUTIONS III: LEADERSHIP

The final method of solution for the prisoners' dilemma pertains to situations in which one player takes on the role of leader in the interaction. In most examples of the prisoners' dilemma, the game is assumed to be *symmetric*. That is, all the players stand to lose (and gain) the same amount from cheating (and cooperation). However, in actual strategic situations, one player may be relatively "large" (a leader) and the other "small." If the size of the payoffs is unequal enough, so much of the harm from cheating may fall on the larger player that he acts cooperatively, even while knowing that the other will cheat. Saudi Arabia, for example, played such a role as the "swing producer" in OPEC for many years; it cut back on its output when one of the smaller producers like Libya expanded, in order to keep oil prices high.

We illustrate the idea of **leadership** using a slightly amended version of the pizza-store game. Suppose the game remains the same as described in Section 2 except that Donna's Deep Dish has a much larger loyal clientele, who guarantee it the sale of 11,000 (rather than 3,000) pies a week. Profit margins are still $12 and $10 for the High and Medium pricing strategies, and the floating demand is still 4,000 pies. Under these new conditions, the payoff table changes to that shown in Figure 8.6. Only Donna's payoffs are changed, but the equilibrium changes also. In the new game with one large and one small store, Medium re-

		PIERCE'S PIZZA PIES	
		High	Medium
DONNA'S DEEP DISH	High	156, 60	132, 70
	Medium	150, 36	130, 50

FIGURE 8.6 Donna's as Leader in the Pizza-Store Prisoners' Dilemma

mains the dominant strategy for Pierce's. Donna's, however, finds that High is now its dominant strategy. The new Nash equilibrium is (High, Medium), and the dilemma has disappeared.

What has happened to change Donna's choice of strategy? The answer lies in the size of the loyal clientele. The captive market for Donna's is so large that if it lowers its price, what it loses on the captive market outweighs what is gained by garnering the floating demand. Switching from High to Medium means giving up $2 in profit on 11,000 pies ($22,000) in order to get $10 in profit on half of the floating demand, or 2,000 pies ($20,000). The decrease in price is not worth it; the gain from the move is exceeded by the accompanying loss. Thus, Donna's newfound role as the "large" player in the pizza pricing game changes the firm's incentives. It is now better to maintain the high price, thereby taking the cooperative stance in the game.

If the players can make side payments and credible promises to each other, they can get an even better outcome. Note that the total profit from (High, High) is 156 + 60 = 216, while that from (High, Medium) is 132 + 70 = 202. If Donna's can promise Pierce's 15 when Pierce's prices high, then their payoffs will be (141, 75), which is better than (132, 70) for both. And 15 is not the only payment that will achieve such a mutually better outcome. Donna's is willing to pay anything less than 156 − 132 = 24, and Pierce's will accept anything more than 70 − 60 = 10. If they were to bargain over the sum that Donna's will pay to Pierces's in exchange for Pierce's pricing high, what would be the outcome? We look at such questions in Chapter 16.

Situations of leadership in what would otherwise be prisoners' dilemma games may occur in many industries and are common in international diplomacy as well. The role of leader often falls naturally to the biggest or most well established of the players, a phenomenon labeled "the exploitation of the great by the small."[4] For many decades after World War II, for instance, the United States carried a disproportionate share of the expenditures of our defense alliances like NATO and maintained a policy of relatively free international trade

---

[4]Mancur Olson, *The Logic of Collective Action*, (Cambridge, Mass.: Harvard University Press, 1965), p. 29.

even when our partners such as Japan and Europe were much more protectionist. In such situations, it might be reasonable to suggest further that a large or well-established player may accept the role of leader because its own interests are closely tied with those of the players as a whole; if the large player makes up a substantial fraction of the whole group, such a convergence of interests would seem unmistakable. The large player would then be expected to act more cooperatively than might otherwise be the case.

5 EXPERIMENTAL EVIDENCE

Numerous people have conducted experiments in which subjects compete in prisoners' dilemma games against each other.[5] Such experiments show that cooperation can and does occur in such games, even in repeated versions of known and finite length. Many players start off by cooperating, and go on cooperating for quite a while, as long as the rival player reciprocates. Only in the last few plays of a finite game does cheating seem to creep in. Although this goes against the reasoning of rollback, it can be "profitable" if sustained for a reasonable length of time. The pairs get higher payoffs than would rational calculating strategists who cheat from the very beginning.

Such observed behavior can be rationalized in different ways. Perhaps the players are not sure that the relationship will actually end at the stated time. Perhaps they believe their reputation for cooperation will carry over to other similar games against the same, or to other opponents. Perhaps they think it possible that their opponent is a naive cooperator and they are willing to risk a little loss in testing this hypothesis out for a couple of plays. If successful, the experiment will lead to higher payoffs for a sufficiently long time.

In some laboratory experiments, players engage in multiple-round games, each round consisting of a given finite number of repetitions. All of the repetitions in any one round are played against the same rival, but each new round is played against a new opponent. Thus there is the opportunity to develop coop-

---

[5]The literature on experiments involving the prisoners' dilemma game is vast. A brief overview is given by Alvin Roth in *The Handbook of Experimental Economics* (Princeton: Princeton University Press, 1995), pp. 26–28. Journals in both psychology and economics can be consulted for additional references. For some examples of the outcomes we describe, see Kenneth Terhune, "Motives, Situation, and Interpersonal Conflict within prisoners' dilemmas," *Journal of Personality and Social Psychology Monograph Supplement,* vol. 8, no. 3 (1968), pp. 1–24; R. Selten and R. Stoecker, "End Behavior in Sequences of Finite Prisoners' Dilemma Supergames," *Journal of Economic Behavior and Organization,* vol. 7 (1986), pp. 47–70. Robert Axelrod's *The Evolution of Cooperation* (New York: Basic Books, 1984), presents the results of his computer simulation tournament for the best strategy in an infinitely repeated dilemma.

eration with an opponent in each round and to "learn" from previous rounds when devising one's strategy against new opponents as the rounds continue. These situations have shown that cooperation lasts longer in early rounds than in later rounds. This suggests that the theoretical argument on the unraveling of cooperation, based on the use of rollback, is being learned from experience of the play itself over time as players begin more fully to understand the benefits and costs of their actions. Another possibility is that players learn simply that they want to be the first to defect, so the timing of the initial defection occurs earlier, as the number of rounds played increases.

Suppose you were playing a game with a prisoners' dilemma structure and found yourself in a cooperative mode with the known end of the relationship approaching. When should you decide to cheat? You do not want to do so too early, while a lot of potential future gains remain. But you also do not want to leave it to too late in the game, because then your opponent might preempt you and leave you with a low payoff for the period in which she cheats. In fact, your decision about when to cheat cannot be deterministic. If it were, your opponent would figure it out and cheat the period before you planned to do so. If no deterministic choice is feasible, then the unwinding of cooperation must involve some uncertainty, such as mixed strategies, for both players. Many thrillers involving tenuous cooperation among criminals, or between informants and police, acquire their suspense precisely because of this uncertainty.

Examples of the collapse of cooperation as players near the end of a repeated game are observed in numerous situations in the real world, as well as in the laboratory. The story of a long-distance bicycle (or foot) race is one such example. There may be a lot of cooperation for most of the race, as players take turns leading and letting others ride in their slipstream; nevertheless, as the finish line looms, each participant will want to make a dash for the tape. Similarly, signs saying "no checks accepted" often appear in stores in college towns each spring near the end of the semester.

Computer simulation experiments have matched a range of very simple to very complex contingent strategies against each other in two-player prisoners' dilemmas. The most famous of these were conducted by Robert Axelrod at the University of Michigan. He invited people to submit computer programs that specified a strategy for playing a prisoners' dilemma repeated a finite but large number (200) of times. There were 14 entrants. Axelrod held a "league tournament" that pitted pairs of these programs against each other, in each case for a run of the 200 repetitions. The point scores for each pairing and its 200 repetitions were kept, and each program's scores over all its runs against different opponents were added up to see which program did best in the aggregate against all other programs. Axelrod was initially surprised when "nice" programs did well; none of the top eight programs were ever the first to defect. The winning strategy turned out to be the simplest program: Tit-for-tat, submitted by the

Canadian game theorist Anatole Rapoport. Programs that were eager to defect in any particular run got the cheating payoff early but then suffered repetitions of mutual defections and poor payoffs. On the other hand, programs that were always nice and cooperative were badly exploited by their opponents. Axelrod explains the success of Tit-for-tat in terms of four properties: it is at once forgiving, nice, provocable, and clear.

In Axelrod's words, one does well in a repeated prisoners' dilemma to abide by these four simple rules: "Don't be envious. Don't be the first to defect. Reciprocate both cooperation and defection. Don't be too clever."[6] Tit-for-tat embodies each of the four ideals for a good, repeated prisoners' dilemma strategy. It is not envious; it does not continually strive to do better than the opponent, only to do well for itself. In addition, Tit-for-tat clearly fulfills the admonitions not to be the first to defect and to reciprocate, defecting only in retaliation to a previous defection of the opponent and always reciprocating in kind. Finally, Tit-for-tat does not suffer from being overly clever; it is simple and understandable to the opponent. In fact, it won the tournament not because it helped players achieve high payoffs in any individual game—the contest was not about "winner takes all"—but because it was always close; it simultaneously encourages cooperation and avoids exploitation while other strategies cannot.

Axelrod then announced the results of his tournament, and invited submissions for a second round. Here people had a clear opportunity to design programs that would beat Tit-for-tat. The result: Tit-for-tat won again! The programs that were cleverly designed to beat it could not beat it by very much, and they did poorly against one another. Axelrod also arranged a tournament of a different kind. Instead of a league where each program met each other program once, he ran a game with a whole population of programs, with a number of copies of each program. Each type of program met an opponent randomly chosen from the population. Those programs that did well were given a larger proportion of the population; those that did poorly had their proportion in the population reduced. This was a game of evolution and natural selection, which we will study in greater detail in Chapter 10. But the idea is simple in this context, and the results are fascinating. At first, nasty programs did well at the expense of nice ones. But as the population became more and more nasty, each nasty program met other nasty programs more and more often, and they began to do poorly and fall in numbers. Then Tit-for-tat started to do well and eventually triumphed.

However, Tit-for-tat has some flaws. Most importantly, it assumes no errors in execution of the strategy. If there is some risk that the player intends to play the cooperative action but plays the cheating action by error, then this can initiate a sequence of retaliatory cheating actions that locks two Tit-for-tat programs playing each other into a bad outcome; it requires another error to rescue them

---

[6]Axelrod, *Evolution of Cooperation*, p. 110.

from this sequence. When Axelrod ran a third variant of his tournament, which provided for such random mistakes, Tit-for-tat could be beaten by even "nicer" programs that tolerated an occasional episode of cheating to see if it was a mistake or a consistent attempt to exploit them, and retaliated only when convinced that it was not a mistake.[7]

6 REAL-WORLD DILEMMAS

Games with the prisoners' dilemma structure arise in a surprisingly varied number of contexts in the world. While we would be foolish to try to show you every possible instance in which the dilemma can arise, we take the opportunity in this section to discuss in detail four specific examples, from a variety of fields of study. One example involves policy setting on the part of state governments, another considers labor arbitration outcomes, still another comes from evolutionary biology (a field which we will study in greater detail in Chapter 10), and the final example describes the policy of "price matching" as a solution to a prisoners' dilemma pricing game.

A. Policy Setting

This example is based on a *Time* magazine article entitled "A No-Win War Between the States."[8] The article discusses the fact that individual states have been using financial incentives to get firms to relocate to their states and/or to get local firms to stay in state. Such incentives, which come primarily in the form of corporate and real-estate tax breaks, can be extremely costly to the states that offer them. They can, however, bring significant benefits if they work. Large manufacturing firms bring with them the promise of increased employment opportunities for state residents and potential increases in state sales-tax revenue. The problem with the process of offering such incentives, however, is that it is a prisoners' dilemma. Each state has a dominant strategy to offer incentives, but the ultimate equilibrium outcome is worse for everyone than if they had refrained from offering incentives in the first place.

Let us consider a game played by two states, Ours and Theirs. Both states currently have the same number of firms and are considering offering identi-

---

[7] For a description and analysis of Axelrod's computer simulations from a biological perspective, see Matt Ridley, *The Origins of Virtue* (New York: Penguin Books, 1997), pp. 61, 75. For a discussion of the difference between computer simulations and experiments using human players, see John K. Kagel and Alvin E. Roth, *Handbook of Experimental Economics* (Princeton, N.J.: Princeton University Press, 1995), p. 29.

[8] John Greenwald, "A No-Win War Between the States," *Time*, April 8, 1996.

cal incentive packages to induce some of the other state's firms to move. Each state has two strategies—to offer incentives or not offer them—which we will call Incentives and None. If no incentive offer is made by either state, the status quo is maintained. Neither state incurs the cost of the incentives, but it does not get new firms to move in either; each state retains its original number of firms. If Ours offers incentives while Theirs offers none, then Ours will encourage firms to move in while Theirs loses firms; Ours gains at the expense of Theirs. (We assume that the states have done enough prior investigation to be able to construct an incentive package whose costs are offset by the benefits of the arrival of the new firms.) The opposite will be true for the situation in which Theirs offers Incentives and Ours offers none; Theirs will gain at the expense of ours. If both states offer incentives, firms are likely to switch states to take advantage of the incentive packages, but the final number of firms in each state will be the same as in the absence of incentives. Each state has incurred a high cost to obtain an outcome equivalent to the original situation.

Given this description of the possible strategies and their payoffs, we can construct a simple payoff table for the states, as done in Figure 8.7. The payoff numbers reflect the rating given each outcome by each state. The highest rating of 4 goes to the state that offers incentives when its rival does not; that state pays the cost of incentives but benefits from gaining new firms. At the other end of the spectrum, the lowest rating of 1 goes to the state that does not offer incentives when its rival does; that state incurs no cost but loses firms to its rival. In between, a rating of 3 goes to the outcome in which the status quo is maintained with no cost, and a rating of 2 goes to the outcome in which the status quo is effectively maintained but the costs of incentives are incurred.

		THEIRS	
		Incentives	None
OURS	Incentives	2, 2	4, 1
	None	1, 4	3, 3

FIGURE 8.7   Payoffs for the Incentive-Package Prisoners' Dilemma

The payoff table in Figure 8.7 is a standard prisoners' dilemma. Each state has a dominant strategy to choose Incentives, and the Nash equilibrium outcome entails incentives being offered by both states. Of course, as in all prisoners' dilemmas, both states would have been better off had they agreed not to offer any incentives at all. Similar situations arise in other policy contexts as well. Governments face the same difficulties when deciding whether to provide export subsidies to domestic firms, knowing that foreign governments are simultaneously making the same choice.

Is there a solution to this dilemma? From the standpoint of the firms receiving the incentives, the equilibrium outcome is a good one, but from the perspective of the state governments and the households that must shoulder an increased tax burden, the no-incentives outcome is preferable. Federal legislation prohibiting the use of such incentive packages could solve this particular dilemma; such a solution would fall into the category of a punishment enforced by a third party.

B. Labor Arbitration

Our second example comes from an intriguingly titled paper by O. Ashenfelter and D. Bloom, "Lawyers as Agents of the Devil in a Prisoners' Dilemma Game."[9] The authors argue that many arbitration situations between labor unions and employers have a prisoners' dilemma structure. Given the choice of whether to retain legal counsel for arbitration, both sides find it in their interests to do so. The presence of a lawyer is likely to help the legally represented side in an arbitration process substantially if the other side has no lawyer. Hence, it is better to use a lawyer if your rival does not. But you are also sure to want a lawyer if your rival uses one. Using a lawyer is a dominant strategy. Unfortunately, Ashenfelter and Bloom go on to claim that the outcomes for these labor arbitration games change little in the presence of lawyers. That is, the probability of the union "winning" an arbitration game is basically the same when no lawyers are used as when both sides use lawyers. But, of course, lawyers are costly, just like the incentive packages states offer to woo new firms. Players always want to use lawyers but do so at great expense and with little reward in the end. The resulting equilibrium with lawyers is worse than if both sides choose No Lawyers because each side has to pay substantial legal fees without reaping any beneficial effect on the probability of winning the game.

Figure 8.8 shows the results of Ashenfelter and Bloom's detailed study of actual arbitration cases. The values in the table are their econometric estimates of the percentage of arbitration cases won by the employer under different combinations of legal representation for the two sides. The Union sees its chances of winning against an unrepresented Employer increase (from 56% to 77%) when it chooses to use a lawyer, and the Employer sees a similar increase (from 44% to 73%). But when both use lawyers, the winning percentages change only slightly, by 2% (with measurement error) compared with the situation in which neither

---

[9] O. Ashenfelter and D. Bloom, "Lawyers as Agents of the Devil in a Prisoners' Dilemma Game," *Rand Journal of Economics*, forthcoming.

		UNION	
		Lawyer	No Lawyer
EMPLOYER	Lawyer	46%	73%
	No Lawyer	23%	44%

FIGURE 8.8   Predicted Percentage of Employer "Wins" in Arbitration Cases

side uses a lawyer. Thus the costs incurred in hiring the lawyers are unlikely to be recouped in increased wins for either side, and each would be better off if both could credibly agree to forgo legal representation.

The search for a solution to this dilemma returns us to the possibility of a legislated penalty mechanism. Federal arbitration rules could be established that preclude the possibility of the sides bringing lawyers to the table. In this game, the likelihood of a continuing relationship between the players also gives rise to the possibility that cooperation could be sustained through the use of contingent strategies. If the union and the employer start out in a cooperative mode, the fear of never being able to return to the No Lawyer outcome might be sufficient to prevent a single occurrence of cheating—hiring a lawyer.

C. Evolutionary Biology

As our third example, we consider a game known as the bowerbirds' dilemma, from the field of evolutionary biology.[10] Male bowerbirds attract females by building intricate nesting spots called bowers, and female bowerbirds are known to be particularly choosy about the bowers built by their prospective mates. For this reason, male bowerbirds often go out on search-and-destroy missions aimed at ruining other males' bowers. When they are out, however, they run the risk of losing their own bower at the beak of another male. The ensuing competition between male bowerbirds and their ultimate choice regarding whether to maraud or guard has the structure of a prisoners' dilemma game.

Ornithologists have constructed a table that shows the payoffs in a two-bird game with two possible strategies, Maraud and Guard. That payoff table is shown in Figure 8.9. GG represents the benefits associated with guarding when the rival bird also Guards; GM represents the payoff from guarding when the rival bird is a marauder. Careful scientific study of bowerbird matings led to the discovery that MG > GG > MM > GM. In other words, the payoffs in the bowerbird game have exactly the same structure as the prisoners' dilemma. The birds'

---

[10] Larry Conick, "Science Classics: The Bowerbird's Dilemma," *Discover*, October 1994.

		BIRD 2	
		Maraud	Guard
BIRD 1	Maraud	MM, MM	MG, GM
	Guard	GM, MG	GG, GG

FIGURE 8.9 Bowerbirds' Dilemma

dominant strategy is to maraud, but when both choose that strategy they end up in an equilibrium worse off than if they had both chosen to guard.

In reality, the strategy used by any particular bowerbird is not actually the result of a process of rational choice on the part of the bird. Rather, in evolutionary games, it is assumed that strategies are genetically "hardwired" into individuals and that payoffs represent reproductive success for the different types. Then equilibria in such games define the type of population that can be expected to be observed—all marauders, for instance, if Maraud is a dominant strategy as in Figure 8.9. This equilibrium outcome is not the best one, however, given the existence of the dilemma. In constructing a solution to the bowerbirds' dilemma, we can appeal to the repetitive nature of the interaction in the game. In the case of the bowerbirds, repeated play against the same or different opponents over the course of several breeding seasons can allow you, the bird, to choose a flexible strategy based on your opponent's last move. Contingent strategies like TFT can be, and often are, adopted in evolutionary games in order to solve exactly this type of dilemma. We will return to the idea of evolutionary games and provide detailed discussions of their structure and equilibrium outcomes in Chapter 10.

D. Price Matching

Finally, we return to a pricing game, in which we consider two specific stores engaged in price competition with each other, using identical price-matching policies. The stores in question, Toys "R" Us and Kmart, are both national chains that regularly advertise prices for name-brand toys (and other items). In addition, each store maintains a published policy that guarantees customers that they will match the advertised price of any competitor on a specific item (model and item numbers must be identical) as long as the customer provides the competitor's printed advertisement.[11]

---

[11] The price-matching policy at Toys "R" Us is printed and posted prominently in all stores. A simple phone call confirmed that Kmart has an identical policy. Similar policies exist at many of the other well known national "discount department store" chains.

For the purposes of this example, we assume that the firms have only two possible prices they can charge for a particular toy (Low or High). In addition, we use hypothetical profit numbers and further simplify the analysis by assuming that Toys "R" Us and Kmart are the only two competitors in the toy market in a particular city—Billings, Montana, for example.

Suppose, then, that the basic structure of the game between the two firms can be illustrated as in Figure 8.10. If both firms advertise low prices, they split the available customer demand and each earns $2,000. If both advertise high prices, they split a market with lower sales, but their markups end up being large enough to let them each earn $3,000. Finally, if they advertise different prices, then the one advertising a high price gets no customers and earns nothing while the one advertising a low price earns $4,000.

		KMART	
		Low	High
TOYS "R" US	Low	2,000, 2,000	4,000, 0
	High	0, 4,000	3,000, 3,000

FIGURE 8.10 Toys "R" Us and Kmart Toy Pricing

The game illustrated in Figure 8.10 is clearly a prisoners' dilemma. Advertising and selling at a low price is the dominant strategy for each firm, although both would be better off if each advertised and sold at the high price. But as we mentioned earlier, each firm actually makes use of a third pricing strategy: a price-matching guarantee to its customers. How does the inclusion of such a policy alter the prisoners' dilemma that would otherwise exist between these two firms?

Consider the effects of allowing firms to choose between pricing low, pricing high, and price matching. The Match strategy entails advertising a high price but promising to match any lower advertised price by a competitor; a firm using Match then benefits from advertising high if the rival firm does so also, but does not suffer any harm from advertising a high price if the rival advertises a low price. We can see this in the payoff structure for the new game, shown in Figure 8.11. In that table, we see that a combination of one firm playing Low while the other plays Match is equivalent to both playing Low, while a combination of one firm playing High while the other plays Match (or both playing Match) is equivalent to both playing High.

		KMART		
		Low	High	Match
TOYS "R" US	Low	2,000, 2,000	4,000, 0	2,000, 2,000
	High	0, 4,000	3,000, 3,000	3,000, 3,000
	Match	2,000, 2,000	3,000, 3,000	3,000, 3,000

FIGURE 8.11 Toys "R" Us and Kmart Price Matching

Using our standard tools for analyzing simultaneous-play games shows that High is weakly dominated by Match for both players and that, once High is eliminated, Low is weakly dominated by Match also. The resulting Nash equilibrium entails both firms using the Match strategy. In equilibrium, both firms earn $3,000—the profit level associated with both firms pricing high in the original game. The addition of the Match strategy has allowed the firms to emerge from the prisoners' dilemma they faced when they had only the choice between two simple pricing strategies, Low or High.

How did this happen? The Match strategy acts as a penalty mechanism. By guaranteeing to match Kmart's low price, Toys "R" Us substantially reduces the benefit Kmart achieves by advertising a low price while Toys "R" Us is advertising a high price. In addition, promising to meet Kmart's low price hurts Toys "R" Us too, because the latter has to accept the lower profit associated with the low price. Thus the price-matching guarantee is a method of penalizing both players whenever either one cheats. This is just like the crime mafia example we discussed in Section 3, except that this penalty scheme—and the higher equilibrium prices that it supports—is observed in markets in virtually every city in the country.

This result should put all customers on alert. Even though stores that match prices promote their policies in the name of competition, the ultimate outcome when all firms use such policies is better for the firms than if there were no price matching at all, and so customers are the ones who are hurt.

SUMMARY

The prisoners' dilemma is probably the most famous game of strategy; each player has a dominant strategy (to Defect), but the equilibrium outcome is worse for all players than when each uses her dominated strategy (to Cooperate). The most well known solution to the dilemma is *repetition of play*. In a finitely lived game, the *present value* of future cooperation is eventually zero and rollback yields an equilibrium with no cooperative behavior. With infinite play (or an uncertain end date), cooperation can be achieved using an appropriate contingent strategy such as *tit-for-tat (TFT)* or the *grim strategy;* in either case, cooperation is possible only if the present value of cooperation exceeds the present value of defecting. More generally, the prospects of "no tomorrow" or of short-term relationships lead to increased competitiveness among players.

The dilemma can also be "solved" with *penalty* schemes that alter the payoffs for players who defect from cooperation when their rivals are cooperating or when others are also defecting. A third solution method arises if a large or strong player's loss from defecting is larger than the available gain from cooperative behavior on that player's part.

Experimental evidence suggests that players often cooperate longer than theory might predict. Such behavior can be explained by incomplete knowledge of the game on the part of the players or on their views regarding the benefits of cooperation. Tit-for-tat has been observed to be a simple, nice, provocable, and forgiving strategy that performs very well on the average in repeated prisoners' dilemmas.

Prisoners' dilemmas arise in a variety of contexts. Specific examples from the areas of policy setting, labor arbitration, evolutionary biology, and product pricing show how to explain actual behavior using the framework of the prisoners' dilemma.

KEY TERMS

actual rate of return (263)
compound interest (261)
contingent strategy (259)
discount factor (262)
effective rate of return (263)
grim strategy (259)
infinite horizon (261)

leadership (269)
penalty (266)
present value (PV) (260)
punishment (259)
repeated play (257)
tit-for-tat (TFT) (259)
trigger strategy (259)

EXERCISES

1. "If a prisoners' dilemma is repeated 100 times, the players are sure to achieve their cooperative outcome." True or false? Explain and give an example of a game that illustrates your answer.

2. Consider the pizza stores introduced in Chapter 4: Donna's Deep Dish and Pierce's Pizza Pies. Suppose they are in a repeated relationship, trying to sus-

tain their joint profit-maximizing prices of $13.50 each. They print new menus each month, and thereby commit themselves to prices for the whole month. In any one month, one of them can cheat on the agreement. If one of them holds the price at the agreed level, what is the best cheating price for the other? What are its resulting profits? For what discount factors will their collusion be sustainable using grim trigger strategies?

3. Consider the bread and cheese stores in Exercise 12 of Chapter 4, and suppose they are in a repeated relationship. Carry out the same kind of analysis and answer the same questions as in Exercise 2 above.

4. A town council consists of three members who vote every year on their own salary increases. Two Yes votes are needed to pass the increase. Each member would like a higher salary, but would like to vote against it herself because that looks good to the voters. Specifically, the payoffs of each are as follows:

Raise passes, own vote is No:	10 points
Raise fails, own vote is No:	5 points
Raise passes, own vote is Yes:	4 points
Raise fails, own vote is Yes:	0 points

Voting is simultaneous. Write down the (three-dimensional) payoff table, and show that in the Nash equilibrium the raise fails unanimously. Examine how a repeated relationship among the members can secure them salary increases every year if (1) every member serves a three-year term, and every year in rotation one of them is up for reelection, and (2) the townspeople have short memories, remembering only the votes on the salary increase motion during the current year and not those of the past years.

5. Consider a two-player game between Child's Play and Kid's Korner, each of whom produce and sell wooden swing sets for children. Each player can set either a high or a low price for a standard two-swing, one-slide set. If they both set a high price, each receives profits of $64,000 (per year). If one sets a low price while the other sets a high price, the low price firm earns profits of $72,000 (per year) while the high price firm earns $20,000. If they both set a low price, each receives profits of $57,000.

(a) Verify that this game has a prisoners' dilemma structure by looking at the ranking of payoffs associated with the different strategy combinations (both cooperate, both defect, one cheats and so on). What are the Nash equilibrium strategies and payoffs in the simultaneous-play game if the players make price decisions only once?

(b) Next suppose the two firms decide to play this game for a fixed number of periods, say for four years, with moves simultaneous within each year but of course sequential from one year to the next. What would each firm's *total* profits be at the end of the game? (Don't discount.) Explain how you arrived at your answer.

(c) Suppose that the two firms play this game repeatedly forever, with the moves still being simultaneous within each period. Let each of them use a grim strategy in which they both price high unless one of them "cheats," in which case they price low for the rest of the game. What is the one-time gain from cheating against an opponent playing such a strategy? How much does each firm lose, in each future period, after it cheats once? If $r = 0.25$ ($d = 0.8$), will it be worthwhile to cooperate? Find the range of values of r (or d) for which this strategy is able to sustain cooperation between the two firms.

(d) Suppose the firms play this game repeatedly year after year, neither expecting any change to their interaction. If the world were to end after four years, without either having anticipated this event, what would each firm's total profits (*not* discounted) be at the end of the game? Compare your answer here with the answer in part (b). Explain why the two answers are different, if they are different, or why they are the same, if they are the same.

(e) Suppose now that the firms know that there is a 10% probability that one of them may go bankrupt in any given year. If this occurs, the repeated game between the two firms ends. Will this knowledge change the firms' actions when $r = 0.25$? What if the probability of a bankruptcy increases to 35% in any year?

6. Two people, Baker and Cutler, play a game where they choose and divide a prize. Baker decides how large the total prize should be; she can choose either $10 or $100. Cutler chooses how to divide the prize chosen by Baker; Cutler can choose either an equal division or a split, where she gets 90% and Baker gets 10%. Write down the payoff table of the game and find its equilibria,

(a) when the moves are simultaneous,

(b) when Baker moves first,

(c) when Cutler moves first.

(d) Is this game a prisoners' dilemma?

7. Consider the game of chicken from Chapter 4, with slightly more general payoffs (Figure 4.11 had $k = 1$):

		DEAN	
		Swerve	Straight
JAMES	Swerve	0, 0	−1, k
	Straight	k, −1	−2, −2

Suppose this game is played repeatedly, every Saturday evening. If $k < 1$, the two players stand to benefit by cooperating to play (Swerve, Swerve) all the time, whereas if $k > 1$ they stand to benefit by cooperating so that one plays Swerve and the other plays Straight, taking turns to go Straight in alternate weeks. Can either type of cooperation be sustained?

8. Consider the following game, which comes from James Andreoni and Hal Varian at the University of Michigan.[12] A neutral referee runs the game. There are two players, Row and Column. The referee gives two cards to each; 2 and 7 to Row and 4 and 8 to Column. This is common knowledge. Then, playing simultaneously and independently, each player is asked to hand over to the referee either his high card or his low card. Depending on the cards he collects, the referee hands out payoffs, which come from a central kitty, not from the players' pockets, and are measured in dollars. If Row chooses his Low card 2, then Row gets $2; if he chooses his High card 7, then Column gets $7. If Column chooses his Low card 4, then Column gets $4; if he chooses his High card 8, then Row gets $8.

   (a) Show that the complete payoff table is as follows:

		COLUMN	
		Low	High
ROW	Low	2, 4	10, 0
	High	0, 11	8, 7

   (b) What is the Nash equilibrium? Verify that it is a prisoners' dilemma.

   Now suppose the game has the following stages. The referee hands out cards as before; who gets what cards is common knowledge. Then at stage I, each player, out of his own pocket, can hand over a sum of money, which the referee is to hold in an escrow account. This amount can be zero, but not negative. The rules for the treatment of these sums are as follows, and are also common knowledge. If Column chooses his high card, then the referee hands over to Column the sum put by Row in the escrow account; if Column chooses Low, then Row's sum reverts back to Row. The disposition of the sum deposited by Column depends similarly on what Row does. At stage II, each player hands over one of his cards, High or Low, to the referee. Then the referee hands out payoffs from the central kitty, according to the table, and also disposes of the sums (if any) in the escrow account, according to the rules just given.

   (c) Find the rollback (subgame-perfect) equilibrium of this two-stage game. Does it resolve the prisoners' dilemma? What is the role of the escrow account?

---

[12] James Andreoni and Hal Varian, "Pre-Play Contacting in the Prisoners' Dilemma," University of Michigan, Ann Arbor, manuscript, October 1993.

Appendix: Infinite Sums

The computation of present values requires us to determine the current value of a sum of money that is paid to us in the future. As we saw in Section 2, the present value of a sum of money, say x, that is paid to us n weeks from now is just $x/(1 + r)^n$, where r is the appropriate weekly rate of return. But the present value of a sum of money that is paid to us next week and every following week in the foreseeable future is more complicated to determine. In that case, the payments continue infinitely, so there is no defined end to the sum of present values that we need to compute. To compute the present value of this flow of payments requires some knowledge of the mathematics of the summation of infinite series.

Consider a player who stands to gain \$10 this week from cheating in a prisoners' dilemma but who will then lose \$10 every week in the future as a result of her choice to continue cheating while her opponent punishes her (using the tit-for-tat, or TFT, strategy). During the first of the future weeks—the first for which there is a loss and the first for which values need to be discounted—the present value of her loss is $10/(1 + r)$; during the second future week, the present value of the loss is $10/(1 + r)^2$; during the third future week, the present value of the loss is $10/(1 + r)^3$. That is, in each of the n future weeks that she incurs a loss from cheating, that loss equals $10/(1 + r)^n$.

We could write out the total present value of all of her future losses as a large sum with an infinite number of components,

$$PV = \frac{10}{1 + r} + \frac{10}{(1 + r)^2} + \frac{10}{(1 + r)^3} + \frac{10}{(1 + r)^4} + \frac{10}{(1 + r)^5} + \frac{10}{(1 + r)^6} + \cdots,$$

or we could use summation notation as a shorthand device and instead write

$$PV = \sum_{n=1}^{\infty} \frac{10}{(1 + r)^n}.$$

This expression, which is equivalent to the previous one, is read as "the sum, from n equals 1 to n equals infinity, of 10 over $(1 + r)$ to the nth power." Because 10 is a common factor—it appears in each term of the sum—it can be pulled out to the front of the expression. Thus we can write the same present value as

$$PV = \sum_{n=1}^{\infty} \frac{1}{(1 + r)^n}.$$

We now need to determine the value of the sum within the present-value expression in order to calculate the actual present value. To do so, we will simplify our notation by switching to the *discount factor d* in place of $1/(1 + r)$. Then the sum that we are interested in evaluating is

$$\sum_{n=1}^{\infty} d^n.$$

It is important to note here that $d = 1/(1 + r) < 1$ because r is strictly positive.

An expert on infinite sums would tell you, after inspecting this last sum, that it converges to the finite value $d/(1 - d)$.[13] Convergence is guaranteed because increasingly large powers of a number less than 1, d in this case, become smaller and smaller, approaching zero as n approaches infinity. The later terms in our present value, then, decrease in size until they get sufficiently small that the series approaches (but technically never exactly reaches) the particular value of the sum. While a good deal of more sophisticated mathematics is required to deduce that the convergent value of the sum is $d/(1 - d)$, providing that this is the correct answer is relatively straightforward.

We use a simple trick to prove our claim. Consider the sum of the first m terms of the series, and denote it by S_m. Thus

$$S_m = \sum_{n=1}^{m} d^n = d + d^2 + d^3 + \cdots + d^{m-1} + d^m.$$

Now we multiply this by $(1 - d)$ to get

$$(1 - d)S_m = d + d^2 + d^3 + \cdots + d^{m-1} + d^m$$
$$- d^2 - d^3 - d^4 - \cdots - d^m - d^{m+1}$$
$$= d - d^{m+1}.$$

Dividing both sides by $(1 - d)$, we have

$$S_m = \frac{d - d^{m+1}}{1 - d}.$$

Finally we take the limit of this as m approaches infinity, in order to evaluate our original infinite sum. As m goes to infinity, the value of d^{m+1} goes to zero because very large and increasing powers of a number less than one get increasingly small but stay nonnegative. Thus as m goes to infinity, the right-hand side of the above equation goes to $d/(1 - d)$, which is therefore the limit of S_m, that is, the limit of the infinite sum, as we set out to prove.

We need only convert back to r to be able to use our answer in the calculation of present values in our prisoners' dilemma games. Because $d = 1/(1 + r)$, it follows that

---

[13] An infinite series *con*verges if the sum of the values in the series approaches a specific value, getting closer and closer to that value as additional components of the series are included in the sum. The series *di*verges if the sum of the values in the series gets increasingly larger (or more negative) with each addition to the sum. Convergence requires that the components of the series get progressively smaller.

$$\frac{d}{1-d} = \frac{1/(1+r)}{r/(1+r)} = \frac{1}{r}.$$

The present value of an infinite stream of $10s earned each week, starting next week, is then

$$10 \sum_{n=1}^{\infty} \frac{1}{(1+r)^n} = \frac{10}{r}.$$

This is the value we use to determine whether a player should cheat forever in Section 2. Notice that incorporating a probability of continuation, $p \leq 1$, into the discounting calculations changes nothing in the summation procedure used here. We could easily substitute R for r in the preceding calculations, and dp for the discount factor, d.

Remember that you need to find present values only for losses (or gains) incurred (or accrued) *in the future*. The present value of $10 lost today is just $10. So if you wanted the present value of a stream of losses, all of them $10, that begins *today*, you would take the $10 lost today and add it to the present value of the stream of losses in the future. We have just calculated that present value as $10/r$. Thus the present value of the stream of lost $10s, including the $10 lost today, would be $10 + 10/r$, or $10[(r + 1)/r]$, which equals $10/(1 - d)$. Similarly, if you wanted to look at a player's stream of profits under a particular contingent strategy in a prisoners' dilemma, you would not discount the profit amount earned in the very first period; you would only discount those profit figures that represent money earned in future periods.

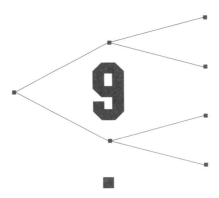

9

Games with Strategic Moves

A GAME IS SPECIFIED BY the choices or moves available to the players; the order, if any, in which they make those moves; and the payoffs that result from all logically possible combinations of all the players' choices. In Chapter 5 we saw how changing the order of moves from sequential to simultaneous or vice versa can alter the game's outcomes. Adding or removing moves available to a player, or changing the payoffs at some terminal nodes or in some cells of the game table, can also change outcomes. Unless the rules of a game are fixed by an outside authority, each player has the incentive to manipulate them to produce an outcome that is more to his own advantage. Devices to manipulate a game in this way are called **strategic moves,** which are the subject of this chapter.

A strategic move changes the rules of the original game to create a new two-stage game. The second stage is the original game, often with some alteration of the order of moves and the payoffs. The first stage specifies how you will act in the second stage. Different first-stage actions correspond to different strategic moves, and we classify them into three types: *commitments, threats,* and *promises.* The aim of all three is to alter the outcome of the second-stage game to your own advantage. Which, if any, suits your purpose depends on the context. But most importantly, any of the three works only if the other player believes that at the second stage you will indeed do what you declared at the first stage. In other words, the **credibility** of the strategic move is open to question; only a credible strategic move will have the desired effect.

You are all probably more familiar with the use and credibility of strategic moves than you might think. Parents, for instance, constantly attempt to influence the behavior of their children using threats ("no dessert unless you finish your vegetables") and promises ("you will get the new racing bike for your next birthday if you maintain at least a B average in school this term"). And children know very well that many of these threats and promises are not credible; much bad behavior can escape the threatened punishment if the child sweetly promises not to do that again, even though the promise itself may not be credible. Furthermore, when the children get older and become concerned with their own appearance, they find themselves making commitments to themselves to exercise and diet; of course, many of these commitments also turn out to lack credibility. All of these devices—commitments, threats, and promises—are examples of strategic moves. Their purpose is to alter the actions of another player, perhaps even your own future self, at a later stage in a game. But they will not achieve this purpose unless they are credible. In this chapter we will use game theory to study systematically how to use such strategies and how to make them credible.

Be warned, however, that credibility is a difficult and subtle matter. We can offer you some general principles and an overall understanding of how strategic moves can work—a science of strategy. But actually making them work depends on your specific understanding of the context, and your opponent may get the better of you by having a better understanding of the concepts or the context or both. Therefore the use of strategic moves in practice retains a substantial component of art. It also involves risk, particularly when using the strategy of *brinkmanship*, which can sometimes lead to disasters. You can have success as well as fun trying to put these ideas into practice, but note our disclaimer and warning: Use such strategies at your own risk.

1 A CLASSIFICATION OF STRATEGIC MOVES

Because the use of strategic moves depends so critically on the order of moves, to study them we need to know what it means "to move first." Thus far we have taken this concept to be self-evident, but now we need to make it more precise. It has two components. First, your action must be **observable** to the other player; second, it must be **irreversible**.

Consider a strategic interaction involving two players, A and B, in which A's move is made first. If A's choice is not observable to B, then B cannot respond to it, and the mere chronology of action is irrelevant. For example, suppose A and B are two companies bidding in an auction. A's committee meets in secret on Monday to determine its bid; B's committee meets on Tuesday; the bids are separately mailed to the auctioneer and opened on Friday. When B makes its decision, it does not know what A has done; therefore the moves are strategically the same as if they were simultaneous.

If A's move is not irreversible, then A might pretend to do one thing, lure B into responding, and then change his own action to his own advantage. Of course, B should anticipate this ruse and not be lured; then he will not be responding to A's choice. Once again, in the true strategic sense A does not have the first move.

Considerations of observability and irreversibility affect the nature and types of strategic moves as well as their credibility. We begin with a taxonomy of strategic moves available to players.

A. Unconditional Strategic Moves

Let us suppose that player A is the one making a strategic observable and irreversible move in the first stage of the game. He can declare: "In the game to follow, I will make a particular move, X." This says that A's future move is unconditional; A will do X irrespective of what B does. Such a statement, if credible, is tantamount to changing the order of the game at stage 2 so that A moves first and B second, and A's first move is X. This strategic move is called a **commitment.**

If the previous rules of the game at the second stage already have A moving first, then such a declaration would be irrelevant. But if the game at the second stage has simultaneous moves, or if A is to move second there, then such a declaration, if credible, can change the outcome, because it changes B's beliefs about the consequences of his actions. Thus a commitment is a simple seizing of the first-mover advantage when it exists.

In the Senate race game from Chapter 3, an incumbent (Gray) and a challenger (Green) play a sequential-move game in which the incumbent first makes a decision about whether to use campaign ads and the challenger then decides whether to enter the race. The rollback equilibrium entails Gray using ads and Green choosing not to run, with payoffs of 3 to each. By making a credible commitment to enter the race, however, Green could alter the outcome of the game. Even though the date for filing her candidacy will not arrive until after Gray makes the decision concerning her ads, Green could "leak" stories of her candidacy to such an extent that her reputation would suffer badly if she did not run. Then Green essentially commits herself to entering the race regardless of what Gray decides to do, and before Gray's decision concerning ads is made. In other words, Green changes the game to one in which she is in effect the first mover. You can easily check that the new rollback equilibrium entails Gray choosing not to advertise and the two politicians getting payoffs of 2 (to Gray) and 4 (to Green)—the equilibrium outcome associated with the game when Green moves first. Several more detailed examples of commitments are given in the following sections.

B. Conditional Strategic Moves

Another possibility for A is to declare at the first stage: "In the game to follow, I will respond to your choices in the following way. If you choose Y_1, I will do Z_1; if you do Y_2, I will do Z_2, . . ." In other words, A can use a move that is conditional on B's behavior; we call this type of move a **response rule** or *reaction function.* A's statement means that in the game to be played at the second stage, A will move second, but how he will respond to B's choices at that point is already predetermined by A's declaration at stage 1. For such declarations to be meaningful, A must be physically able to wait to make his move at the second stage until after he has observed what B has irreversibly done. In other words, at the second stage B should have the true first move in the double sense just explained.

Conditional strategic moves take different forms depending on what they are trying to achieve and how they set about achieving it. When A wants to stop B from doing something, we say that A is trying to deter B, or to achieve **deterrence;** when A wants to induce B to do something, we say that A is trying to compel B, or to achieve **compellence.** We return to this distinction later. Of more immediate interest is the method used in pursuit of either of these aims. If A declares, "*Unless* your action (or inaction, as the case may be) conforms to my stated wish, I will respond in a way that will *hurt* you," that is, a **threat.** If A declares, "*If* your action (or inaction, as the case may be) conforms to my stated wish, I will respond in a way that will *reward* you," that is, a **promise.** Of course, "hurt" and "reward" are measured in terms of the payoffs in the game itself. When A hurts B, A does something that lowers B's payoff; when A rewards B, A does something that leads to a higher payoff for B. Threats and promises are the two conditional strategic moves upon which we focus our analysis.

To understand the nature of these strategies, consider the dinner game mentioned earlier. In the natural chronological order of moves, first the child decides whether to eat its vegetables, and then the parent decides whether to give the child dessert. Rollback analysis tells us the outcome: the child refuses to eat the vegetables, knowing that the parent, unwilling to see the child hungry and unhappy, will give it the dessert. The parent can foresee this outcome, however, and can try to alter it by making an initial move, namely by stating a conditional response rule of the form "no dessert unless you finish your vegetables." This declaration constitutes a threat. It is a first move in a pregame, which fixes how you will make your second move in the actual game to follow. If the child believes the threat, that alters the child's rollback calculation. The child "prunes" that branch of the game tree in which the parent serves dessert even if the child has not finished its vegetables. This may alter the child's behavior; the parent hopes that it will make the child act as the parent wants it to. Similarly, in the "study game," the promise of the bike may induce a child to study harder.

2 CREDIBILITY OF STRATEGIC MOVES

We have already seen that payoffs to the other player can be altered by one player's strategic move, but what about the payoffs for the player making that move? Of course, A gets a higher payoff when B acts in conformity with A's wishes. But A's payoff may also be affected by his own response. In the case of a threat, A's threatened response if B does not act as A would wish may have consequences for A's own payoffs: the parent may be made unhappy by the sight of the unhappy child who has been denied dessert. Similarly, in the case of a promise, rewarding B if he does act as A would wish can affect A's own payoff: the parent who rewards the child for studying hard has to incur the monetary cost of the gift, but is happy to see the child's happiness on receiving the gift and even happier about the academic performance of the child.

This effect on A's payoffs has an important implication for the efficacy of A's strategic moves. Consider the threat. If A's payoff is actually increased by carrying out the threatened action, then B reasons that A will carry out this action even if B fulfills A's demands. Therefore B has no incentive to comply with A's wishes, and the threat is ineffective. For example, if the parent is a sadist who enjoys seeing the child go without dessert, then the child thinks, "I am not going to get dessert anyway, so why eat the vegetables?"

Therefore an essential aspect of a threat is that it should be costly for the threatener to carry out the threatened action. In the dinner game, the parent must prefer to give the child dessert. Threats in the true strategic sense have the innate property of imposing some cost on the threatener too; they are threats of *mutual harm.*

In technical terms, a threat fixes your strategy (response rule) in the subsequent game. A strategy must specify what you will do in each eventuality along the game tree. Thus, "no dessert if you don't finish your vegetables" is an incomplete specification of the strategy; it should be supplemented by "and dessert if you do." Threats generally don't specify this latter part. Why not? Because the second part of the strategy is automatically understood; it is implicit. And for the threat to work, this second part of the strategy—the implied promise in this case—has to be automatically credible too.

Thus the threat "no dessert if you don't finish your vegetables" carries with it an implicit promise of "dessert if you do finish your vegetables." This promise should also be credible if the threat is to have the desired effect. In our example, the credibility of the implicit promise is automatic when the parent prefers to see the child get and enjoy its dessert. In other words, the implicit promise is automatically credible precisely when the threatened action is costly for the parent to carry out.

To put it yet another way, a threat carries with it the stipulation that you will do something if your wishes are not met that, if those circumstances actually arise, you will regret having to do. Then why make this stipulation at the first stage? Why tie your own hands in this way when it might seem that leaving one's options open would always be preferable? Because in the realm of game theory, having more options is not always preferable. In the case of a threat, your lack of freedom in the second stage of the game has strategic value. It changes other players' expectations about your future responses, and you can use this to your advantage.

A similar issue arises with a promise. If the child knows that the parent enjoys giving him gifts, he may expect to get the racing bike anyway on some occasion in the near future, for example, an upcoming birthday. Then the promise of the bike has little effect on the child's incentive to study hard. To have the intended strategic effect, the promised reward must be so costly to provide that the other player would not expect you to hand over that reward anyway. (This is a useful lesson in strategy you can point out to your parents: the rewards they promise must be larger and more costly than what they would give you just for the pleasure of seeing you happy.)

The same is true of unconditional strategic moves (commitments too). In bargaining, for example, others know that when you have the freedom to act, you also have the freedom to capitulate, so a "no concessions" commitment can secure you a better deal. If you hold out for 60% of the pie, and the other party offers you 55%, you may be tempted to take it. But if you can credibly assert in advance that you will not take less than 60%, then this temptation does not arise and you can do better than you otherwise would.

Thus it is in the very nature of strategic moves that after the fact—that is, when the stage 2 game actually requires it—you do not want to carry out the action you had stipulated you would take. This is true for all types of strategic moves and it is what makes credibility so problematic. You have to do something that creates credibility—something that convincingly tells the other player that you will not give in to the temptation to deviate from the stipulated action when the time comes—in order for your strategic move to work. That is why giving up your own freedom to act can be strategically beneficial. Alternatively, credibility can be achieved by changing your own payoffs in the second-stage game in such a way that it becomes truly optimal for you to act as you declare.

Thus there are two general ways of making your strategic moves credible: remove from your own set of future choices the other moves that may tempt you, or reduce your own payoffs from those temptation moves so that the stipulated move becomes the actual best one. In the sections that follow, we first elucidate the mechanics of strategic moves, assuming them to be credible. We make some comments about credibility as we go along, but postpone our general analysis of credibility to the last section of the chapter.

3 COMMITMENTS

We studied the game of chicken in Chapter 4, and found two pure-strategy Nash equilibria. Each player prefers the equilibrium in which he goes straight and the other person swerves.[1] We saw in Chapter 6 that if the game were to have sequential rather than simultaneous moves, the first mover would choose Straight, leaving the second to make the best of the situation by settling for Swerve rather than causing a crash. Now we can consider the same matter from another perspective. Even if the game itself has simultaneous moves, if one player can make a strategic move—create a *first stage* in which he makes a credible declaration about his action in the chicken game itself, which is to be played at the *second stage*—then he can get the same advantage afforded a first mover by making a commitment to act tough (choose Straight).

While the point is simple, we outline the formal analysis to develop your understanding and skill, which will be useful for later, more complex, examples. Remember our two players, James and Dean. Suppose James is the one who has the opportunity to make a strategic move. Figure 9.1 shows the tree for the two-stage game. At the first stage, James has to decide whether to make a commitment. Along the upper branch emerging from the first node, he does not make the commitment. Then at the second stage the simultaneous-move game is played, and its payoff table is the familiar one shown in Figures 4.11 or 6.14. This second-stage game has multiple equilibria, and James gets his best payoff in only one of them. Along the lower branch, James makes the commitment. Here we interpret this to mean giving up his freedom to act in such a way that Straight is the only action available to James at this stage. Therefore the second-stage game table has only one row for James, corresponding to his declared choice of Straight. In this table, Dean's best action is Swerve, so the equilibrium outcome gives James his best payoff. Therefore at the first stage, James finds it optimal to make the commitment; this strategic move ensures his best payoff, while not committing leaves the matter uncertain.

How can James make this commitment credibly? Like any first move, the commitment move must be (1) irreversible, and (2) visible to the other player. People have suggested some extreme and amusing ideas. James can disconnect the steering wheel of the car and throw it out of the window, so Dean can see that James can no longer Swerve. (James could just tie the wheel so that it could no longer be turned, but it would be more difficult to demonstrate to Dean that the wheel was truly tied, and that the knot was not a trick one that could be undone quickly.) These devices simply remove the Swerve option from the set of choices

[1] We saw in Chapter 5, and will see again in Chapter 10, that the game has a third equilibrium, in mixed strategies, in which both players do quite poorly.

available to James in the stage 2 game, leaving Straight as the only thing he *can* do.

More plausibly, if such games are played every weekend, James can acquire a general reputation for toughness that acts as a guarantee of his action on any one day. In other words, James can alter his own payoff from swerving, by subtracting an amount that represents the loss of reputation. If this amount is large enough, say 3, then the second-stage game when James has made the commitment has a different payoff table. The complete tree for this version of the game is shown in Figure 9.2.

Now in the second stage with commitment, Straight has become truly optimal for James; in fact, it is his dominant strategy in that stage. Dean's optimal

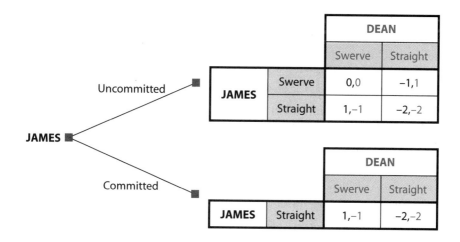

FIGURE 9.1   Chicken: Commitment by Restricting Freedom to Act

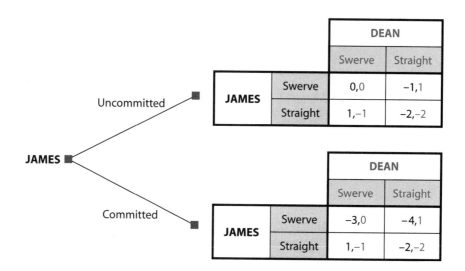

FIGURE 9.2   Chicken: Commitment by Changing Payoffs

strategy is then Swerve. Looking ahead to this outcome at stage 1, James sees that he gets 1 by making the commitment (changing his own stage 2 payoffs), while without the commitment he cannot be sure of 1 and may do much worse. Thus a rollback analysis shows that James should make the commitment.

Of course, both (or all) can play the game of commitment, so success may depend on both the speed with which you can seize the first move and on the credibility with which you can make that move. If there are lags in observation, the two may even make incompatible simultaneous commitments: each disconnects his steering wheel and tosses it out of the window just as he sees the other's wheel come flying out, and then the crash is unavoidable.

Even if one of the players has the advantage in making a commitment, the other player can defeat the first player's attempt to do so. The second player could demonstrably remove his ability to "see" the other's commitment, for example, by cutting off communication.

Games of chicken may be a 1950s anachronism, but our second example is perennial and familiar. In a class, the teacher's deadline enforcement policy can be Weak or Tough, and the students' work can be Punctual or Late. Figure 9.3 shows this game in the strategic form. The teacher does not like being tough; for him the best outcome (a payoff of 4) is when students are punctual even when he is weak; the worst (1) is when he is tough but students are still late. Of the two intermediate strategies, he recognizes the importance of punctuality, and rates (Tough, Punctual) better than (Weak, Late). The students most prefer the outcome (Weak, Late), where they can party all weekend without suffering any penalty for the late assignment. (Tough, Late) is the worst for them, just as it is for the teacher. Between the intermediate ones, they prefer (Weak, Punctual) over (Tough, Punctual) because they have better self-esteem if they can think they acted punctually on their own volition rather than because of the threat of a penalty.[2]

		STUDENT	
		Punctual	Late
TEACHER	Weak	4, 3	2, 4
	Tough	3, 2	1, 1

FIGURE 9.3 Payoff Table for Class Deadline Game

---

[2] You may not regard these specific rankings of outcomes as applicable either to you or to your own teachers. We ask you to accept them for this example, whose main purpose is to convey some *general ideas* about commitment in a simple way. The same disclaimer also applies to all the examples that follow.

If this is played as a simultaneous-move game, or if the teacher moves second, Weak is dominant for the teacher, and then the student chooses Late. The equilibrium outcome is (Weak, Late), and the payoffs are (2, 4). But the teacher can achieve a better outcome by committing at the outset to the policy of Tough. We do not draw a tree as we did in Figures 9.1 and 9.2. The tree would be very similar to that for the chicken case above, so we leave it for you to draw. Without the commitment, the second-stage game is as before, and the teacher gets a 2. When the teacher is committed to Tough, the students find it better to respond with Punctual at the second stage, and the teacher gets a 3.

The teacher commits to a move *different from* what he would do in simultaneous play, or indeed, his best-second move if the students moved first. This is where the strategic thinking enters. The teacher has nothing to gain by declaring that he will have a Weak enforcement regime; the students expect that anyway in the absence of any declaration. To gain advantage by making a strategic move, he must commit *not* to follow what would be his equilibrium strategy of the simultaneous-move game. This strategic move changes the students' expectations, and therefore their action. Once they believe the teacher is really committed to tough discipline, they will choose to turn in their assignments punctually. Of course, if they tested this out by being late, the teacher would like to forgive them, maybe with an excuse to himself, such as "just this once." The existence of this temptation to shift away from your commitment is what makes its credibility problematic.

Even more dramatically, in this instance the teacher benefits by making a strategic move that commits him to a dominated strategy. He commits to choosing Tough, which is dominated by Weak. The choice of Tough gets the teacher a 3 if the student chooses Punctual and a 1 if the student chooses Late, whereas if the teacher had chosen Weak, his corresponding payoffs would have been 4 and 2. If you think it paradoxical that one can gain by choosing a dominated strategy, you are extending the concept of dominance beyond the proper scope of its validity. Dominance involves either of two calculations: (1) After the other player does something, how do I respond, and is some choice best (or worst), given all possibilities? (2) If the other player is simultaneously doing action X, what is best (or worst) for me, and is this the same for all the X actions the other could be choosing? Neither is relevant when you are moving first. Instead, you must look ahead to how the other will respond. Therefore the teacher does not compare his payoffs in vertically adjacent cells of the table (taking the possible actions of the students one at a time). Instead, he calculates how the students will react to each of his moves. If he is committed to Tough, they will be Punctual, but if he is committed to Weak (or uncommitted), they will be Late, so the only pertinent comparison is that of the top right cell with the bottom left, of which the teacher prefers the latter.

To be credible, the teacher's commitment must be everything a first move has to be. First, it must be made before the other side makes its move. The teacher must establish the ground rules of deadline enforcement before the assignment is due. Next, it must be observable—the students must know the rules by which they must abide. Finally, and perhaps the most important, it must be irreversible—the students must know that the teacher cannot, or at any rate will not, change his mind and forgive them. A teacher who leaves loopholes and provisions for incompletely specified emergencies is merely inviting imaginative excuses accompanied by fulsome apologies and assertions that "it won't happen again."

The teacher might achieve credibility by hiding behind general university regulations; this simply removes the Weak option from his set of available choices at stage 2. Or, as is true in the chicken game, he might establish a reputation for toughness, changing his own payoffs from Weak by creating a sufficiently high cost of loss of reputation.

4 THREATS AND PROMISES

We emphasize that threats and promises are *response rules:* your actual future action is conditioned on what the other players do in the meantime, but your freedom of future action is constrained to following the stated rule. Once again, the aim is to alter the other players' expectations and therefore their actions in a way favorable to you. Tying yourself to a rule, which you would not want to follow if you were completely free to act at the later time, is an essential part of this process. Thus the initial declaration of intention must be credible. Once again we will elucidate some principles for achieving credibility of these moves, but we remind you that their actual implementation remains largely an art.

Remember the taxonomy given in Section 1: A *threat* is a response rule that leads to a bad outcome for the other players if they act contrary to your interests. A *promise* is a response rule by which you offer to create a good outcome for the other players if they act in a way that promotes your own interests. Each of these may aim either to stop the other players doing something that they would otherwise do (*deterrence*), or to induce them to do something that they would otherwise not do (*compellence*). We consider these features in turn.

A. Example of a Threat: U.S.–Japanese Trade Relations

Our example comes from a hardy perennial of U.S. international economic policy, namely trade friction with Japan. Each country has the choice of keeping its own markets open or closed to the other's goods. They have somewhat different preferences over the outcomes.

Figure 9.4 shows the payoff table for the trade game. For the United States, the best outcome (a payoff of 4) comes when both markets are open; this is partly because of its overall commitment to the market system and free trade, and partly because of the benefit of trade with Japan itself—U.S. consumers get high-quality cars and consumer electronics products, and U.S. producers can export their agricultural and high-tech products. Similarly its worst outcome (payoff 1) occurs when both markets are closed. Of the two outcomes when only one market is open, the U.S. would prefer its own market to be open, because the Japanese market is smaller, and loss of access to it is less important than the loss of access to Hondas and Walkmen.

As for Japan, for the purpose of this example we accept the protectionist, producer-oriented picture of Japan, Inc. Their best outcome is when the U.S. market is open and their own is closed, their worst when matters are the other way around. Of the other two outcomes, they prefer that both markets be open, because their producers then have access to the much larger U.S. market.[3]

		JAPAN	
		Open	Closed
UNITED STATES	Open	4, 3	3, 4
	Closed	2, 1	1, 2

FIGURE 9.4   Payoff Table for the United States–Japan Trade Game

Both sides have dominant strategies. No matter how the game is played—simultaneously or sequentially with either move order—the equilibrium outcome is (Open, Closed), and the payoffs are (3, 4). This outcome also fits well the common American impression of how the actual trade policies of the two countries work.

Japan is already getting its best payoff in this equilibrium, and so has no need to try any strategic moves. The United States, however, can try to get a 4 instead of a 3. But in this case an ordinary unconditional commitment will not work. Japan's best response, no matter what commitment the United States makes, is to keep its market closed. Then the United States does better for itself by committing to keep its own market open, which is the equilibrium without any strategic moves anyway.

But suppose the United States can choose the conditional response rule "We

---

[3] Again we ask you to accept this payoff structure as a vehicle for conveying the ideas. You can experiment with the payoff tables to see what difference that would make to the role and effectiveness of the strategic moves.

will close our market if you close yours." The situation then becomes the two-stage game shown in Figure 9.5. If the United States does not use the threat, the second stage is as before and leads to the equilibrium in which the U.S. market is open and it gets a 3, while the Japanese market is closed and it gets a 4. If the United States does use the threat, then at the second stage only Japan has freedom of choice; given what Japan does, the United States then merely does what its response rule dictates. Therefore along this branch of the tree we show only Japan as an active player and write down the payoffs to the two parties: If Japan keeps its market closed, the United States closes its own, and the United States gets a 1 and Japan gets a 2. If Japan keeps its market open, then the United States threat has worked, it is happy to keep its own market open, and it gets a 4 while Japan gets a 3. Of these two possibilities, the second is better for Japan.

Now we can use the familiar rollback reasoning. Knowing how the second stage will work in all eventualities, it is better for the United States to deploy its threat at the first stage. This threat will result in an open market in Japan, and the United States will get its best outcome.

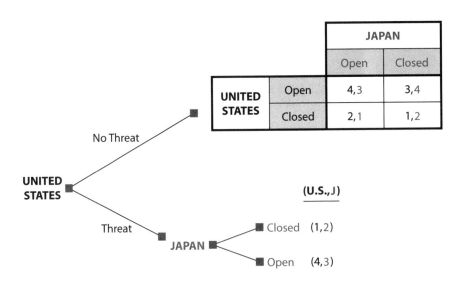

FIGURE 9.5 Tree for the United States–Japan Trade Game with Threat

Having described the mechanics of the threat, we now point out some of its important features.

1. When the United States deploys its threat credibly, Japan doesn't follow its dominant strategy Closed. Again, the idea of dominance is relevant only in the context of simultaneous moves or when Japan moves second. Here Japan knows that the United States will take actions that depart from *its* dominant strategy. In the payoff table, Japan is looking at a choice between just two cells, the top left and the bottom right, and of those two, it prefers the former.

2. Credibility of the threat is problematic because, if Japan puts it to the test by keeping its market closed, the United States faces the temptation to refrain from carrying out the threat. In fact, if the threatened action were the best U.S. response after the fact, then there would be no need to make the threat in advance (but the United States might issue a *warning* just to make sure the Japanese understand the situation). The strategic move has a special role exactly because it locks a player into doing something other than what it would have wanted to do after the fact. As we explained earlier, a threat in the true strategic sense is necessarily costly for the threatener to carry out; the threatened action would inflict *mutual* harm.

3. The conditional rule "We will close our market if you close yours" does not completely specify the U.S. *strategy*. To be complete, it needs an additional clause indicating what the United States will do in response to an open Japanese market: "and we will keep our market open if you keep yours open." This additional clause, the implicit promise, is really part of the threat, but it does not need to be stated explicitly because it is automatically credible. Given the payoffs of the second-stage game, it is in the best interests of the United States to keep its market open if Japan keeps its market open. If that were not the case, if the United States would respond by keeping its market closed even when Japan kept its own market open, then the implicit promise would have to be made explicit and somehow made credible. Otherwise the U.S. threat would become tantamount to the unconditional commitment "We will keep our market closed," and that would not draw forth the desired response from Japan.

4. The threat, when credibly deployed, results in a change in Japan's action. Depending on the status quo, we can regard this as deterrence or compellence. If the Japanese market is initially open, and they are considering a switch to protectionism, then the threat deters them from that action. But if the Japanese market is initially closed, then the threat compels them to open it. Thus whether a strategic move is deterrent or compellent depends on the status quo. The distinction may seem to be a matter of semantics, but in practice the credibility of a move and the way it works are importantly affected by this distinction. We return to these issues later in the chapter.

5. Here are a few ways the United States can make its threat credible. (a) It can enact a law that mandates the threatened action under the right circumstances. This removes the temptation action from the set of available choices at stage 2. Some reciprocity provisions in the World Trade Organization agreements have this effect, but the procedures are very slow and uncertain. (b) It can delegate fulfillment to an agency like the U.S. Commerce Department that is captured by U.S. producers who would like to keep our markets closed and so reduce the competitive pressure on themselves. This changes the U.S. payoffs at stage 2—replacing the true U.S. payoffs by those of the Commerce Department—with the result that the threatened action becomes truly optimal. (The

danger is that the Commerce Department will then retain a protectionist stance even if Japan opens its market; gaining credibility for the threat may lose credibility for the implied promise.)

6. If a threat works, it doesn't have to be carried out. So its cost to you is immaterial. In practice, the danger that you may have miscalculated or the risk that the threatened action will occur by error even if the other player complies are strong reasons to refrain from using threats more severe than necessary. To make the point starkly, the United States could threaten to pull out of defensive alliances with Japan if they didn't buy our rice and semiconductors, but that threat is "too big" and too risky for the United States to ever carry out; therefore it is not credible.

But sometimes a range of threats is not available from which a player can choose one that is, on the one hand, sufficiently big that the other player fears it and alters his action in the way the first player desires, and on the other hand, not so big as to be too risky for the first player to ever carry out and therefore lacking credibility. If the only available threat is too big, then a player can reduce its size by making its fulfillment a matter of chance. Instead of saying, "If you don't open your markets, we will refuse to defend you in the future," the United States can say to Japan, "If you don't open your markets, the relations between our countries will deteriorate to the point where Congress may refuse to allow us to come to your assistance if you are ever attacked, even though we do have an alliance." In fact, the United States can deliberately foster sentiments that raise the probability that Congress will do just that, so the Japanese will feel the danger more vividly. A threat of this kind, which creates a *risk* but not a certainty of the bad outcome, is called **brinkmanship.** It is an extremely delicate and even dangerous variant of the strategic move. We will study brinkmanship in greater detail in Chapter 13.

7. Japan gets a worse outcome when the United States deploys its threat than it would without this threat; so it would like to take strategic actions that defeat or disable the United States's attempts to use the threat. For example, suppose their market is currently closed, and the United States is attempting compellence. The Japanese can accede in principle but stall in practice, pleading necessary delays for assembling the necessary political consensus to legislate the market opening, then delays for writing the necessary administrative regulations to implement the legislation, and so on. Since the United States does not want to go ahead with its threatened action, at each point it has the temptation to accept the delay. Or Japan can claim that their domestic politics makes it difficult for them to open all markets fully; will the United States accept the outcome if Japan keeps just a few of its industries protected? It gradually expands this list, and at any point the extra small step is not enough cause for the United States to unleash a trade war. This device of defeating a compellent threat by small steps, or "slice by slice," is called **salami tactics.**

B. Example of a Promise: The Pizza Pricing Game

We now illustrate a promise using the pizza-store prisoners' dilemma from Chapters 4 and 8. We consider only the two strategies of High and Medium prices, and reproduce the payoff table as Figure 9.6. Without any strategic moves, the game has the usual equilibrium in dominant strategies in which both stores charge the medium price, and both get lower profits than they would if they both charged the high price.

		PIERCE'S PIZZA PIES	
		High	Medium
DONNA'S DEEP DISH	High	60, 60	36, 70
	Medium	70, 36	50, 50

FIGURE 9.6   Payoff Table for the Pizza Sellers' Prisoners' Dilemma

If either side can make the credible promise "I will charge a high price if you do," the cooperative outcome is achieved. For example, if Pierce's makes the promise, then Donna's knows that its choice of High will be reciprocated, leading to the payoff shown in the upper left cell of the table, and that its choice of Medium will bring forth Pierce's usual action, namely Medium, leading to the lower right cell. Between the two, Donna's prefers the first, and therefore chooses High.

The analysis can be done more properly by drawing a tree for the two-stage game in which Pierce's has the choice of making or not making the promise at the first stage. We omit the tree, partly so you can improve your understanding of the process by constructing it yourself, and partly to show how such detailed analysis becomes unnecessary as one becomes familiar with the ideas.

The credibility of Pierce's promise is open to doubt. In order to respond to what Donna's does, Pierce's must arrange to move second in the second stage of the game; correspondingly, Donna's must move first in stage 2. Remember that a first move is an irreversible and observable action. Therefore, if Donna's moves first and prices high, it leaves itself vulnerable to Pierce's cheating, and Pierce's is very tempted to renege on its promise to price high when it sees Donna's in this vulnerable position. Pierce's must somehow convince Donna's that it will not give in to the temptation to charge a medium price when Donna's charges a high price.

How can it do so? Perhaps Pierce's owner can leave the pricing decision in the hands of a local manager, with clear written instructions to reciprocate with the high price if Donna's charges the high price. Pierce's owner can invite Donna's to inspect these instructions, after which he leaves on a solo round-the-world sail-boat trip so he cannot rescind them. (Even then, Donna's management may be doubtful—Pierce might secretly carry a telephone or a laptop computer on board.) This scenario is tantamount to removing the cheating action from the choices available to Pierce's at stage 2.

Or Pierce's store can develop a reputation for keeping its promises, in business and in the community more generally. In a repeated relationship, the promise may work because reneging on the promise once may cause future cooperation to collapse. In essence, an ongoing relationship means splitting the game into smaller segments, in each of which the benefit from reneging is too small to justify the costs. In each such game then, the payoff from cheating is altered by the cost of collapse of future cooperation.

We saw earlier that every threat has an implicit attached promise. Similarly every promise has an implicit attached threat. In this case it is "I will charge the medium price if you do." It does not have to be stated explicitly because it is automatically credible—it describes Pierce's best response to Donna's medium price.

There is also an important difference between a threat and a promise. If a threat is successful, it doesn't have to be carried out, and is then costless to the threatener. Therefore a threat can be bigger than what is needed to make it effective (although making it too big may be too risky, even to the point of losing its credibility as we suggested earlier). If a promise *is* successful in altering the other's action in the desired direction, then the promisor has to deliver what he had promised, so it is costly. In the above example the cost is simply giving up the opportunity to cheat and get the highest payoff; in other instances where the promisor offers an actual gift or an inducement to the other, the cost may be more tangible. In either case, the player making the promise has a natural incentive to keep its size small—just big enough to be effective.

C. Example Combining Threat and Promise: Joint U.S.–European Military Operations

When we considered threats and promises one at a time, the explicit statement of a threat included an implicit clause of a promise that was automatically credible, and vice versa. There can, however, be situations in which the credibility of both aspects is open to question; then the strategic move has to make both aspects explicit and make them both credible.

Our example of an explicit-threat-and-promise combination comes from a context, unfortunately all too frequent these days, in which the United States and Western European countries must contemplate military intervention in some other country, usually closer to Europe than to the United States, in order

		UNITED STATES	
		Yes	No
EUROPE	Yes	3, 3	2, 4
	No	4, 1	1, 2

FIGURE 9.7   Payoff Table for the Military Intervention Problem

to prevent a civil war or other regional conflict. We consider a plausible scenario of this kind and show in Figure 9.7 the payoff table for the United States and Europe.

Each party would like the other to take on the whole burden of intervention, so the top right cell has the best payoff for the United States (4), and the bottom left cell is best for Europe. The worst situation for the United States is where it takes on the whole burden. The Europeans are closer to the scene of the conflict, however, so their worst outcome arises when neither party gets involved. Both regard a joint involvement as the second-best (a payoff of 3). The United States assigns a payoff of only 2 to the situation in which neither side intervenes, because this weakens NATO (The North Atlantic Treaty Organization) and reflects badly on the United States's reputation as a superpower in other parts of the world. For Europe, a payoff of 2 is assigned to their own sole involvement in the intervention.

Without any strategic moves, the intervention game is dominance solvable. No is the dominant strategy for the United States, and then Yes is Europe's best choice. The equilibrium outcome is the top right cell, with payoffs of 2 for Europe and 4 for the United States. Since the United States gets its best outcome, it has no reason to try any strategic moves. But Europe can try to do better than a 2.

What strategic move will work to improve Europe's equilibrium payoff? An unconditional move (commitment) will not work, because the United States will respond "No" to either first move by Europe. A threat alone ("We won't intervene unless you do") does not work because the implied promise ("We will if you do") is not credible—if the United States does intervene, Europe has the temptation to back off and leave everything to the United States, getting a payoff of 4 instead of the 3 that would come from fulfilling the promise. A promise alone won't work: since the United States knows that Europe will intervene if the United States does not, a European promise of "We will intervene if you do" becomes tantamount to a simple commitment to intervene; then the United States can stay out and get its best payoff of 4.

In this game, Europe's explicit promise must carry the implied threat "We won't intervene if you don't," but that threat is not automatically credible. Similarly, Europe's explicit threat must carry the implied promise "We will intervene if you do," but that is also not automatically credible. Therefore Europe has to make both the threat and the promise explicit. It must issue the combined threat-*cum*-promise "We will intervene if, and only if, you do." Of course, it needs to make both clauses credible. Usually such credibility has to be achieved by means of a treaty that covers the whole relationship, not just with agreements negotiated separately when each incident arises.

5 SOME ADDITIONAL ISSUES

A. When Do Strategic Moves Help?

We have seen several examples in which a strategic move brings a better outcome to one player or another, as compared with the original game without such moves. What can be said in general about the desirability of such moves?

An unconditional move—a commitment—need not always be advantageous to the player making it. In fact, if the original game gives the advantage to the second mover, then it is a mistake to commit oneself to move in advance, thereby effectively becoming the first mover.

The availability of a conditional move—threat or promise—can never be an actual disadvantage. At the very worst, one can commit to a response rule that would have been optimal after the fact. However, if such moves bring one an actual gain, it must be because one is choosing a response rule that in some eventualities specifies an action different from what one would find optimal at that later time. Thus whenever threats and promises bring a positive gain, they do so precisely when (one might say precisely because) their credibility is inherently questionable, and must be achieved by some specific credibility "device." We have mentioned some such devices in connection with each earlier example, and later discuss the topic of achieving credibility in greater generality.

What about the desirability of being on the receiving end of a strategic move? It is never desirable to let the other player threaten you. If this seems likely, you can gain by looking for a different kind of advance action—one that makes the threat less effective or less credible. We will consider some such actions shortly. However, it is often desirable to let the other player make promises to you. In fact, both players may benefit when one can make a credible promise, as in the prisoners' dilemma example of pizza pricing earlier in this chapter, in which a promise achieved the cooperative outcome. Thus it may be in the players' mutual interest to facilitate the making of promises by one or both of them.

B. Deterrence Versus Compellence

In principle, either a threat or a promise can achieve either deterrence or compellence. For example, a parent who wants a child to study hard (compellence) can promise a reward (a new racing bike) for good performance in school, or threaten a punishment (a strict curfew the following term) if the performance is not sufficiently good. Similarly, a parent who wants the child to keep away from bad company (deterrence) can try either a reward (promise) or a punishment (threat). In practice, the two types of strategic moves work somewhat differently, and that will affect the ultimate decision regarding which to use. Generally, deterrence is better achieved by a threat and compellence by a promise. The reason is an underlying difference of timing and initiative.

A deterrent threat can be passive—you don't need to do anything so long as the other player doesn't do what you are trying to deter. And it can be static—you don't have to impose any time limit. Thus you can set a trip wire and then leave things up to the other player. So the parent who wants the child to keep away from bad company can say, "If I ever catch you with X again, I will impose a 7 P.M. curfew on you for a whole year." Then the parent can sit back to wait and watch; only if the child acts contrary to its wishes does the parent have to act on its threat. Trying to achieve the same deterrence by a promise would require more complex monitoring and continual action: "At the end of each month that I know you did not associate with X, I will give you $25."

Compellence must have a deadline or it is pointless—the other side can defeat your purpose by procrastinating or by eroding your threat in small steps (salami tactics). This makes a compellent threat harder to implement than a compellent promise. The parent who wants the child to study hard can simply say, "Each term you get an average of B or better, I will give you CDs or games worth $500." The child will then take the initiative in showing the parent each time he has fulfilled the conditions. Trying to achieve the same thing by a threat—"Each term your average falls below B, I will take away one of your computer games"—will require the parent to be much more vigilant and active. The child will postpone bringing the grade report, or will try to hide the games.

Of course, the concepts of reward and punishment are relative to those of some status quo. If the child has a perpetual right to the games, then taking one away is a punishment; if the games are temporarily assigned to the child on a term-by-term basis, then renewing the assignment for another term is a reward. Therefore you can change a threat into a promise, or vice versa, by changing the status quo. You can use this to your own advantage when making a strategic move. If you want to achieve compellence, try to choose a status quo such that what you do when the other player acts to comply with your demand becomes a reward, so you are using a compellent promise. To give a rather dramatic example, a mugger can convert the threat "If you don't give me your wallet, I will take out my knife and cut your throat" into the promise "Here is a knife at your throat; as soon as you give me your wallet I will take it away." But if you want to achieve deterrence, try to choose a status quo such that if the other player acts contrary to your wishes, what you do is a punishment, so you are using a deterrent threat.

6 ACQUIRING CREDIBILITY

We have throughout emphasized the importance of credibility of strategic moves, and we accompanied each example with some brief remarks about how credibility could be achieved in that particular context. Devices for achieving credibility are indeed often context-specific, and there is a lot of art to discovering or developing such devices. Some general principles can help you organize your search.

We pointed out two broad approaches to credibility: reducing your own future freedom of action in such a way that you have no choice but to carry out the action stipulated by your strategic move, and changing your own future payoffs in such a way that it becomes optimal for you to do what you stipulate in your strategic move. We now elaborate some practical methods for implementing each of these approaches.

A. Reducing Your Freedom of Action

I. AUTOMATIC FULFILLMENT Suppose at stage 1 you relinquish your choice at stage 2, and hand it over to a mechanical device or similar procedure or mechanism that is programmed to carry out your committed, threatened, or promised action under the appropriate circumstances. You demonstrate to the other player that you have done so. Then he will be convinced that you have no freedom to change your mind, and your strategic move will be credible. The **doomsday device,** a nuclear explosive device that would detonate and contaminate the whole world's atmosphere if the enemy launched a nuclear attack, is the best known example of this, popularized by the early-1960s movies *Fail Safe* and *Dr. Strangelove*. Luckily, it remained in the realm of fiction. But automatic procedures that retaliate with import tariffs if another country tries to subsidize its exports to your country (countervailing duties) are quite common in the arena of trade policy.

II. DELEGATION A fulfillment device does not even have to be mechanical. You could delegate the power to act to another human or to an organization that is required to follow certain preset rules or procedures. In fact, that is how the countervailing duties work. They are set by two agencies of the U.S. government—the

Commerce Department and the International Trade Commission—whose operating procedures are laid down in the general trade laws of the country.

An agent should not have his own objectives that defeat the purpose of his strategic move. For example, if one player delegates to an agent the task of inflicting threatened punishment, and the agent is a sadist who enjoys inflicting punishment, then he may act even when there is no reason to, that is, even when the second player has complied. If the second player suspects this, then the threat loses its effectiveness, because the punishment becomes a case of "damned if you do and damned if you don't."

Delegation devices are not complete guarantees of credibility. Even the doomsday device may fail to be credible if the other side suspects that you control an override button in order to avoid the risk of a catastrophe. And delegation and mandates can always be altered; in fact, the U.S. government has often set aside the stipulated countervailing duties and reached other forms of agreements with other countries so as to avoid costly trade wars.

III. BURNING BRIDGES Many invaders, from Xenophon in ancient Greece to William the Conqueror in England to Cortes in Mexico, are supposed to have deliberately cut off their own army's avenue of retreat to ensure that it will fight hard. Some of them literally burned bridges behind them, while others burned ships, but the device has become a cliche. Its most recent users in military contexts may have been the Japanese *kamikaze* pilots in World War II, who took only enough fuel to reach the United States naval ships into which they were to ram their airplanes. The principle even appears in the earliest known treatise on war, in a commentary attributed to Prince Fu Ch'ai: "Wild beasts, when they are at bay, fight desperately. How much more is this true of men! If they know there is no alternative they will fight to the death."[4]

Related devices are used in other high-stakes games. Although the European Monetary Union could have retained separate currencies and merely fixed the exchange rates between them, a common currency was adopted precisely to make the process irreversible and thereby give the member countries a much greater incentive to make the union a success. (In fact, it is the extent of the necessary commitment that has kept some nations, Great Britain in particular, from agreeing to be part of the European Monetary Union.) Of course, it is not totally impossible to abandon a common currency and go back to separate national ones; it is just inordinately costly. If things get really bad inside the Union, one or more countries may yet choose to get out. As with automatic devices, the credibility of burning bridges is not an all-or-nothing matter, but one of degree.

---

[4]Sun Tzu, *The Art of War*, trans. Samuel B. Griffith (Oxford, England: Oxford University Press, 1963), p. 110.

IV. CUTTING OFF COMMUNICATION   If you send the other player a message demonstrating your commitment, and at the same time cut off any means for him to communicate with you, then he cannot argue or bargain with you to reverse your action. The danger in cutting off communication is that if both players do this simultaneously, then they may make mutually incompatible commitments that can cause great mutual harm. Also, cutting off commuication is harder to do with a threat, because you have to remain open to the one message that tells you whether the other player has complied, and therefore whether you need to carry out your threat. Of course, in this age it is also quite difficult for an individual to cut himself off from all contact.

But players who are large teams or organizations can try variants of this device. Consider a labor union that makes its decisions at mass meetings of members. To convene such a meeting takes a lot of planning—reserving a hall, communicating with members, and so forth—and several weeks of time. A meeting is convened to decide on a wage demand. If management does not meet the demand in full, the union leadership is authorized to call a strike and then it must call a new mass meeting to consider any counteroffer. This process puts management under a lot of time pressure in the bargaining; it knows that the union will not be open to communication for several weeks at a time. Here we see that cutting off communication for extended periods can establish some degree of, but not absolute, credibility. The union's device does not make communication totally impossible, it only creates several weeks of delay.

B. Changing Your Payoffs

I. REPUTATION   You can acquire a **reputation** for carrying out threats and delivering on promises. Such a reputation is most useful in a repeated game against the same player. It is also useful when playing different games against different players, if each of them can observe your actions in the games you play with others. The circumstances favorable to the emergence of such a reputation are the same as those for achieving cooperation in the prisoners' dilemma, and for the same reasons. The greater the likelihood that the interaction will continue, and the greater the concern for the future relative to the present, the more likely the players will be to sacrifice current temptations for the sake of future gains. The players will therefore be more willing to acquire and maintain reputations.

In technical terms, this device links different games, and the payoffs of actions in one game are altered by the prospects of repercussions in other games. If you fail to carry out your threat or promise in one game, your reputation suffers and you get a lower payoff in other games. Therefore when you consider any one of these games, you should adjust your payoffs in it to reflect such repercussions on your payoffs in the linked games.

The benefit of reputation in ongoing relationships explains why your regular car mechanic is less likely to cheat you and do an unnecessary or excessively costly or shoddy repair than is a random garage you go to in an emergency. But what does your regular mechanic actually stand to gain from acquiring this reputation if competition forces him to charge a price so low that he makes no profit on any deal? His integrity in repairing your car must come at a price—you have to be willing to let him charge you a little bit more than the rates the cheapest garage in the area might advertise.

The same reasoning also explains why, when you are away from home, you might settle for the known quality of a restaurant chain instead of taking the risk of going to an unknown local restaurant. And a department store that expands into a new line of merchandise can use the reputation it has acquired in its existing lines to promise its customers the same high quality in the new line.

In games where credible promises by one or both parties can bring mutual benefit, the players can agree and even cooperate in fostering the development of reputation mechanisms. But if the interaction ends at a known finite time, there is always the problem of the endgame.

In the Middle East peace process, the endgame difficulty arises when Israel has transferred power in the West Bank to the Palestinian Authority, and the latter has to make good on its promises to continue to respect Israel's integrity and safety. If Israel doubts the credibility of this promise, it will be reluctant to perform the earlier steps of territorial transfer. Indeed, in 1998 the peace process stalled. The community of other nations may find it necessary to offer to both parties sufficiently attractive rewards beyond the land-for-peace exchange—for example, contingent offers of economic aid or prospects of expanded commerce—to keep the process going.

II. DIVIDING THE GAME INTO SMALL STEPS   Sometimes a single game can be divided into a sequence of smaller games, thereby allowing the reputation mechanism to come into effect. In home construction projects, it is customary to pay by installments as the work progresses. In the Middle East peace process, Israel would never have agreed to a complete transfer of the West Bank to the Palestinian Authority in one fell swoop in return for a single promise to recognize Israel and cease the terrorism. Proceeding in steps has enabled the process to go at least part of the way. But this again illustrates the difficulty of sustaining the momentum as the endgame approaches.

III. TEAMWORK   Teamwork is yet another way to embed one game into a larger game to enhance the credibility of strategic moves. It requires that a group of players monitor each other. If one fails to carry out a threat or a promise, others are required to inflict punishment on him; failure to do so makes them in turn vulnerable to similar punishment by others, and so on. Thus a player's payoffs in the larger game are altered in a way that makes adhering to the team's creed credible.

Many universities have academic honor codes that act as credibility devices for students. Examinations are not proctored by the faculty; instead students are required to report to a student committee if they see any cheating. Then the committee holds a hearing and hands out punishment, as severe as suspension for a year or outright expulsion, if it finds the accused student guilty of cheating. Of course, students are very reluctant to place their fellow students in such jeopardy. To stiffen their resolve, such codes include the added twist that failure to report an observed infraction is itself an offense against the code. Even then, the general belief is that the system works only imperfectly. A poll conducted at Princeton University last year found that only a third of students said they would report an observed infraction, especially if they knew the guilty person.

IV. IRRATIONALITY Your threat may not be credible because the other player knows that you are rational and that it is too costly for you to follow through with your threatened action. Therefore others believe you will not carry out the threatened action if you are put to the test. You can counter this problem by claiming to be irrational, so others will believe that your payoffs are different from what they originally perceived. Apparent irrationality can then turn into strategic rationality when the credibility of a threat is in question. Similarly, apparently irrational motives like honor or saving face may make it credible that you will deliver on a promise even though tempted to renege.

Of course, the other player may see through such **rational irrationality.** Therefore if you attempt to make your threat credible by claiming irrationality, he will not readily believe you. You will have to acquire a reputation for irrationality, for example, by acting irrationally in some related game, or do something that is a credible signal of irrationality. We will study the requirements of credible signaling in detail in Chapter 12.

V. CONTRACTS You can make it costly to yourself to fail to carry out a threat or to deliver on a promise by signing a **contract** under which you have to pay a sufficiently large sum in that eventuality. If such a contract is written with sufficient clarity that it can be enforced by a court or some outside authority, the change in payoffs makes it optimal to carry out the stipulated action, and the threat or the promise becomes credible.

In the case of a promise, the other player can be the other party to the contract. It is in his interest that you deliver on the promise, so he will hold you to the contract if you fail to fulfill the promise. A contract to enforce a threat is more problematic. The other player does not want you to carry out the threatened action and will not enforce the contract unless he gets some longer-term benefit in associated games from being subject to a credible threat in this one. Therefore in the case of a threat, the contract has to be with a third party. But when you bring in a third party and a contract merely to ensure that you will carry out your threat if put to the test, the third party does not actually benefit from your *failure* to act as stipulated. The contract thus becomes vulnerable to any renegotiation that would provide the third-party enforcer with some positive benefits. If the other player puts you to the test, you can say to the third party, "Look, I don't want to carry out the threat. But I am being forced to do so by the prospect of the penalty in the contract, and you are not getting anything out of all this. Here is a real dollar in exchange for releasing me from the contract." Thus the contract itself is not credible, and therefore neither is the threat. The third party must have its own longer-term reasons for holding you to the contract, such as wanting to maintain its reputation, if the contract is to be renegotiation-proof and therefore credible.

Written contracts are usually more binding than verbal ones, but even verbal ones may constitute commitments. When George Bush said, "Read my lips; no new taxes," in the presidential campaign of 1988, the American public took this promise to be a binding contract; when Bush reneged on it in 1990, the public held that against him in the election of 1992.

VI. BRINKMANSHIP   In the United States–Japan trade-policy game, we found that a threat might be too "large" to be credible. If a smaller but effective threat cannot be found in a natural way, the size of the large threat can be reduced to a credible level by making its fulfillment a matter of chance. The United States cannot credibly say to Japan, "If you don't keep your markets open to U.S. goods, we will not defend you if the Russians or the Chinese attack you." But it can credibly say, "If you don't keep your markets open to U.S. goods, the relations between our countries will deteriorate, and this will create the risk that if you are faced with an invasion, Congress at that time will not sanction U.S. military involvement in your aid." As mentioned earlier, such deliberate creation of risk is called brinkmanship. This is a subtle idea, difficult to put into practice. Brinkmanship is best understood by seeing it in operation, and the detailed case study of the Cuban missile crisis in Chapter 13 serves just that purpose.

We have described several devices for making one's strategic moves credible, and examined how well they work. In conclusion we want to emphasize a feature common to the entire discussion. Credibility in practice is not an all-or-nothing matter but one of degree. Even though the theory is stark—rollback analysis shows either that a threat works or that it does not—practical application must recognize that between these polar extremes lies a whole spectrum of possibility and probability.

7 COUNTERING YOUR OPPONENT'S STRATEGIC MOVES

If your opponent can make a commitment or a threat that works to your disadvantage, then before he actually does so, you may be able to make a strategic countermove of your own. You can do so by making his future strategic move less effective, for example, by removing its irreversibility or undermining its credibility. In this section we examine some devices that can help achieve this purpose. Some are similar to devices the other side can use for its own needs.

A. Irrationality

Irrationality can work for the would-be receiver of a commitment or a threat just as well as it does for the other player. If you are known to be so irrational that you will not give in to any threat and will suffer the damage that befalls you when your opponent carries out that threat, then he may as well not make the threat in the first place because having to carry it out will only end up hurting him too. Of course, everything we said earlier about the difficulties of credibly convincing the other side of your irrationality holds true here as well.

B. Cutting Off Communication

If you make it impossible for the other side to convey to you the message that it has made a certain commitment or a threat, then your opponent will see no point in doing so. Thomas Schelling illustrates this possibility with the story of a child who is crying too loudly to hear his parent's threats.[5] Thus it is pointless for the parent to make any strategic moves; communication has effectively been cut off.

C. Leaving Escape Routes Open

If the other side can benefit by burning bridges to prevent its retreat, you can benefit by dousing those fires or perhaps even by constructing new bridges or roads by which your opponent can retreat. This device was also known to the ancients. Sun Tzu said, "To a surrounded enemy, you must leave a way of escape." Of course, the intent is not actually to allow the enemy to escape. Rather, "show him there is a road to safety, and so create in his mind the idea that there is an alternative to death. Then strike."[6]

D. Undermining Your Opponent's Motive to Uphold His Reputation

If the person threatening you says, "Look, I don't want to carry out this threat, but I must because I want to maintain my reputation with others," you can respond, "It is not in my interest to publicize the fact that you did not punish me. I am only interested in doing well in this game. I will keep quiet; both of us will avoid the mutually damaging outcome; and your reputation with others will stay intact."

[5] Thomas C. Schelling, *The Strategy of Conflict* (Oxford University Press, 1960), p. 146.
[6] Sun Tzu, *The Art of War,* pp. 109–110.

Similarly, if you are a buyer bargaining with a seller, and he refuses to lower his price on the grounds that "if I do this for you, I would have to do it for everyone else," you can point out that you are not going to tell anyone else. Of course, this may not work; the other player may suspect that you would tell a few friends who would tell a few others, and so on.

E. Salami Tactics

Salami tactics are devices used to whittle down the other player's threat the way a salami is cut—one slice at a time. You fail to comply with the other's wishes (whether for deterrence or compellence) to a very small degree, so that it is not worth the other's while to carry out the comparatively more drastic and mutually harmful threatened action just to counter that small transgression. If that works, you transgress a little more, and a little more again, and so on.

You know this perfectly well from your own childhood. Schelling[7] gives a wonderful description of the process:

> Salami tactics, we can be sure, were invented by a child. . . . Tell a child not to go in the water and he'll sit on the bank and submerge his bare feet; he is not yet "in" the water. Acquiesce, and he'll stand up; no more of him is in the water than before. Think it over, and he'll start wading, not going any deeper. Take a moment to decide whether this is different and he'll go a little deeper, arguing that since he goes back and forth it all averages out. Pretty soon we are calling to him not to swim out of sight, wondering whatever happened to all our discipline.

Salami tactics work particularly well against compellence, because they can take advantage of the *time* dimension. When your mother tells you to clean up your room "or else," you can put off the task for an extra hour by claiming you have to finish your homework, then for a half day because you have to go to football practice, then for an evening because you can't possibly miss the Simpsons on TV, and so on.

To counter the countermove of salami tactics you must make a correspondingly graduated threat. There should be a scale of punishments that fits the scale of noncompliance or procrastination. This can also be achieved by gradually raising the *risk* of disaster, another application of brinkmanship.

SUMMARY

Actions taken by players to fix the rules of later play are known as *strategic moves*. These first moves must be *observable* and *irreversible* to be true first moves, and

---

[7]Thomas C. Schelling, *Arms and Influence,* (New Haven, Conn.: Yale University Press, 1966), pp. 66–67.

they must be credible if they are to have their desired effect in altering the equilibrium outcome of the game. *Commitment* is an unconditional first move used to seize a first-mover advantage when one exists. Such a move usually entails committing to a strategy that would not have been one's equilibrium strategy in the original version of the game.

Conditional first moves such as threats and promises are *response rules* designed either to *deter* rivals' actions and preserve the status quo or to *compel* rivals' actions and alter the status quo. *Threats* carry the possibility of mutual harm but cost nothing if they work; threats that create only the risk of a bad outcome fall under the classification of *brinkmanship. Promises* are costly only to the maker and only if they are successful. Threats can be arbitrarily large, although excessive size compromises credibility, but promises are usually kept just large enough to be effective. If the implicit promise (or threat) that accompanies a threat (or promise) is not credible, players must make a move that combines both a promise and a threat and see to it that both components are credible.

Credibility must be established for any strategic move. There are a number of general principles to consider in making moves credible and a number of specific devices that can be used to acquire credibility. These generally work either by reducing your own future freedom to choose or by altering your own payoffs from future actions. Specific devices of this kind include establishing a *reputation*, using teamwork, demonstrating apparent irrationality, burning bridges, and making *contracts*, although the acquisition of credibility is often context-specific. Similar devices exist for countering strategic moves made by rival players.

KEY TERMS

brinkmanship (302)

commitment (290)

compellence (291)

contract (312)

credibility (288)

deterrence (291)

doomsday device (308)

irreversible (289)

observable (289)

promise (291)

rational irrationality (312)

reputation (310)

response rule (291)

salami tactics (302)

strategic moves (288)

threat (291)

EXERCISES

1. For each of the following three games, answer these questions: (i) What is the equilibrium if neither player can use any strategic moves? (ii) Can one player improve his payoff by using a strategic move (commitment, threat, or

promise) or a combination of such moves? If so, which player makes what strategic move(s)?

(a)

		COLUMN	
		Left	Right
ROW	Up	0, 0	2, 1
	Down	1, 2	0, 0

(b)

		COLUMN	
		Left	Right
ROW	Up	4, 3	3, 4
	Down	2, 1	1, 2

(c)

		COLUMN	
		Left	Right
ROW	Up	4, 1	2, 2
	Down	3, 3	1, 4

2. In Exercise 6 from Chapter 8, Baker and Cutler played a game in which they divided a prize. Which player could benefit from making a strategic move in that game? What kind of move should she make to get a better outcome?

3. Consider a game between a parent and a child. The child can choose to be good (G) or bad (B); the parent can punish the child (P) or not (N). The child gets enjoyment worth a 1 from bad behavior, but hurt worth –2 from punishment. Thus a child that behaves well and is not punished gets a 0; one who behaves badly and is punished gets $1 - 2 = -1$; and so on.) The parent gets –2 from the child's bad behavior and –1 from inflicting punishment.

   (a) Set up this game as a simultaneous-move game, and find the equilibrium.

   (b) Next, suppose that the child chooses G or B first, and the parent chooses its P or N after having observed the child's action. Draw the game tree and find the subgame-perfect equilibrium.

   (c) Now suppose that before the child acts, the parent can commit to a strategy, for example, the threat "P if B" ("If you behave badly, I will punish you"). How many such strategies does the parent have? Write down the table for this game. Find all pure-strategy Nash equilibria.

(d) How do your answers to parts (b) and (c) differ? Explain the reason for the difference.

4. The following is an interpretation of the rivalry between the United States and the Soviet Union for geopolitical influence in the 1970s and 1980s.[8] Each side has the choice of two strategies: Aggressive and Restrained. The Soviet Union wants to achieve world domination, so being Aggressive is their dominant strategy. The United States wants to prevent the Soviet Union from achieving world domination; they will match Soviet aggressiveness with aggressiveness, and restraint with restraint. Specifically, the payoff table is:

		SOVIET UNION	
		Restrained	Aggressive
UNITED STATES	Restrained	4, 3	1, 4
	Aggressive	3, 1	2, 2

For each player, 4 is best and 1 is worst.

(a) Consider this game when the two countries move simultaneously. Find the Nash equilibrium.

(b) Next consider three different and alternative ways in which the game could be played with sequential moves: (i) The United States moves first and the Soviet Union moves second. (ii) The Soviet Union moves first and the United States moves second. (iii) The Soviet Union moves first, the United States moves second, but the Soviet Union has a further move in which they can change their first move. For each case, draw the game tree and find the subgame-perfect equilibrium.

(c) What are the key strategic issues (commitment, credibility, and so on) for the two countries?

5. In a scene from the movie *Manhattan Murder Mystery*, Woody Allen and Diane Keaton are at a hockey game in Madison Square Garden. She is obviously not enjoying herself, but he tells her: "Remember our deal. You stay here with me for the entire hockey game, and next week I will come to the opera with you and stay until the end." Later we see them coming out of the Met into a deserted Lincoln Center square, while inside the music is still playing. Keaton is visibly upset: "What about our deal? I stayed to the end of the hockey game, and so you were supposed to stay till the end of the opera."

---

[8] We thank political science professor Thomas Schwartz at UCLA for the idea for this exercise.

Allen answers: "You know I can't listen to too much Wagner. At the end of the first act I already felt the urge to invade Poland." Comment on the strategic choices made here using your knowledge of the theory of strategic moves and credibility.

6. Consider the following games. In each case, (i) identify which player can benefit from making a strategic move, (ii) identify the nature of the strategic move appropriate for this purpose, (iii) discuss the conceptual and practical difficulties that will arise in the process of making this move credible, and (iv) discuss whether and how the difficulties can be overcome.

   (a) The other countries of the European Monetary Union (France, Germany, and so on) would like Britain to join the common currency and the common central bank.

   (b) The United States would like North Korea to stop exporting missiles and missile technology to countries like Iran, and would like China to join the United States in working toward this aim.

   (c) The United Auto Workers would like U.S. auto manufacturers not to build plants in Mexico, and would like the U.S. government to restrict imports of autos made abroad.

   (d) The students at your university or college want to prevent the administration from raising tuition.

   (e) Most participants, as well as outsiders, want to achieve a durable peace in situations like Northern Ireland and the Middle East.

7. Write a brief description of a game in which you have participated, involving strategic moves, such as a commitment, threat, or promise, paying special attention to the essential aspect of credibility. Provide an illustration of the game if possible, and explain why the game you describe ended as it did. Did the players use sound strategic thinking in making their choices?

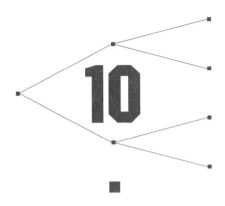

10

Evolutionary Games

W E HAVE SO FAR STUDIED GAMES with many different features—simultaneous and sequential moves, zero-sum and non-zero-sum payoffs, strategic moves to manipulate rules of games to come, one-shot and repeated play, and so on. However, one ground rule has remained unchanged in all of our discussions, namely that all the players in all these games are rational—each player has an internally consistent value system, can calculate the consequences of her strategic choices, and makes the choice that best favors her interests.

In using this rule, we merely follow the route taken by most of game theory, which was developed mainly by economists. Economics was founded on the dual assumptions of rational behavior and equilibrium. Indeed, these assumptions have proved useful in game theory. We have obtained quite a good understanding of games in which the players participate sufficiently regularly to have learned what their best choices are by experience. The assumptions ensure that a player does not attribute any false naiveté to her rivals and thus does not get exploited by these rivals. The theory also gives some prescriptive guidance to players as to how they *should* play.

However, other social scientists are much more skeptical of the rationality assumption, and therefore of a theory built upon such a foundation. Economists too should not take rationality for granted, as we pointed out in Chapter 7. The trouble is finding a viable alternative. While we may not wish to impose conscious and perfectly calculating rationality on players, we do not want to abandon the idea that some strategies are better than others. We want good strategies to be rewarded with higher payoffs; we want players to observe or imitate success, and to experiment with new strategies; we want good strategies to be used more often, and bad strategies less often, as players gain experience playing the game. We find one possible alternative to rationality in the biological theory of evolution and evolutionary dynamics, and devote this chapter to studying its lessons.

1 THE FRAMEWORK

The process of evolution in biology offers a particularly attractive parallel to the theory of games used by social scientists. This theory rests on three fundamentals: heterogeneity, fitness, and selection. The starting point is that a significant part of animal behavior is genetically determined; a complex of one or more genes (**genotype**) governs a particular pattern of behavior, called a behavioral **phenotype.** Natural diversity of the gene pool ensures a heterogeneity of phenotypes in the population. Some behaviors are better suited than others to the prevailing conditions, and the success of a phenotype is given a quantitative measure called its **fitness.** People are used to thinking in terms of the common but misleading phrase "survival of the fittest"; however, the ultimate test of biological fitness is not mere survival, but reproductive success. That is what enables an animal to pass on its genes to the next generation and perpetuate its phenotype. The fitter phenotypes then become relatively more numerous in the next generation than the less fit phenotypes. This process of **selection** is the dynamic that changes the mix of genotypes and phenotypes, and perhaps leads eventually to a stable state.

From time to time, chance produces new genetic **mutations.** Many of these mutations produce behaviors (that is, phenotypes) that are ill suited to the environment, and they die out. But occasionally a mutation leads to a new phenotype that is fitter. Then such a mutant gene can successfully **invade** a population, that is, spread to become a significant proportion of the population.

Biologists call a particular phenotype **evolutionary stable** if its population cannot be invaded successfully by any mutant. The strategy played by an evolutionary stable phenotype—a strategy that will persist in the population if played by all the existing phenotypes—is called an **evolutionary-stable strategy (ESS).** This is a static test, but often a more dynamic criterion is applied: starting from any mixture of phenotypes, a particular phenotype is evolutionary stable if the population evolves to a state in which this phenotype predominates.[1]

---

[1]Underlying these changes in the proportions of the phenotypes are changes in genotypes, but this branch of biology often focuses its analysis at the phenotype level and conceals the genetic aspects of evolution.

The fitness of a phenotype depends on the relationship of the individual animal to its environment; for example, the fitness of a particular bird depends on the aerodynamic characteristics of its wings. It also depends on the whole complex of the proportions of different phenotypes that exist in the environment, how aerodynamic its wings are, relative to the rest of its species. Thus the fitness of a particular animal, with its behavioral traits, like aggression and sociability, depends on whether other members of its species are predominantly aggressive or passive, crowded or dispersed, and so on. For our purpose, this **interaction** between phenotypes within a species is the most interesting aspect of the story. Of course, sometimes an individual interacts with members of another species; then the fitness of a particular type of sheep may depend on the traits that prevail in the local population of wolves. We consider this type of interaction as well, but only after we have covered the within-species case.

The basic idea of evolutionary games in biology is that the strategy of an animal in such an interaction—for example, whether to fight or to retreat—is not calculated but rather predetermined by its phenotype. Since the population is a mix of phenotypes, different pairs selected from it will bring to their interactions different combinations of such strategies. Those animals whose strategies are better suited for these interactions, on average against all the phenotypes they might encounter, will have greater evolutionary success. The eventual outcome of the population dynamics, or a strategy that, when used by all players, cannot be upset by any successful invasion of another, will be an evolutionary-stable strategy.

Biologists have used this approach very successfully. Combinations of aggressive and cooperative behavior, locations of nesting sites, and many more phenomena that elude more conventional explanations can be understood as the stable outcomes of an evolutionary process of selection of fitter strategies. Interestingly, biologists developed the idea of evolutionary games using the pre-existing body of game theory, drawing from its language but modifying the assumption of conscious maximizing to suit their needs. Now game theorists are in turn using insights from the research on biological evolutionary games to enrich their own subject.[2]

Indeed, the theory of evolutionary games seems a ready-made framework for a new approach to game theory, relaxing the assumption of rational behav-

[2] Robert Pool, "Putting Game Theory to the Test," *Science*, vol. 267, (March 17, 1995), pp. 1591–1593, is a good recent article for general readers and has many examples from biology. John Maynard Smith deals with such games in biology in his *Evolutionary Genetics* (Oxford: Oxford University Press, 1989), chap. 7, and *Evolution and the Theory of Games*, (Cambridge: Cambridge University Press, 1982); the former also gives much background on evolution. Recommended for advanced readers are Peter Hammerstein and Reinhard Selten, "Game Theory and Evolutionary Biology," in *Handbook of Game Theory*, vol. 2, ed. R. J. Aumann and S. Hart (Amsterdam: North Holland, 1994), pp. 929–993; and Jorgen Weibull, *Evolutionary Game Theory* (Cambridge, Mass.: MIT Press, 1995).

ior.[3] According to this view of games, individual players have no freedom to choose their strategy at all. Some are "born" to play one strategy, others another. The idea of inheritance of strategies is interpreted more broadly in game theory than in biology. In human interactions, a strategy may be embedded in a player's mind for a variety of reasons—due not merely to genetics but also (and probably more importantly) to socialization, cultural background, education, or a rule of thumb based on past experience. The population can consist of a mixture of different individuals with different backgrounds or experiences that embed different strategies into them. Thus some politicians may be motivated to adhere to certain moral or ethical codes even at the cost of electoral success, while others are mainly concerned with their own reelection; similarly, some firms may pursue profit alone, while others are motivated by social or ecological objectives.

From a population with its heterogeneity of embedded strategies, pairs of phenotypes are repeatedly selected to interact (play the game), with others of the same or different "species." In each interaction the payoff of each player depends on the strategies of both; this is governed by the usual "rules of the game" and reflected in the game table or tree. The fitness of a particular strategy is defined as its aggregate or average payoff in its matchings with all the strategies in the population. Some strategies have a higher level of fitness than others, and in the next generation—that is, the next round of play—these will be used by more players and will proliferate; strategies with lower fitness will be used by fewer players and will decay or die out. The central question is whether this process of selective proliferation or decay of certain strategies in the population will have an evolutionary-stable outcome, and if so, what it will be. In terms of the examples just cited, will society end up with a situation in which all politicians are concerned with reelection and all firms with profit? In this chapter we develop the framework and methods for answering such questions.

Although we use the biological analogy, the reason why the fitter strategies proliferate and the less fit ones die out in socioeconomic games differs from the strict genetic mechanism of biology: players who fared well in the last round will transmit the information to their friends and colleagues playing the next round, those who fared poorly in the last round will observe which strategies succeeded better and will try to imitate them, and some purposive think-

---

[3] Indeed, applications of the evolutionary perspective need not stop with game theory. The following joke offers an "evolutionary theory of gravitation" as an alternative to Newton's or Einstein's physical theories:

Question: Why does an apple fall from the tree to the earth?

Answer: Originally, apples that came loose from trees went in all directions. But only those that were genetically predisposed to fall to the earth could reproduce.

ing and revision of previous rules of thumb will occur between successive rounds. Such "social" and "educational" mechanisms of transmission are far more important in most strategic games than any biological genetics, and indeed, this is how the reelection orientation of legislators and the profit-maximization motive of firms are reinforced. Finally, conscious experimentation with new strategies substitutes for the accidental mutation in biological games.

Outcomes of biological games include an interesting new possibility. A single phenotype need not prevail completely in the eventual outcome of evolutionary dynamics. Two or more phenotypes may be equally fit, coexisting in certain proportions. Then the population is said to exhibit **polymorphism,** that is, a multiplicity (*poly*) of forms (*morph*). Such a state will be stable if no new phenotype or feasible mutant can achieve a higher fitness against such a population than the fitness of the types that are already present in the polymorphic population.

Polymorphism comes close to the game-theoretic notion of a mixed strategy. However, there is an important difference. To get polymorphism, no individual player need follow a mixed strategy. Each can follow a pure strategy, but the population exhibits a mixture because different individuals pursue different pure strategies.

In this chapter, we develop some of these ideas, as usual through a series of illustrative examples. We begin with symmetric games, in which the two players are on similar footing, for example, two members of the same species competing with each other for food or mates; in a social science interpretation, they could be two elected officials competing for the right to continue in public office. In terms of the payoff table for the game, each can be designated as the row player or the column player with no difference in outcome.

2 THE PRISONERS' DILEMMA

Suppose a population is made up of two phenotypes. One type consists of players who are natural-born cooperators; they do not confess under questioning. The other type consists of the defectors; they confess readily. The payoffs of each type in a single play of the dilemma are given by the familiar table of the husband-wife game, Figure 8.1, reproduced here as Figure 10.1 with the now-familiar warning that high numbers in this game are undesirable. Here we call the players simply Row and Column because they can be any individual in the population who is randomly arrested and paired against another random rival.

		COLUMN	
		Confess	Not
ROW	Confess	10 yr, 10 yr	1 yr, 25 yr
	Not	25 yr, 1 yr	3 yr, 3 yr

FIGURE 10.1   The Prisoners' Dilemma (years in jail)

Remember that under the evolutionary scenario, no one has the choice between confessing and not confessing; each is "born" with one trait or the other predetermined. Which is the more successful (fitter) trait in the population?

A confessing type gets 10 years if matched against another confessing type, and one year when matched against a nonconfessing type. A nonconfessing type gets 25 years if matched against a confessing type, and three if matched against another nonconfessing type. No matter what the type of the matched rival, the confessing type does better than the nonconfessing type. Therefore the confessing type has a better (numerically lower) expected payoff (and thus is fitter) than the nonconfessing type, irrespective of the proportions of the two types in the population.

A little more formally, let x be the proportion of cooperators in the population. Consider any one particular cooperator. In a random draw, the probability that she will meet another cooperator (and get three years) is x, and that she will meet a defector (and get 25 years) is $(1 - x)$. Therefore a typical cooperator's expected sentence is $3x + 25(1 - x)$. For a defector, the probability of meeting a cooperator (and getting one year) is x, and that of meeting another defector (and getting 10 years) is $(1 - x)$. Therefore a typical defector's expected sentence is $x + 10(1 - x)$. Now it is immediately apparent that

$$x + 10\,(1 - x) < 3x + 25\,(1 - x) \quad \text{for all } x \text{ between 0 and 1.}$$

Therefore a defector has a shorter expected sentence (better payoff) and is fitter than a cooperator. This will lead to an increase in the proportion of defectors (a decrease in x) from one "generation" of players to the next, until the whole population consists of defectors.

However, the converse is not true. If the population initially consists of all defectors, then no mutant (experimental) cooperator will survive and multiply to take over the population; in other words, the defector population cannot be invaded successfully by mutant cooperators. Even for a very small value of x— that is, when the proportion of cooperators in the population is very small—the cooperators remain less fit than the prevailing defectors and their population proportion will not increase but will be driven to zero; the mutant strain will die out.

Thus Confess must be the evolutionary-stable strategy for this population engaged in this dilemma game. This is a general proposition: if a game has a dominant strategy, that strategy will also be the ESS.

A. The Twice-Played Prisoners' Dilemma

We saw in Chapter 8 how a repetition of the prisoners' dilemma permitted consciously rational players to sustain cooperation for their mutual benefit. Let us see if a similar possibility exists in the evolutionary story. Suppose each chosen pair of players plays the dilemma twice in succession. The overall payoff of a player from such an interaction is the sum of what she gets in the two rounds.

Each individual is still programmed to play just one strategy, but in a game with two moves, a strategy has to be a complete plan of action, so a strategy can stipulate an action in the second play that depends on what happened in the first play. For example, "I will always cooperate no matter what" and "I will always defect no matter what" are valid strategies. But, "I will begin by cooperating and continue to cooperate on the second play if you cooperated on the first but defect otherwise" is also a valid strategy; it is just tit-for-tat (TFT).

Suppose these are the three types of strategies that can possibly exist in the population. Then we have three types of individuals, those who always confess, those who never confess, and those who play TFT. Figure 10.2 shows the outcomes (jail years) summed over the two meetings, of each type when matched against rivals of each type. For brevity we have labeled the strategy of always confessing as A, that of never confessing as N, and tit-for-tat as T.[4]

To see how these numbers arise, consider a couple of examples. When two T players meet each other, neither confesses the first time, and therefore neither

		COLUMN		
		A	T	N
ROW	A	20, 20	11, 35	2, 50
	T	35, 11	6, 6	6, 6
	N	50, 2	6, 6	6, 6

FIGURE 10.2 Outcomes in the Twice-Played Prisoners' Dilemma (years in jail)

[4] When each pair plays twice, the maximum possible sentence is 50 years! The story makes more sense if we now think of the sentences as being in months rather than years.

confesses the second time; both get 3 each time, for a total of six years in jail each. When a T player meets an A player, the latter does well the first time (one year versus 25 for the T player), but then the T player also confesses the second time, and each gets 10 (for totals of 11 years for A and 35 years for T).

What do these numbers tell us about the fitness of the different strategies? The first thing to note is that the naively cooperative strategy of never confessing (N) is not going to do well. It does exactly as well as T when the rival is either N or T. But it does worse than T against an A; it lets itself get fooled twice (shame on it). Therefore in any mixed population, T will be fitter than N, and will eventually dominate N. If the population initially consists of only N and A, a mutant T player can invade successfully and eventually dominate N. If the population initially happens to consist entirely of N players, then a mutant A player does better and can invade successfully. From either point of view, N cannot be an ESS.

So let us confine our attention to the other two strategies, A and T. Their relative fitnesses depend on the composition of the population. If the population is almost wholly A type, then A is fitter than T (because A types meeting mostly other A types get 20 years of jail most of the time, while T types most often get 35 years). But if the population is almost wholly T type, then T is fitter than A (because T types get only six years when they meet mostly other Ts, but A types get 11 years in such a situation). Each type is fitter when it already predominates in the population. Therefore T cannot invade successfully when the population is all A, and vice versa. Each of the two strategies, A and T, is an ESS.

Now consider the case in which the population is made up of a mixture of the two types. What will the evolutionary dynamics in the population entail? That is, how will the composition of the population evolve over time? Suppose a fraction x of the population is T type and the rest, $(1 - x)$, is A type.[5] An individual A player, pitted against various opponents chosen from such a population, gets 11 when confronting a T player, which happens a fraction x of the times, and 20 against another A player, which happens a fraction $(1 - x)$ of the times. This gives an average expected sentence of

$$11x + 20(1 - x) = 20 - 9x$$

for each A player. Similarly, an individual T player gets an average expected sentence of

$$6x + 35(1 - x) = 35 - 29x$$

[5] Literally, the fraction of any particular type in the population is finite and can only take on values like 1/1,000,000, 2/1,000,000, and so on. But if the population is sufficiently large, and we show all such values by points on a straight line, as in Figure 10.3, then these points are very tightly packed together, and we can regard them as forming a continuous line. This amounts to letting the fractions take on any real value between 0 and 1. We can then talk of the population *proportion* of a certain behavioral type. By the same reasoning, if one individual goes to jail and is removed from the population, her removal does not change the population's proportions of the various phenotypes.

Then a T player is fitter than an A player if the former serves less jail time on average, that is, if

$$35 - 29x < 20 - 9x$$
$$20x > 15$$
$$x > 0.75.$$

In other words, if more than 75% of the population is already T type, then T players are fitter and their proportion will grow until it reaches 100%. If the population starts with less than 75% T, then A players will be fitter, and the proportion of T players will go on declining until there are 0% of them, or 100% of the A players. Both extremes are evolutionary-stable strategies (ESS).

Thus we have identified two equilibria. In each one the population is all of one type, and so each equilibrium is said to be **monomorphic.** Both of these equilibria are evolutionary stable. For example, if the population is initially 100% T, then even after a small number of experimenting mutant A types arise the population mix will still be more than 75% T; T will remain the fitter type, and the mutant A strain will die out. Similarly, if the population is initially 100% A, then a small number of experimenting T-type mutants will leave the population mix with less than 75% T, so the A types will be fitter and the mutant T strain will die out. And, as we saw earlier, experimenting mutants of type N can never succeed in a population of A and T types that is either largely T or largely A.

What if the initial population has exactly 75% T players (and 25% A players)? Then the two types are equally fit. This population proportion is a polymorphic equilibrium but an unstable one; the population can sustain this delicately balanced outcome only until a mutant of either type surfaces. By chance, such a mutant must arise sooner or later. The mutant's arrival will tip the fitness calculation in favor of the mutant type, and the advantage will accumulate until the ESS with 100% of that type is reached.

This reasoning can be shown in a simple graph that closely resembles the graphs we drew when calculating the equilibrium proportions in a mixed-strategy equilibrium with consciously rational players. The only difference is that in the evolutionary context the proportion in which the separate strategies are played is not a matter of choice by any individual player but a property of the whole population; the diagram for this case is shown in Figure 10.3.

Figure 10.3 shows the proportion x of T players measured from 0 to 1 from left to right in the horizontal direction. The vertical direction measures unfitness; since these are jail times, higher numbers indicate greater unfitness. Each line shows the unfitness of one type. The line for the T type starts higher (at 35 as against 20 for the A type line) and ends lower (6 against 11). The two lines cross when $x = 0.75$. To the right of this point, the T type is fitter, so its population proportion increases over time and x *increases* toward 1. Similarly, to the left of this point, the A type is fitter, so its population proportion increases over time and x

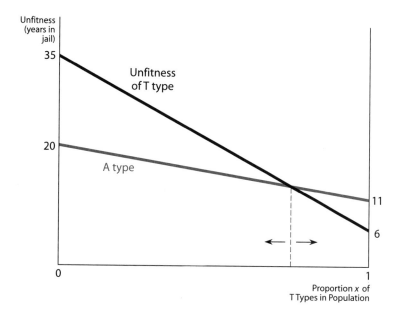

FIGURE 10.3 Fitness Graphs and Equilibria for the Twice-Played Prisoners' Dilemma

decreases toward 0. Such diagrams often prove useful as visual aids, and we will use them extensively.

B. Multiple Repetitions

What if each pair plays more than two repetitions of the game? Let us focus on a population comprised only of A and T types in which interactions between random pairs occur n times (where $n > 2$). The table of the total outcomes from playing n repetitions is shown in Figure 10.4. When two A types meet, they always cheat and get 10 years every time, so each gets $10n$ years over n plays. When two T types meet, they begin by cooperating, and no one is the first to defect, so they get three years every time, for a total of $3n$ years. When an A type meets a T type, on the first play the T type cooperates and the A type defects, so the A type gets one and the other gets 25; thereafter the T type retaliates against the previous defection of the A type for all remaining plays, and each gets 10 years in all of the remaining $(n - 1)$ plays. Thus the A type gets a total of $1 + 10(n - 1) = 10n - 9$ years in n plays against a T type, while the T type gets $25 + 10(n - 1) = 10n + 15$ years in n plays against an A type. (Please accept these as a metaphor, even though for large values of n the sentences run longer than lifetimes.)

If the proportion of T types in the population is x, then a typical A type gets $x(10n - 9) + (1 - x) 10n$ on average, and a typical T type gets $x3n + (1 - x) (10n + 15)$ on average. Therefore the T type is fitter if

		COLUMN	
		A	T
ROW	A	10n, 10n	10n – 9, 10n + 15
	T	10n + 15, 10n – 9	3n, 3n

FIGURE 10.4 Outcomes in the n-Fold Repeated Dilemma

$$x(10n-9) + (1-x)\,10n > x\,3n + (1-x)\,(10n+15)$$
$$x(7n+6) > 15$$
$$x > \frac{15}{7n+6}.$$

Once again we have two stable monomorphic ESSs, one all-T [or $x = 1$, to which the process converges starting from any $x > 15/(7n+6)$] and the other all-A [or $x = 0$, to which the process converges starting from any $x < 15/(7n+6)$], and an unstable polymorphic equilibrium at the balancing point $x = 15/(7n+6)$.

Notice that the proportion of T at the balancing point depends on n; it is smaller when n is larger. When $n = 10$, it is 15/76, or approximately 0.20. So if the population initially is 20% T players, in a situation where each pair plays 10 repetitions, their numbers will grow until they reach 100%. Recall that when pairs played only two repetitions ($n = 2$), the T players needed an initial strength of 75% or more to achieve this outcome. Remember too that a population consisting of all T players achieves cooperation. Thus cooperation emerges from a larger range of the initial conditions when the game is repeated more times. In this sense, with more repetition, cooperation becomes more likely. What we are seeing is the result of the fact that the value of establishing cooperation increases as the length of the interaction increases.

C. Comparing the Evolutionary and Rational-Player Models

Finally, let us return to the twofold repeated game, and instead of using the evolutionary model, consider it played by two consciously rational players. What are the Nash equilibria? There are two in pure strategies, one in which both play A, and the other in which both play T. There is also an equilibrium in mixed strategies, in which T is played 75% of the time and A 25% of the time. The first two are just the monomorphic ESSs we found, and the third is the unstable polymorphic evolutionary equilibrium. In other words, there is a close relationship between evolutionary and consciously rational perspectives on games.

That is not a coincidence. An ESS must be a Nash equilibrium of the game played by consciously rational players with the same payoff structure. To see this, suppose the contrary for the moment. If all players using some strategy, call it S, is not a Nash equilibrium, then some other strategy, call it T, must yield a higher payoff for one player when played against S. A mutant playing T will achieve greater fitness in a population playing S, and so will invade successfully. Thus S cannot be an ESS. In other words, if all players' using S is not a Nash equilibrium, then S cannot be an ESS. This is the same as saying that if S is an ESS, it must be a Nash equilibrium for all players to use S.

Thus the evolutionary approach provides a backdoor justification for the rational approach. Even when players are not consciously maximizing, if the more successful strategies get played more often and the less successful ones die out, and if the process converges eventually to a stable strategy, then the outcome must be the same as that resulting from consciously rational play.

There is one limitation of our analysis of the repeated game. At the outset we allowed just three strategies: A, T, and N. Nothing else was supposed to exist or arise as a mutation. In biology, the kinds of mutations that arise are determined by genetic considerations. In social or economic or political games, the genesis of new strategies is presumably governed by history, culture, and experience of the players; the ability of people to assimilate and process information and to experiment with different strategies must also play a role. However, the restrictions we place on the set of strategies that can possibly exist in a particular game have important implications for which of these strategies (if any) can be evolutionary stable. In the twice-repeated prisoners' dilemma example, if we had allowed for a strategy S that cooperated on the first play and defected on the second, then S-type mutants could have successfully invaded an all-T population, so T would not have been an ESS. We develop this possibility further in Exercise 3, at the end of this chapter.

3 CHICKEN

Remember our 1950s youths racing their cars toward each other and seeing who will be the first to swerve to avoid a collision? Now we suppose the players have no choice in the matter: each is genetically hardwired to be either a Wimp (always swerve) or a Macho (always go straight). The population consists of a mixture of the two types. Pairs are picked at random every week to play the game. Figure 10.5 shows the payoff table for any two such players, say A and B. (The numbers replicate those we used before in Figures 4.11 and 5.7).

		B	
		Wimp	Macho
A	Wimp	0, 0	−1, 1
	Macho	1, −1	−2, −2

FIGURE 10.5 Payoff Table for Chicken

How will the two types fare? The answer depends on the initial population proportions. If the population is almost all Wimps, then a Macho mutant will win and score 1 lots of times while all the Wimps meeting their own types get mostly zeroes. But if the population is mostly Macho, then a Wimp mutant scores −1, which may look bad but is better than the −2 that all the Machos get. You can think of this appropriately in terms of the biological context and the sexism of the 1950s: in a population of Wimps, a Macho newcomer will show all the rest to be chickens and so will impress all the girls. But if the population consists mostly of Machos, they will be in the hospital most of the time and the girls will have to go for the few Wimps that are healthy.

In other words, each type is fitter when it is relatively rare in the population. Therefore each can invade successfully a population consisting of the other type. We should expect to see both types in the population in equilibrium; that is, we should expect an ESS with a mixture, or polymorphism.

To find the proportions of Wimps and Machos in such an ESS, let us calculate the fitness of each type in a general mixed population. Write x for the fraction of Machos and $(1 - x)$ for the proportion of Wimps. A Wimp meets another Wimp and gets 0 for a fraction $(1 - x)$ of the time, and meets a Macho and gets −1 for a fraction x of the time. Therefore the fitness of a Wimp is $0 \times (1 - x) - 1 \times x = -x$. Similarly, the fitness of a macho is $1 \times (1 - x) - 2x = 1 - 3x$. The macho type is fitter if

$$1 - 3x > -x$$
$$2x < 1$$
$$x < 1/2.$$

If the population is less than half Macho, then the Machos will be fitter and their proportion will increase. On the other hand, if the population is more than half Macho, then the Wimps will be fitter and the Macho proportion will fall. Either way, the population proportion of Machos will tend toward 1/2, and this 50-50 mix will be the stable polymorphic ESS.

Figure 10.6 shows this outcome graphically. Each straight line shows the fit-

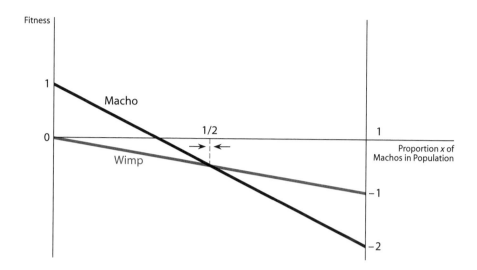

FIGURE 10.6   Fitness Graphs and Polymorphic Equilibrium for Chicken

ness (the expected payoff in a match against a random member of the population) for one type, in relation to the proportion x of Machos. For the Wimp type, this functional relation showing their fitness as a function of the proportion of the Machos is $-x$, as we saw two paragraphs above. This is the gently falling line, which starts at the height 0 when $x = 0$, and goes to -1 when $x = 1$. The corresponding function for the Macho type is $1 - 3x$. This is the rapidly falling line, which starts at height 1 when $x = 0$, and falls to -2 when $x = 1$. The Macho line lies above the Wimp line for $x < 1/2$ and below it for $x > 1/2$, showing that the Macho types are fitter when the value of x is small and the Wimps are fitter when x is large.

Now we can compare and contrast the evolutionary theory of this game with our earlier theory of Chapters 4 and 5, which was based on the assumption that the players were conscious rational calculators of strategies. There we found three Nash equilibria, two in pure strategies, where one player goes straight and the other swerves, and one in mixed strategies, where each player goes straight with a probability of $1/2$ and swerves with a probability of $1/2$.

In one sense we still have all three of those equilibria. If the population is truly 100% Macho, then all players are equally fit (or equally unfit) and the situation can persist. Similarly, a population of nothing but Wimps can continue. So we still have two equilibria of the pure type. But they are unstable. In an all-Macho population, a Wimp mutant will outscore them and invade successfully.[6] Once some Wimps get established, no matter how few, our analysis shows that their proportion will rise inexorably toward $1/2$. Similarly, an all-Wimp popula-

---

[6]*The Invasion of the Mutant Wimps* could be an interesting science fiction movie.

tion is vulnerable to a successful invasion of mutant Machos, and the process again goes to the polymorphic ESS.

Most interesting is the connection between the mixed-strategy equilibrium of the rationally played game and the polymorphic ESS of the evolutionary game. The mixture proportions in the equilibrium strategy of the former are *exactly the same* as the population proportions in the latter: a 50-50 mixture of Wimp and Macho. But the interpretations differ: in the rational framework, each player mixes his own strategies; in the evolutionary framework, every individual uses a pure strategy, but different individuals use different strategies, so we see a mixture in the population.[7]

This correspondence between Nash equilibria of a rationally played game and stable outcomes of a game with the same payoff structure when played according to the evolutionary rules is a very general proposition, and we see it in its generality later, in Section 6. Indeed, evolutionary stability provides an additional rationale for choosing one of the multiple Nash equilibria in such rationally played games.

When we looked at chicken from the rational perspective, the mixed-strategy equilibrium seemed puzzling. It left open the possibility of costly mistakes. Each player went straight one time in two, so one time in four they collided. The pure-strategy equilibria avoided the collisions. At that time this may have led you to think that there was something undesirable about the mixed-strategy equilibrium, and you may have wondered why we were spending time on it. Now you see the reason. The seemingly strange equilibrium emerges as the stable outcome of a natural dynamic process in which each player tries to improve his payoff against the population he confronts.

4 THE ASSURANCE GAME

Among the important classes of strategic games we introduced in Chapter 4, we have studied prisoners' dilemma and chicken from the evolutionary perspective. That leaves the assurance game. We illustrated this type of game in Chapter 4, using the story of two superpowers deciding whether to build more nuclear arms. In the evolutionary context it makes no sense to have pairs chosen from a large population of superpowers playing at arms races time and again, so we use a different example.

[7]There can also be evolutionary-stable mixed strategies. Each player has a mixed strategy. The population is polymorphic and, by the law of large numbers, the proportion playing each pure strategy equals the probability that an individual chooses it in her mixture. We leave this for more advanced treatments.

Suppose pairs of players are drawn from the U.S. population. The members of each pair, whose identities are unknown to each other, are supposed to meet somewhere in New York City, but the exact location is not specified for them. There are two types in the population: theater fans and movie buffs. When the theater fans think of New York City, they think of nothing but the Theater District and Times Square, more specifically, and that is where they will go. The movie buffs, remembering a triad of famous film rendezvous (or attempted rendezvous)—in *An Affair to Remember, Sleepless in Seattle,* and, of course, *King Kong*—will go to what for them is the most obvious meeting place: the top of the Empire State Building.

If the two people in a particular pair go to different places and fail to meet, each of them will get a payoff of 0. If both are theater fans and go to Times Square, each will get 1, while if both are movie buffs and go to the top of the Empire State Building, each will get 2. The payoffs are higher when both go to the top of the Empire State Building because it is a smaller location and meeting there is easier, not to mention safer. Figure 10.7 shows the payoff table for a random pairing in this game. As before, we call the two players A and B. For the two types and their strategies, we let T denote the theater type and its *Times Square* strategy, and M denote the movie type and its *Empire* State Building strategy.

		B	
		T	M
A	T	1, 1	0, 0
	M	0, 0	2, 2

FIGURE 10.7 Payoff Matrix for the Meeting Game

If this were a game played by rational strategy-choosing players, there would be two equilibria in pure strategies: (T, T) and (M, M). The latter is better for both players. If they communicate and coordinate explicitly, they can settle on it quite easily. But if they are making the choices independently, they need tacit coordination through a convergence of expectations—that is, a focal point.

The game has a third equilibrium, in mixed strategies, although we did not consider this equilibrium in our analysis of the arms race in Chapter 4. The equilibrium mixtures are easy to find though, using the method of keeping the opponent indifferent. If A chooses T with probability p and M with probability $(1 - p)$, B gets the expected payoff p from choosing T and $2(1 - p)$ from choosing

M. Her indifference requires $p = 2(1 - p)$, or $p = 2/3$. The expected payoff for each player is then $1 \times 2/3 + 0 \times 1/3 = 2/3$. This is worse than the payoff associated with the less attractive of the two pure-strategy equilibria, (T, T). The mixed-strategy equilibrium here yields a lower payoff for a reason similar to what we found for the mixed-strategy equilibrium of chicken in Chapter 5. When the two players mix their strategies independently, they make clashing or bad choices quite a lot of the time. Here the bad outcome (a payoff of 0) occurs when A chooses T and B chooses M or the other way around; the probability of this occurring is $2 \times 2/3 \times 1/3 = 4/9$: the two players go to different meeting places almost half the time.

What happens when this is an evolutionary game? In the large population, each individual is hardwired, some to choose T and the others to choose M. Randomly chosen pairs of such people are assigned to go to New York City and to attempt a meeting. Suppose x is the proportion of T types in the population and $(1 - x)$ that of M types. Then the fitness of a particular T type—her expected payoff in a random encounter of this kind—is $x \times 1 + (1 - x) \times 0 = x$. Similarly, the fitness of each M type is $x \times 0 + (1 - x) \times 2 = 2(1 - x)$. Therefore the T type is fitter when $x > 2(1 - x)$, or for $x > 2/3$. The M type is fitter when $x < 2/3$. At the balancing point $x = 2/3$, the two types are equally fit.

As in chicken, once again the probabilities associated with the mixed-strategy equilibrium that would obtain under rational choice seem to reappear under evolutionary rules as the population proportions in a polymorphic equilibrium. But now this mixed equilibrium is not stable. The slightest chance departure of the proportion x from the balancing point 2/3 will set in motion a cumulative process that takes the population mix farther away from the balancing point. If x increases from 2/3, the T type becomes fitter and propagates faster, increasing x even more. If x falls from 2/3, the M type becomes fitter and propagates faster, lowering x even more. Depending on which disturbance occurs, eventually x will either rise all the way to 1 or fall all the way to 0. The difference is that in chicken each type was fitter when it was rarer, so the population proportions tended to move away from the extremes and toward a midrange balancing point. In contrast, in the assurance game each type is fitter when it is more numerous; the risk of failing to meet falls when more of the rest of the population are the same type as you—so population proportions tend to move toward the extremes.

Figure 10.8 illustrates the fitness graphs and equilibria for the meeting version of the assurance game; this diagram is very similar to Figure 10.6. The two lines show the fitness of the two types in relation to the population proportion. The intersection of the lines gives the balancing point. The only difference is that away from the balancing point, the more numerous type is the fitter, whereas in Figure 10.6 it was the less numerous type.

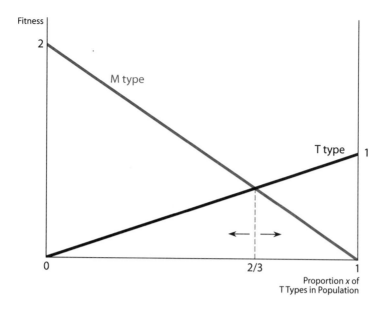

FIGURE 10.8 Fitness Graphs and Equilibria for the Meeting Game

Since each type is less fit when it is rare, only the two extreme monomorphic equilibria survive. It is easy to check that both are evolutionary stable: an invasion by a small mutant population of the other type will die out because the mutants, being rare, will be less fit.

Thus in assurance or coordination games, unlike in chicken, the evolutionary process does not preserve the bad equilibrium, where the players choose clashing strategies. But the process cannot guarantee that it will lead to the better of the two extreme pure-strategy equilibria. That depends on the initial mix in the population.

5 INTERACTIONS ACROSS SPECIES

The battle-of-the-two-cultures game of Chapter 4 (Figure 4.12) looks similar to the assurance game in some respects. In our example, a university faculty was voting on the use of some vacant space. The Humanities Faculty prefers to have a theater, and the Science Faculty prefers a lab. If the two agree on one use, that use will be chosen. If each faculty insists on its preferred use, the result will be an impasse, nothing will be done, and that will be the worst outcome for both faculties. Thus there is a premium on taking mutually consistent actions, just as in the assurance example. But the consequences of the two possible mutually consistent actions differ. The types in the assurance game do not differ in their preferences; both prefer (M, M) to (T, T). The faculty in the battle game differ in theirs: the theater gives a payoff of 2 to the humanists and 1 to the scientists, and the lab the other way around. These preferences distinguish the two types. In the language of biology, they can no longer be considered random draws from a homogeneous population of animals. Effectively, they belong to different species (as indeed humanities and science faculties often believe of each other).

To study such games from an evolutionary perspective, we must extend our methodology to the case in which the matches are between randomly drawn members of different species or populations. We develop the battle-of-the-two-cultures example to illustrate how this is done.

Suppose there is a large population of scientists and a large population of humanists. One of each "species" is picked, and the two are asked to cast secret ballots for the lab or the theater. The votes are compared, and the comparison determines the outcome—lab, theater, or impasse—according to the rule stated earlier.

All scientists agree among themselves about the valuation (payoffs) of the lab, the theater, and the impasse. Likewise, all humanists are agreed among themselves. But within each population, some individuals are hard-liners and others are compromisers. A hard-liner will always vote for her species' preferred use. A compromiser recognizes that the other species wants the opposite, and votes for that, to get along.

If the random draws happen to have picked a hard-liner of one species and a compromiser of the other, the outcome is that preferred by the hard-liner's species. We have an impasse if two hard-liners meet, and strangely, also if two compromisers meet, because they vote for each other's preferred use. (Remember, they have to vote in secret and cannot negotiate. Perhaps even if they did meet, they would reach an impasse of "No, I insist on giving way to your preference.")

We alter the payoff table in Figure 4.12 as shown in Figure 10.9; of course, what were choices of votes are now interpreted as actions predetermined by your type (hard-liner or compromiser).

In comparison with all the evolutionary games studied so far, the new feature here is that the row player and the column player come from different species. While each species is a heterogeneous mixture of hard-liners and compromisers, there is no reason why the proportions of the types should be the same in both species. Therefore we must introduce two variables to represent the two mixtures, and study the dynamics of both.

We let x be the proportion of hard-liners among the scientists, and y that among the humanists. Consider a particular hard-liner scientist. She meets a hard-liner humanist a proportion y of the time and gets a 0, and meets a com-

		HUMANITIES FACULTY	
		Hard-liner	Compromiser
SCIENCE FACULTY	Hard-liner	0, 0	2, 1
	Compromiser	1, 2	0, 0

FIGURE 10.9 Payoffs in the Battle-of-the-Two-Cultures Game

promising humanist the rest of the time and gets a 2. Therefore her expected payoff (fitness) is $y \times 0 + (1 - y) \times 2 = 2(1 - y)$. Similarly, a compromising scientist's fitness is $y \times 1 + (1 - y) \times 0 = y$. Among scientists, therefore, the hard-liner type is fitter when $2(1 - y) > y$, or $y < 2/3$. The hard-liner scientists will reproduce faster when they are fitter; that is, x increases when $y < 2/3$. Note the new, and at first sight surprising, feature of the outcome: the fitness of each type within a given species depends on the proportion of types found in *other* species. Of course, there is nothing surprising here—after all, now the games each species plays are all against the members of the other species.[8]

Similarly, considering the other species, we have the result that the hard-liner humanists are fitter, so y increases when $x < 2/3$. To understand the result intuitively, note that it says that the hard-liners of each species do better when the other species does not have too many hard-liners of its own, because then they meet compromisers of the other species quite frequently.

Figure 10.10 shows the dynamics of the configurations of the two species. Each of x and y can range from 0 to 1, so we have a graph with a unit square and x and y on their usual axes. Within that, the vertical line AB shows all points where $x = 2/3$, the balancing point at which y neither increases or decreases. If the current population proportions lie to the left of this line (that is, $x < 2/3$), y is increasing (moving the population proportion of hard-liner humanists in the vertically upward direction). If the current proportions lie to the right of AB ($x > 2/3$), then y is decreasing (motion vertically downward). Similarly, the horizontal line CD shows all points where $y = 2/3$, which is the balancing point for x. For population proportions of hard-liner humanists below this line (that is, for $y < 2/3$) the proportion of hard-liner scientists, x, increases (motion horizontal and rightward) and decreases for population proportions above it, when $y > 2/3$ (motion horizontal and leftward).

---

[8] And this finding supports and casts a different light on the property of mixed-strategy equilibria, that each player's mixture keeps the *other* player indifferent among her pure strategies. Now we can think of it as saying that in a polymorphic evolutionary equilibrium of a two-species game, the proportion of each species' type keeps all the surviving types of the other species equally fit.

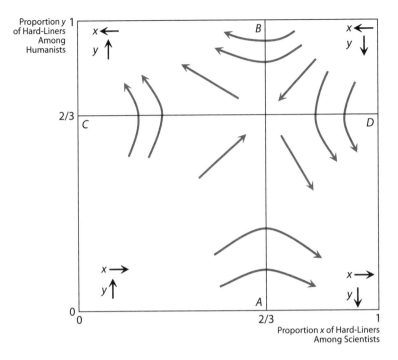

FIGURE 10.10 Population Dynamics in the Battle of the Two Cultures

When we combine the motions of x and y, we can follow their dynamic paths to determine the location of the population equilibrium. From a starting point in the bottom left quadrant of Figure 10.10, for example, the dynamics entail y and x both increasing. This joint movement (to the northeast) continues until either $x = 2/3$ and y begins to decrease (motion now to the southeast) or $y = 2/3$ and x begins to decrease (motion now to the northwest). Similar processes in each quadrant yield the curved dynamic paths shown in the diagram. The vast majority of these lead to either the southeast or northwest corners of the diagram; that is, they converge either to (1, 0) or (0, 1). Thus evolutionary dynamics will lead in most cases to a configuration in which one species is entirely hard-line and the other is entirely compromising. Which species will be which type depends on the initial conditions. Note that the population dynamics starting from a situation with a small value of x and a larger value of y are more likely to cross the CD line first and head for (0, 1)—all hard-line humanists, $y = 1$—than to hit the AB line first and head for (1, 0); similar results follow for a starting position with a small y but a larger x. The species that starts out with more hard-liners will have the advantage of ending up all hard-line and getting the payoff of 2.

If the initial proportions are balanced just right, the dynamics may lead to the polymorphic point (2/3, 2/3). But unlike the polymorphic ESS in chicken, the polymorphic equilibrium in the battle of the two cultures is unstable. Any chance departure will set in motion a cumulative process that leads to one of the two extreme equilibria; those are the two ESSs for this game.

6 THE HAWK-DOVE GAME

The **hawk-dove game** was the first example studied by biologists in their development of the theory of evolutionary games. It has instructive parallels with our analyses so far of the prisoners' dilemma and chicken, so we describe it here to reinforce and improve your understanding of the concepts.

The game is played not by birds of these two species, but by two animals of the same species, and Hawk and Dove are merely the names for their strategies. The context is competition for a resource. The Hawk strategy is aggressive, and fights to try to get the whole resource of value V. The Dove strategy is to offer to share but to shirk from a fight. When two Hawk types meet each other, they fight. Each animal is equally likely (probability 1/2) to win and get V, or to lose, be injured, and get $-C$. Thus the expected payoff for each is $(V - C)/2$. When two Dove types meet, they share without a fight, so each gets $V/2$. When a Hawk type meets a Dove type, the latter retreats and gets a 0, while the former gets V. Figure 10.11 shows the payoff table.

		B	
		Hawk	Dove
A	Hawk	$\frac{V-C}{2}, \frac{V-C}{2}$	$V, 0$
	Dove	$0, V$	$\frac{V}{2}, \frac{V}{2}$

FIGURE 10.11   Payoff Table for the Hawk-Dove Game

The analysis of the game is similar to that for the prisoners' dilemma and chicken games, except that the numerical payoffs have been replaced by algebraic symbols. We will compare the equilibria of this game when the players rationally choose to play Hawk or Dove, and the outcomes when players are acting mechanically and success is being rewarded with faster reproduction.

A. Rational Strategic Choice and Equilibrium

1. If $V > C$, then the game is a prisoners' dilemma. Hawk is the dominant strategy for each, but (Dove, Dove) is the jointly better outcome.

2. If $V < C$, then it's a game of chicken. There are two pure-strategy Nash equilibria: (Hawk, Dove) and (Dove, Hawk). There is also a mixed-strategy equilibrium, where B's probability p of choosing Hawk is such as to keep A indifferent:

$$p \frac{V-C}{2} + (1-p) V = p \times 0 + (1-p) \frac{V}{2}$$

$$p = V/C.$$

B. Evolutionary Stability for $V > C$

We start with an initial population predominantly of Hawks, and test whether it can be invaded by mutant Doves. Following the convention used in analyzing such games, we could write the population proportion of the mutant phenotype as m, for mutant, but for clarity in our case we will use d for mutant Dove. The population proportion of Hawks is then $(1 - d)$. Then in a match against a randomly drawn opponent, a Hawk will meet a Dove a proportion d of the time and get V on each of those occasions, and meet another Hawk a proportion $(1 - d)$ of the time and get $(V - C)/2$ on each of those occasions. Therefore the fitness of a Hawk is $[dV + (1 - d) (V - C)/2]$. Similarly, the fitness of one of the mutant doves is $[d(V/2) + (1 - d) \times 0]$. Since $V > C$, it follows that $(V - C)/2 > 0$. Also, $V > 0$ implies that $V > V/2$. Then for any value of d between 0 and 1, we have

$$dV + (1-d) \frac{V-C}{2} > d \frac{V}{2} + (1-d) \times 0,$$

so the Hawk type is fitter. The Dove mutants cannot invade, and the Hawk strategy is evolutionary stable.

The same holds true for any population proportion of Doves, for all values of d. Therefore, from any initial mix, the proportion of Hawks will grow and they will predominate. In addition, if the population initially is all Doves, mutant Hawks can invade and take over. Thus the Hawk strategy is the only ESS. This confirms and extends our earlier finding for the prisoners' dilemma.

C. Evolutionary Stability for $V < C$

If the initial population is again predominantly Hawks, with a small proportion d of Dove mutants, then each has the same fitness function derived in Section 6.B. When $V < C$, however, $(V - C)/2 < 0$. We still have $V > 0$, so $V > V/2$. But since d is very small, the comparison of the terms with $(1 - d)$ is much more important than that of the terms involving d, so

$$d \frac{V}{2} + (1-d) \times 0 > dV + (1-d) \frac{V-C}{2}.$$

Thus the Dove mutants have higher fitness than the predominant Hawks and can invade successfully.

But if the initial population is almost all Doves, then we must consider whether a small proportion h of Hawk mutants can invade. (Note that because the mutant is now a Hawk, we have followed convention and used h instead of m for the proportion of the mutant Hawk type.) The Hawk mutants have a fitness of $[h(V-C)/2 + (1-h)V]$ compared with $[h \times 0 + (1-h)(V/2)]$ for the Doves. Again $V < C$ implies that $(V-C)/2 < 0$, and $V > 0$ implies that $V > V/2$. But when h is small, we get

$$h\frac{V-C}{2} + (1-h)V > h \times 0 + (1-h)\frac{V}{2}.$$

This shows that Hawks have greater fitness and will successfully invade a Dove population. Thus mutants of each type can invade populations of the other type; there is no pure ESS. This also confirms our earlier finding for chicken.

What happens in the population then when $V < C$? There are two possibilities. In one, every individual follows a pure strategy, but the population has a stable mix of players following different strategies. This is the polymorphic equilibrium we developed for Chicken in Section 10.3. The other possibility is that every individual uses a mixed strategy. We begin with the polymorphic case.

D. $V < C$: Stable Polymorphic Population

When the population proportion of Hawks is h, the fitness of a Hawk is $h(V-C)/2 + (1-h)V$, and the fitness of a Dove is $h \times 0 + (1-h)(V/2)$. The Hawk type is fitter if

$$h\frac{V-C}{2} + (1-h)V > (1-h)\frac{V}{2},$$

which simplifies to:

$$h\frac{V-C}{2} + (1-h)\frac{V}{2} > 0$$
$$V - hC > 0$$
$$h < \frac{V}{C}.$$

The Dove type is then fitter when $h > V/C$, or when $(1-h) < 1 - V/C = (C-V)/C$. Thus each type is fitter when it is rarer. Therefore we have a stable polymorphic equilibrium at the balancing point, where the proportion of Hawks in the population is $h = V/C$. This is exactly the probability with which each individual plays the Hawk strategy in the mixed-strategy Nash equilibrium of the game under the assumption of rational behavior, as we calculated in Section 6.A.

Again we have an evolutionary "justification" for the mixed-strategy outcome in chicken.

We leave it to you to draw a graph similar to Figure 10.6 for this case. Doing so will require that you determine the dynamics by which the population proportions of each type converge to the stable equilibrium mix.

E. *V* < *C*: Each Individual Mixes Strategies

Recall the equilibrium mixed strategy of the rational-play game calculated earlier in Section 6.A in which $p = V/C$ was the probability of choosing to be a Hawk while $(1 - p)$ was the probability of choosing to be a Dove. Let us examine whether this strategy, call it M, is an ESS. Now there are three types to consider: H types who play the pure Hawk strategy, D types who play the pure Dove strategy, and M types who play the mixture in the proportions $p = V/C$ and $1 - p = 1 - V/C = (C - V)/C$.

When an H or a D meets an M, their expected payoffs depend on p, the probability that M is playing H, and on $(1 - p)$, the probability that M is playing D. Then each player gets p times her payoff against an H, plus $(1 - p)$ times her payoff against a D. So when an H type meets an M type, she gets the expected payoff

$$p\,\frac{V - C}{2} + (1 - p)\,V = \frac{V}{C}\,\frac{V - C}{2} + \frac{C - V}{C}\,V$$

$$= -\frac{1}{2}\,\frac{V}{C}(C - V) + \frac{V}{C}(C - V)$$

$$= V\,\frac{(C - V)}{2C}\,.$$

And when a D type meets an M type, she gets

$$p \times 0 + (1 - p)\,\frac{V}{2} = \frac{C - V}{C}\,\frac{V}{2} = \frac{V\,(C - V)}{2C}\,.$$

The two fitnesses are equal. This should not be a surprise; the proportions of the mixed strategy are determined to achieve exactly this equality. Then an M type meeting another M type also gets the same expected payoff. For brevity of future reference, we call this common payoff K, where $K = V(C - V)/2C$.

But these equalities create a problem when we test M for evolutionary stability. Suppose the population consists entirely of M types and that a few mutants of the H type, comprising a very small proportion h of the total population, invade. Then the typical mutant gets the expected payoff

$$h\,\frac{V - C}{2} + (1 - h)K\,.$$

To calculate the expected payoff of an M type, note that she faces another M type a fraction $(1 - h)$ of the time and gets K in each of these. She then faces an H type for a fraction h of the interactions; in these she plays H a fraction p of the time and gets $(V - C)/2$, and plays d a fraction $(1 - p)$ of the time and gets 0. Thus her total expected payoff (fitness) is

$$hp \frac{(V - C)}{2} + (1 - h)K.$$

Since h is very small, the fitnesses of the M types and the mutant H types are almost equal. The point is that when there are very few mutants, both the H type and the M type meet only M types most of the time, and in this interaction the two have equal fitness as we just saw.

Therefore evolutionary stability hinges on the distinction of whether the original population M type is fitter than the mutant H when each is matched against one of the few mutants, that is, whether $pV(C - V)/2C = pK > (V - C)/2$. This is true in this case, because $V < C$, so $(V - C)$ is negative, while K is positive. In words, an H-type mutant will always do badly against another H-type mutant because of the high cost of fighting, but the M type fights only part of the time and therefore suffers this cost only a fraction p of the time.

Similarly, the success of a Dove invasion against the M population depends on the comparison between the mutant Dove's fitness: $[dV/2 + (1 - d) K]$ and the fitness of an M type: $d \times [pV + (1 - p)V/2] + (1 - d)K$. As earlier, the mutant faces another D a fraction d of the time and faces an M a fraction $(1 - d)$ of the time. An M type also faces another M type a fraction $(1 - d)$ of the time; but a fraction d of the time, the M faces a D and plays H a fraction p of these times, thereby gaining pV, and plays D a fraction $(1 - p)$ of these times, thereby gaining $(1 - p)V/2$. Then a Dove invasion is successful only if $V/2$ is greater than $pV + (1 - p)V/2$. This condition does not hold, since the latter expression, being a weighted average of V and $V/2$, must exceed the former whenever $V > 0$. Thus the Dove invasion cannot succeed either.

This analysis tells us that M is an ESS. The case of $V < C$ can have two equilibria. One entails a mixture of types (a stable polymorphism) and the other entails a single type that mixes its strategies in the same proportions that define the polymorphism.

7 SOME GENERAL THEORY

We now generalize the ideas illustrated in Section 6 to get a theoretical framework and set of tools that can then be applied further. This unavoidably requires some slightly abstract notation and a bit of algebra. Therefore we cover only the case of monomorphic equilibria in a single species. Readers who are adept at this level of mathematics can readily develop the cases of polymorphism and two species by analogy. Readers who are not prepared for or interested in this material can omit this section without loss of continuity.[9]

We consider random matchings from a single species whose population has available strategies I, J, . . . Some of these may be pure strategies; some may be mixed. Each individual is hardwired to play just one of these strategies. We let $E(I, J)$ denote the payoff to an I player in single encounter with a J player. The payoff of an I player meeting another of her own type is $E(I, I)$ in the same notation. We write $W(I)$ for the fitness of an I player. This is just her expected payoff in encounters with randomly picked opponents, when the probability of meeting a type is just the proportion of this type in the population.

Suppose the population is all I type. We consider whether this can be an evolutionary-stable configuration. To do so, we imagine that the population is invaded by a few J-type mutants, so the proportion of mutants in the population is a very small number, m. Now the fitness of an I type is

$$W(I) = m\, E(I, J) + (1 - m)\, E(I, I),$$

and the fitness of a mutant is

$$W(J) = m\, E(J, J) + (1 - m)\, E(J, I).$$

Therefore, the difference in the fitness of the population's main type and the mutant type is

$$W(I) - W(J) = m[E(I, J) - E(J, J)] + (1 - m)[E(I, I) - E(J, I)].$$

Since m is very small, we will have

$$W(I) > W(J) \quad \text{if } E(I, I) > E(J, I),$$

which guarantees that the second half of the $W(I) - W(J)$ expression is positive; the fact that m is small guarantees that the second term determines the sign of the overall inequality. Then the main type in the population cannot be invaded; it is fitter than the mutant type when each is matched against a member of the main type. This forms the **primary criterion** for evolutionary stability. Conversely, if $W(I) < W(J)$, due to $E(I, I) < E(J, I)$, the J-type mutants will invade successfully and an all-I population cannot be evolutionary stable.

However, it is possible that $E(I, I) = E(J, I)$, as indeed happens if the population initially is mixing between strategies I and J (a monomorphic equilibrium with a mixed strategy), as was the case in our final variant of the Hawk-Dove game (Section 6.E). Then the difference between $W(I)$ and $W(J)$ is governed by

---

[9]Conversely, readers who want more details can find them in Maynard Smith, *Evolution and the Theory of Games*, especially pp. 14–15. Maynard Smith is a pioneer of the theory of evolutionary games.

how each type fares against the mutants.[10] When $E(I, I) = E(J, I)$, we get $W(I) > W(J)$ if $E(I, J) > E(J, J)$. This is the **secondary criterion** for the evolutionary stability of I, to be invoked only if the primary one is inconclusive—that is, only if $E(I, I) = E(J, I)$.

The primary criterion carries a punch. It says that if the strategy I is evolutionary stable, then for all other strategies J that a mutant might try, $E(I, I) \geq E(J, I)$. This means that I is the best response to itself. In other words, if the members of this population suddenly started playing as rational calculators, everyone playing I would be a Nash equilibrium. *Evolutionary stability thus implies Nash!*

This is a remarkable result. If you were dissatisfied with the rational-behavior assumption underlying the theory of Nash equilibria given in earlier chapters, and you came to the theory of evolutionary games looking for a better explanation, you find it yields the same results. The very appealing biological description—fixed nonmaximizing behavior, but selection in response to resulting fitness—does not yield any new outcomes. If anything, it provides a backdoor justification for Nash equilibrium. When a game has several Nash equilibria, the evolutionary dynamics may even provide a good argument for choosing among them.

However, your reinforced confidence in Nash equilibrium should be cautious. Our definition of *evolutionary stable* is static rather than dynamic. It only checks whether the configuration of the population (monomorphic, or polymorphic in just the right proportions) that we are testing for equilibrium cannot be successfully invaded by a small proportion of mutants. It does not test whether, starting from an arbitrary initial population mix, all the unwanted types will die out and the equilibrium configuration will be reached. And the test is carried out for those particular classes of mutants that are deemed logically possible; if the theorist has not specified this classification correctly and some type of mutant she overlooked could actually arise, that mutant might invade successfully and destroy the supposed equilibrium. Our remark at the end of the twice-played prisoners' dilemma in Section 2.A warned of this possibility, and Exercise 2 below shows how it can arise. Finally, in the next section we show how evolutionary dynamics can fail to converge at all.

8 DYNAMICS WITH THREE TYPES IN THE POPULATION

If there are only two possible phenotypes (strategies), say I and J, we can show the dynamics of the population in an evolutionary game using figures similar to

[10] If the initial population is polymorphic and m the proportion of J types, then m may not be "very small" any more. The size of m is no longer crucial, however, since the second term in $W(I) - W(J)$ is now assumed to be zero.

10.3, 10.6, and 10.8. If there are three or more possible phenotypes, matters can get more complicated. As a final illustration that will lead you into more advanced study of evolutionary games, we show the dynamics of the Rock–Paper–Scissors (RPS) game from an evolutionary viewpoint.

In Chapter 7 we considered mixed strategies in RPS from a rational-choice perspective. Figure 10.12 reproduces the relevant part of the payoff table from Figure 7.10. In the evolutionary context, players are not choosing mixtures, but there can be a mixture of types in the population.

		PLAYER 2			
		R	P	S	q-Mix
PLAYER 1	R	0	−1	1	$-q_2 + (1 - q_1 - q_2)$
	P	1	0	−1	$q_1 - (1 - q_1 - q_2)$
	S	−1	1	0	$-q_1 + q_2$

FIGURE 10.12   Payoffs in the Evolutionary RPS Game

Suppose q_1 is the proportion of types in the population that are playing Rock, q_2 the proportion of Paper-playing types, and the rest, $(1 - q_1 - q_2)$ play Scissors. The last column of the table shows each Row type's payoffs when meeting this mixture of strategies; that is, just her fitness. Suppose the proportion of each type in the population grows when its fitness is positive and declines when it is negative.[11] Then

$$q_1 \text{ increases} \quad \text{if and only if} \quad -q_2 + (1 - q_1 - q_2) > 0, \text{ or } q_1 + 2q_2 < 1.$$

The proportion of Rock-playing types in the population increases when q_2, the proportion of Paper-playing types, is small or when $(1 - q_1 - q_2)$, the proportion of Scissors-playing types, is large. This makes sense; Rock players do poorly against Paper players but well against Scissors players. Similarly, we see that

$$q_2 \text{ increases} \quad \text{if and only if} \quad q_1 - (1 - q_1 - q_2) > 0, \text{ or } 2q_1 + q_2 > 1.$$

Paper players do better when the proportion of Rock players is large or the proportion of Scissors players is small.

Figure 10.13 shows graphically the population dynamics and resulting equilibria for this game. The triangular area defined by the axes and the line $q_1 + q_2 = 1$

---

[11]A little more care is necessary to ensure that the three proportions sum to 1, but that can be done, and we hide the mathematics so as to convey the ideas in a simple way.

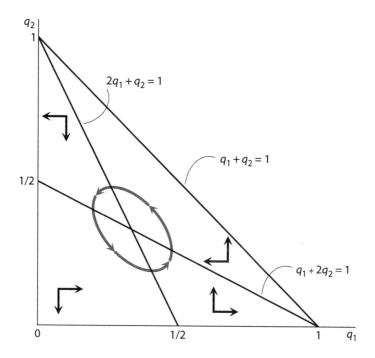

FIGURE 10.13   Population Dynamics in the Evolutionary RPS Game

contains all the possible equilibrium combinations of q_1 and q_2. There are also two straight lines within this area. The first is $q_1 + 2q_2 = 1$ (the flatter one), which is the balancing line for q_1; for combinations of q_1 and q_2 below this line, q_1 (the proportion of Rock players) increases while for combinations above this line, q_1 decreases. The second, steeper line is $2q_1 + q_2 = 1$, which is the balancing line for q_2. To the right of this line (when $2q_1 + q_2 > 1$), q_2 increases, while to the left of the line (when $2q_1 + q_2 < 1$), q_2 decreases. Arrows on the diagram show directions of motion of these population proportions; curves labeled by arrows show typical dynamic paths. The general idea is the same as that of Figure 10.10.

Each of the two straight lines consists of points where one of q_1 and q_2 neither increases nor decreases. Therefore the intersection of the two lines represents the point where q_1, q_2, and therefore also $(1 - q_1 - q_2)$, are all constant; this point thus corresponds to a polymorphic equilibrium. It is easy to check that here $p_1 = p_2 = 1 - p_1 - p_2 = 1/3$. Thus its proportions are the same as the probabilities in the rational mixed-strategy equilibrium we found in Section 4.A of Chapter 7.

Is this polymorphic outcome stable? We cannot say. The dynamics indicate paths (shown in Figure 10.13 as a single ellipse) that wind around it. Whether these paths wind in a decreasing spiral toward the intersection (in which case we have stability) or in an expanding spiral away from the intersection (indicating instability) depends on the precise response of the population proportions to the fitnesses. It is even possible that the paths circle as drawn, neither approaching nor departing from the equilibrium.

9 EVOLUTION OF COOPERATION AND ALTRUISM

Evolutionary game theory rests on two fundamental ideas: first, that individuals are engaged in games with others in their own species or with members of other species, and second, that the genotypes that lead to higher-payoff (fitter) strategies proliferate while the rest decline in their proportions of the population. These ideas suggest a vicious struggle for survival like that depicted by some interpreters of Darwin, who understood "survival of the fittest" in a literal sense and who conjured up images of a "nature red in tooth and claw." In fact, nature shows many instances of cooperation (where individual animals behave in a way that yields greater benefit to everyone in a group) and even altruism (where individuals incur significant costs in order to benefit others). Beehives and ant colonies are only the most obvious examples. Can such behavior be reconciled with the perspective of evolutionary games?

The behavior of ants and bees is probably the easiest to understand in these terms. All the individuals in an ant colony or a beehive are closely related and therefore share genes to a substantial extent. All worker ants in a colony are full sisters and therefore share half their genes; therefore the survival and proliferation of one ant's genes is helped just as much by the survival of two of its sisters as by its own survival. All worker bees in a hive are half sisters and therefore share a quarter of their genes. Of course, an individual ant or bee does not make a fine calculation of whether it is worthwhile to risk its own life for the sake of two or four sisters, but those groups whose members exhibit such behavior (phenotype) will see the underlying genes proliferate. The idea that evolution ultimately operates at the level of the gene has had profound implications for biology, although it has been misapplied by many people, just as Darwin's original idea of natural selection was misapplied.[12] The interesting idea is that a "selfish gene" may prosper by behaving unselfishly in a larger organization of genes, such as a cell. Similarly a cell and its genes may prosper by participating cooperatively and accepting their allotted tasks in a body. The next step goes beyond biology and

---

[12] In this very brief account we cannot begin to do justice to all the issues and the debates. An excellent popular account, and the source of many examples we cite in this section, is Matt Ridley, *The Origins of Virtue* (New York: Penguin, 1996). We should also point out that we do not examine the connection between genotypes and phenotypes in any detail, or the role of sex in evolution. Another book by Ridley, *The Red Queen* (New York: Penguin, 1995), gives a fascinating treatment of this subject.

into sociology: a body (and its cells and ultimately its genes) may benefit by behaving cooperatively in a collection of bodies, namely a society.

This suggests that we should see some cooperation even among individuals in a group who are not close relatives, and indeed we do find instances of this. Groups of predators like wolves are a case in point; groups of apes often behave like extended families. Even among species of prey, cooperation arises when individuals in a school of fish take turns to look out for predators. Cooperation also extends across species. Some small fish and shrimp thrive on parasites that collect in the mouths and gills of some large fish; the large fish let the small ones swim unharmed through their mouths for this "cleaning service."

Many of these situations can be seen as resolutions of prisoners' dilemmas through repetition. The mutually beneficial outcome is sustained by strategies that are remarkably like tit-for-tat. It is not to be supposed that each animal consciously calculates whether it is in its individual interest to continue the cooperation or to defect. Instead, the behavior is instinctive. Reciprocity seems to come automatically to many animals, including humans.

An instinct is hardwired into an individual's brain by genetics, but reciprocity and cooperation can also arise from more purposive thinking or experimentation within the group, and spread by socialization—by way of explicit instruction or observation of the behavior of elders—instead of genetics. The relative importance of the two channels—nurture and nature—will differ from one species to another and from one situation to another. One would expect socialization to be relatively more important among humans, but there are instances of its role among other animals. We cite a remarkable one. The expedition that Robert F. Scott led to the South Pole in 1911–1912 used teams of Siberian dogs. This group of dogs, brought together and trained for this specific purpose, developed within a few months a remarkable system of cooperation and sustained it using punishment schemes. "They combined readily and with immense effect against any companion who did not pull his weight, or against one who pulled too much . . . their methods of punishment always being the same and ending, if unchecked, in what they probably called justice, and we called murder."[13]

While this is an encouraging account of how cooperative behavior can be compatible with evolutionary game theory, you should not conclude that all animals have overcome dilemmas of selfish actions in this way. "Compared to nepotism, which accounts for the cooperation of ants and every creature that cares for its young, reciprocity has proved to be scarce. This, presumably, is due to the fact that reciprocity requires not only repetitive interactions, but also the ability to recognize other individuals and keep score."[14] In other words, precisely

---

[13] Apsley Cherry-Garrard, *The Worst Journey in the World* (London: Constable, 1922; reprinted New York: Carroll and Graf, 1989), pp. 485–486.

[14] Ridley, *Origins of Virtue*, p. 83.

the conditions that our theoretical analysis in Section 2.D of Chapter 8 identified as being necessary for a successful resolution of the repeated prisoners' dilemma are seen to be relevant in the context of evolutionary games. In the next chapter we will consider when and how the purposive design of institutions, organizations, and incentive mechanisms can help humans overcome the numerous dilemmas of collective or cooperative action that our societies face.

SUMMARY

The biological theory of evolution parallels the theory of games used by social scientists. Evolutionary games are played by behavioral *phenotypes* with genetically predetermined, rather than rationally chosen, strategies. In an evolutionary game, phenotypes with higher *fitness* survive repeated *interactions* with others to reproduce and to increase their representation in the population. If a phenotype maintains its dominance in the population when faced with an *invading mutant* type, that phenotype's strategy is called *evolutionary stable*. If two or more types are equally fit, the population exhibits *polymorphism*.

When the theory of evolutionary games is used more generally for nonbiological games, the strategies followed by individuals are understood to be standard operating procedures or rules of thumb, instead of being genetically fixed. The process of reproduction stands for more general methods of transmission including socialization, education, and imitation; and *mutations* represent experimentation with new strategies.

Evolutionary games may have payoff structures similar to those analyzed in Chapters 4 and 5, including the prisoners' dilemma and chicken. In each case, the *evolutionary-stable strategy* mirrors either the pure-strategy Nash equilibrium of a game with the same structure played by rational players, or the proportions of the equilibrium mixture in such a game. In a prisoners' dilemma, "always defect" is evolutionary stable; in chicken, types are fitter when rare, so there is a polymorphic equilibrium; in the assurance game, types are less fit when rare, so the polymorphic outcome is unstable and the equilibria are at the extremes. When play occurs between two different types of members of each of two different species, a more complex but similarly structured analysis is used to determine equilibria.

The *hawk-dove game* is the classic biological example. Analysis of this game parallels that of the prisoners' dilemma and chicken versions of the evolutionary game; evolutionary-stable strategies depend on the specifics of the payoff structure. The analysis can also be done when more than two types interact or in very general terms. This theory shows that the requirements for evolutionary stability yield an equilibrium strategy that is equivalent to the Nash equilibrium obtained by rational players.

evolutionary stability (321)

evolutionary-stable strategy (ESS) (321)

fitness (321)

genotype (321)

hawk-dove game (341)

interaction (322)

invasion by a mutant (321)

monomorphism (328)

mutation (321)

phenotype (321)

polymorphism (324)

primary criterion (346)

secondary criterion (347)

selection (321)

━━━━━━━━━━ **EXERCISES** ━━━━━━━━━━

1. Prove the following statement: "If a strategy is strongly dominated in the payoff table of a game played by rational players, then in the evolutionary version of the same game it will die out no matter what the initial population mix. If a strategy is weakly dominated, it may coexist with some other types but not in a mixture involving all types."

2. Suppose that a single play of a prisoners' dilemma has the following payoffs:

		PLAYER 2	
		Cooperate	Defect
PLAYER 1	Cooperate	3, 3	1, 4
	Defect	4, 1	2, 2

In a large population of individuals where each individual's behavior is genetically determined, each player will be either a defector (that is, always defects in any play of a prisoners' dilemma game) or a tit-for-tat player (in multiple rounds of a prisoners' dilemma, she cooperates on the first play, and on any subsequent play does whatever her opponent did on the previous play). Pairs of randomly chosen players from this population will play "sets" of n single plays of this dilemma (where $n \geq 1$). The payoff to each player over one whole set (of n plays) is the sum of her payoffs over the n plays.

Let the population proportion of defectors be p and the proportion of tit-for-tat players be $(1 - p)$. Each individual in the population plays sets of dilemmas repeatedly, matched against a new randomly chosen opponent for each new set. A tit-for-tat player always begins each new set by cooperating on its first play.

(a) Show in a two-by-two table the payoffs to an individual of each type when, over one set of plays, each player meets an opponent of each of the two types.

(b) Find the fitness (average payoff over one set against a randomly chosen opponent) for a defector.

(c) Find the fitness for a tit-for-tat player.

(d) Use the answers to parts (b) and (c) to show that when $p > (n-2)/(n-1)$, the defector type has greater fitness, and that when $p < (n-2)/(n-1)$, the tit-for-tat type has greater fitness.

(e) If evolution leads to a gradual increase in the proportion of the fitter type in the population, what are the possible eventual equilibrium outcomes of this process for the population described above? (That is, what are the possible equilibria, and which are evolutionary stable?) Use a diagram with the fitness graphs to illustrate your answer.

(f) In what sense does greater repetition (larger values of n) facilitate the evolution of cooperation?

3. Consider the twice-repeated prisoners' dilemma of Figure 10.2. Suppose that as well as the three strategy types A, T, and N present there, a fourth type S can also exist. This type cooperates on the first play and cheats on the second play of each episode of two successive plays against the same opponent.

(a) Show the four-by-four fitness table for this game.

(b) If the population initially has some of each of the four types, show that the types N, T, and S will die out in that order.

(c) Show that once a strategy (N, T, or S) has died out, as in part (b), if a small proportion of mutants of that type reappear in the future, they cannot successfully invade the population at that time.

(d) What is the ESS in this game? Why is it different from that found in the text?

4. In the assurance (meeting-place) game in this chapter, the payoffs were meant to describe the value of something material gained by the players in the various outcomes; they could be prizes given for a successful meeting, for example. Then other individuals in the population might observe the expected payoffs (fitness) of the two types, see which was higher, and gradually imitate the fitter strategy. Thus the proportions of the two types in the population would change. But a more biological interpretation can be made. Suppose the row players are always female and the column players always male. When two of these players meet successfully, they pair off, and their children are of the same type as the parents. Therefore the types would proliferate or die off as a result of meetings. The formal mathematics of this new version of the game makes it a "two-species game" (although the biology of it does not, of course). Thus the proportion of T-type females in the population—call this

proportion x—need not equal the proportion of T-type males—call this proportion y.

Examine the dynamics of x and y using methods similar to those used in the text for the battle-of-the-two-cultures game. Find the stable outcome or outcomes of this dynamic process.

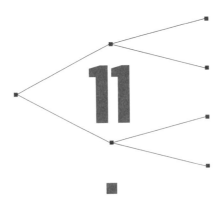

11

Collective-Action Games

THE GAMES AND STRATEGIC SITUATIONS we considered in the preceding chapters have usually involved only two or three players interacting with each other. Such games are also common in our own academic, business, political, and personal lives and so are important to understand and analyze. But many social, economic, and political interactions are strategic situations that involve numerous players at the same time. Strategies for career paths, investment plans, rush-hour commuting routes, and even studying have associated benefits and costs that depend on the actions of many other people. If you have been in any of these situations, you likely thought something was wrong—too many students, investors, and commuters crowding just where you want to be, for example. If you have tried to organize fellow students or your community in some worthy cause, you probably faced frustration of the opposite kind—too few willing volunteers. In other words, multiple-person games in society often seem to produce outcomes that are not deemed satisfactory by many or even all of the people in that society. In this chapter we will examine such games from the perspective of the theory we have already developed. We present an understanding of what goes wrong in such situations, and what can be done about it.

In the most general form, such many-player games concern problems of **collective action.** The aims of the whole society or collective are best served if its members take some particular action or actions, but these actions are not in the best private interests of those individuals. In other words, the socially optimal outcome is not automatically achievable as the Nash equilibrium of the game.

Therefore we must examine how the game can be modified to lead to the optimal outcome, or at least to improve upon the unsatisfactory Nash equilibrium. To do so, we must first understand the nature of such games. We find they come in three forms, all of them familiar to you by now: the prisoners' dilemma, chicken, and assurance games. Although our main focus in this chapter is on situations where numerous players play such games at the same time, we build on familiar ground by beginning with games between just two players.[1]

1 COLLECTIVE-ACTION GAMES WITH TWO PLAYERS

Imagine that you are a farmer. Your neighboring farmer and you can both benefit by constructing an irrigation and flood-control project. The two of you can join together to do this, or one might do so on his own. However, once the project has been constructed, the other automatically gets some benefit from it. Therefore each is tempted to shirk. That is the essence of their strategic interaction, and the difficulty of securing collective action.

In Chapter 4 we encountered a game of this kind: three neighbors were each deciding whether to contribute to a street garden they would all enjoy. That game became a prisoners' dilemma in which all three shirked; our analysis here will include an examination of a more general range of possible payoff structures. Also, in the street-garden game we rated the outcomes on a scale of 1 to 6; when we describe more general games, we will have to consider more general forms of benefits and costs for each player.

Our irrigation project has two important characteristics. First, its benefits are **nonexcludable:** a person who has not contributed to paying for it cannot be prevented from enjoying the benefits. Second, its benefits are **nonrival:** your benefits are not diminished by the mere fact that someone else is also getting the benefit. Economists call such a project a **pure public good;** national defense is often given as an example. By contrast, a pure *private* good is fully excludable and rival: nonpayers can be excluded from its benefits, and if one person gets the benefit, no one else does. A loaf of bread is a good example of a pure private good. Most goods fall somewhere on the two-dimensional spectrum of varying degrees of excludability and rivalness. We do not need this taxonomy, but men-

---

[1] You may think that games with many players should be treated like the evolutionary games discussed in Chapter 10. In fact, there are two important differences. First, although there is a large population in evolutionary games, any one interaction is between just two individuals selected from that population; in collective-action games the whole population participates in the game at the same time. Second, in evolutionary games the players mechanically follow strategies hardwired into them (genetically or otherwise); in multiplayer games the players are rational maximizers. However, there are some similarities in the methods of analysis, especially in the graphical tools we use.

tion it to help you relate our discussion to what you may encounter in other courses and books.[2]

A. Numerical Examples

The costs and the benefits associated with building the irrigation project can depend on which players participate. Suppose each of you acting alone could complete the project in seven weeks, whereas if the two of you acted together, it would take only four weeks of time from each. The two-person project is also of better quality; each farmer gets benefits worth six weeks of work from a one-person project (whether constructed by you or by your neighbor), and eight weeks' worth of benefit from a two-person project.

In the game, each farmer has to decide whether to work toward the construction of the project. (Presumably there is a short window of time in which the work can be done, and you could pretend to be called away on some very important family matter at just that time, as could your neighbor.) Figure 11.1 shows the payoff table of the game, where the numbers measure the values in weeks of work. Given the payoff structure of this table, your best response if your neighbor does not participate is not to participate either: your benefit from completing the project by yourself (6) is less than your cost (7), whereas you can get 0 by not participating. Similarly, if your neighbor does participate, then you can reap the benefit (6) from his work at no cost to yourself; this is better for you than working yourself to get the larger benefit of the two-person project (8) while incurring the cost of the work (4). In this case, you are said to be a **free rider** on your neighbor's effort if you let the other do all the work and then reap the benefits all the same. The more general lesson of the game is that it is better for you not to participate no matter what your neighbor does; the same logic holds for him. Thus the game is a prisoners' dilemma: not building is the dominant strategy for each, but both would be better off if the two were to work together to build.

		YOUR NEIGHBOR	
		Build	Not
YOU	Build	4,4	−1,6
	Not	6,−1	0,0

FIGURE 11.1 Collective Action as a Prisoners' Dilemma: Version I

[2]Public goods are studied in more detail in textbooks on "public economics" such as Harvey Rosen, *Public Finance*, 4th ed. (Chicago: Richard D. Irwin, 1995), or Joseph Stiglitz, *Economics of the Public Sector*, 2nd ed. (New York: Norton, 1988).

We see in this prisoners' dilemma one of the main difficulties that arises in games of collective action. Individually optimal choices—in this case, not to build regardless of what the other farmer chooses—may not be optimal from the perspective of society as a whole, even if the society is made up of just two farmers. The "social" optimum in a collective-action game occurs when the sum total of the players' payoffs is maximized; in this prisoners' dilemma, the social optimum is the (Build, Build) outcome. Nash equilibrium behavior of the players does not regularly bring about the socially optimal outcome, however. Hence, the study of collective-action games has focused on methods to improve upon observed (generally Nash) equilibrium behavior in order to move outcomes toward the socially best outcomes. As we will see, the divergence between Nash equilibrium and socially optimum outcomes appears in every version of collective-action games.

Now consider what the game would look like if the numbers were to change slightly. Suppose the two-person project yields benefits that are not much better than those in the one-person project: 6.3 weeks' worth of work to each farmer. Then each of you gets $6.3 - 4 = 2.3$ when both of you build. The resulting payoff table is shown in Figure 11.2. The game is still a prisoners' dilemma, and leads to the equilibrium (Not, Not). However, when both farmers build, the total payoff for both of you is only 4.6. The total payoff is maximized when one of you builds and the other does not, in which case together you get a grand total of 5—and there are two possible ways to get this outcome. Achieving the social optimum in this case then poses a new problem: who should build and suffer the payoff of −1, while the other is allowed to be a free rider and enjoy the payoff of 6?

		YOUR NEIGHBOR	
		Build	Not
YOU	Build	2.3, 2.3	−1, 6
	Not	6, −1	0, 0

FIGURE 11.2 Collective Action as a Prisoners' Dilemma: Version II

Now consider a different variation in the numbers of the original prisoners' dilemma game of Figure 11.1. Suppose the cost of the work is reduced so that it becomes better for you to build your own project if the neighbor does not. Specifically, suppose the one-person project requires four weeks of work, and the two-person project takes three weeks from each; the benefits are the same as before. Figure 11.3 shows the payoff matrix resulting from these changes. Now

		YOUR NEIGHBOR	
		Build	Not
YOU	Build	5,5	2,6
	Not	6,2	0,0

FIGURE 11.3 Collective Action as Chicken: Version I

your best response is to shirk when your neighbor works and work when he shirks. Formally, this is just like a game of chicken, where shirking is the Straight strategy (tough or uncooperative), while working is the Swerve strategy (conciliatory or cooperative).

If this game results in one of its pure-strategy equilibria, the two payoffs sum to 8; this total is less than the total outcome both players could get if both of them build. That is, neither of the Nash equilibria provides as much benefit to society as a whole as the coordinated outcome entailing both farmers choosing to build. If the chicken game results in a mixed-strategy equilibrium, the outcome for this two-farmer society will be even worse: the expected payoffs will add to something less than 8.

The collective-action chicken game also has another possible structure if we make some additional changes to the benefits associated with the project. As with version II of the prisoners' dilemma, suppose the two-person project is not much better than the one-person project. Then each farmer's benefit from the two-person project is only 6.3, while each still gets a benefit of 6 from the one-person project. We ask you to practice your skill by constructing the payoff table for this game. You will find that it is still a game of chicken—call it chicken II—and that the two Nash equilibria are the same, but the sum of the payoffs when both build is only 6.6, while the sum when only one builds is 8. The social optimum is for only one farmer to build. This outcome occurs in each of the two Nash equilibria of the game, but each farmer prefers the equilibrium in which the other builds. Then the outcome in which each waits for the other to build may arise, or the farmers might use mixed strategies, which, as we know from the analysis of chicken in Chapter 5, Section 4A, result in low expected payoffs for both.

We have supposed thus far that the project yields equal benefits to both you and your neighbor farmer. But what if one's activity causes harm to the other, as would happen if the only way to prevent one farm from being flooded is to divert the water to the other? Then each player's payoffs would be lower than those shown in Figure 11.3 in the case that your neighbor chose Build. Thus a third variant of chicken could arise in which each of you wants to build when the other does not, whereas it would be collectively better if neither of you did.

		YOUR NEIGHBOR	
		Build	Not
YOU	Build	4,4	−4,3
	Not	3,−4	0,0

FIGURE 11.4   Collective Action as an Assurance Game

Finally, let us change the payoffs of the prisoners' dilemma case in a different way altogether, leaving the benefits of the two-person project and the costs of building as originally set out and reducing the benefit of a one-person project to 3. This change reduces your benefit as a free rider so much that now, if your neighbor chooses Build, your best response is also Build. Figure 11.4 shows the payoff table for this version of the game. This is now an assurance game with two pure-strategy equilibria: one where both of you participate, the other where neither of you does.

As in the chicken II version of the game, the socially optimal outcome here is one of the two Nash equilibria. But there is a difference. In chicken II, the two players differ in their preferences between the two equilibria, whereas in the assurance game both of them prefer the same equilibrium. Therefore achieving the social optimum should be easier in the assurance game than in chicken.

B. Generalization of the Two-Person Case

We now gather all of our examples of the two-person collective-action game "to build an irrigation project" within a single general statement. Suppose a one-person project yields benefits b_1 to each farmer and imposes costs c_1 on the builder, while a two-person project has benefits b_2 and costs c_2 for each. Then the general payoff table is as shown in Figure 11.5.

		YOUR NEIGHBOR	
		Build	Not
YOU	Build	$b_2 - c_2, b_2 - c_2$	$b_1 - c_1, b_1$
	Not	$b_1, b_1 - c_1$	0,0

FIGURE 11.5   General Form of a Two-Person Collective-Action Game

The game illustrated in the figure is a prisoners' dilemma if the inequalities

$$b_1 > b_2 - c_2, \quad 0 > b_1 - c_1, \quad \text{and} \quad b_2 - c_2 > 0$$

all hold at the same time. The first says that the best response to Build is Not, the second says that the best response to Not is also Not, and the third says that (Build, Build) is jointly preferred to (Not, Not).

The game is one of chicken if

$$b_1 > b_2 - c_2, \quad \text{and} \quad 0 < b_1 - c_1.$$

These inequalities indicate that you want to shirk when your neighbor works and work when he shirks; this is true for both versions I and II of chicken.

For both the prisoners' dilemma and the chicken cases, we get version I (in which it is socially optimal for both of them to build) if

$$2(b_2 - c_2) > 2b_1 - c_1,$$

and version II (where it is jointly best for either one but not both players to build) if

$$2(b_2 - c_2) < 2b_1 - c_1.$$

Finally, the game is one of assurance if

$$b_1 < b_2 - c_2, \quad 0 > b_1 - c_1, \quad \text{and} \quad b_2 - c_2 > 0.$$

Here the inequalities say that Build is the best response to Build, that Not is the best response to Not, and that (Build, Build) yields higher payoffs to both players than does (Not, Not). For the assurance game, the first of the above inequalities implies that

$$2(b_2 - c_2) > 2b_1 > 2b_1 - c_1,$$

and so it is socially optimal for both to build; the version II that was possible for the prisoners' dilemma and chicken games cannot arise for the assurance game.

Just as the problems pointed out in these examples are familiar, the various alternative ways of tackling the problems also follow the general principles we discussed in earlier chapters. Before turning to solutions, let us see how the problems manifest themselves in the more realistic setting where several players interact simultaneously in such games.

2 COLLECTIVE-ACTION PROBLEMS IN LARGE GROUPS

In this section, we extend our irrigation-project example to a situation in which a population of N farmers must each decide whether to participate. If n of them participate in the construction of the project, each of the participants incurs a cost c that depends on the number n, so we write it as the function $c(n)$. Also, each person in the population, whether a contributor to the building of the project or not, enjoys a benefit from its completion that is also a function of n; we write the benefit function as $b(n)$. Thus each participant gets the payoff $p(n) = b(n) - c(n)$, while each nonparticipant, or shirker, gets the payoff $s(n) = b(n)$.

Suppose you are contemplating whether to participate or to shirk. Your decision will depend on what the other $(N-1)$ farmers in the population are doing. In general, you will have to make your decision when the other $(N-1)$ players consist of n participants and $(N-1-n)$ shirkers. If you decide to shirk, the number of participants in the project is still n, so you get a payoff of $s(n)$. If you decide to participate, the number of participants becomes $n + 1$, so you get $p(n + 1)$. Therefore your final decision depends on the comparison of these two payoffs; you will participate if $p(n + 1) > s(n)$, and you will shirk if $p(n + 1) < s(n)$. This comparison holds true for every version of the collective-action game we analyzed in Section 1; differences in behavior in the different versions arise because the changes in the payoff structure alter the values of $p(n + 1)$ and $s(n)$.

We can use the payoff functions $p(n)$ and $s(n)$ to construct a third function showing the total payoff to society as a function of n, which we write as $T(n)$. The total payoff to society consists of the value $p(n)$ for each of the n participants and the value $s(n)$ for each of the $(N-n)$ shirkers:

$$T(n) = n\, p(n) + (N-n)\, s(n)$$

Now we can ask: What allocation of people between participants and shirkers maximizes the total payoff $T(n)$? To get a better understanding of this, it is convenient to write $T(n)$ differently, as

$$T(n) = N\, s(n) - n\, [s(n) - p(n)]$$

This derivation of the total social payoff shows that we can calculate it as if we gave every one of the N people the shirker's payoff, but then removed the shirker's extra benefit $[s(n) - p(n)]$ from each of the n participants. We normally expect $s(n)$ to increase as n increases; therefore the first term in this expression, $N\, s(n)$, also increases as n increases. If the second term does not increase too fast as n increases—as would be the case if the shirker's extra benefit, $[s(n)-p(n)]$, is small and does not increase—then the effect of the first term dominates in determining the value of $T(n)$; $T(n)$ increases steadily with n in this case and is maximized at $n = N$. But otherwise and more generally, $T(n)$ can be maximized for some value of n less than N. That is, society's aggregate payoff may be maximized by allowing some shirking. We saw this earlier in the two-person examples, and we will encounter more examples in the multiperson situations to be examined shortly.

Now we can use graphs of the $p(n + 1)$ and $s(n)$ functions to help us determine which type of game we have encountered, its Nash equilibrium, and its socially optimal outcome. We draw the graphs showing n over its full range from 0 to $(N-1)$ along the horizontal axis and payoffs along the vertical axis. Two separate curves show $p(n+1)$ and $s(n)$. Actually, n takes on only integer values, and therefore each function $p(n+1)$ and $s(n)$ technically consists only of a discrete set of points rather than a continuous set as implied by our smooth lines. But when N is large, the discrete points are sufficiently close together that we can join up the successive points and show each payoff function as a continuous curve. This is similar to what we did in Chapter 10 on evolutionary games, where the horizontal axis showed the proportion of the population following one strategy or the other. We show the $p(n+1)$ and $s(n)$ functions as straight lines to bring out the basic issues most simply, and will discuss more complicated possibilities later.

The first of our graphs, Figure 11.6, illustrates the case in which the curve $s(n)$ lies entirely above the curve $p(n+1)$; this is the prisoners' dilemma. The implication of this type of relationship is that no matter how many others participate (that is, no matter how large n gets), your payoff is higher if you shirk than if you participate; shirking is your dominant strategy. This is true for everyone, therefore the equilibrium of the game involves everyone shirking. But both curves are rising as n increases: for each action you take, you are better off if more of the others participate. And the left intercept of the $s(n)$ curve is below the right intercept of the $p(n+1)$ curve, or $s(0) < p(N)$. This says that if everyone including you shirks, your payoff is less than if everyone including you participates: everyone would be better off than they are in the Nash equilibrium of the game if the outcome in which everyone participates could be sustained.

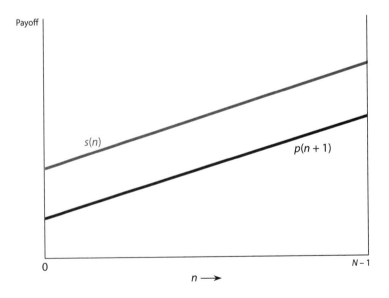

FIGURE 11.6 Multiperson Prisoners' Dilemma

However, as we saw earlier, the fact that each individual would be better off if everyone participated does not automatically imply that full participation is the best thing for society; it is not automatic that the total payoff function is maximized when n is as large as possible in the prisoners' dilemma case. In fact, if the gap between $s(n)$ and $p(n)$ widens sufficiently fast as n increases, then the negative impact of the second term in the expression for $T(n)$ outweighs the positive impact of the first term as n approaches N; then it may be best to let some people shirk—that is, the socially optimal value for n may be less than N. Of course, this type of outcome creates an inequality in the payoffs—the shirkers fare better than the participants—which adds another dimension of difficulty to society's attempts to resolve the dilemma.

Figure 11.7 shows the case for chicken. Here for small values of n, $p(n + 1) > s(n)$, so if few others are participating, your choice is to participate; and for large values of n, $p(n + 1) < s(n)$, so if many others are participating, your choice is to shirk. Note the equivalence of these two statements to the idea in the two-person game that "you shirk if your neighbor works and work if he shirks." If the two curves intersect at a point corresponding to an integer value of n, then that is the Nash equilibrium number of participants. If that is not the case, then strictly speaking the game has no Nash equilibrium. But in practice, if the current value of n in the population is the integer just to the left of the point of intersection, then one person will just want to participate, whereas if the current value of n is the integer just to the right of the point of intersection, one person will want to switch to shirking. Therefore the number of participants will stay in a small

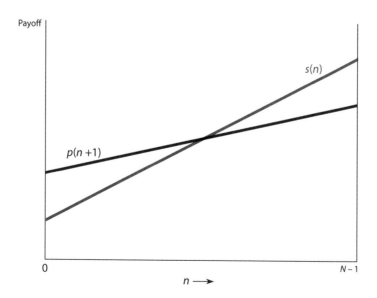

FIGURE 11.7   Multiperson Chicken

neighborhood of the point of intersection, and we can justifiably speak of the intersection as the equilibrium in some approximate sense.

The equilibrium illustrated in Figure 11.7 looks very much like the polymorphic equilibrium of the evolutionary chicken game (Figure 10.6). There the two curves had negative slopes. Here we have shown both positively sloped, although they don't have to be; it is conceivable that the benefit for each person is smaller when more people participate. The important feature common to the two games is that when there are few people taking one action, it is better for any one individual to take that action, and when there are many people taking one action, it is better for any one individual to take the other action. Thus the two games have similar equilibria for a common reason. The difference remains that only two players selected from the large population play the evolutionary game at any one time, whereas here the whole population plays at once.

What is the socially optimal outcome in the chicken form of the collective-action problem? If each participant's payoff $p(n)$ increases as the number of participants increases, and if each shirker's payoff $s(n)$ does not become too much greater than the $p(n)$ of each participant, then the total social payoff is maximized when everyone participates. But more generally it may be better to let some shirk. This is exactly the difference between versions I and II of chicken in our earlier numerical example in Section 1. As we pointed out earlier in Section 2, in the context of the prisoners' dilemma, where the social optimum can have some participants and some shirkers, payoffs can be unequally distributed in the population, making it harder to implement the optimum. In more general games of this kind, the optimal number of participants could even be smaller than that in the Nash equilibrium. We return to examine the question of the social optimum of all of these versions of the game in greater detail in Section 5.

Figure 11.8 shows the assurance case. Here $s(n) > p(n + 1)$ for small values of n, so if few others are participating, then you want to shirk too, whereas $p(n + 1) > s(n)$ for large values of n, so if many others are participating, then you want to participate too. This game has two Nash equilibria at the two extremes: either everyone shirks or everyone participates. When the curves are rising—so each person is better off if more people participate—then clearly the right-hand extreme equilibrium is the better one for society, and the question is how to bring it about.

When the total number of people in the group, N, is very large, and any one person makes only a small difference, then $p(n + 1)$ is almost the same as $p(n)$. Thus any one person will choose to shirk if $p(n) < s(n)$. Expressing this in terms of the benefits and costs of the common project in our example, namely $p(n) = b(n) - c(n)$ and $s(n) = b(n)$, we see that $p(n)$ is *always* less than $s(n)$, so individuals will always want to shirk when N is very large even when the game appears to have the structure of an assurance game. And that is why problems of collective provi-

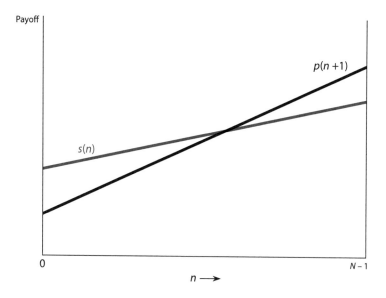

Payoff

$p(n+1)$

$s(n)$

0

$n \longrightarrow$

$N-1$

FIGURE 11.8 Multiperson Assurance Game

sion of public projects in a large group almost always manifest themselves as prisoners' dilemmas. But as we saw in the two-person examples discussed in Section 1, this result is not necessarily true for small groups. Morever, collective-action problems cover a broader range than that of participating in the construction of a public project whose benefits are equally available to shirkers. We will see several other examples of such games later in Section 5.

Therefore, in general, we must allow for a broader interpretation of the payoffs $p(n)$ and $s(n)$ than we did in the specific case involving the benefits and the costs of a project; we must allow $p(n)$ and $s(n)$ to be any functions of n. Then there is no automatic presumption about which of the two payoff functions is larger, and all three kinds of games—prisoners' dilemma, chicken, and assurance—and their associated graphs deserve our attention. In fact, in the most general case, $p(n)$ and $s(n)$ do not even have to be straight lines and can intersect many times. Then there can be several equilibria, although each can be thought of as representing one of the three types we have described so far.[3] When we make this more general interpretation, we will speak of two actions labeled P and S, which have no necessary connotation of "participation" and "shirking" but allow us to continue with the same symbols for the payoffs. Thus, when there are n players taking the action P, $p(n)$ becomes the payoff of each player taking the action P, and $s(n)$ that of each player taking the action S.

[3] Exercises 2, 3, and 5 at the end of this chapter construct some simple situations with nonlinear curves and multiple equilibria. For a more general theoretical analysis of such equilibria, see Thomas Schelling, *Micromotives and Macrobehavior* (New York: Norton, 1978), chap. 7.

3 A BRIEF HISTORY OF IDEAS

The problem of collective action has been recognized by social philosophers and economists for a very long time. The 17th-century British philosopher Thomas Hobbes argued that society would break down in a "war of all against all" unless it was ruled by a dictatorial monarch or *Leviathan* (the title of his book). One hundred years later, the French philosopher Jean-Jacques Rousseau described the problem of a prisoners' dilemma in his *Discourse on Inequality*. A stag hunt needs the cooperation of the whole group of hunters to encircle and kill the stag, but any individual hunter who sees a hare may find it better for himself to leave the circle to chase the hare. But Rousseau thought that such problems were the product of civilization, and that people in the natural state lived harmoniously as "noble savages." At about the same time, two Scots pointed out some dramatic solutions to such problems: David Hume in his *Treatise on Human Nature* argued that the expectations of future returns of favors can sustain cooperation. Adam Smith's *Wealth of Nations* developed a grand vision of an economy in which production of goods and services motivated purely by private profit could result in an outcome that was best for society as a whole.[4]

The optimistic interpretation persisted, especially among many economists and even several political scientists, to the point where it was automatically assumed that if an outcome was beneficial to a group as a whole, the actions of its members would bring this about. This belief received a necessary rude shock in the mid-1960s when Mancur Olson published *The Logic of Collective Action*. He pointed out that the best outcome collectively would not prevail unless it was in each individual's private interest to perform his assigned action, that is, unless it was a Nash equilibrium. However, he did not specify the collective action game very precisely. Although it looked like a prisoners' dilemma, Olson insisted that it was not necessarily so, and we have already seen that the problem can also take the form of a chicken game or an assurance game.[5]

Another major class of collective-action problems, namely those involving the depletion of common-access resources, received attention at about the same time. If a resource like a fishery or a meadow is open to all, each user will exploit

[4]The great old books cited in this paragraph have been reprinted many times in many different versions. For each, we list the year of original publication, and details of one relatively easily accessible reprint. In each case, the editor of the reprinted version provides an introduction that conveniently summarizes the main ideas. Thomas Hobbes, *Leviathan, or the Matter, Form, and Power of Commonwealth Ecclesiastical and Civil*, 1651 (Everyman Edition, London: J. M. Dent, 1973). David Hume, *A Treatise of Human Nature*, 1739 (Oxford: Clarendon Press, 1976). Jean-Jacques Rousseau, *A Discourse on Inequality*, 1755 (New York: Penguin Books, 1984). Adam Smith, *An Inquiry into the Nature and Causes of the Wealth of Nations*, 1776 (Oxford: Clarendon Press, 1976).

[5]Mancur Olson, *The Logic of Collective Action* (Cambridge, Mass.: Harvard University Press, 1965).

it as much as he can because any self-restraint on his part will merely make more available for the others to exploit. Garrett Hardin wrote a famous article on this subject entitled "The Tragedy of the Commons."[6] The problem in the common-resource game is just the reverse of our example of participation in the construction of a project that yields benefits mostly to the others. In our example, each individual has a strong private incentive to refrain from participation and to enjoy a free rider's benefits. In the case of a common resource, each individual has a strong private incentive to exploit it to the full, making everyone else pay the social cost that results from the degradation of the resource.

Until recently, many social scientists, and most physical scientists, took a Hobbesian line on the common-resource problem, arguing that it can only be solved by a government that forces everyone to behave cooperatively. Others, especially economists, retained their Smithian optimism. They argued that placing the resource in proper private ownership, where its benefits can be captured in the form of profit by the owner, will induce the owner to restrain its use in a socially optimal manner. He will realize that the value of the resource (fish or grass, for example) may be higher in the future because there will be less available, and therefore he can make more profit by saving some of it for that future.

Nowadays, thinkers from all sides have begun to recognize that collective-action problems come in diverse forms, and that there is no uniquely best solution to all of them. They also understand that groups or societies do not stand helpless in the face of such problems and they devise various ways to cope with them. Much of this work has been informed by game-theoretic analysis of repeated prisoners' dilemmas and similar games. Michael Taylor's book *The Possibility of Cooperation* gives a theoretical overview. Our classification of collective-action problems in Section 1 builds on his work. Elinor Ostrom's *Governing the Commons* is probably the best account of how common resources are managed in practice; Matt Ridley's *The Origins of Virtue* offers a fascinating biological perspective on the emergence of cooperation.[7]

4 SOLVING COLLECTIVE-ACTION PROBLEMS

In this section we offer a brief discussion of conceptual and practical solutions to collective-action problems, organized around the threefold classification of collective-action games as prisoners' dilemma, chicken or assurance games. The

---

[6] Garrett Hardin, "The Tragedy of the Commons," (*Science*, vol. 162 [1968], pp. 1243–1248).

[7] Michael Taylor, *The Possibility of Cooperation* (New York: Cambridge University Press, 1987). Elinor Ostrom, *Governing the Commons* (New York: Cambridge University Press, 1990). Matt Ridley, *The Origins of Virtue* (New York: Viking Penguin, 1996).

feature common to all three types is the need to induce individuals to act cooperatively, or in a manner that would be best for the group, even though the individual's interests may best be served by doing something else, in particular, taking advantage of the others' cooperative behavior.

The problem of the need to attain cooperation and its solutions are not unique to human societies; nature provides numerous puzzles and examples pertaining to cooperation. Striking examples are the cooperative behavior of ant and bee colonies, where individuals sacrifice seemingly important activities like reproduction—even their lives—in the interests of the collective. Biologists generally explain these behaviors in terms of the advantage to the gene, as we saw in Section 9 of Chapter 10; all the individuals in an ant colony or a beehive are genetically so closely related that the genes of any one of them stand the best chance of propagation if it abides by its allotted task in the division of labor that best serves the interests of the group. In a sense, every individual is a collective of cells that abide by just such a division of labor.

But we also see cooperation in groups of animals in which the individuals are not so close in their genetic composition. Groups of fish, birds, foxes, apes, and so forth exhibit remarkable instances of cooperation in gathering and sharing food, rearing their young, and avoiding predators. Biologists explain this type of behavior as the evolution of **instincts.** If the advantages of cooperation are so large that it is better to be a cooperative member of a group of cooperative individuals than to be a selfish member of a group of selfish individuals, then the cooperative groups and their members likewise will thrive. If cooperation is an instinct hardwired in the brain and transmitted by genetic inheritance, then such behavior can emerge triumphant in the evolutionary process.[8]

Humans also show many cooperative instincts; for example, the act of reciprocating gifts and skills at detecting cheating are so common in all societies and throughout history that there is good reason to think they are instincts.[9] But human societies rely more on purposive social and cultural customs, norms, and sanctions in inducing cooperative behavior from their individual members. These methods are more conscious, deliberate attempts to design the game in order to solve the collective-action problem than are observed in animal societies, and we focus on them here.[10] We approach the issue of solution methods from the perspective of the type of game being played.

---

[8] See Ridley, *Origins of Virtue,* especially chap. 3.

[9] See Ridley, ibid., chaps. 6 and 7.

[10] The social sciences do not have precise and widely accepted definitions of terms like *custom* and *norm*, nor are the distinctions among such terms always clear and unambiguous. We set out some definitions in this section, but be aware that you may find a different usage in other books. Our approach is similar to that in David Kreps, "Intrinsic Motivation and Extrinsic Incentives," *American Economic Review,* Papers and Proceedings, vol. 87, no. 2 (May 1997), pp. 359–364, but he uses the term "norm" for all the concepts that we classify under different names.

Sociologists have a different taxonomy of norms than economists; it is based on the importance

A. Analysis

A solution is easiest if the collective-action problem takes the form of an assurance game. Then it is in every individual's private interest to take the socially best action if he expects all other individuals to do likewise. In other words, the socially optimal outcome is a Nash equilibrium; the only problem is that the same game has other, socially less good, Nash equilibria. But then all that is needed to achieve the best Nash equilibrium, and thereby the social optimum, is to make it a focal point, that is, to ensure the convergence of the players' expectations upon it. Such a convergence can result from a social **custom** or **convention**—namely, a mode of behavior that finds automatic acceptance because it is in everyone's interest to follow it so long as others are expected to do likewise. For example, if all the farmers, herders, weavers, and other producers in an area want to get together to trade their wares, all they need is the assurance of finding others with whom to trade. Then the custom that the market is held in village X on day Y of every week makes it optimal for everyone to be there on that day.

But our analysis in Section 2 suggested that individual payoffs are often configured in such a way that collective-action problems, particularly of large groups, take the form of a prisoners' dilemma. Not surprisingly, the methods for coping with such problems have received the most attention. In Chapter 8 we described in detail three methods for achieving a cooperative outcome in prisoners' dilemma games: repetition, penalties (and/or rewards), and leadership. In that discussion, we were mainly concerned with two-person dilemmas. But the same methods apply to collective-action problems in large groups, with some important modifications or innovations.

Society can inflict punishment on "defectors" (or shirkers) in several different ways. One is through **sanctions** imposed by other members of the group. Sanctions often take the form of disqualification from future games played by the group. Society can also create **norms** of behavior that change individual payoffs so as to induce cooperation. A norm changes the private payoff scale of each player, by adding an extra cost in the form of shame, guilt, or dislike of the mere disapproval of others. Society establishes norms through a process of education or culture. Norms differ from customs in that an individual would not automatically follow the norm merely because of the expectation that others would follow it; the extra cost is essential. Norms also differ from sanctions in that others do

---

of the issue (trivial situations like table manners are called *folkways*, and more weighty matters are called *mores*), and on whether the norms are formally codified as *laws*. They also maintain a distinction between *values* and norms, recognizing that some norms may run counter to individuals' values, and therefore require sanctions to enforce them. This corresponds to our distinction between customs and norms; the conflict between individual values and social goals arises for norms but not for *conventions,* as we label them. See Donald Light and Suzanne Keller, *Sociology,* 4th ed. (New York: Knopf, 1987) pp. 57–60.

not have to take any explicit actions to hurt you if you violate the norm; the extra cost becomes internalized in your own payoff scale.[11]

Society can also reward desirable actions just as it can punish undesirable ones. Again the rewards, financial or otherwise, can be given externally to the game (like sanctions) or individuals' payoffs can be changed so that they take pleasure in doing the right thing (norms). The two types of rewards can interact; for example the peerages and knighthoods given to British philanthropists and others who do good deeds for British society are external rewards, but individuals value them only because respect for knights and peers is a British social norm.

Norms are reinforced by observation of society's general adherence to them, and lose their force if they are frequently seen to be violated. Before the advent of the welfare state, when those who fell on hard economic times had to rely on help from family or friends or their immediate small social group, the work ethic constituted a norm that held in check the temptation to slacken one's own efforts and become a free rider on the support of others. As the government took over the supporting role and unemployment compensation or welfare became an entitlement, this norm of the work ethic weakened. After the sharp increases in unemployment in Europe in the late 1980s and early 1990s, a significant fraction of the population became users of the official support system, and the norm weakened even further.[12]

Different societies or cultural groups may develop different conventions and norms to achieve the same purpose. At the relatively trivial level, each culture has its own set of good manners—ways of greeting strangers, indicating approval of food, and so on. When two people from different cultures meet, misunderstandings can arise. More importantly, each company or office has its own ways of getting things done. The differences between these customs and norms are subtle and difficult to pin down, but many mergers fail because of a clash of these "corporate cultures."

The external penalties of sanctions, or the internalized ones of norms, can in principle work even in single plays of collective-action games, although the force of norms is more likely to emerge and be sustained through repetition. Now let us consider repeated plays more explicitly. We saw in Chapter 8 how repetition can achieve cooperative outcomes as equilibria of individual actions in a repeated two-person prisoners' dilemma by holding up the prospect that cheating will lead to a breakdown of cooperation. More generally, what is needed to maintain cooperation is the expectation in the minds of each player that his personal benefits of cheating are transitory, and that they will quickly be replaced by a

---

[11] See Ostrom, *Governing the Commons*, p. 35. Our distinction between norms and sanctions is similar to Kreps's distinction between functions (iii) and (iv) of norms (Kreps, "Intrinsic Motivation and Extrinsic Incentives," p. 359).

[12] Assar Lindbeck, "Incentives and Social Norms in Household Behavior," *American Economic Review*, vol. 87, no. 2 (May 1997), pp. 370–377.

payoff lower than that associated with cooperative behavior. For players to believe that cheating is not beneficial from a long-term perspective, cheating should be detected quickly, and the punishment that follows (reduction in future payoffs) should be sufficiently swift, sure, and painful.

Detection and punishment are never easy. In most real situations, payoffs are not completely determined by the players' actions, but are subject to some random fluctuations. One person's cheating may get lost in this uncertainty, so the other players may not be aware that cheating has occurred. Even if they are sure, they may not know who has cheated. And inflicting severe punishment on a player without being sure of his guilt beyond a reasonable doubt is not only morally repulsive; it is counterproductive—the incentive to cooperate gets blunted if cooperative actions are susceptible to punishment by mistake. Finally, punishment can bring its own collective problems. For example, if a trading group punishes a cheater by ostracism—by denying him the benefits of trading with the other members of the group—then to punish one member, the others must forgo the benefits they get from future dealings with him.

Successful solution of collective-action problems clearly hinges on success in detection and punishment. As a rule, small groups are more likely to have better information about their members and the members' actions, and therefore are more likely to be able to detect cheating. They are also more likely to be able to organize when inflicting punishment on a cheater. One sees many instances of successful cooperation in small village communities that would be unimaginable in a large city or state.

This argues for letting a small group solve its own collective-action problem. An external enforcer of cooperation may not be able to detect cheating or impose punishment with sufficient clarity and swiftness. Thus the frequent reaction that centralized or government policy is needed to solve collective-action problems is often proved wrong. For example, village communities or "communes" in late 19th-century Russia solved many collective-action problems of irrigation, crop rotation, management of woods and pastures, and road and bridge construction and repair, in just this way. "The village . . . was not the haven of communal harmony. . . . It was simply that the individual interests of the peasants were often best served by collective activity." Reformers of early 20th-century czarist governments and Soviet revolutionaries of the 1920s alike failed, partly because the old system had such a hold on the peasants' minds that they resisted anything new, but also because the reformers failed to understand the role that some of the prevailing practices played in solving collective-action problems, and thus failed to replace them by equally effective alternatives.[13]

---

[13] Orlando Figes, *A People's Tragedy: The Russian Revolution 1891–1924* (New York: Viking Penguin, 1997), pp. 89–90, 240–241, 729–730. See also Ostrom, *Governing the Commons*, p. 23, for other instances where external government-enforced attempts to solve common-resource problems actually made them worse.

Next, consider the chicken form of collective-action games. Here the nature of the remedy depends on whether the largest total social payoff is attained when everyone participates (what we called "chicken version I" in Section 1) or when some cooperate and others are allowed to shirk (chicken II). For chicken I, where everyone has the individual temptation to shirk, the problem is much like that of sustaining cooperation in the prisoners' dilemma, and all the earlier remarks for that game apply here too. Chicken II is different—easier in one respect and harder in another. Once an assignment of roles between participants and shirkers is made, no one has the private incentive to switch: if the other driver is assigned the role of going straight, then you are better off swerving and the other way around. Therefore if a custom creates the expectation of an equilibrium, it can be maintained without further social intervention such as sanctions. However, in this equilibrium the shirkers get higher payoffs than the participants, and this inequality can create its own problems for the game; the conflicts and tensions, if they are major, can threaten the whole fabric of the society. Often the problem can be solved by repetition. The roles of participants and shirkers can be rotated to equalize payoffs over time.

Sometimes the problem of differential payoffs in versions II of the prisoners' dilemma or chicken is "solved," not by restoring equality, but by **oppression** or **coercion,** which forces a dominated subset of society to accept the lower payoff and allows the dominant subgroup to enjoy the higher payoff. In many societies throughout history, the work of handling animal carcasses was forced on particular groups or castes in this way. The history of the maltreatment of racial and ethnic minorities and of women provides vivid examples of such practices. Once such a system becomes established, no one member of the oppressed group can do anything to change the situation: in the chicken game it is better for you to swerve if you know that the other player is going straight. The oppressed must get together as a group and act to change the whole system, itself another problem of collective action.

Finally, consider the role of leadership in solving collective-action problems. In Chapter 8 we pointed out that if the players are of very unequal "size," the prisoners' dilemma may disappear because it may be in the private interests of the larger player to continue cooperation and to accept the cheating of the smaller player. Here we recognize the possibility of a different kind of bigness, namely having a "big heart." People in most groups differ in their preferences, and many groups have one or a few who take genuine pleasure in expending personal effort to benefit the whole. If there are enough such people for the task at hand, then the collective-action problem disappears. Most schools, churches, local hospitals, and other worthy causes rely on the work of such willing volunteers. This solution, like others before it, is more likely to work in relatively small groups, where the fruits of their actions are more closely and immediately visible to the benefactors, who are therefore encouraged to continue.

B. Applications

Elinor Ostrom, in her book *Governing the Commons,* discusses several examples of resolution of common-resource problems at local levels. Most of these involve taking advantage of features specific to the context in order to set up systems of detection and punishment. A fishing community on the Turkish coast, for example, assigns and rotates locations to its members; the person who is assigned a good location on any given day will naturally observe and report any intruder who tries to usurp his place. Many other users of common resources, including the grazing commons in medieval England, actually restricted access and controlled overexploitation by allocating complex, tacit but well-understood rights to individuals. In one sense, this solution bypasses the common-resource problem by dividing up the resource into a number of privately owned subunits.

The most striking feature of Ostrom's range of cases is their immense variety. Some of the prisoners' dilemmas of the exploitation of common-property resources that she examined were solved by private initiative by the group of people actually involved in the dilemma, others by external public or governmental intervention. In some instances the dilemma was not resolved at all, and the group remained trapped in the all-shirk outcome. Despite this variety, Ostrom identifies several common features that make it easier to solve prisoners' dilemmas of collective action: (1) It is essential to have an identifiable and stable group of potential participants. (2) The benefits of cooperation have to be large enough to make it worth paying all the costs of monitoring and enforcing the rules of cooperation. And (3) it is very important that the members of the group can communicate with one another. This last feature accomplishes several things. First, it makes the norms clear—everyone knows what behavior is expected, what kind of cheating will not be tolerated, and what sanctions will be imposed on cheaters. Next, it spreads information about the efficacy of the detection of the cheating mechanism, thereby building trust and removing the suspicion each participant might hold that he is abiding by the rules while others are getting away with breaking them. Finally, it enables the group to monitor the effectiveness of the existing arrangements, and to improve upon them as necessary. All these requirements look remarkably like those we identified in Chapter 8 from our theoretical analysis of the prisoners' dilemma (Sections 2.D and 3) and from the observations of Axelrod's tournaments (Section 5).

Ostrom's study of the fishing village also illustrates what can be done if the collective optimum requires different individuals to do different things, so that some get higher payoffs than others. In a repeated relationship, the advantageous position can rotate among the participants, thereby maintaining some sense of equality over time. We analyze other instances of this problem of inequality and other resolutions of it in the following section.

The idea that small groups are more successful at solving collective-action problems forms the major theme of Olson's *The Logic of Collective Action*, and has led to an insight important in political science. In a democracy all voters have equal political rights, and the majority's preference should prevail. But we see many instances in which this does not happen. The effects of policies are generally good for some groups and bad for others. To get its preferred policy adopted, a group has to take political action—lobbying, publicity, campaign contributions, and so on. To do these things the group must solve a collective-action problem, because each member of the group may hope to shirk and enjoy the benefits that the others' efforts have secured. If small groups are better able to solve this problem, then the policies resulting from the political process will reflect *their* preferences even if other groups who fail to organize are more numerous and suffer greater losses than the successful groups' gains.

The most dramatic example of this comes from the arena of trade policy. A country's import restrictions help domestic producers whose goods compete with these imports, but they hurt the consumers of the imported goods and the domestic competing goods alike, because prices for these goods are higher than they would be otherwise. The domestic producers are few in number, and the consumers are almost the whole population; the total dollar amount of the consumers' losses is typically far bigger than the total dollar amount of the producers' gains. Political considerations based on constituency membership numbers and economic considerations of dollar gains and losses alike would lead us to expect a consumer victory in this policy arena; we would expect to see at least a push for the idea that import restrictions should be abolished, but we don't. The smaller and more tightly knit associations of producers are better able to organize for political action than the numerous dispersed consumers.

Over 60 years ago, the American political scientist E. E. Schattschneider provided the first extensive documentation and discussion of how pressure politics drives trade policy. He recognized that "the capacity of a group for organization has a great influence on its activity," but did not develop any systematic theory of what determines this capacity.[14] The analysis of Olson and others has improved our understanding of the issue, but the triumph of pressure politics over economics persists in trade policy to this day. For example, in the late 1980s the U.S. sugar policy cost each of the 240 million people in the United States about $11.50 per year for a total of about $2.75 billion, while it increased the incomes of about 10,000 sugar-beet farmers by about $50,000 each, and the incomes of a thousand sugarcane farms by as much as $500,000 each, for a total of about $1 billion. The net loss to the U.S. economy was $1.75 billion.[15] Each of the unorganized con-

---

[14] E. E. Schattschneider, *Politics, Pressures, and the Tariff,* (New York: Prentice-Hall, 1935), see especially pp. 285-86.

[15] Stephen V. Marks, "A Reassessment of the Empirical Evidence on the U.S. Sugar Program," in *The Economics and Politics of World Sugar Policies,* ed. Stephen V. Marks and Keith E. Maskus (Ann Arbor, Mich.: University of Michigan Press, 1993), pp. 79–108.

sumers continues to bear his small share of the costs in silence; many of them are not even aware that each is paying $11.50 a year too much for his sweet tooth.

You might be tempted to come away from this discussion of collective-action problems with the impression that individual freedom leads to harmful outcomes that can and must be improved by social norms and sanctions. Remember, however, that societies face problems other than those of collective action; some of these are better solved by individual initiative than by joint efforts. Societies can often get hidebound and autocratic, becoming trapped in their norms and customs, and stifling the innovation that is so often the key to economic growth. Collective action can become collective inaction.[16]

5 SPILLOVERS, OR EXTERNALITIES

Now we return to the comparison of the Nash equilibrium and the social optimum in collective-action games. The two questions we want to answer are: what are the socially optimal numbers of people who should choose the one strategy or the other, and how will these numbers differ from the ones that emerge in a Nash equilibrium?

Remember that the total social payoff $T(n)$, when n of the N people choose the action P and $(N - n)$ choose the action S, is given by the formula

$$T(n) = np(n) + (N - n)s(n)$$

Suppose that there are initially n people who have chosen P, and one person switches from S to P. Then the number choosing P increases by 1 to $(n + 1)$, and the number choosing S decreases by 1 to $(N - n - 1)$, so the total social payoff becomes

$$T(n + 1) = (n + 1)p(n + 1) + [N - (n + 1)]s(n + 1)$$

The increase in the total social payoff is the difference between $T(n)$ and $T(n + 1)$:

$$\begin{aligned}
T(n + 1) - T(n) &= (n + 1)p(n + 1) + [N - (n + 1)]s(n + 1) - np(n) - (N - n)s(n) \\
&= [p(n + 1) - s(n)] + n[p(n + 1) - p(n)] \\
&\quad + [N - (n + 1)][s(n + 1) - s(n)]
\end{aligned} \tag{11.1}$$

after collecting and rearranging terms.

Equation (11.1) has a very useful interpretation. It classifies the various different effects of one person's switch from S to P. Since this one person is a small part of the whole group, the change due to his switch of strategies is small, or *marginal*. The equation shows how the overall marginal change in the whole

---

[16] David Landes, *The Wealth and Poverty of Nations* (New York: Norton, 1998), chaps. 3 and 4, makes a spirited case for this effect.

group's or whole society's payoffs is divided into the marginal change in payoffs for various subgroups of the population.

The first part of Eq. (11.1), namely $[p(n+1) - s(n)]$, is the change in the pay-off of the person who switched. We encountered the two components of this expression in Section 2. This difference in the person's individual or private payoffs between what he receives when he chooses S, namely $s(n)$, and what he receives when he chooses P, namely $p(n+1)$, is what drives his choice, and all such individual choices then determine the Nash equilibrium.

Next, when one person switches from S to P, there are now $(n+1)$ people choosing P; the increase from n "participants" to $(n+1)$ changes the payoffs of all the other individuals in society, both those who play P and those who play S. For example, if one commuting driver switches from route A to route B, this decreases the congestion on route A and increases that on route B, reducing the time taken by all the other drivers on route A (improving their payoffs) and increasing the time taken by all the other drivers on route B (decreasing their payoffs). To be sure, the effect that one driver's switch has on the time for any one driver on either route is very small, but, when there are numerous other drivers (that is, when N is large), the total effect can be substantial.

The other two groups of terms in Eq. (11.1) are just the quantifications of these effects on others. For the n other people choosing P, each sees his payoff change by the amount $[p(n+1) - p(n)]$ when one more individual switches to P; this effect is seen in the second group of terms in Eq. (11.1). There are also $N - (n+1)$ (or $N - n - 1$) others still choosing S after the one person switches and each of these players sees his payoff change by $[s(n+1) - s(n)]$; this effect is shown in the third group of terms in the equation.

As an example of these various effects, suppose you are one of 6,000 commuters who drive every day from a suburb to the city and back; each of you takes either the expressway (action P) or a network of local roads (action S). The route via the local roads takes a constant 45 minutes, no matter how many cars are going that way. The expressway takes only 15 minutes when uncongested. But every car beyond 2,000 increases the time for each driver on the expressway by 0.01 minute (about half a second).

Let payoffs be measured in minutes of time saved—by how much the commute time is less than one hour, for instance. Then the payoff to drivers on the local roads, $s(n)$, is a constant $60 - 45 = 15$ regardless of the value of n. But the payoff to drivers on the expressway, $p(n)$, does depend on n; in particular, $p(n) = 60 - 15 = 45$ for all $n \leq 2,000$, but $p(n)$ decreases by $1/100$ for every commuter over 2,000. Thus, $p(n) = 45$ for values of n up to 2,000 and $p(n) = 45 - (n - 2,000)/100$ for values of n beyond 2,000.

Suppose that initially there are 4,000 cars on the expressway. With this many cars on that road, it takes each of them $15 + (4,000 - 2,000)/100 = 15 + 20 = 35$ minutes to commute to work; each gets a payoff of $p(n) = 25$ (which is $60 - 35$ or

$p(4,000)$). If you switch from a local road to the expressway, the time for each of the now 4,001 drivers there (including you) will be 35 and 1/100, or 35.01 minutes; each expressway driver now gets a payoff of $p(n + 1) = p(4,001) = 24.99$. This payoff is higher than the 15 from driving on the local roads. But the 4,000 other drivers on the expressway now take 0.01 minute more each; their total payoff has decreased by a total of $4,000 \times 0.01 = 40$ (minutes). Thus you have a private incentive to make the switch because for you, $p(n + 1) > s(n)$ (24.99 > 15). But the overall social effect of your switch is bad; the social payoff is reduced by a total of 40 (minutes lost by expressway commuters) − 9.99 (minutes gained by person who switched) = 30.01 minutes; it follows that $T(n + 1) - T(n) < 0$ in this case. You should be able to verify these outcomes using the payoff functions $p(n)$ and $s(n)$ that we derived earlier for this example.

It helps now to introduce some terminology that captures the ideas behind the numbers in our example. As we noted above, the effect of one person's switch in strategies on the total social payoff is called the marginal change in social payoff—**marginal social gain** for short. This consists of two parts, as we saw in our discussion of Eq. (11.1). One part is the change in the payoff of the person choosing the action; we call this the **marginal private gain** because it is "privately held" by one individual. The rest of the marginal social gain comes from the effect this person's action has on everyone else's payoffs. When one person's action affects others, it is called a **spillover effect, external effect,** or **externality.** Since we are considering a small change, we should call it the *marginal spillover effect.* In this language, we can rewrite Eq. (11.1) for a general switch of one person from either S to P or P to S as

Marginal social gain = marginal private gain + marginal spillover effect.

Note that the marginal private gain includes the equivalent of the first set of terms from Eq. (11.1), while the marginal spillover effect includes the equivalents of both the second and third set of terms from that equation. For an example (such as our driving story) in which one person switches from S to P, we have

Marginal social gain = $T(n + 1) - T(n)$,
Marginal private gain = $p(n + 1) - s(n)$, and
Marginal spillover effect = $n[p(n + 1) - p(n)] + [N - (n + 1)] [s(n + 1) - s(n)]$.

In the commuting example, we called driving on the expressway the action P and taking the local road the action S; we showed earlier that $s(n) = 15$ (minutes) for all n and that $p(n)$ is 45 for $n < 2,000$, and then falls by 0.01 minute for each unit increase in n. This gave us a gain of $p(n + 1) - s(n) = 9.99$ minutes in the payoff of the person who switched and which represents the marginal private gain from the switch. Next, we had 4,000 other commuters choosing P, for each of whom $p(4,001) - p(4,000) = -0.01$, and 1,999 others choosing S, for each of whom $s(4,001) - s(4,000) = 0$. The cumulative effect on these drivers is

– 4,000 × 0.01 + 1,999 × 0 = –40; this is the marginal spillover effect. The overall marginal social gain is 9.99 – 40 = –30.01 minutes.

In this example, the marginal spillover effect is negative, in which case we say there is a negative externality. A negative externality exists when the action of one person lowers others' payoffs; it imposes some extra costs on the rest of society. But this person is motivated only by his own payoffs. (Remember that any guilt he may suffer from harming others should already be reflected in his payoffs). Therefore he does not take the spillover into account when making his choice. He will change his action from S to P as long as this has a positive marginal *private* gain. Of course, society would be better off if the individual's decision were governed by the marginal *social* gain. When the marginal spillover effect is negative, as in our example, the marginal social gain is less than the marginal private gain and it could be negative even if the marginal private gain is positive. In the event that the marginal social gain *is* negative but the marginal private gain is positive, the individual makes the switch even though society as a whole would be better off if he did not do so. More generally, in situations with negative externalities the marginal social gain will be smaller than the marginal private gain. People will make decisions based on a cost-benefit calculation that is the wrong one from society's perspective; individuals will choose the actions with negative spillover effects more often than society would like them to do.

Similarly, when judged from a social perspective, people choose actions with positive spillover effects less often than society would want them to do. Such actions include the maintenance and conservation of natural resources, basic research whose benefits are freely available to all who can read scientific journals, and so on. We will consider actions of this type and their positive spillovers later in this section.

A. The Calculus of the General Case

Before examining some spillover situations in more detail to see what can be done to achieve socially better outcomes, we restate the general concepts of the analysis in the language of calculus. If you do not know this language, you can omit this section without loss of continuity; but if you do know it, you will find the alternative statement much simpler to grasp and use than the algebra employed earlier.

If the total number N of people in the group is very large, say in the hundreds or thousands, then one person can be regarded as a very small, or "infinitesimal," part of this whole. This allows us to treat the number n as a continuous variable. If $T(n)$ is the total social payoff, we calculate the effect of changing n by considering an increase of an infinitesimal marginal quantity dn, instead of a full unit increase from n to $(n + 1)$. To the first order, the change in payoff is $T'(n) \, dn$, where $T'(n)$ is the derivative of $T(n)$ with respect to n. Using the expression for the total social payoff,

$$T(n) = np(n) + (N - n)s(n),$$

and differentiating, we have

$$T'(n) = p(n) + np'(n) - s(n) + (N - n)s'(n)$$
$$= [p(n) - s(n)] + np'(n) + (N - n)s'(n) \qquad (11.2)$$

This is the calculus equivalent of Eq. (11.1). $T'(n)$ represents the marginal social gain. The marginal private gain is $p(n) - s(n)$, which is just the change in the payoff of the person making the switch from S to P. In Eq. (11.1) we had $p(n + 1) - s(n)$ for this change in payoff; now we have $p(n) - s(n)$. This is because the infinitesimal addition of dn to the group of the n people choosing P does not change the payoff to any one of them by a significant amount. However, the total change in their payoff, $np'(n)$, is sizable and is recognized in the calculation of the spillover effect—it is the second term in Eq. (11.2)—as is the change in the payoff of the $(N - n)$ people choosing S, namely $(N - n)s'(n)$, the third term in Eq. (11.2). These last two terms comprise the marginal-spillover-effect portion of Eq. (11.2).

In the driving problem, for $n > 2,000$, when the expressway starts to get congested, we have $p(n) = 45 - (n - 2,000) \times 0.01 = 65 - 0.01n$, and $s(n) = 15$. Then the private marginal gain for each driver who switches to the expressway when there are already n drivers using it is $p(n) - s(n) = 50 - 0.01n$. Since $p'(n) = -0.01$ and $s'(n) = 0$, the spillover effect is $n \times (-0.01) + (N - n) \times 0 = -0.01n$, which equals -40 when $n = 4,000$. The answers are the same as before, but calculus simplifies the derivation.

B. Negative Spillovers

In the commuting example, if there are already n people using the expressway and another driver is contemplating switching from the local roads to the expressway, he stands to gain from this if $p(n + 1) > s(n)$, whereas the total social payoff increases if $T(n + 1) - T(n) > 0$. Using our formulas (11.1) and (11.2), we can calculate the precise conditions under which a switch will be beneficial for a particular individual versus for society as a whole. The private gain is positive if

$$45 - (n + 1 - 2,000) \times 0.01 > 15$$
$$64.99 - 0.01n > 15$$
$$n < 100 \, (64.99 - 15) = 4,999,$$

whereas the condition for the social gain to be positive is

$$45 - (n + 1 - 2,000) \times 0.01 - 15 - 0.01 \, n > 0$$
$$49.99 - 0.02 \, n > 0$$
$$n < 2499.5.$$

Thus, if given the free choice, commuters will crowd onto the expressway until there are almost 5,000 of them, but all crowding beyond 2,500 reduces the total social payoff. Society as a whole would be best off if the number of commuters on the expressway were kept down to 2,500.

We show this result graphically in Figure 11.9. On the horizontal axis from left to right we measure the number of commuters using the expressway. On the vertical axis we measure the payoffs to each commuter when there are n others using the expressway. The payoff to each driver on the local road is constant and equal to 15 for all n; this is shown by the horizontal line $s(n)$. The payoff to each driver who switches to the expressway is shown by the line $p(n + 1)$. It stays constant at 45 while the expressway is uncongested, that is for $n < 2,000$, and then starts to fall by 0.01 for each unit increase in n. The two lines meet at $n = 4,999$, that is, at the value of n for which $p(n + 1) = s(n)$ or for which the marginal private gain is just zero. Everywhere to the left of this value of n, any one driver on the local roads calculates that he gets a positive gain by switching to the expressway. As some drivers make this switch, the numbers on the expressway increase—the value of n in society rises. Conversely, to the right of the intersection point (that is, for $n > 4,999$), $s(n) > p(n + 1)$, so each of the $(n + 1)$ drivers on the expressway stands to gain by switching to the local road. As some do this, the numbers on the expressway decrease and n falls. From the left of the intersection, this process converges to $n = 4,999$ and from the right it converges to 5,000.

If we had used the calculus approach, we would have regarded 1 as a very small increment in relation to n, and graphed $p(n)$ instead of $p(n + 1)$. Then the intersection point would have been at $n = 5,000$ instead of at 4,999. As you can

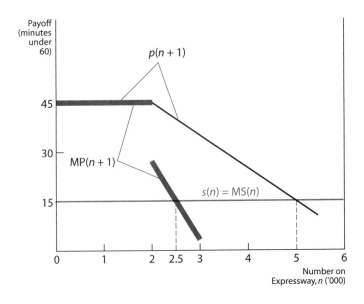

FIGURE 11.9 Equilibrium and Optimum in Route-Choice Game

see, it makes very little difference in practice. What this means is that we can call $n = 5,000$ the Nash equilibrium of the route-choice game when choices are governed by purely individual considerations. Given a free choice, 5,000 of the 6,000 total commuters will choose the expressway and only 1,000 will drive on the local roads.

But we can also interpret the outcome in this game from the perspective of the whole society of commuters. Society benefits from an increase in the number of commuters, n, on the expressway when $T(n + 1) - T(n) > 0$ and loses from an increase in n when $T(n + 1) - T(n) < 0$. To figure out how to show this on the graph, we express the idea somewhat differently; we take Eq. (11.1) and rearrange it into two pieces, one depending only on p and the other depending only on s:

$$T(n + 1) - T(n) = (n + 1) p(n + 1) + [N - (n + 1)] s(n + 1) - np(n) - [N - n] s(n)$$
$$= \{p(n + 1) + n[p(n + 1) - p(n)]\} - \{s(n) + [N - (n + 1)][s(n + 1) - s(n)]\}.$$

The expression in the first set of braces is the effect on the payoffs of the set of commuters who choose P; this expression includes the $p(n + 1)$ of the switcher and the spillover effect, $n[p(n + 1) - p(n)]$, on all the other n commuters who choose P. We call this the marginal social payoff for the P-choosing subgroup, when their number increases from n to $n + 1$, or MP$(n + 1)$ for short. Similarly, the expression in the second set of braces is the marginal social payoff for the S-choosing subgroup, or MS(n) for short. Then, the full expression for $T(n + 1) - T(n)$ tells us that the total social payoff increases when one person switches from S to P (or decreases if the switch is from P to S) if MP$(n + 1) >$ MS(n), and the total social payoff decreases when one person switches from S to P (or increases when the switch is from P to S) if MP$(n + 1) <$ MS(n).

Using our expressions for $p(n + 1)$ and $s(n)$ in the commuting example, we have MP$(n + 1) = 45$ for $n < 2,000$, and for values of n beyond that,

$$MP(n + 1) = 45 - (n + 1 - 2,000) \times 0.01 + n \times (-0.01) = 64.99 - 0.02n$$

while MS$(n) = 15$ for all values of n. Figure 11.9 includes graphs of the relationships MP$(n + 1)$ and MS(n), shown as the thick curves. Note that the MP$(n + 1)$ curve coincides with the $p(n + 1)$ curve for $n < 2,000$ (while the expressway is uncongested), and MS(n) coincides with $s(n)$ everywhere because the local roads are always uncongested. But for $n > 2,000$ the MP$(n + 1)$ curve lies below the $p(n + 1)$ curve—because of the negative spillover, the social gain from switching one person to the expressway is less than the private gain to the switcher. Also, as n crosses 2,000, the MP$(n + 1)$ curve suffers a discontinuous vertical drop of 20, because the 2001st person on the expressway reduces the payoff to all the previous 2,000 by 0.01 minutes each.

The MP$(n + 1)$ and MS(n) curves meet at $n = 2,499$, or approximately 2,500. To the left of this intersection, MP$(n + 1) >$ MS(n) and society stands to gain by allowing one more person on the expressway; to the right, the opposite is true and society stands to gain by shifting one person from the expressway to the local roads. Thus the socially optimal allocation of drivers is 2,500 on the expressway and 3,500 on the local roads.

If you wish to use calculus, you can write the total payoff for the expressway drivers as $np(n) = n [45 - 0.01 \times (n - 2,000)] = 65n - 0.01 n^2$, and then the $MP(n + 1)$ is the derivative of this with respect to n, namely $65 - 0.01 \times 2n = 65 - 0.02 n$. The rest of the analysis can then proceed as before.

How might this society achieve the optimum allocation of its drivers? Different cultures and political groups use different systems, and each has its own merits and drawbacks. The society could simply restrict access to the expressway to 2,500 drivers. But how would it choose those 2,500? It could adopt a first-come, first-served rule, but then drivers would race each other to get there early and waste a lot of time. A bureaucratic society may set up criteria based on complex calculations of needs and merits as defined by civil servants; then everyone will undertake some costly activities to meet these criteria. In a politicized society, the important "swing voters" or organized pressure groups or contributors may be favored. In a corrupt society, those who bribe the officials or the politicians may get the preference. A more egalitarian society could allocate the rights to drive on the expressway by lottery, or rotate them from one month to the next. A scheme that lets you drive only on certain days depending on the last digit of your car's license plate is an example, but this is not as egalitarian as it seems because the rich can have two cars and choose license plate numbers that will allow them to drive every day.

Many economists prefer a more open system of charges. Suppose each driver on the expressway is made to pay a tax t, measured in units of time. Then the private benefit from using the expressway becomes $p(n) - t$, and the number n in the Nash equilibrium will be determined by $p(n) - t = s(n)$. (Here we are ignoring the tiny difference between $p(n)$ and $p(n+1)$, which is possible when N is very large.) We know that the socially optimal value of n is 2,500. Using the expressions $p(n) = 65 - 0.01 n$ and $s(n) = 15$, and plugging in 2,500 for n, we find that $p(n) - t = s(n)$—that is, drivers are indifferent between the expressway and the local roads—when $65 - 25 - t = 15$, or $t = 25$. If we value time at the minimum wage of about $5 an hour, 25 minutes comes to a little over $2. This is the tax or toll which, when charged, will keep the numbers on the expressway down to what is socially optimal.

Note that when there are 2,500 drivers on the expressway, the addition of one more increases the time spent by each of them by 0.01 minute, for a total of 25 minutes. This is exactly the tax each driver is being asked to pay. In other words, each driver is made to pay the cost of the negative spillover he creates on the rest of society. This "brings home" to each driver the extra cost of his action, and therefore induces him to take the socially optimal action; economists say the individual is being made to **internalize the externality.** This idea, that people whose actions hurt others are made to pay for the harm they cause, adds to the appeal of this approach. But the proceeds from the tax are not used to compensate the others directly. If they were, then each expressway user would count on receiving from others just what he pays, and the whole purpose would be defeated. Instead, the proceeds of the tax go into general government revenues, where they may or may not be used in a socially beneficial manner.

Those economists who prefer to rely on markets argue that if the expressway is owned by a private individual, his profit motive will induce him to charge for its use just enough to reduce the number of users to the socially optimal level. An owner knows that if he charges a tax t for each user, the number of users n will be determined by $p(n) - t = s(n)$. His revenue will be $tn = n[p(n) - s(n)]$, and he will act in such a way as to maximize this. In our example, the revenue is $n [65 - 0.01 n - 15] = n[50 - 0.01 n] = 50n - 0.01 n^2$; it is easy to see this is maximized when $n = 2,500$. But in this case, the revenue goes into the owner's pocket; most people regard this as a bad solution.

C. Positive Spillovers

Many matters pertaining to positive spillovers or positive externalities can be understood simply as mirror images of those for negative spillovers. Individuals' private benefits from undertaking activities with positive spillovers are less than society's marginal benefits from such activities. Therefore such actions will be underutilized and their benefits underprovided in the Nash equilibrium. A better outcome can be achieved by augmenting people's incentives; providing those individuals whose actions provide positive spillovers with a reward just equal to the spillover benefit will achieve the social optimum.

Indeed, the distinction between positive and negative spillovers is to some extent a matter of semantics. Whether a spillover is positive or negative depends on which choice you call P and which you call S. In the commuting example, suppose we called the local roads P and the expressway S. Then one commuter's switch from S to P will reduce the time taken by all the others who choose S, so this action will convey a positive spillover to them. As another example, consider vaccination against some infectious disease. Each person getting vaccinated reduces his own risk of catching the disease (marginal private gain) and also reduces the risk of others getting the disease through him (spillover). If being unvaccinated is called the S action, then getting vaccinated has a positive spillover effect. If, instead, remaining unvaccinated is called the P action, then the act of getting the vaccine has a negative spillover effect. This has implications for the design of policy to bring individual action into conformity with the social optimum. Society can either reward those who get vaccinated or penalize those who fail to do so.

But actions with positive spillovers can have one very important new feature that distinguishes them from actions with negative spillovers, namely **positive feedback.** Suppose the spillover effect of your choosing P is to increase the payoff to the others who are also choosing P. Then your choice increases the attraction of that action (P), and may induce some others to take it also, setting in train a process that culminates in everyone taking that action. Conversely, if very few people are choosing P, then it may be so unattractive that they too give it up, leading to a situation in which everyone chooses S. In other words, positive feedback can give rise to multiple Nash equilibria. We now illustrate this using a very real example.

Consider the choice of which home computer to purchase: an Apple Macintosh or an IBM PC clone. This game has the necessary payoff structure. As the number of Macintosh owners rises, the better it will be to purchase a Mac—there will be more software available, more experts available to help with any problems that arise, and potentially fewer problems in the first place; the same is true of purchasing a PC clone. In addition, many computing aficionados would argue that the "all Mac" outcome provides higher benefits to computer users as a whole, due to efficiency issues, than does the all-PC outcome.[17]

A diagram similar to Figures 11.6 to 11.8 can be used to show the payoffs to an individual computer purchaser of the two strategies, "purchase a Mac" and "purchase a PC." As shown in Figure 11.10, the Mac payoff rises as the number of Mac owners rises and the PC payoff rises as the number of Mac owners falls. Note that the diagram is drawn so that the payoff to Mac purchasers when everyone in the population is a Mac owner is higher than the payoff to PC purchasers when everyone in the population is a PC owner. That is, we have assumed that the all-Mac outcome is better for everyone than the all-PC outcome. You may disagree, but bear with us; our purpose is to ask the question: Will individual choice lead to the socially best outcome?

If the current population has only a small number of Mac owners, then the computer distribution in society lies to the left of the intersection of the two payoff lines at *I*, and each individual finds it better to choose a PC. When there is a larger number of Mac owners in the population, to the right of *I*, it is better for each person to choose a Mac. Thus a mixed population of Mac and PC owners is sustainable as an equilibrium only when the current population has exactly *I* Mac owners; only then will no member of the population have any incentive to switch platforms. But if one more person by accident makes a different decision, his choice will push the population either to the left of *I*, in which case there will

---

[17]In common parlance, positive feedback is often thought to result in a good thing, but in technical language the term merely characterizes the process and entails no general value judgment about the outcome. In this example, the same positive feedback mechanism could lead to either an all-Mac outcome or an all-PC outcome; one could be worse than the other.

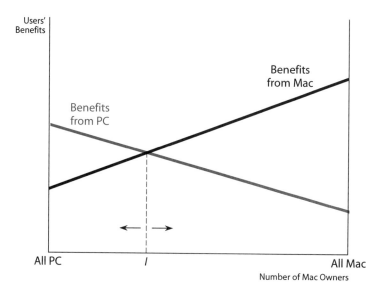

FIGURE 11.10   Payoffs in Computer Choice Game

then be an incentive for people to buy PCs, or to the right of I, for which an incentive will emerge for more people to buy Macs. A mixed population with I Mac owners is an equilibrium but an unstable one. The game's stable equilibria are found at the endpoints, where everyone is either a Mac owner or a PC owner.

But which of the two stable equilibria will occur in this game? The answer depends on where the game starts. If you look at the configuration of today's computer users, you will see a heavily PC-oriented population. Thus it seems that, since there are so few Mac users (or so many PC users), the world is moving toward the all-PC equilibrium. Schools, businesses, and private users have become **locked in** to this particular equilibrium as a result of an accident of history. If it is indeed true that the Macintosh and its operating system provide more benefits to society when used by everyone, then the all-Mac equilibrium should be preferred over the all-PC one we are approaching. Unfortunately, although society as a whole might be better off with the change, no individual computer user has an incentive to make a change from the current situation. Only coordinated action can swing the pendulum toward Macintosh. A critical mass of individuals, more than I in Figure 11.10, must own and use Macs before it becomes individually rational for others to choose the same machine.

There are many examples of similar choices of convention being made by different groups of people. The most famous cases are those in which it has been argued, in retrospect, that wrong choice was made. Advocates claim that steam power could have been developed for greater efficiency than gasoline; it certainly would have been cleaner. Proponents of the Dvorak typewriter/computer key configuration claim it would be better than the QWERTY keyboard if used everywhere. Many engineers agree that Betamax had more going for it than VHS in the video recorder market. In such cases, the whims of the public or the genius of advertisers help determine the ultimate equilibrium and may lead to a "bad" or "wrong" outcome from society's perspective. Other situations do not suffer from such difficulties. Few people concern themselves with fighting for a reconfiguration of traffic light colors, for example.[18]

The ideas of positive feedback and lock-in find an important application in macroeconomics. Production is more profitable the higher the level of demand in the economy, which happens when national income is higher. In turn, income is higher when firms are producing more and therefore hiring more workers. This positive feedback creates the possibility of multiple equilibria, of which the high-production, high-income one is better for society, but individual decisions may lock the economy into the low-production, low-income equilibrium. In Exercise 6 at the end of the chapter we ask you to show this in a figure. The better equilibrium could be turned into a focal point by public declaration—"the only thing we have to fear is fear itself"—but a better way is for the government to inject demand into the economy to the extent necessary to move it to the better equilibrium. In other words, the possibility of unemployment due to a deficiency of aggregate demand, as discussed in the supply-and-demand language of economic theory by the British economist John Maynard Keynes in his famous 1936 book *Employment, Interest, and Money,* can be seen from a game-theoretic perspective as the result of a failure to solve a collective-action problem.[19]

6 "HELP!"—A GAME OF CHICKEN WITH MIXED STRATEGIES

In the chicken variant of collective-action problems we discussed in earlier sections, we looked only at the pure-strategy equilibria. But we know from Section 4.A in Chapter 5 and Section 6.A in Chapter 10 that such games have mixed-strategy equilibria too, and although they yield poor expected payoffs, they have some claim to attention because they are evolutionary stable. In collective-action problems, where each participant is thinking, "It is better if I wait for enough others to participate so I can shirk; but then again maybe they won't, in

[18] Not everyone agrees that the Dvorak keyboard and the Betamax video recorder were clearly superior alternatives. See two articles by S. J. Liebowitz and Stephen E. Margolis, "Network Externality: An Uncommon Tragedy," *Journal of Economics Perspectives,* vol. 8 (Spring 1994), pp. 146–149, and "The Fable of the Keys," *Journal of Law and Economics,* vol. 33, (April 1990), pp. 1–25.

[19] John Maynard Keynes, *Employment, Interest, and Money* (London: Macmillan, 1936). See also John Bryant, "A Simple Rational-Expectations Keynes-type Model," *Quarterly Journal of Economics,* vol. 98 (1983), pp. 525–528, and Russell Cooper and Andrew John, "Coordination Failures in a Keynesian Model," *Quarterly Journal of Economics,* vol. 103, (1988), pp. 441–463, for formal game-theoretic models of unemployment equilbria.

which case I should participate," mixed strategies nicely capture the spirit of such vacillation. Our last story is a dramatic, even chilling application of such a mixed-strategy equilibrium.

In 1964 in New York City (in Kew Gardens, Queens), a woman named Kitty Genovese was killed in a brutal attack that lasted over half an hour. She screamed through it all, and although her screams were heard by many people, and more than 30 actually watched the attack taking place, not one went to help her, nor even called the police.

The story created a sensation, and found several ready theories to explain it. The press, and most of the public, saw this as a confirmation of their belief that New Yorkers—or big-city dwellers, or Americans, or people more generally—were just apathetic or didn't care about their fellow human beings.

However, even a little introspection or observation will convince you that people do care about the well-being of other humans, even strangers. Social scientists offered a different explanation for what happened, which they labeled **pluralistic ignorance.** The idea behind this is that no one can be sure about what is happening, whether help is really needed and how much. They look to each other for clues or guidance about these matters, and try to interpret other people's behavior in this light. If they see no one else is doing anything to help, they interpret this as meaning that help is probably not needed, so they don't do anything either.

This has some intuitive appeal, but is unsatisfactory in the Kitty Genovese context. There is a very strong presumption that a screaming woman needs help. What did the onlookers think—that a movie was being shot in their obscure neighborhood? If so, where were the lights, the cameras, the director, other crew?

A better explanation would recognize that while each onlooker may experience strong personal loss from Kitty's suffering, and get genuine personal pleasure if she were saved, each must balance that against the cost of getting involved. You may have to identify yourself if you call the police; you may then have to appear as a witness, and so on. Thus, we see that each person may prefer to wait for someone else to call and hope to get for himself the free rider's benefit of the pleasure of a successful rescue.

Social psychologists have a slightly different version of this idea of free riding, which they label **diffusion of responsibility.** Here the idea is that everyone might agree that help is needed, but they are not in direct communication with each other and so cannot coordinate on who should help. Each person may believe that help is someone else's responsibility. And the larger the group, the more likely it is that each person will think that someone else would probably help, and therefore he can save himself the trouble and the cost of getting involved.

Social psychologists conducted some experiments to test this hypothesis. They staged situations in which someone needed help of different kinds in different places and with different-sized crowds. Among other things, they found that the larger the size of the crowd, the less likely was help to come forth.

The concept of diffusion of responsibility seems to explain this, but not quite. It says that the larger the crowd, the less likely is any one person to help. But there are more people, and it needs only one to act and call the police to secure help. To make it less likely that even one person helps, the chance of any one person helping has to go down sufficiently fast to offset the increase in the total number of potential helpers. To find whether it does so requires game-theoretic analysis, which we now supply.[20]

We consider only the aspect of diffusion of responsibility in which action is not consciously coordinated, and leave aside all other complications of information and inference. Thus we assume that everyone believes the action is needed and is worth the cost.

Suppose there are N people in the group. The action brings each of them a benefit B. Only one person is needed to take the action; more are redundant. Anyone who acts bears the cost C. We assume that $B > C$, so it is worth the while of any one individual to act even if no one else is acting. Thus the action is justified in a very strong sense.

The problem is that anyone who takes the action gets the value B and pays the cost C for a net payoff of $(B - C)$, while he would get the higher payoff B if someone else took the action. Thus each has the temptation to let someone else go ahead, and to become a free rider on their effort. When all N people are thinking thus, what will be the equilibrium or outcome?

If $N = 1$, the single person has a simple decision problem rather than a game. He gets $B - C > 0$ if he takes the action, and 0 if he does not. Therefore he goes ahead and helps.

If $N > 1$, we have a game of strategic interaction with several equilibria. Let us begin by ruling out some possibilities. With $N > 1$, there cannot be a pure-strategy Nash equilibrium in which all people act, because then any one of them would do better by switching to free-ride. Likewise, there cannot be a pure-strategy Nash equilibrium in which no one acts, because *given that no one else is acting* (remember that under the Nash assumption each player takes the others' strategies as given), it pays any one person to act.

There *are* Nash equilibria where exactly one person acts; in fact, there are N such equilibria, one corresponding to each member. But when everyone is taking the decision individually in isolation, there is no way to coordinate and designate who is to act. Even if they were to attempt such coordination, they might

[20] For a fuller account of the Kitty Genovese story, and the analysis of such situations from the perspective of social psychology, see John Sabini, *Social Psychology*, 2nd ed., (New York: Norton, 1995), pp. 39–44. Our game-theoretic model is based on Thomas Palfrey and Howard Rosenthal, "Participation and the Provision of Discrete Public Goods," *Journal of Public Economics*, vol. 24 (1984), pp. 171–193.

try to negotiate over the responsibility, and not reach a conclusion, or at least not in time to be of help. Therefore it is of interest to examine equilibria in which all members have identical strategies.

We already saw that there cannot be an equilibrium in which all N people follow the same pure strategy. Therefore we should see whether there can be an equilibrium in which they all follow the same mixed strategy. Actually, mixed strategies are quite appealing in this context. The people are isolated, and each is trying to guess what the other will do. Each is thinking, Perhaps I should call the police . . . but maybe someone else will . . . but what if they don't . . . ? Each breaks off this process at some point and does the last thing he thought of in this chain, but we have no good way of predicting what that last thing is. A mixed strategy carries the flavor of this idea of a chain of guesswork being broken at a random point.

So suppose P is the probability that any one person will *not* act. If one particular person is willing to mix strategies, he must be indifferent between the two pure strategies of acting and not acting. Acting gets him $(B - C)$ for sure. Not acting will get him 0 if none of the other $(N - 1)$ people act, and B if at least one of them does act. Since the probability that any one fails to act is P, and they are deciding independently, the probability that none of the $(N - 1)$ others acts is P^{N-1}, and the probability that at least one does act is $(1 - P^{N-1})$. Therefore the expected payoff of the one person, when he does not act, is

$$0 \times P^{N-1} + B\,(1 - P^{N-1}) = B\,(1 - P^{N-1}).$$

And that one person is indifferent between acting and not acting when

$$B - C = B(1 - P^{N-1}) \quad \text{or when} \quad P^{N-1} = \frac{C}{B} \quad \text{or} \quad P = \left(\frac{C}{B}\right)^{1/(N-1)}.$$

Note how this **indifference condition** of *one* selected player serves to determine the probability with which the *other* players mix their strategies.

Having obtained the equilibrium mixture probability, we can now see how it changes as the group size N changes. Remember that $C/B < 1$. As N increases from 2 to infinity, the power $1/(N-1)$ decreases from 1 to 0. Then C/B raised to this power, namely P, increases from C/B to 1. Remember that P is the probability that any one person does not take the action. Therefore the probability of action by any one person, namely $(1 - P)$, falls from $1 - C/B = (B - C)/B$ to 0.

In other words, the more people there are, the less likely is any one of them to act. This is of course intuitively true, and in good conformity with the idea of diffusion of responsibility.

But it does not yet give us the conclusion that help is less likely to be forthcoming in a larger group. As we said before, help requires action by only one person. As there are more and more people each of whom is less and less likely to act, we cannot conclude immediately that the probability of *at least one* of them acting gets smaller. More calculation is needed to see whether this is the case.

Since the N individuals are randomizing independently in the Nash equilibrium, the probability Q that *not even one* of them helps is

$$Q = P^N = \left(\frac{C}{B}\right)^{N/(N-1)}.$$

As N increases from 2 to infinity, $N/(N-1)$ decreases from 2 to 1, and then Q increases from $(C/B)^2$ to C/B. Correspondingly, the probability that *at least one* person helps, namely $(1 - Q)$, decreases from $1 - (C/B)^2$ to $1 - C/B$.

So our exact calculation does bear out the hypothesis: the larger the group, the *less* likely is help to be given at all. The probability of provision does not, however, reduce to zero even in very large groups; instead it levels off at a positive value, namely $(B - C)/B$, which depends on the benefit and cost of action to each individual.

We see how game-theoretic analysis sharpens the ideas from social psychology with which we started. The diffusion of responsibility theory takes us part of the way, namely to the conclusion that any one person is less likely to act when he is part of a larger group. But the desired conclusion—that larger groups are less likely to provide help at all—needs further and more precise probability calculation based on the analysis of individual mixing and the resulting interactive (game) equilibrium.

And now we ask, did Kitty Genovese die in vain? Do the theories of pluralistic ignorance, diffusion of responsibility, and free-riding games still play out in the decreased likelihood of individual action within increasingly large cities? Perhaps not. John Tierney of the *New York Times* has publicly extolled the virtues of "urban cranks."[21] These are individuals who encourage the civility of the group through prompt punishment of those who exhibit unacceptable behavior—including litterers, noise polluters, and the generally obnoxious boors of society. Such people act essentially as enforcers of a cooperative norm for society. And, as Tierney surveys the actions of known "cranks," he reminds the rest of us that "[n]ew cranks must be mobilized! At this very instant, people are wasting time reading while norms are being flouted out on the street. . . . You don't live alone in this world! Have you enforced a norm today?" In other words, we need social norms and some people who get some innate payoff from enforcing these norms.

SUMMARY

Many-person games generally concern problems of *collective action*. The general structure of collective-action games may be manifested as a prisoners'

---

[21] John Tierney, "The Boor War: Urban Cranks, Unite—Against All Uncivil Behavior. Eggs Are a Last Resort." *New York Times Magazine*, January 5, 1997.

dilemma, chicken, or an assurance game. The critical difficulty with such games, in any form, is that the Nash equilibrium arising from individually rational choices may not be the socially optimal outcome—the outcome that maximizes the sum of the payoffs of all the players.

Problems of collective action have been recognized for many centuries and discussed by scholars from diverse fields. Several early works professed no hope for the situation, while others offered up dramatic solutions; the work of Olson in the 1960s put the problems in perspective by noting that socially optimal outcomes would not arise unless each individual had a private incentive to perform his assigned task. The most recent treatments of the subject acknowledge that collective-action problems arise in diverse areas and that there is no single optimal solution.

Some general solution methods come from a variety of fields. Biologists point to the role of *instinct*. More social scientific analysis suggests that social *custom*, or *convention*, can lead to cooperative behavior. Other possibilities for solutions come from the use of *sanctions* for the uncooperative, or the creation of *norms* of acceptable behavior. Much of the literature agrees that small groups are more successful at solving collective-action problems than large ones.

In collective-action games, when an individual's action has some effect on the payoffs of all the other players, we say there are *spillovers*, or *externalities*. These can be positive or negative and lead to individually driven outcomes that are not socially optimal; when actions create negative spillovers, they are overused from the perspective of society, and when actions create positive spillovers, they are underused. Mechanisms that induce individuals to incur the cost (or reap the benefit) of their actions can be used to achieve the social optimum. The additional possibility of *positive feedback* exists when there are positive spillovers; in such a case, there may be multiple Nash equilibria of the game.

In large-group games, *diffusion of responsibility* can lead to behavior in which individuals wait for others to take action and *free-ride* off the benefits of that action. If help is needed, it is less likely to be given at all as the size of the group available to provide it grows.

KEY TERMS

coercion (374)

collective action (356)

convention (371)

custom (371)

diffusion of responsibility (389)

external effect (379)

externality (379)

free rider (358)

indifference condition (391)

instincts (370)

internalize the externality (385)

locked in (387)

marginal private gain (379)

marginal social gain (379)

nonexcludable benefits (357)

nonrival benefits (357)

norm (371)

oppression (374)

pluralistic ignorance (389)

positive feedback (386)

pure public good (357)

sanction (371)

spillover effect (379)

EXERCISES

1. A group has 100 members. If n of them participate in a common project, then each participant derives the benefit $p(n) = n$, and each of the $(100 - n)$ shirkers derives the benefit $s(n) = 4 + 3n$.
   (a) What kind of a game is this?
   (b) Write the expression for the total benefit of the group.
   (c) Show that $n = 74$ yields the maximum total benefit for the group.
   (d) What difficulties will arise in trying to get exactly 74 participants and allowing the remaining 26 to shirk?
   (e) How might the group try to overcome these difficulties?

2. Consider a small geographic region with a total population of 1 million people. There are two towns, Alphaville and Betaville, in which each person can choose to live. For each individual, the benefit from living in a town increases for a while with the size of the town (because larger towns have more amenities, etc.), but after a point it decreases (because of congestion and so on). If x is the fraction of the population that lives in the same town as you do, your payoff is given by

$$x \quad \text{if} \quad x < 0.4$$
$$0.6 - 0.5 x \quad \text{if} \quad 0.4 < x < 1$$

   (a) Draw a graph of this relationship. Regard x as a continuous variable that can take any real value between 0 and 1.
   (b) Equilibrium occurs **either** when both towns are populated and their residents have equal payoffs, **or** when one town, say Betaville, is totally depopulated, and the residents of the other town (Alphaville) get a higher payoff than would the very first person who seeks to populate Betaville. Use your graph to find all such equilibria.
   (c) Now consider a dynamic process of adjustment whereby people gradually move toward the town whose residents currently enjoy a larger payoff than do the residents of the other town. Which of the equilibria identified in part (b) will be stable with these dynamics, and which ones unstable?

3. There are two routes for driving from A to B. One is a freeway, and the other consists of local roads. The benefit of using the freeway is constant and equal

to 1.8, irrespective of the number of people using it. Local roads get congested when too many people use this alternative, but if too few people use it, the few isolated drivers run the risk of becoming victims of crimes. Suppose that when a fraction x of the population is using the local roads, the benefit of this mode to each driver is given by

$$1 + 9x - 10x^2.$$

(a) Draw a graph showing the benefits of the two driving routes as functions of x, regarding x as a continuous variable that can range from 0 to 1.

(b) Identify all possible equilibrium traffic patterns from your graph in (a). Which equilibria are stable and which ones unstable, and why?

(c) What value of x maximizes the total benefit to the whole population?

4. Suppose an amusement park is being built in a city with a population of 100. Voluntary contributions are being solicited to cover the cost. Each citizen is being asked to give $100. The more people contribute, the larger will be the park and the greater the benefit to each citizen. But it is not possible to keep out the noncontributors; they get their share of this benefit anyway. Suppose that when there are n contributors, where n can be any whole number between 0 and 100, the benefit to each citizen in monetary unit equivalents is n^2 dollars.

(a) Suppose that initially no one is contributing. You are the mayor of the city. You would like everyone to contribute, and can use persuasion on some people. What is the minimum number you need to persuade before everyone else will join in voluntarily?

(b) Find the Nash equilibria of the game where each citizen is deciding whether to contribute.

5. Suppose a class of 100 students is comparing the choice between two careers—lawyer or engineer. An engineer gets a take-home pay of $100,000 per year, irrespective of the numbers who choose this career. Lawyers make work for each other, so as the total number of lawyers increases, the income of each lawyer increases—up to a point. Ultimately the competition between them drives down the income of each. Specifically, if there are N lawyers, each will get $100N - N^2$ thousand dollars a year. The annual cost of running a legal practice (office space, secretary, paralegals, access to online reference services, and so forth) is $800,000. So each lawyer takes home $100N - N^2 - 800$ thousand dollars a year when there are N of them.

(a) Draw a graph showing the take-home income of each lawyer on the vertical axis and the number of lawyers on the horizontal axis. (Plot a few points, say for 0, 10, 20, . . . , 90, 100 lawyers, and fit a curve to the points, or use a computer graphics program if you have access to one.)

(b) When career choices are made in an uncoordinated way, what are the possible equilibrium outcomes?

(c) Now suppose the whole class decides how many should become lawyers, aiming to maximize the total take-home income of the whole class. What will be the number of lawyers? (Use calculus, regarding N as a continuous variable, if you can. Otherwise you can use graphical methods, or a spreadsheet.)

6. Put the idea of Keynesian unemployment described at the end of Section 5.C into a properly specified game, and show the multiple equilibria in a diagram. Show the level of production (national product) on the vertical axis, as a function of a measure of the level of demand (national income) on the horizontal axis. Equilibrium occurs when national product equals national income, that is, when the function relating the two cuts the 45° line. For what shapes of the function can there be multiple equilibria, and why might you expect such shapes in reality? Suppose that income increases when current production exceeds current income, and income decreases when current production is less than current income. With this dynamic process, which equilibria are stable and which ones unstable?

7. Write a brief description of a strategic game that you have witnessed or participated in that involves a large number of players, in which individual players' payoffs depend on the number of other players and their actions. Try to illustrate your game with a graph if possible. Discuss the outcome of the actual game in light of the fact that many such games involve inefficient outcomes. Do you see evidence of such an outcome in your game? Explain.

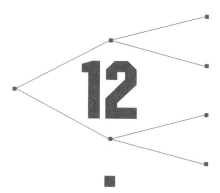

Uncertainty and Information

UNCERTAINTY ABOUT OUTCOMES EXISTS in many of the games we have studied so far. Sometimes this uncertainty is the result of the players' mixed strategies, but at other times it is something intrinsic to the game. In tennis or football, for example, we supposed quite realistically that the players know only the probabilities of success of the various strategy combinations; they cannot forecast the exact result on any one occasion when those strategies are used. Then we computed a player's payoff from each strategy combination as the expected value, or the probability-weighted average, of the payoffs of success and failure of the shot or the play. In Section 3.B of Chapter 2, we pointed out that this meant assuming the players were indifferent to risk. We also said that dislike of risk can be accommodated by taking the expected value of a nonlinear payoff scale called utility; we provided a detailed discussion of this methodology in the Appendix to Chapter 5. However, we have not made much use of it thus far.

Allowing different attitudes toward risk introduces some interesting new aspects of strategy that have to do with manipulation of risk itself. When a player is averse to risk, he benefits from devising strategies that reduce the risk he faces. Sometimes this can be done if two or more such players can get together and combine their risk. In some games, particularly ones in which the players are pitted against each other in a contest, their strategies toward risk interact in interesting ways. In this chapter we study some of these matters.

Different people may have different amounts of information about an uncertain prospect, or may be able to control it to different extents. When such differently informed people interact as players in a game, the choices of one affect the expected payoff of the other. This adds yet another dimension to their strategy, namely manipulating each other's information. We spoke briefly of some such strategies, namely screening and signaling, in Chapters 1 and 2. Here we study them in a little more detail.

The study of the topic of information and its manipulation in games has been very active and important in the last 25 years. It has shed new light on many previously puzzling matters like the nature of incentive contracts, the organization of companies, markets for labor and for durable goods, government regulation of business, and myriad others. This chapter introduces you to some of these, and prepares the way for you to read about these fascinating ideas in economics and politics. Many of the lessons apply to strategies in your own life too.

1 CONTROLLING AND MANIPULATING RISK

A. Strategies to Reduce Risk

Imagine you are a farmer subject to the vagaries of weather. If the weather is good for your crops, you will have an income of $15,000. If it is bad, your income will be only $5,000. The two possibilities are equally likely (probability 1/2 or 0.5, or 50% each). Therefore your average or expected income is $10,000, but there is considerable risk around this average value.

What can you do to reduce the risk you face? You might try a crop that is less subject to the vagaries of weather, but suppose you have already done all such things that are under your individual control. Now you can reduce your income risk further only by combining it with other people's risks. There are two ways to do this, pooling and trading.

We begin with **pooling risks.** Suppose your neighbor faces a similar risk, but gets good weather exactly when you get bad weather and vice versa. (Suppose you live on opposite sides of an island, and rain clouds visit one side or the other but not both.) In technical jargon, *correlation* is a measure of alignment between any two uncertain quantities—in this discussion, between one person's risk and another's. Thus we would say that your neighbor's risk is totally **negatively correlated** with yours. Then your combined income is $20,000, no matter what the weather: it is totally risk-free. You can enter into a contract that gets each of you $10,000 for sure: you promise to give him $5,000 in years when you are lucky, and he promises to give you $5,000 in years when he is lucky. You have eliminated your risks by pooling them together.

Even without such perfect correlation, risk pooling has some benefit. Suppose the farmers' risks are independent, as if the rain clouds could toss a separate coin to decide whether to visit each of you. Then there are four possible outcomes, each with a probability of 1/4. In the first case, both of you are lucky, and your combined incomes total $30,000. In the second and third cases, one of you is lucky and the other is not (the two cases differ only as to which one is which); in both of these situations your combined incomes total $20,000. In the fourth outcome, both of you are unlucky, and your combined incomes amount to only $10,000. Facing these possibilities, if the two of you make a contract to share and share alike, each of you will have $15,000 with a probability of 1/4, $5,000 with a probability of 1/4, and $10,000 with a probability of 1/2. Without the trade, each would have $15,000 or $5,000 with probabilities of 1/2 each. Thus for each of you, the contract has reduced the probabilities of the two extreme outcomes from 1/2 to 1/4 and increased the probability of the middle outcome from 0 to 1/2. In other words, the contract has reduced the risk for each of you.

In fact, so long as the two farmers' incomes are not totally **positively correlated**—that is, their luck does not move in perfect tandem—they can reduce mutual risk by pooling. And if there are more than two farmers with some degree of independence in their risks, then even greater reduction in the risk of each is made possible by the law of large numbers.

This risk reduction is the whole basis of insurance: by pooling together the similar but independent risks of many people, an insurance company is able to compensate any one of them when he suffers a large loss. It is also the basis of portfolio diversification: by dividing your wealth among many different assets with different kinds and degrees of risk, you can reduce your total exposure to risk.

However, several complexities limit the possibility of risk reduction by pooling. At the simplest, the pooling promise may not be credible: the farmer who gets the good luck may then refuse to share it. This is relatively easily overcome if a legal system enforces contracts, provided its officials can observe the farmers' outcomes in order to decide who should give how much to whom. The premiums people pay to the insurance company are advance guarantors of their promise, like setting aside money in an escrow account. Then the insurance company's desire to maintain its reputation in an ongoing business ensures that it will pay out when it should. However, if outcomes cannot be observed by others, then people may try to defraud the company by pretending to have suffered a loss. Or, if people's risks can be influenced by their actions, they may become careless knowing they are insured; this is called **moral hazard.** Or, if the insurance company cannot know as much about the risks (say, of health) as do the insured, then those people who face the worst risks may choose to insure more, thereby worsening the average risk pool and raising premiums for all; this is called **adverse selection.** We will study some such problems in more detail soon. For now, we merely point out that to cope with these problems, insurance companies usually provide only partial insurance, requiring you to bear part of the loss yourself. The portion you have to bear consists of a *deductible* (an initial amount of the loss) and *coinsurance* (an additional fraction of the loss beyond the deductible).

Next consider **trading risks.** Suppose you are the farmer facing the same risk as before. But now your neighbor has a sure income of $10,000. You face a lot of risk and he faces none. He may be willing to take a little of your risk, for a price that is agreeable to both of you.

To understand and calculate this price, we must allow for *risk aversion.* Suppose both of you are risk-averse. As we saw in the Appendix to Chapter 5, your attitudes to risk can be captured by using a concave scale to convert your money incomes into "utility" numbers. In the appendix we used the square root as a simple example of such a scale; let us continue to do so here. Suppose you pay your neighbor $100 up front, and in exchange he agrees to the following deal with you: after your crop is harvested and if you have been unlucky, he will give you $2,000, but if you have proved to be lucky, *you* will give *him* $2,000. In other words, he agrees to assume $2,000 of your risk in exchange for a sure payment of $100. The end result is that his income will be $10,000 + $100 + $2,000 = $12,100 if you are lucky, and $10,000 + $100 − $2,000 = $8,100 if you are unlucky. His expected utility is

$$0.5 \times \sqrt{12,100} + 0.5 \times \sqrt{8,100} = 0.5 \times 110 + 0.5 \times 90 = 100.$$

This is the same as the utility he would get if he did not trade with you: $\sqrt{10,000} = 100$. Thus he is just willing to make this trade with you. Would you be willing to pay him the $100 up front to have him assume $2,000 of your risk?

Your expected utility from the risky income prospect you face is

$$0.5 \times \sqrt{5,000} + 0.5 \times \sqrt{15,000} = 0.5 \times 70.7 + 0.5 \times 122.5 = 96.6.$$

If you pay your neighbor $100 and he gives you $2,000 when your luck is bad, (so that he is left with $8,100), you will have $6,900. In return, when your luck is good, you are to give him $2,100 ($100 of initial payment plus $2,000 from your high income), so he has the $12,100 he needs to compensate him for the risk, and you are left with $12,900. Your expected utility with risk trading is

$$0.5 \times \sqrt{6,900} + 0.5 \times \sqrt{12,900} = 0.5 \times 83.0 + 0.5 \times 113.6 = 98.3.$$

Thus the arrangement has given you a greater expected utility, while leaving your neighbor just as happy.

Your neighbor might suggest another arrangement, which would give him a higher expected utility while keeping yours at the original 96.6. There are other deals in between where both of you have a higher expected utility than you would without trading risk. Thus there is a whole range of contractual arrangements, some favoring one party and some the other, and you can bargain over them. The precise outcome of such bargains depends on the parties' bargaining power; we will study bargaining in some detail in Chapter 16. If there are many people, then an impersonal market (examined in Chapter 17) can determine the price of risk. Common to all such arrangements is the idea that mutually beneficial deals can be struck whereby, for a suitable price, someone facing less risk takes some of the risk off the shoulders of someone else who faces more.

In fact, this idea that there exists a price and a market for risk is the basis for almost all the financial arrangements in a modern economy. Stocks and bonds, and all the complex financial instruments like derivatives, are just ways of spreading risk to those who are willing to bear it for the least asking price. Many people think these markets are purely forms of gambling. In a sense, they are. But the gambles are taken by those who start out with the least risk, perhaps because they have already diversified in the way we saw earlier. And the risk is sold or shed by those who are initially most exposed to it. This enables the latter to be more adventurous in their enterprises than they would be if they had to bear all of the risk themselves. Thus financial markets promote entrepreneurship by facilitating risk trading.

Of course, these markets are subject to similar kinds of moral hazard, adverse selection, and even outright fraud as is risk pooling. Therefore these markets are not able to do a really full job of spreading risk. For example, managers are often required to bear some of the risk created by their decisions, through equity participation (stock ownership) or some other mechanism, in order to give them the right incentive to make an effort. We will see this in more detail later in this chapter.

B. Using Risk

Risk is not always bad; in some circumstances it can be used to one's advantage. While the general theory would take us too far afield, we offer a brief example of a prominent instrument that uses risk, namely an **option.**

An option gives the holder the right but not the obligation to take some stipulated action. For example, a call option on a stock allows the holder to buy the stock at a preset price, say 110. Suppose the stock now trades at 100, and may go to 150 or 50 with equal probability. If the former, the holder will exercise the option and make a profit of 40; if the latter, he will let the option lapse and make 0. Therefore the expected profit is

$$0.5 \times 40 + 0.5 \times 0 = 20.$$

If the uncertainty is larger, so the stock can go to 200 or 0 with equal probabilities, the expected profit from the option is

$$0.5 \times 90 + 0.5 \times 0 = 45.$$

Thus the option is more valuable the greater the uncertainty in the underlying stock. Of course, the market recognizes this, and the price for which one can buy such an option itself fluctuates as the volatility of the market changes.

C. Manipulating Risk in Contests

Our farmers in Section 1.A faced risk, but it was due to the weather rather than to any actions of their own or of other farmers. If instead the players in a game can affect the risk faced by themselves or others, then they can use such manipulation of risk strategically. A prime example is contests such as R&D races between companies to develop and market new information technology or biotech products; many sporting contests have similar features.

The outcome of sporting and related contests is determined by a mixture of skill and chance. You win if

$$\text{Your skill} + \text{your luck} > \text{rival's skill} + \text{rival's luck}$$

or

$$\text{Your luck} - \text{rival's luck} > \text{rival's skill} - \text{your skill}.$$

Your *luck surplus,* or L (which represents the difference between your luck and your rival's luck), is uncertain. Suppose the distribution of L can be shown using a normal, or bell curve, as in Figure 12.1. At any point on the horizontal axis, the height of the curve represents the probability that L takes on that value. Thus the area under this curve between any two points on the horizontal axis equals the probability that L lies between those points. Suppose your rival has more skill, so you are an underdog. Your *skill deficit,* which equals the difference between your rival's skill and your skill, is therefore positive, as shown by the point S. You win if your luck surplus—your value of L—exceeds your skill deficit, S. Therefore the area under the curve to the right of the point S, which is shaded in the figure, represents your probability of winning. If you make the situation chancier, the bell curve will flatten because the probability of rela-

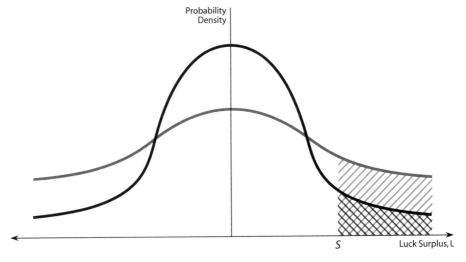

FIGURE 12.1   The Effect of Greater Risk on Chances of Winning

tively high and low values of L increases while the probability of moderate values decreases. Then the area under the curve to the right of S also increases. As the underdog, you should therefore adopt a strategy that flattens the curve. Conversely, if you are the favorite, you should try to reduce the element of chance in the contest.

Thus we should see underdogs or those who have fallen behind in a long race try unusual or risky strategies, because "it is their only chance to get level or ahead," while favorites or those who have stolen a lead will play it safe. A practical piece of advice based on this principle: if you want to challenge someone who is a better player than you to a game of tennis, choose a windy day.

You may be able to manipulate not just the amount of risk in your strategy, but also, as mentioned earlier, the correlation between the risks. The player who is ahead will try to choose a correlation as high and positive as possible: then, whether his own luck is good or bad, the luck of his opponent will be the same and his lead protected. Conversely, the player who is behind will try to find a risk as uncorrelated with that of his opponent as possible. It is well known that in a two-sailboat race, the boat that is behind should try to steer differently from the boat ahead, and the boat ahead should try to imitate all the tackings of the one behind.[1]

2 SOME STRATEGIES TO MANIPULATE INFORMATION

Our analysis of games thus far has assumed that all information about the game is common knowledge among the players. There may be risk or uncertainty about the consequences of actions, but the probabilities of the various alternatives are known to all. In reality, games have a lot of **information asymmetry.** One player, say A, often has some information that affects other players' decisions. For example, the preferences of A regarding the possible outcomes (such as risk tolerance in a game of brinkmanship, or patience in bargaining) are not likely to be known to others. The same is true for some innate characteristic of A (such as the skill of an employee, or the risk class of an applicant for auto or health insurance). Or the actions available to A (for example, the weaponry and readiness of a country), are not fully known to other players. Even when the possible actions are known, the actual actions taken by A may not be observable to others; for example, a manager may not know very precisely the effort or diligence with which the employees in his group are carrying out their tasks. Finally, some actual outcomes (like the actual dollar value of loss to an insured home-

[1]See Avinash Dixit and Barry Nalebuff, *Thinking Strategically* (New York: Norton, 1991), pp. 10–11, for this example. For more general theoretical analysis, see Luis Cabral, "Football, Sailing, Tennis, and R&D: Dynamic competition with Strategic Choice of Variance and Covariance," discussion paper, London Business School, 1997.

owner in a flood or an earthquake) may be observed by A but not by others. Since these things matter to others, and in turn their mattering matters to A, we have a new game of manipulating information.

We can easily list the various possibilities. The better-informed player, A, may want to do one of the following:

1. *Conceal information* or *reveal misleading information:* When mixing moves in a zero-sum game, you don't want the other player to see what you have done; you bluff in poker to mislead others about your cards.

2. *Reveal selected information truthfully:* When you make a strategic move, you want others to see what you have done so they may respond in the way you desire. For example, if you are in a tense situation but your intentions are not hostile, you want others to know this credibly so there will be no unnecessary fight.

Similarly, the less-informed player may want to do one of the following:

1. *Elicit information* or *filter truth from falsehood:* Examples: An employer wants to find out the skill of a prospective employee and the effort of a current employee. An insurance company wants to know an applicant's risk class, the amount of loss of a claimant, and any contributory negligence by the claimant that would reduce its liability.

2. *Remain ignorant:* Being unable to know your opponent's strategic move can immunize you against his commitments and threats. For example, top-level politicians or managers often benefit from having "credible deniability."

The general principle governing all such situations in which players' information and interests differ is that actions speak louder than words. The others should watch what A does, not what he says. And, knowing that the others will interpret actions in this way, A will in turn try to manipulate his actions for their information content.

When you are playing a strategic game, you may have information that is "good" (for yourself), in the sense that if the other players knew this information, they would alter their actions in a way that would increase your payoff. Or, you may have "bad" information, which would cause others to act in a way that would hurt you. You know that others will infer your information from your actions. Therefore you try to think of, and take, actions that will induce them to believe your information is good. Such actions are called **signals,** and the strategy of using them is called **signaling.** Conversely, if others are likely to conclude that your information is bad, you may be able to stop them from making this inference by confusing them. This strategy, called **signal jamming,** typically involves a mixed strategy, because the randomness of mixed strategies make inferences imprecise.

If other players know more than you do, or take actions that you cannot directly observe, you can use strategies that reduce your informational disadvantage. The strategy of making another player act so as to reveal his information is called **screening,** and specific methods used for this purpose are called **screen-**

ing devices.[2] A strategy that attempts to influence an unobservable action of another player, by giving him some reward or penalty based on an observable outcome of that action, is called an **incentive scheme.** We will elaborate on and illustrate all the concepts introduced here later on, in Sections 3 through 5.

Later in the chapter we give some examples that explain these concepts and teach you how to use them. But first we need to clear up a mistaken notion you may have regarding your ability to elicit information from your opponents.

A. Literal and Strategic Liars

You might think that truth can be elicited by asking cleverly designed questions. A well-known puzzle shows how. A traveler approaches a split in the road. One branch leads to the village of the truth tellers, the other to the village of the liars, but the traveler does not know which branch is which. A local inhabitant is sitting nearby. The traveler must find out which branch leads to which village by asking just one question. Asking straightforward questions does not work; for example, if the traveler asks, "Which road leads to the village of the truth tellers?" and the bystander points to the one going leftward, the traveler does not know whether the person is a truth teller pointing to the correct road or a liar pointing to the wrong one. But more complicated questions can elicit the truth. One such question is: "If you were a member of the other village, how would you answer the question, 'Does the left branch lead to the village of the truth tellers?'"

To see that this question allows the traveler to determine which village lies where, consider the possible combinations of the reality and the answers to the question. We show these in Figure 12.2. Just one example will show you how the entries arise. If the truth is No and the local is a liar, he knows that if he were a truth teller he would answer "No," and so of course being a liar he answers "Yes." Now we see that the answers are independent of the respondent's type—if the truth is Yes, both types will say "No," and the other way around. The answers are

		TRUTH IS	
		Yes	No
LOCAL INHABITANT IS	Truthful	No	Yes
	Liar	No	Yes

FIGURE 12.2   The Literal Liar's Responses

[2]A word of warning: Don't confuse screening with signal jamming. The alternative definition of screening, namely "concealment," is not the sense in which the term is used in game theory; rather, it means testing or scrutinizing. Thus a less-informed player uses screening to find out what a more-informed player knows. In contrast, a more-informed player uses signal jamming to confuse the less-informed player when the latter's finding out the truth would harm the former in the game.

		TRUTH: STOCK WILL GO	
		Up	Down
STOCKBROKER IS	Honest	Down	Up
	Literal Liar	Down	Up
	Strategic Liar	Up	Down

FIGURE 12.3   The Strategic Liar's Responses

always just the opposite of the truth, but that is perfectly fine for the questioner to draw the correct inference, just as if weather forecasts were always exactly wrong so we could interpret a forecast of Rain to mean Sun and vice versa.

But this mechanism works only because the members of the liars' village are *literal liars*: they will always tell the opposite of the literal truth, just for the sake of doing so and not for the sake of any profit they will get by misleading you. If liars were instead *strategic*, they would choose a response deliberately meant to deceive.

To see what difference this makes, consider a seemingly equivalent version of the story. A certain stock may go up or down; the stockbroker knows which. You know that there are two types of stockbrokers: honest and dishonest. The latter want to profit by getting you to buy the stock when it is going to go down and to sell it when it is going to go up. Now suppose you try asking, "If you were a stockbroker of the opposite kind, how would you answer the question 'Will this stock go up or down?'" You would get the answers shown in Figure 12.3.

When the stock is going to go up, the strategic liar does not lie mechanically; instead he gives the answer that, given the way you intend to interpret it, would lead you to infer that the stock will go down and so you should sell it. And similarly if it is going to go down. Of course, if you recognize you might be talking to a strategic liar, you will realize that your question will not get you the truth.

Therefore mere clever questions don't work against strategic liars. You must devise an *action* that will uncover strategic lying; for example, you could ask the stockbroker to use a commission scheme in which he takes a share of your loss. Such screening mechanisms also require some subtlety, as the next example shows.

B. King Solomon's Unwisdom

Two women came before King Solomon, disputing who was the true mother of a child. The King James Bible (I Kings 3:24–28) takes up the story:

> And the king said, Bring me a sword. And they brought a sword before the king.

And the king said, Divide the living child in two, and give half to the one, and half to the other.

Then spake the woman whose the living child was, unto the king, for her bowels yearned upon her son, and she said, O my lord, give her the living child, and in no wise slay it. But the other said, Let it be neither mine nor thine, but divide it.

Then the king answered and said, give her [the first woman] the living child, and in no wise slay it; she is the mother thereof.

And all Israel heard of the judgment which the king had judged; and they feared the king: for they saw that the wisdom of God was in him, to do judgment.

Given our expertise in strategy, we can see that the second woman made a strategic blunder. She should have given the same answer as the first or else she was sure to reveal herself as the false claimant; the child's true mother would never have consented to its death. But also the king was lucky that his ploy worked; its success came only from the second woman's error. Therefore "all Israel" should have flunked Solomon on strategy, fearing him perhaps for his luck but not for his wisdom.

What strategic device could Solomon have used that would have worked even against a pair of strategically savvy women? Game theorists have found several screening devices that would work under slightly different circumstances; we describe a simple one here. Call the two women Anna and Bess. Then suppose Solomon sets up the game as follows:

Move 1: Solomon decides on a fine or punishment F; this need not be monetary and need not be drastic.

Move 2: Anna is chosen to go first. *Either* she gives up her claim to the child (in which case Bess gets the child and the game ends) *or* she asserts her claim, in which case the game goes on to move 3.

Move 3: Bess *either* accepts Anna's claim (in which case Anna gets the child and the game ends) *or* challenges her claim. In the latter case, Bess must put in a bid, B, of her own choosing for the child, and Anna must pay the fine, F, to Solomon. The game goes on to move 4.

Move 4: Anna now *either* matches Bess's bid (in which case Anna gets the child, Anna pays B to Solomon, and Bess pays the fine, F, to Solomon) *or* chooses not to match (in which case Bess gets the child and pays her bid, B, to Solomon).

We show the extensive form for this game in Figure 12.4. At the terminal nodes, we show what the two players receive: C_A represents the value of getting the child for Anna and C_B the value to Bess; $-B$ means the player pays her bid, and $-F$ means the fine is paid. These payoffs cannot be explicitly ranked because

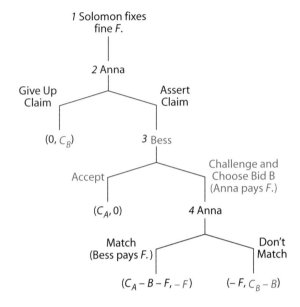

Payoffs are shown as (Anna, Bess).

FIGURE 12.4 King Solomon's Game

the true mother is likely to value the child more highly than the false claimant, but we don't know which woman is which. Thus we do not know whether C_A or C_B is larger. However, we can now consider the possible outcomes of the game.

Suppose the true mother does, indeed, value the child more and that Anna is the true mother; then $C_A > C_B$. Bess knows at move 3 that the only way she can end up with the child after move 4 is if she bids enough to force Anna's payoffs to satisfy $C_A - B - F < -F$; then Anna would prefer not to match Bess's bid. This outcome would arise only if $C_A < B$, or if $B > C_A > C_B$: Bess gets the child only if she bids more than the child is worth to her. So knowing this, Bess will not bid at move 3. Stepping back to move 2, Anna can look ahead and reason that Bess will not bid and so Anna will claim the child and get it. Thus, the ploy works when Anna is the true mother. And what if Bess is the true mother, so that $C_B > C_A$? Then Bess at move 3 will bid as long as $C_A - B - F < -F$ which implies that $B > C_A$; since we know $C_B > C_A$, Bess is assured of being able to find an appropriate bid, B. Then Anna knows in move 2 that Bess in move 3 will choose a bid that is not worth Anna's while to match in move 4, so she (Anna) is simply going to end up paying the fine, F, and not getting the child. So in move 2 Anna does best for herself by giving up her claim to the child. The ploy works in this case also.[3]

[3] For a full discussion of this problem, see John Moore, "Implementation, Contracts, and Renegotiation," in *Advances in Economic Theory*, vol. 1, ed. Jean-Jacques Laffont (Cambridge: Cambridge University Press, 1992), pp. 184–185 and 190–194.

Thus as long as the true mother values the child more than the false claimant, in the rollback (subgame-perfect) equilibrium the true mother gets the child. Solomon does not have to know the women's values. No fines or bids are actually paid along the equilibrium path of play through the tree; their sole purpose is to avoid any false claims by either woman, by affecting the consequences of deviating from the equilibrium path.

You may think it sordid to introduce money into such a human situation. But remember that no money actually changes hands in the equilibrium. It serves only as a backstop of threats to keep the women on the path of truth. In this it is no different than Solomon's original device of threatening to cut the child in two; and while money may be sordid, cutting up a child is gruesome.

There is only one problem. If the two women's material circumstances are sufficiently different that the false claimant is able to pay a lot to get the child while the true mother cannot afford that much to keep it, then in monetary terms the false claimant may value the child higher than does the true mother, which violates the condition under which the scheme works. But in this case the two women seem to be in quite similar circumstances (the Bible says they were both "harlots"), so the difficulty does not seem serious here.

Now let us move on to a more systematic study of strategies used to manipulate information.

3 INCENTIVES TO INDUCE EFFORT

Suppose you are the owner of a firm, looking for a manager to undertake a project. The outcome of the project is uncertain, and the probability of success depends on the quality of the manager's effort. If successful, the project will earn $600,000. The probability of success is 60% (0.6) if the manager's effort is of routine quality but rises to 80% if it is of high quality.

Putting out high-quality effort entails a subjective cost to the manager. For example, he may have to think about the project night and day, and his family may suffer. You have to pay him $100,000 even for the routine effort, but he requires an extra $50,000 for making the extra effort.

Is the extra payment in return for the extra effort worth your while? Without it, you have a 0.6 chance of $600,000 (that is, an expected value of $360,000), and you are paying the manager $100,000, for an expected net profit of $260,000. With it, the expected profit becomes $0.8 \times \$600,000 - \$150,000 = \$480,000 - \$150,000 = \$330,000$. Therefore the answer is yes; paying the manager $50,000 for his extra effort is to your benefit.

How do you implement such an arrangement? The manager's contract could specify that he gets a basic salary of $100,000 and an extra $50,000 if his effort is of high quality. But how can you tell? The manager might simply take the extra

$50,000 and make the routine effort anyway. Much of the effort may involve thinking, which may be done at home during evenings and weekends. The manager can always claim to have done this; you cannot check on him. If the project fails, he can attribute that to bad luck; after all, even with the high-quality effort the chances of failure are 1 in 5. And if a dispute arises between you and the manager about the quality of his effort, how can a court or an arbitrator get any better information to judge the matter?

If effort cannot be observed and, in case of a dispute, verified in a court of law, then the manager's contract must be based on something that can be observed and verified. In the present instance, it's the project's success or failure. Because success or failure is *probabilistically* related to effort, it gives some information, albeit imperfect, about effort, and can serve to motivate effort.

Consider a compensation package consisting of a base salary, s, along with a bonus, b, that is paid if the project succeeds. Therefore the manager's expected earnings will be $(s + 0.6\,b)$ if he makes the routine effort, and $(s + 0.8\,b)$ if he makes the high-quality effort. His expected extra earnings from making the better effort will be $(s + 0.8\,b) - (s + 0.6\,b)$ or $(0.8 - 0.6)\,b = 0.2\,b$. For the extra effort to be worth his while, it must be true that

$$0.2\,b \geq \$50,000$$
$$b \geq \$250,000.$$

The interpretation is simple: the bonus, multiplied by the *increase* in the probability of getting the bonus if he makes the extra effort, equals the manager's expected extra earnings from the extra effort. This expected increase in earnings should be at least enough to compensate him for the cost to him of that extra effort.

Thus a sufficiently high bonus for success creates enough of an incentive for the manager to put out the extra effort. The expression just given is called the **incentive-compatibility condition** or **constraint,** on your compensation package.

There is one other condition. The package as a whole must be good enough to get the manager to work for you at all. When the incentive compatibility condition is met, the manager will make the high-quality effort if he works for you, and his expected earnings will be $(s + 0.8\,b)$. He also requires at least $150,000 to work for you with high-quality effort. Therefore your offer has to satisfy the relation

$$s + 0.8\,b \geq \$150,000.$$

This is called the **participation condition** or **constraint.**

You want to maximize your own profit. Therefore you want to keep the manager's total compensation as low as possible, consistent with the incentive compatibility and the participation constraints. To hold it down to its floor of $150,000 means choosing the smallest value of s that satisfies the participation condition, or choosing $s = \$150,000 - 0.8\,b$. But since b must be at least $250,000, s can be no more than $\$150,000 - 0.8 \times \$250,000 = \$150,000 - \$200,000 = -\$50,000$.

What does a negative base salary mean? There are two possible interpretations. In one case, the negative amount might represent the amount of capital that the manager must put up for the project. Then his bonus can be interpreted as the payout from an equity stake or partnership in the enterprise. The second possibility is that the manager does not put up any capital but is fined if the project fails.

Sometimes neither of these can be done. Qualified applicants for the managerial job may not have enough capital to invest, and the law may prohibit the fining of employees. In that case, the total compensation cannot be held down to the minimum that satisfies the participation constraint. For example, suppose the basic salary, s, must be nonnegative. Even if the bonus is at the lowest level needed to meet the incentive compatibility constraint, namely $b = \$250,000$, the expected total payment to the manager when s takes on the smallest possible nonnegative value (zero) is $0 + 0.8 \times \$250,000 = \$200,000$. You are forced to over-fulfill the participation constraint.

If effort could be observed and verified directly, you would write a contract where you paid a base salary of $100,000 and an extra $50,000 when the manager was actually seen making the high-quality effort, for a total of $150,000. When effort is not observable or verifiable, you expect to pay $200,000 to give the manager enough incentive to make the extra effort. The additional $50,000 is an extra cost caused by this observation problem, that is, by the information asymmetry in the game. Such a cost generally exists in situations of asymmetric information. Game theory shows ways of coping with the asymmetry, but at a cost that someone, often the less-informed player, must bear.

Is this extra cost of circumventing the information problem worth your while? In this instance, by paying it you get a 0.8 probability of $600,000, for an expected net profit of

$$0.8 \times \$600,000 - \$200,000 = \$480,000 - \$200,000 = \$280,000.$$

You could have settled for the low effort, for which you need pay the manager only the basic salary of $100,00, but get only a 60% probability of success, for an expected net profit of

$$0.6 \times \$600,000 - \$100,000 = \$360,000 - \$100,000 = \$260,000.$$

Thus, even with the extra cost caused by the information asymmetry, you get more expected profit by using the incentive scheme.

But that need not always be the case. Other numbers in the example can lead to different results; if the project when successful yields only $400,000 instead of $600,000, then you do better to settle for the low-pay, low-effort situation.

The nature of the problem here is analogous to that faced by an insurance company whose clients run the risk of a large loss—theft or fire—but can also affect the probability of the loss by taking good or poor precautions, or being more or less careful. If you are not sure you double-locked your front door when you left your house, it is tempting not to bother to go back to double-check if you are insured. As we mentioned earlier, the insurance industry calls this problem moral hazard. Sometimes there may be deliberate fraud, for example, setting fire to your own house to collect insurance, but in many situations there is no necessary judgment of morality involved.

Economic analyses of situations of asymmetric information and incentive problems have accepted this terminology. Most generally, *any* strategic interaction where one person's action is unobservable or unverifiable is termed **moral hazard.** In the insurance context, moral hazard is controlled by requiring the insured to retain a part of the risk, using deductibles or coinsurance. This acts as an incentive scheme; in fact, it is just a mirror image of our contract for the manager. The manager gets a bonus when the outcome, whose probability is affected by his effort, is good; the insured has to carry some loss when the outcome, whose probability is affected by his care, is bad.

4 SCREENING AND SIGNALING

Many of you expect that when you graduate you will be among the elite of America's workers, the so-called symbolic analysts. Employers of such workers want them to possess the appropriate qualities and skills—capacity for hard work, numeracy, logic, and so on. You know your own qualities and skills far better than your prospective employer does. He can test and interview you, but what he can find out by these methods is limited by the available time and resources. Your mere assertions as to your qualifications are not credible; more objective evidence is needed.

What items of evidence can the employer seek, and what can you offer? Recall from Section 2 that the former are called *screening devices,* and their use by the prospective employer to identify your qualities and skills is called *screening.* Devices that you use at your initiative are called *signals,* and when you do this, you are said to engage in *signaling.* Sometimes similar or even identical devices can be used for either signaling or screening.

In this instance, if you have selected (and passed) particularly tough and quantitative courses in college, your course choices can be credible evidence of your capacity for hard work in general, and your skills in symbolic analysis in particular. Let us consider the role of course choice as a screening device.

To keep things simple, suppose college students are of just two types when it comes to the qualities most desired by employers: A (able) and C (challenged). Potential employers are willing to pay $150,000 a year to a type A, and $100,000 to a type C (and, in the equilibrium of a competitive job market, they have to pay this much). Since employers cannot directly observe any particular job applicant's type, they have to look for other credible means to distinguish between them.

Suppose the types differ in their tolerance for taking a tough course rather than an easy one in college. Each type must sacrifice some party time or other activities in order to take a tougher course, but this sacrifice is less, or easier to bear, for the A types than it is for the C types. Suppose the A types regard the cost of each such course as equivalent to $6,000 a year of salary, while the C types regard it as $9,000 a year of salary. Can an employer use this differential to screen his applicants and tell the A types from the C types?

Consider the following policy: anyone who has taken a certain number, n, or more of the tough courses will be regarded as an A and paid $150,000, and anyone who has taken less than n will be regarded as a C and paid $100,000. The aim of this policy is to create natural incentives whereby only the A types will take the tough courses, while the C types will not. Of course, neither wants to take more of these than he has to, so the choice is between taking n to qualify as an A, or giving up and settling for being regarded as a C, in which case one may as well not take any of the tougher courses and just coast through college.

The criterion employers devise to distinguish an A from a C—namely the number of tough courses taken—should be sufficiently strict that the C types do not bother to meet it, but not so strict as to discourage even the A types from attempting it. The correct value of n must be such that the true C types prefer to settle for being revealed as such, rather than incur the extra cost of imitating the A type's behavior. That is, we need

$$100,000 \geq 150,000 - 9,000n$$
$$9n \geq 50$$
$$n \geq 6.[4]$$

Similarly, the condition that the true A types prefer to prove their type by taking the n tough courses is

$$150,000 - 6,000n \geq 100,000$$
$$6n \leq 50$$
$$n \leq 8.[5]$$

These are called the incentive compatibility constraints for this problem. What makes it possible to meet both the conditions is the *difference* between the two types: the cost is sufficiently lower for the "good" type the employers wish to identify. When the constraints are met, the employer can use a policy to which the two types will respond differently, thereby revealing their types. This is called **separation of types** based on **self-selection.**

[4] Note that $50/9 = 5.56$. For n to take on integer values at least this large, we require that $n \geq 6$.

[5] Note again that $50/6 = 8.33$, but this time we need n to take on integer values less than this, so require $n \leq 8$.

We did not assume here that the tough courses actually imparted any additional skills or work habits that might convert C types into A types. In our scenario the tough courses serve only the purpose of identifying the individuals who already possess these attributes. In other words, they have a purely screening function.

In reality, of course, education does increase productivity. But it also has the additional screening or signaling function of the kind described here. In our example we found that education might be undertaken solely for the latter function; in reality the corresponding outcome is that education is carried further than is justified by the extra productivity alone. This extra education carries an extra cost—the cost of the information asymmetry.

When the requirement of taking enough tough courses is used for screening, the A types bear the cost. Assuming that only the minimum needed to achieve separation is used, namely $n = 6$, the cost to each A type has the monetary equivalent of $6 \times \$6,000 = \$36,000$. This is the cost, in this context, of the information asymmetry. It would not exist if a person's type could be directly and objectively identified. Nor would it exist if the population consisted solely of A types. The A types have to bear this cost *because* there are some C types in the population, from whom they (or their prospective employers) seek to distinguish themselves. In the language of Chapter 11 (Section 5), this is a negative external effect inflicted by the C types on the A types.

Rather than having the A types bear this cost, might it be better not to bother with the separation of types at all? With the separation, A types get a salary of $150,000 but suffer a cost the monetary equivalent of $36,000 in taking the tough courses; thus their net money-equivalent payoff is $114,000. And C types get the salary of $100,000. What happens to the two types if they are not separated?

If employers do not use screening devices, they have to treat every applicant as a random draw from the population and pay all the same salary. This is called **pooling of types.**[6] In a competitive job market, the common salary under pooling will be the population average of what the types are worth to an employer, and this average will depend on the proportions of the types in the population.

For example, if 20% of the population is type A and 80% is type C, then the common salary with pooling will be

$$0.2 \times \$150,000 + 0.8 \times \$100,000 = \$110,000.$$

---

[6] Note that pooling of types is quite different from *pooling of risks*, which we met earlier, in Section 1. The latter occurs when each person in a group faces some uncertain prospect, so they agree to put these assets into one pool from which they share the returns. As long as their risks are not perfectly correlated, such an arrangement reduces the risk for each person. In *pooling of types*, players who are different in some relevant innate characteristic (such as ability) nevertheless get the same outcome or treatment (such as allocation to a job) in the equilibrium of the game. Note that this concept also contrasts with *separation of types*, in which players of each different type would get a different outcome, and so the outcome would serve to reveal the type perfectly.

The A types will then prefer the situation with separation because it yields $114,000 instead of the $110,000 with pooling. But if the proportions are 50-50, then the common salary with pooling will be $125,000, and the A types will be worse off under separation than they would be under pooling. The C types are always better off under pooling. The existence of the A types in the population means that the common salary with pooling will always exceed the C type's separation salary of $100,000.

However, even if both types prefer the pooling outcome, it cannot be an equilibrium. Suppose the population proportions are 50-50, and there is an initial equilibrium with pooling where both types are paid $125,000. An employer can announce that he will pay $132,000 for someone who takes just one tough course. Relative to the initial situation, the A types will find this worthwhile because their cost of taking the course is only $6,000 and it raises their salary by $7,000, while C types will not find it worthwhile because their cost, $9,000, exceeds the benefit, $7,000. Since this particular employer selectively attracts the A types, each of whom is worth $150,000 to him but is paid only $132,000, he makes a profit by deviating from the pooling salary package.

But his deviation starts a process that causes the old pooling situation to collapse. As A types flock to work for him, the pool available to the other employers is of lower average quality, and eventually they cannot afford to pay $125,000 anymore. As the salary in the pool is lowered, the differential between that salary and the $132,000 offered by the deviating employer widens to the point where the C types also find it desirable to take that one tough course. But then the deviating employer must raise his requirement to two courses, and increase the salary differential, to the point where the C types again choose not to act like the A types. Other employers who would like to hire some A types must use similar policies if they are to attract them. This process continues until the job market reaches the separating equilibrium described earlier.

Even if the employers did not take the initiative to attract As rather than Cs, a type A earning $125,000 in a pooling situation might take a tough course, take his transcript to a prospective employer, and say: "I have a tough course on my transcript, and I am asking for a salary of $132,000. This should be convincing evidence that I am type A—no type C would make you such a proposition." Given the facts of the situation, the argument is valid, and the employer should find it very profitable to agree: the employee, being type A, will generate $150,000 for the employer but get only $132,000 in salary. Other A types can do the same. This starts the same kind of cascade that leads to the separating equilibrium. The only difference is who takes the initiative. Now the type A workers choose to get the extra education as credible proof of their type; it becomes a case of signaling rather than screening.

The general point is that even though the pooling outcome may be better for all, they are not choosing the one or the other in a cooperative binding process. They are pursuing their own individual interests, which lead them to the separating equilibrium. This is like a multiperson prisoners' dilemma, and therefore there is something unavoidable about the cost of the information asymmetry.

We have considered only two types, but the idea generalizes immediately. Suppose there are several types: A, B, C, ... , ranked in an order that is at the same time decreasing in their worth to the employer, and increasing in the costs of extra education. Then it is possible to set up a sequence of requirements of successively higher and higher levels of education, such that the very worst type needs none, the next-worst type needs the lowest level, the type second from the bottom needs the next higher level, and so on, and the types will self-select the level that identifies them.

In this example we developed the idea of a tough course requirement as an evidence of skills seen mainly from the perspective of the employer, that is, as an example of a screening device. But at times we also spoke of a student initiating the same action as a signal. As mentioned earlier, there are indeed many parallels between signaling and screening, although in some situations the equilibrium can differ depending on who initiates the process, and on the precise circumstances of the competition among several people on the one side or the other of the transaction. We will leave such details for more advanced treatises on game theory. Here we make a further point specific to signaling.

Suppose you are the informed party and have available an action that would credibly signal your good information (one whose credible transmission would work to your advantage). If you fail to send that signal, you will be assumed to have bad information. In this respect, signaling is like playing chicken: if you refuse to play, you have already played and lost.

You should keep this in mind when you have the choice between taking a course for a letter grade or on a pass-fail basis. The whole population in the course spans the whole spectrum of grades; suppose the average is B. A student is likely to have a good idea of his own abilities. Those reasonably confident of getting an A+ have a strong incentive to take the course for a letter grade. When they have done so, the average of the rest is less than B, say B−, because the top end has been removed from the distribution. Now among the rest, those expecting an A have a strong incentive to switch. That in turn lowers the average of the rest. And so on. Finally, the pass-fail option is chosen by only the Cs and Ds. And that is how the reader of a transcript (a prospective employer or the admissions officer for a professional graduate school) will interpret it.

5 SEPARATION AND POOLING

We explained the general concepts of screening and signaling, and the possible outcomes of separation and pooling that can arise when these strategies are being used. We also saw how these would operate in an environment where many employers and employees meet each other, so the salaries are determined by conditions of competition in the market. However, we have not specified, nor found the equilibrium of, a game in which just two players with differential information confront each other. Now we develop an example to show how that can be done. Some of the concepts and the analysis gets a little intricate, and beginning readers can omit this section without loss of continuity.

Suppose the old established firm Oldstar faces a challenge from Nova, a newcomer to their market. Oldstar must then decide whether to fight a price war. Eventually the market will be dominated by only one firm: Oldstar if Nova decides not to challenge at all, Nova if it challenges and Oldstar gives up without a fight, and the winner of the fight if Nova challenges and Oldstar fights.

Oldstar is set in its ways. If Nova has a superior product or a more efficient technology, then Oldstar cannot hope to compete, and would do better to give up and go out of business, leaving the field to Nova. However, Nova may just be pretending to be a superior hotshot, hoping to frighten Oldstar into giving up and then enjoying monopoly profits itself. In our jargon, Nova can be of one of two "types": Strong or Weak.

If there is a fight, Oldstar will beat a Weak-type Nova, but a Strong-type Nova will beat Oldstar. The possible outcomes of an actual encounter are shown in Figure 12.5. The idea is that a Strong-type Nova can make a profit of 4 if it is left alone in the market, Oldstar can make 3, and a Weak-type Nova can make 2. A fight costs 2. In the bottom left cell in Figure 12.5, a Weak-type Nova mounts a challenge, and Oldstar fights and wins. Therefore Nova simply loses the 2 for the fight (payoff –2), and Oldstar gets 3 after it is left to enjoy sole possession of the market, but pays 2 to fight (payoff 3 – 2 = 1). You can verify the payoffs in the other cells by similar arguments.

Nova knows which type it is, but Oldstar does not know this about Nova. Oldstar does, however, know the general circumstances of the industry and the technology, and can therefore figure out the probability that Nova is Weak. We let w denote this probability.[7] Oldstar has to decide whether to fight, without knowing

		OLDSTAR	
		Fight	Retreat
NOVA	Strong	2,–2	4,0
	Weak	–2,1	2,0

FIGURE 12.5 Payoffs of Encounter Between Rival Firms

---

[7] To be more precise, Oldstar has to know w, Nova has to know that Oldstar knows, and so on; in other words, w has to be *common knowledge* between the two. See the discussion of common knowledge of rules of the game in Chapter 2, Section 3.D.

the actual type of a challenger when it appears. In turn, Nova has to decide whether to challenge, knowing that Oldstar will respond in this way.

If faced with a challenge and offered no other basis for information, Oldstar will simply calculate the expected payoff from fighting, which is

$$w \times 1 + (1 - w)(-2) = 3w - 2.$$

If this is positive (that is, if $w > 2/3$), Oldstar will fight; otherwise it will retreat. This makes intuitive sense: if Oldstar's prior belief is that Nova is likely to be weak, it makes sense for Oldstar to fight.

Now suppose Nova can present some evidence of its strength—for example, by displaying prototypes of advanced products even though it does not yet have the ability to deliver large quantities at the same quality specifications. We call such activities a display, much like the display animals put on before a fight when they are trying to convince other animals of their strength. If Nova is in fact Strong, it puts on this display automatically and costlessly. A Weak Nova might try to pass as Strong by displaying, but such behavior comes only at a cost. We let c be the dollar figure of the cost incurred by the Weak-type Nova when it tries to imitate the Strong type. We suppose that c is common knowledge, just as w is.

The game tree is shown in Figure 12.6. It has some special features that take

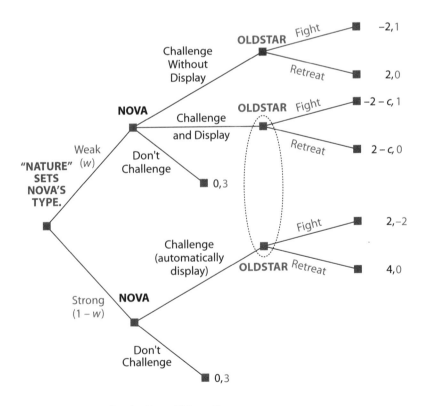

FIGURE 12.6 Game Tree for the Nova-Oldstar Game

into account the information asymmetry. First, Nova's type is a matter of chance; therefore we begin the game at a node where a fictitious player, Nature, makes this choice. The next nodes are Nova's moves. Knowing its own type, it has to decide whether to challenge, and if it is the Weak type, whether to try to imitate the Strong type by making a display at cost, c. Then come the nodes for Oldstar's decisions. If faced with a challenge, it must decide whether to fight. When doing so, it does not know Nova's true type but can observe a display, if any. Therefore Oldstar cannot tell apart the node representing a challenge with display by a Weak-type Nova, from the node for a challenge by a Strong-type Nova (which, remember, displays automatically and thus without cost). Oldstar responds the same way—either Fight or Retreat—at both nodes because both are in the same information set.[8]

The concept of equilibrium for this game must also recognize the information aspects. We require that

1. at each node, the player acting there takes the best action in light of the available information, and

2. players draw the correct inferences from their observations, as specified by Bayes' rule for drawing inferences from observations.[9]

These criteria constitute a **Bayesian perfect equilibrium.** Whether the equilibrium entails separation, pooling, or a combination of the two depends on the probability w of Nova being Weak, and the cost, c, of display for a Weak-type Nova. We consider all of the possibilities in the following discussion.

A. Separating Equilibrium

If the cost of display for the Weak-type Nova is too high, then the display serves to separate the two types because it will not be optimal for the Weak-type Nova to incur this cost. Therefore Oldstar can correctly infer that a challenger who makes the display is Strong, and so Oldstar will respond by retreating. The precise condition is that $c > 2$. For any cost at this level or higher, the Weak-type gets a negative payoff if it tries to challenge. We assume for now that this condition holds, and verify that it leads to a separating equilibrium.

To be more precise about how the game works in this case, we note that the strategies and inferences are as follows:

1. Nova challenges if and only if it is Strong (the Weak type does not use the display).

2. If Oldstar sees a challenge with display, it infers that the challenger is Strong, and it retreats; if it observes a challenge without display, it infers that the challenger is Weak, and it fights.

---

[8] See the discussion of information sets in Chapter 6, Section 1.
[9] See the Appendix to Chapter 5 for an explanation of Bayes' rule.

Note that we have to specify Oldstar's strategy at all nodes, even though they do not actually occur (because in equilibrium a Weak-type Nova does not challenge). We also have to make sure that these actions at off-equilibrium nodes would be optimal for Oldstar if by some mischance that node were to be reached. This ensures that Oldstar's strategy does not rely on a noncredible threat, so the equilibrium is subgame perfect.

Now we proceed with the verification of the equilibrium. Given Nova's strategies (item 1 above), Oldstar is correct to infer that a display indicates strength and a lack thereof indicates weakness. We know from the payoff table of Figure 12.5 that Oldstar does better to fight a Weak-type Nova (a payoff of 1 versus a 0 for retreating) and better to retreat from a strong-type Nova (a payoff of 0 as against –2 from fighting); therefore Oldstar's actions (item 2) are optimal given Nova's actions specified in situation 1. Conversely, given that Oldstar is making this interpretation, a Weak-type Nova stands to gain $(2 - c)$ if it challenges with display and 0 if it does not challenge (and even worse, –2, if it challenges without display). Since $c > 2$ in this case, it is optimal not to challenge. For a Strong-type Nova, it is always optimal to challenge. Thus the actions specified in situation 1 are optimal for Nova given Oldstar's inferences and actions specified in item 2.

B. Pooling Equilibrium

If the cost of display is small, we should expect a Weak-type Nova to use this device and pretend to be the Strong type. Therefore we should look for a pooling equilibrium if $c < 2$. However, there is a further condition for such an equilibrium, namely that the probability of being Weak is sufficiently small. Otherwise, if Nova was quite likely to be Weak, Oldstar would find it optimal to fight any randomly picked type of challenger, and then the Weak type would not find it optimal to challenge at all. The precise condition turns out to be $w < 2/3$, as we now show.

We start by supposing that $c < 2$ and $w < 2/3$, and verify that all the requirements of a pooling equilibrium are met. The precise specification of strategies and inferences is as follows:

1. Both types of Nova challenge, and the weak one uses the display.
2. If a challenger with a display appears, Oldstar will retreat. (If a challenger without a display were to appear, Oldstar would fight.)

Given Nova's strategies (item 1), Oldstar cannot infer anything by observing the display, and its estimate of the probability that it faces a Weak-type Nova is w. Therefore Oldstar's expected payoff from fighting is

$$w \times 1 + (1 - w)(-2) = 3w - 2 .$$

Since $w < 2/3$ under the premises of this case, this expected payoff is negative, and it is optimal for Oldstar to retreat and get zero. From Oldstar's perspective, it is sufficiently probable that a randomly chosen opponent is strong to make retreat better than taking a chance at losing a fight. Thus an actually Weak-type Nova is successful in its deception.

Given Oldstar's strategies (item 2), a Weak-type Nova gets $(2 - c)$ by challenging with display and 0 by not challenging (and $-c$ by challenging without display). Since $c < 2$ in this case, it is optimal to challenge.

C. Semiseparating Equilibrium

Finally, what kind of an equilibrium do we get if $c < 2$ and $w > 2/3$? We cannot get a separating equilibrium. If there were an equilibrium of this kind, then Oldstar would be drawing the inference that a challenge with display was evidence of a Strong type. Then a truly Weak-type Nova would find it optimal to exploit this belief: by challenging with display, it would get mistaken for a Strong type, Oldstar would retreat, and then the Weak Nova would get the payoff of $(2 - c)$. Thus a separating equilibrium cannot last; a Weak-type Nova would switch to imitating the Strong type. Nor can there be a pooling equilibrium. With $w > 2/3$, the probability of Nova's type being Weak is so high that if everyone were challenging with display, it would be optimal for Oldstar to fight every time, and then the Weak-type Nova would get $(-2 - c)$. In that case, the Weak-type Novas should not display, and pooling cannot last either.

This analysis shows that in this case it cannot be optimal for the Weak-type Nova *never* to challenge; nor can it be optimal for it *always* to challenge. We know from our extensive experience of such situations that the equilibrium should then involve mixing strategies. Therefore let us look for an equilibrium with mixed strategies in which:

1. a Weak-type Nova challenges with probability p.
2. Oldstar draws inferences from its observations using Bayes' rule, and responds to a challenge with display by fighting with probability, q, (and to a challenge without display by fighting for sure).

Of course, we have to find the values of p and q compatible with the requirements of a mixed-strategy equilibrium. We will find that

$$p = 2\,\frac{1 - w}{w} \quad \text{and} \quad q = \frac{2 - c}{4}.$$

First we must see how Bayes' rule gives Oldstar its inference from observing a challenge with display. As explained in the Appendix to Chapter 5, we must draw up a table of probabilities of various combinations of the true condition (Nova's type) and the observation (display). Figure 12.7 shows this. The first two rows correspond to the two types, Strong and Weak, and the first two columns to the display, Yes or No. The last column gives the sum in each row.

These are just the probabilities of each type of Nova as estimated by Oldstar in the absence of any observation (display): $(1 - w)$ for Strong and w for Weak. If Nova is strong, it displays automatically, therefore in the first row the Yes column gets the whole quantity $(1 - w)$ and the No column gets 0. We have supposed that a Weak-type Nova mixes strategies, and displays with probability, p; therefore in the second row, the Yes column gets wp (probability that Nova is Weak times the probability that Nova displays given that it is Weak), and the No column gets $w(1 - p)$.

		DISPLAY		Sum of Row
		Yes	No	
NOVA'S TRUE TYPE	Strong	$1 - w$	0	$1 - w$
	Weak	wp	$w(1 - p)$	w
Sum of Column		$1 - w + wp$	$w(1 - p)$	

FIGURE 12.7 Probabilities of Nova's Types and Displays

The last row gives the sums of the entries in each column. In the Yes column we have $(1 - w + wp)$; this is the total probability of display, comprising all of the probability that Nova is Strong but only the displaying part of the probability that Nova is Weak. Similarly, in the last row of the No column we have $w(1 - p)$, the total probability of nondisplay. Therefore if Oldstar knows that $c < 2$ and $w > 2/3$ and it observes a display, it recalculates the probability of Nova's type being Weak, now conditioned on the fact that it has observed the display. By Bayes' rule, this recalculated probability is $wp/(1 - w + wp)$, using the probability that a Weak Nova has displayed and the probability of observing a display from the entries in Figure 12.7. Similarly, the probability of Nova being Strong, conditioned on the display, is $(1 - w)/(1 - w + wp)$.

Then Oldstar also recalculates the expected payoff from fighting using its recalculated probabilities of facing a Weak versus a Strong Nova. The expected payoff from fighting is now

$$1 \times \frac{wp}{1 - w + wp} + (-2) \times \frac{1 - w}{1 - w + wp} = \frac{wp - 2(1 - w)}{1 - w + wp}$$

If Oldstar is to follow a mixed strategy, this must equal 0, so that Oldstar's expected payoffs from its two pure strategies are identical. Therefore, we need the numerator of the fraction to equal 0:

$$wp - 2(1 - w) = 0$$
$$p = \frac{2(1 - w)}{w}.$$

Note that $w > 2/3$ ensures $p < 1$. Moreover, as w increases from $2/3$ to 1, p gradually falls from 1 to 0. In the context of our example, this means that, as the probability of Nova being Weak increases, the mixing Weak-type Nova deploys a lower probability of putting on a display to try to mimic the Strong type if it is to avoid a fighting response from Oldstar.

Next, given Oldstar's strategies and inferences (item 2 above), the expected payoff from challenging with display for a weak-type Nova is

$$q(-2 - c) + (1 - q)(2 - c) = 2 - c - 4q.$$

If the Weak-type Nova is to mix strategies, this expected payoff must equal 0, the payoff from not challenging. Therefore $q = (2 - c)/4$.

To review, in this case we could not get a separating equilibrium; this is because, given the expectation that in such an equilibrium the Weak type does not display, the first one to try would succeed. Nor could we get a completely pooling equilibrium (because with everyone displaying and more than a 2/3 probability of Nova being Weak, Oldstar would respond by fighting, when the Weak-type Nova would find it better not to challenge). What we have is a **semiseparating equilibrium,** where having no display identifies Nova as Weak, but using a display leaves some ambiguity about its type.

Figure 12.8 shows how the three types of equilibria can arise for different possible combinations of the cost of display, c, and the probability, w, of a challenger being Weak. If the cost of imitation is too high, we get a separating equilibrium regardless of the probability of being Weak. If the cost is sufficiently low, then the type of equilibrium obtained depends on the probabilities of being Weak versus Strong. If the probability of being Weak is sufficiently low, then a Weak challenger can exploit the fact that a random challenger is quite likely to be Strong. He does this by acting like one, that is, by displaying. Otherwise, when the probability of being Weak is high but the cost of display is low, we get a semiseparating equilibrium.

In contrast to the usual idea that a player with good information looks for a signal to convey this information credibly, here the player with bad information takes an action that he hopes would get him mistaken for one with good information. This can be thought of as a kind of signal jamming: the Weak type's randomization muddles up Oldstar's process of drawing inferences from observations. This is somewhat like bluffing in poker.

		PROBABILITY OF WEAK TYPE, w	
		$w < 2/3$	$w > 2/3$
COST OF DISPLAY, c	$c < 2$	Pooling	Semiseparating
	$c > 2$	Separating	

FIGURE 12.8 Classification of Equilibria in the Nova-Oldstar Game

One more fact about the semiseparating equilibrium: Oldstar responds in a way that leaves the expected payoff for the Weak-type Nova the same as if the latter had not troubled to display or challenge. But now the Strong-type Nova has to fight with a positive probability, and then it receives the payoff of 2 instead of 4. Thus its expected payoff goes down by $(4 - 2)$ times the probability, p, of the fight. In other words, the fact that Nova may be Weak exerts a kind of negative external effect on the Strong type.

If the Strong type could find a way of demonstrating its strength that the Weak type could not mimic (or mimic only at excessive cost), then the Strong type would use that as a signal to separate themselves. Likewise, if Oldstar could find some effective way to screen its challengers and sort the Strong from the Weak, they could benefit too.

6 SOME EVIDENCE FOR SIGNALING AND SCREENING

As we saw in Section 5, characterization and solution of Bayesian perfect equilibria for games of signaling and screening involve subtle concepts and computations. Should we expect players to perform such calculations correctly? How realistic are these equilibria as descriptions of the outcomes of these games?

There is ample evidence that people are very bad at performing calculations involving probabilities, and especially bad at conditioning probabilities on new information.[10] These are exactly the calculations that must be performed to update one's information on the basis of observed actions. Therefore we should be justifiably suspicious of equilibria that depend on the players' doing so. Relative to this expectation, the findings of economists who have conducted laboratory experiments of signaling games are encouraging. Surprisingly subtle forms (refinements) of Nash equilibrium have been successfully observed, even though these require players not only to update their information by observing actions along the equilibrium path, but also to decide how they would infer information from off-equilibrium actions that should never have been taken in the first place. However, the verdict of the experiments is not unanimous; much seems to depend on the precise details of the "market institution" simulated in the laboratory, that is, on the design of the experiment.[11]

---

[10]Deborah J. Bennett, *Randomness* (Cambridge, Mass,: Harvard University Press, 1998), pp. 2–3 and Chap. 10. See also Paul Hoffman, *The Man Who Loved Only Numbers* (New York: Hyperion, 1998), pp. 233–240, for an entertaining account of how several probability theorists, as well as the brilliant and prolific mathematician Paul Erdős, got a very simple probability problem wrong, and even failed to understand their error when it was explained to them.

[11]See Douglas D. Davis and Charles A. Holt, *Experimental Economics* (Princeton, N.J.: Princeton University Press, 1995), chap. 7, for review and discussion of these experiments.

While the equilibria of signaling and screening games can be quite subtle and complex, the basic idea of the role of signaling or screening to elicit information is very simple. Players of different "types"—that is, possessing different information about their own characteristics or about the game and its payoffs more generally—should find it optimal to take different actions, so their actions truthfully reveal their types. One can point to numerous practical applications of this idea. Here are some examples; once you start thinking along these lines, you should be able to come up with dozens more.

1. Insurance companies usually offer a spectrum or menu of policies, with different provisions for deductible amounts and coinsurance. Those customers who know themselves to be bad risks prefer the policies with low deductibles and coinsurance, while those who know themselves to be good risks are more willing to take the policies in which they have to bear more of the losses. Thus the different risk classes have different optimal actions, and insurance companies use their clients' self-selection to screen and elicit the risk class of any particular client.

2. Venture capitalists seek inventors with good ideas, but how are they to judge the quality of any particular idea? They look for an inventor's willingness to "put his money where his mouth is"—that is, for someone who will take up as large an equity participation in the venture as his financial situation permits. Thus the willingness to bear a portion of any loss from the project becomes a credible indicator of the inventor's belief that his project will turn a profit.

3. The quality of durable goods such as cars and computers is hard for consumers to evaluate, but each manufacturer has a much better idea of the quality of his own product than do his customers. And if a product is of higher quality, it is less costly for the maker to offer a longer or better warranty on it. Therefore warranties can serve as signals of quality, and consumers are intuitively quite aware of this when they make their purchase decisions.

4. A somewhat unusual idea comes from political economy: Governments sell bonds to the public to finance their budget deficits. They can sell bonds that are *indexed*—whose interest rate in any year is adjusted by the rate of inflation that year—or not. An indexed bond, because it has to pay more interest when there is more inflation, is more costly to issue for a government that intends to create more inflation. Thus the issuing of indexed bonds should serve as a credible signal of an anti-inflation resolve. The government in effect is saying, "Look, if we intended to pursue inflationary policies, we wouldn't find it optimal to issue these indexed bonds." Very few governments issue indexed bonds; we as citizens should draw the obvious conclusion from that. This example illustrates a general principle. Suppose it is known that a signal is available to a player whereby he can credibly convey information about himself, but is not doing so. Information can be good or bad (in the sense defined in Section 2 in conjunction with the definition of signaling): information about one player is good if other players

would respond to it in such a way that the equilibrium of the game became more favorable to the first player, and bad if the opposite were true. The other players assume that, if the information were good, surely he would be sending that signal. They therefore can infer from the signal's *absence* that his information about himself must be bad.

5. Finally, we supply an example from biology. In many species of birds, the males have elaborate and heavy plumage that females find attractive. One should expect the females to seek genetically superior males so that their offspring will be better equipped to survive and attract mates in their turn. But why does heavy plumage indicate such desirable genetic qualities? One would think it would be a handicap, making the bird more visible to predators (including human hunters), less mobile, and therefore less able to evade getting caught. Why do females choose these handicapped males? The answer comes from the conditions for credible signaling. While heavy plumage is a handicap, it is less of one to a male who is sufficiently superior genetically in qualities like strength and speed. The weaker the male, the harder it is for him to produce and maintain plumage of a given quality. Thus it is precisely the heaviness of the plumage that makes it a credible signal of the male's quality.[12] Of course, as in all biological games, we do not claim that females have an explicit awareness of this connection; merely that such behavior will result in a fitter progeny and will therefore constitute an evolutionary stable equilibrium.

7 ADDITIONAL READING ON INFORMATION MANIPULATION

The theory of games involving information manipulation has developed rapidly over the last 20 years, and its ideas have come to dominate much new thinking in economics, business, and political science. We expect many of you will want to acquire more knowledge of this topic than an elementary textbook can accommodate. Therefore we list some further reading. First, two classics:

Erving Goffman, *The Presentation of Self in Everyday Life*, rev. ed., (New York: Anchor Books, 1959). (A classic from psychology)

A. Michael Spence, *Market Signaling*, (Cambridge, Mass.: Harvard University Press, 1974). (A classic from economics)

Next, two intermediate-level textbooks:

Ken Binmore, *Fun and Games* (Heath, 1992), chaps. 10–12 (*Warning:* The title is misleading; the book is actually quite challenging.)

---

[12]Matt Ridley, *The Red Queen* (New York: Penguin Books, 1995), p. 148.

Eric Rasmusen, *Games and Information*, (Cambridge, Mass.: Basil Blackwell, 1989), part II.

Finally, two advanced graduate-level textbooks:

Drew Fudenberg and Jean Tirole, *Game Theory* (Cambridge, Mass.: MIT Press, 1991), chaps. 6–8.

David Kreps, *A Course in Microeconomic Theory* (Princeton, N.J.: Princeton University Press, 1990), chaps. 13, 16, 17.

SUMMARY

When game players have different attitudes toward risk or different amounts of information, there is scope for strategic behavior in the control and manipulation of risk and information. Players can reduce their risk through *trading* or *pooling* (combining), although the latter is complicated by issues of *moral hazard* and *adverse selection*. Risk may be useful in some contexts, and it can be manipulated to a player's benefit depending on the circumstances within the game.

Players with private information may want to conceal or reveal that information, while those without the information try to elicit it or avoid it. Actions speak louder than words in the presence of *asymmetric information*. To reveal information, a credible *signal* is required. *Signaling* works only if the signal action entails different costs to players with different information. To obtain information, when questioning is not sufficient to elicit truthful information, a *screening* scheme that looks for a specific action may be required. Screening works only if the *screening device* induces others to reveal their types truthfully; there must be *incentive compatibility* to get *separation*. Information asymmetries may be costly to resolve, and external effects may arise as a result of the resolution process.

In the equilibrium of a game with imperfect information, players must not only use their best actions, given their information; they must also draw correct inferences (update their information) by observing the actions of others. This type of equilibrium is known as a *Bayesian perfect equilibrium*. The outcome of such a game may entail pooling, separation, or partial separation depending on the specifics of the payoff structure.

The evidence on players' abilities to achieve Bayesian perfect equilibria seems to suggest that, despite the difficult probability calculations necessary, such equilibria are often observed. Different experimental results appear to depend largely on the design of the experiment. Many examples of signaling and screening games can be found in ordinary situations like the provision of insurance or the issuing of government bonds.

KEY TERMS

adverse selection (399)
Bayesian perfect equilibrium (419)
incentive-compatibility condition
  (constraint) (410)
incentive scheme (405)
information asymmetry (403)
moral hazard (399)
negatively correlated risks (398)
option (401)
participation condition
  (constraint) (410)
pooling of types (414)

pooling risks (398)
positively correlated risk (399)
screening (404)
screening devices (404)
self-selection (413)
semiseparating equilibrium
  (423)
separation of types (413)
signal jamming (404)
signaling (404)
signals (404)
trading risks (400)

EXERCISES

1. In our discussion of risk trading in Section 1.A, you had a risky income that was $15,000 in good weather (a probability of 0.5) and $5,000 in bad weather (a probability of 0.5). When your neighbor had a sure income of $10,000, we derived a scheme in which you could keep him indifferent while raising your own expected utility; you simply paid your neighbor $100 with the understanding that he would give you $2,000 when the weather was bad, and you would pay him an additional $2,000 when the weather was good. Construct a similar type of arrangement in which your neighbor keeps you at your original level of expected utility but he achieves a higher level of expected utility than he would in the absence of the arrangement.

2. The managerial-effort example given in Section 3 assumes that a successful project would earn $600,000 for the firm. Using the same values for the rest of the example—that the probability of success for the project is 60% with routine-quality managerial effort, the manager requires $50,000 extra payment for extra effort, and so on—show that the firm does better to settle for the low-pay, low-effort situation when the value of the successful project is only $400,000.

3. An economy has two types of jobs, Good and Bad, and two types of workers, Qualified and Unqualified. The population consists of 60% Qualified and 40% Unqualified. In a Bad job, either type of worker produces 10 units of output. In a Good job, a Qualified worker produces 100 and an Unqualified worker produces 0. Companies have numerous job openings of each type, and must pay for each type of job what they expect the appointee will produce.

Companies cannot directly observe a worker's type before hiring, but Qualified workers can signal their qualification by getting educated. The cost of getting educated to level n for a Qualified worker is $n^2/2$, and for an Unqualified worker, is n^2. These costs are measured in the same units as output, and n must be an integer.

(a) What is the minimum level n that will achieve separation?

(b) Now suppose the signal is made unavailable. Which kinds of jobs will be filled by which kinds of workers, and at what wages? Who will gain and who will lose from this change?

4. Mictel Corporation has a world monopoly on the production of personal computers. It can make two kinds of computers, low-end and high-end. The total population of prospective buyers is P. There are two types of prospective buyers: casual users and intensive users. The casual ones comprise a fraction, c, of the population, and the rest, $(1 - c)$, are intensive users.

The costs of production of the two kinds of machines, and the benefits gained from the two by the two types of prospective buyers, are given in the following table (all figures in thousands of dollars):

		COST	BENEFIT FOR USER TYPE	
			Casual	Intensive
PC TYPE	Low-End	1	4	5
	High-End	3	5	8

Each type of buyer calculates the net payoff (benefit minus price) he would get from each type of machine, and buys the type that would give the higher net payoff provided this is nonnegative. If both types give equally non-negative net payoffs for a buyer, he goes for the high end; if both types have negative net payoff for a buyer, he does not purchase. Mictel wants to maximize its expected profit.

(a) If Mictel were omniscient, then knowing the type of a prospective customer, the company could offer to sell him just one type of machine at a stated price, on a take-it-or-leave-it basis. What kind of machine would Mictel offer, and at what price, to what kind of buyer?

In fact, Mictel does not know the type of any particular buyer. It just makes its "catalog" available for all buyers to choose from.

(b) First, suppose the company produces just the low-end machines and sells them for price x. What value of x will maximize its profit? How does the answer depend on c, the proportion of casual users in the population?

(c) Next, suppose Mictel produces just the high-end machines and sells them for price y. What value of y will maximize its profit, and how does the answer depend on c?

(d) Finally, suppose the company produces both types of machines, selling the low-end ones for price x and the high-end ones for price y. What are the incentive-compatibility constraints on x and y that the company must satisfy if it wants the casual users to buy the low-end machines and the intensive users to buy the high-end ones? What is the company's expected profit from this policy? What values of x and y will maximize the expected profit? How does the answer depend on c?

(e) Putting all of this together, what production and pricing policy should the company pursue? How does the answer depend on c?

5. Rosencrantz and Guildenstern pass the time on their voyage to England by playing the following game. Each makes an initial bet of 8 ducats. Then each separately tosses a fair coin (probability 1/2 each of heads and tails). Each sees the outcome of his own toss but not that of the other; Hamlet acts as the impartial referee and observes and records both outcomes to prevent cheating.

Then Rosencrantz decides whether to pass or to bet an additional 4 ducats. If he chooses to pass, the two coin tosses are compared. If the outcomes are different, the one with the heads collects the whole pot. The pot has 16 ducats, of which the winner himself contributed 8, so his winnings are 8 ducats; the loser's payoff is –8 ducats. If the outcomes are the same, the pot is split equally and each gets his 8 ducats back (a payoff of 0).

If Rosencrantz bets, then Guildenstern has to decide whether to concede, or to match with his own additional 4 ducats. If Guildenstern concedes, then Rosencrantz collects the pot irrespective of the numbers tossed. If Guildenstern matches, then the coin tosses are compared. The procedure is the same as that in the previous paragraph, with heads again the winner, but the pot is now bigger.

(a) Show the game in extensive form. (Be careful about information sets.)

(b) If the game is instead written in the normal form, Rosencrantz has four strategies: "pass always" (PP), "bet always" (BB), "bet if his own coin comes up heads/pass if tails" (BP), and "pass on heads/bet on tails" (PB). Similarly, Guildenstern has four strategies: "concede always" (CC), "match always" (MM), "match on heads/concede on tails" (MC), and the other way around (CM). Show that the table of payoffs to Rosencrantz is as follows:

		GUILDENSTERN			
		CC	MM	MC	CM
ROSEN-CRANTZ	PP	0	0	0	0
	BB	8	0	1	7
	BP	2	1	0	3
	PB	6	−1	1	4

(For each of the 16 cells, you will have to take an expected value by averaging over the consequences for each of the four possible combinations of the coin-toss outcomes.)

(c) Eliminate dominated strategies as far as possible. Find the mixed-strategy equilibrium in the remaining table, and the expected payoff to Rosencrantz in the equilibrium.

(d) Use your knowledge of the theory of signaling and screening to explain intuitively why the equilibrium has mixed strategies.

6. [For Discussion] The design of a health care system involves issues of information and strategy at several points. The users—potential and actual patients—have better information about their own state of health, lifestyle, and so forth, than the insurance companies can find out. The providers—doctors, hospitals, and so on—know more about what the patients need than do either the patients themselves or the insurance companies. Doctors also know more about their own skills and efforts, and hospitals about their own facilities. Insurance companies may have some statistical information about outcomes of treatments or surgical procedures from their past records. But outcomes are affected by many unobservable and random factors, so the underlying skills, efforts, or facilities cannot be inferred perfectly from observations of the outcomes. The pharmaceutical companies know more about the efficacy of drugs than do the others. As usual, the parties do not have natural incentives to share their information fully or accurately with others. The design of the overall scheme must try to face these issues and find the best feasible solutions.

From this strategic perspective, consider the relative merits of various payment schemes: fee for service versus capitation fees to doctors, comprehensive premiums per year versus payment for each visit for patients, and so on. Which are likely to be most beneficial to those seeking health care? To those providing health care? Think also about the relative merits of private insurance and coverage of costs from general tax revenues.

7. Michael Lewis, in reviewing Peter Robinson's memoir of becoming an M.B.A., writes:[13]

> Mr. Robinson pretty much concludes that business schools are a sifting device—M.B.A. degrees are union cards for yuppies. But perhaps the most important fact about the Stanford business school is that all meaningful sifting occurs before the first class begins. No messy weeding is done within the walls. "They don't want you to flunk. They want you to become a rich alum who'll give a lot of money to the school." But one wonders: If corporations are abdicating to the Stanford admissions office the responsibility for selecting young managers, why don't they simply replace their personnel departments with Stanford admissions officers, and eliminate the spurious education? Does the very act of throwing away a lot of money and two years of one's life demonstrate a commitment to business that employers find appealing?

What answer to Lewis's question can you give based on our analysis of strategies in situations of asymmetric information?

8. In a television commercial for a well-known brand of instant cappuccino, a gentleman is entertaining a lady friend at his apartment. He wants to impress her and offers her cappuccino with dessert. When she accepts, he goes into the kitchen to make the instant cappuccino—simultaneously tossing takeout boxes into the trash and faking the noises made by a high-class (and expensive) cappuccino maker. As he is doing this, a voice comes from the other room: "I want to see the machine. . . . "

Use your knowledge of games of asymmetric information to comment on the actions of these two people. Pay attention to their attempts to use signaling and screening, and point out specific instances of each strategy. Offer an opinion about which player is the better strategist.

9. [Optional] A teacher wants to find out how confident the students are about their own abilities. He proposes the following scheme: "After you answer this question, state your estimate of the probability that you are right. I will then check your answer to the question. Suppose you have given the probability estimate x. If your answer is actually correct, your grade will be $\log(x)$. If incorrect, it will be $\log(1 - x)$." Show that the scheme will elicit the students' own estimates truthfully; that is, if the truth is p, show that a student's stated estimate is $x = p$.

---

[13] Michael Lewis, "Boot Camp for Yuppies," review of *Snapshots from Hell: The Making of an MBA*, by Peter Robinson, *New York Times*, May 8, 1994; Book Review section.

PART FOUR

■

Applications to Specific Strategic Situations

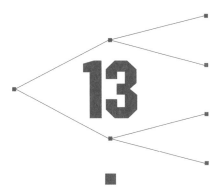

13

Brinkmanship: The Cuban Missile Crisis

I N CHAPTER 1, we explained that our basic approach was neither pure theory, nor pure case study, but a combination, where theoretical ideas were developed using features of particular cases or examples. Thus we ignored those aspects of each case that were incidental to the conceptual issue being developed. However, once you have learned the theoretical ideas, a richer mode of analysis becomes available to you, in which factual details of a particular case are more closely integrated with game-theoretic analysis to achieve a fuller understanding of what has happened and why. Such *theory-based case studies* have begun to appear in diverse fields—business, political science, and economic history.[1]

Here we offer an example from political and military history, namely nuclear brinkmanship in the Cuban missile crisis of 1962. Our choice is motivated by the sheer drama of the episode, the wealth of factual information that has become available recently, and the applicability of an important concept from game theory. You may think the risk of nuclear war died with the dissolution of the Soviet Union and that therefore our case is a historical curiosity. But nuclear arms races continue in many parts of the world, and rivals like India and Pakistan, or Iran and Israel, may find use for the lessons we take from the Cuban crisis. More importantly for many of you, brinkmanship must be practiced in many everyday

---

[1] Two excellent recent examples of theory-based case studies are Pankaj Ghemawat, *Games Businesses Play: Cases and Models* (Cambridge, Mass.: MIT Press, 1997), and Robert H. Bates, Avner Greif, Margaret Levi, Jean-Laurent Rosenthal, and Barry Weingast, *Analytic Narratives* (Princeton, N.J.: Princeton University Press), 1998.

situations, for example, strikes and marital disputes. Although the stakes in such games are lower than those in a nuclear confrontation between superpowers, the same principles of strategy apply.

In Chapter 9 we introduced the concept of brinkmanship as a strategic move; here is a quick reminder. A *threat* is a response rule, and the threatened action inflicts a cost on both the player making the threat and the player whose action the threat is intended to influence. However, if the threat succeeds in its purpose, this action is not actually carried out. Therefore there is no apparent upper limit to the cost of the threatened action. But the risk of *errors*—that is, the risk that the threat may fail to achieve its purpose, or the threatened action may occur by accident—forces the strategist to use the minimal threat that achieves its purpose. If a smaller threat is not naturally available, a large threat can be scaled down by making its fulfillment probabilistic. You do something in advance that creates a probability, but not certainty, that the mutually harmful outcome will happen if the opponent defies you. If the need actually arose, you would not take that bad action if you had the full freedom to choose. Therefore you must arrange in advance to let things get out of your control to some extent. *Brinkmanship* is the creation and deployment of such a probabilistic threat; it consists of a deliberate loss of control.

The word *brinkmanship* is often used in connection with nuclear weapons, and the Cuban missile crisis of 1962, when the world came as close to an unaccidental nuclear war as it ever has, is often offered as the classic example of brinkmanship. We will use the crisis as an extended case study to explicate the concept. In the process we will find that many popular interpretations and analyses of the crisis are simplistic. A deeper analysis reveals brinkmanship to be a subtle and dangerous strategy. It also shows that many conflicts in business and personal interactions—such as strikes and breakups of relationships—are examples of brinkmanship gone wrong. Therefore a clear understanding of the strategy, and of its limitations and risks, is very important to all game players, which includes just about everyone.

1 A BRIEF NARRATION OF EVENTS

We begin with a brief story of the unfolding of the crisis. Our account draws on several books, including some that were written recently with the benefit of documents and statements released since the collapse of the Soviet Union.[2] We can-

[2]Our sources include Robert Smith Thompson, *The Missiles of October* (New York: Simon & Schuster, 1992); James G. Blight and David A. Welch, *On the Brink: Americans and Soviets Reexamine the Cuban Missile Crisis* (New York: Hill and Wang, 1989); Richard Reeves, *President Kennedy—Profile of Power* (New York: Simon & Schuster, 1993); Donald Kagan, *On the Origins of War and the Preserva-*

not hope to do justice to the detail, let alone the drama, of the events. President Kennedy said at the time of the crisis: "This is the week when I earn my salary," and much more than a president's salary stood in the balance. We urge you to read the books that tell the story in vivid detail, and also to talk to any relatives who lived through it to get their firsthand memories.

In late summer and early fall of 1962, the Soviet Union (USSR) started to place medium- and intermediate-range ballistic missiles (MRBMs and IRBMs) in Cuba. The MRBMs had a range of 1,100 miles and could hit Washington; the IRBMs, with a range of 2,200 miles, could hit most of the major U.S. cities and military installations. The missile sites were guarded by the latest Soviet SA-2-type surface-to-air missiles (SAMs), which could shoot down U.S. high-altitude U-2 reconnaissance planes. There were also IL-28 bombers, and tactical nuclear weapons called Luna by the Soviets and FROG (free rocket over ground) by the United States, which could be used against invading troops.

This was the first time the Soviets had ever attempted to place their missiles and nuclear weapons outside Soviet territory. Had they been successful, it would have increased their offensive capability against the United States manyfold. It is now believed that the Soviets had less than 20, and perhaps as few as "two or three," operational intercontinental ballistic missiles (ICBMs) in their own country capable of reaching the United States (*War*, 464, 509–510). Their initial placement in Cuba had about 40 MRBMs and IRBMs, which was a substantial increase. But the United States would still have retained vast superiority in the nuclear balance between the superpowers. Also, as the Soviets built up their submarine fleet, the relative importance of land-based missiles near the United States would have decreased. But the missiles had more than mere direct military value to the Soviets. Successful placement of missiles so close to the United States would have been an immense boost to Soviet prestige around the world, especially in Asia and Africa, where the superpowers were competing for political and military influence. Finally, the Soviets had come to think of Cuba as a "poster child" for socialism. The opportunity to deter a feared U.S. invasion of Cuba and to counter Chinese influence in Cuba weighed importantly in the cal-

---

tion of Peace (New York: Doubleday, 1995); Aleksandr Fursenko and Timothy Naftali, *One Hell of a Gamble: The Secret History of the Cuban Missile Crisis* (New York: Norton, 1997); and last, latest, and most direct, *The Kennedy Tapes: Inside the White House During the Cuban Missile Crisis*, ed. Ernest R. May and Philip D. Zelikow (Cambridge, Mass.: Harvard University Press, 1997). Graham T. Allison's *Essence of Decision: Explaining the Cuban Missile Crisis* (Boston: Little Brown, 1971) remains important not only for its narrative, but also for its analysis and interpretation. Our view differs from his in some important respects, but we remain in debt to his insights. We follow and extend the ideas in Avinash Dixit and Barry Nalebuff, *Thinking Strategically* (New York: Norton, 1991), chapter 8.

When we cite these sources to document particular points, we do so in parentheses in the text, in each case using a key word from the title of the book, and the appropriate page number or page range. In the list above, the key words have been underlined.

culations of the Soviet leader and Premier, Nikita Khrushchev. (See *Gamble*, 182–183 for an analysis of Soviet motives.)

U.S. surveillance of Cuba and of shipping lanes during the late summer and early fall of 1962 had indicated some suspicious activity. When questioned about this by U.S. diplomats, the Soviets denied any intentions to place missiles in Cuba. Later, faced with irrefutable evidence, they said their intention was defensive, to deter the United States from invading Cuba. It is hard to believe this, although we know that an offensive weapon *can* serve as a defensive deterrent threat.

An American U-2 "spy plane" took photographs over western Cuba on Sunday and Monday, October 14 and 15. When developed and interpreted, these showed unmistakable signs of construction of MRBM launching sites. (Evidence of IRBMs was found later, on October 17.) These photographs were shown to President Kennedy the following day (October 16). He immediately convened an ad hoc group of advisers, which later came to be called the Executive Committee of the National Security Council (ExComm), to discuss the alternatives. At the first meeting (on the morning of October 16), he decided to keep the matter totally secret until he was ready to act, mainly because if the Soviets knew that the Americans knew, they might speed up the installation and deployment of the missiles before the Americans were ready to act, but also because spreading the news without announcing a clear response would create panic in the United States.

Members of ExComm who figured most prominently in the discussions were the Secretary of Defense, Robert McNamara; the National Security Adviser, McGeorge Bundy; the Chairman of the Joint Chiefs of Staff, General Maxwell Taylor; the Secretary of State, Dean Rusk, and Undersecretary George Ball; the Attorney General Robert Kennedy (who was also the President's brother); the Secretary of the Treasury, Douglas Dillon (also the only Republican in the Cabinet); and Llewellyn Thompson, who had recently returned from being U.S. Ambassador in Moscow. During the two weeks that followed, they would be joined by, or would consult with, several others, including the U.S. Ambassador to the United Nations, Adlai Stevenson; the former Secretary of State and a senior statesman of U.S. foreign policy, Dean Acheson; and the Chief of the U.S. Air Force, General Curtis Lemay.

During the rest of that week (October 16 to 21) the ExComm met numerous times. To preserve secrecy, the President continued his normal schedule, including travel to speak for Democratic candidates in the midterm Congressional elections that were to be held in November 1962. Of course, he kept in constant touch with ExComm. He dodged press questions about Cuba, and persuaded one or two trusted media owners or editors to preserve the facade of business as usual. ExComm's own attempts to preserve secrecy in Washington sometimes verged on the comic, as when once almost a dozen of them had to pile into one limo, because the sight of several government cars going from the White House to the State Department in a convoy could cause speculation in the media.

Different members of ExComm had widely differing assessments of the situation and supported different actions. The military Chiefs of Staff thought that the missile placement changed the balance of military power substantially; Defense Secretary McNamara thought it changed "not at all" but regarded the problem as politically important nonetheless (*Tapes*, 89). President Kennedy pointed out that the first placement, if ignored by the United States, could grow into something much bigger, and that the Soviets could use the threat of missiles so close to the United States to try to force the withdrawal of the U.S., British, and French presence in West Berlin. Kennedy was also aware that this was a part of the *geopolitical* struggle between the United States and the Soviet Union (*Tapes*, 92).

It now appears that he was very much on the mark in this assessment. The Soviets planned to expand their presence in Cuba into a major military base (*Tapes*, 677). They expected to complete the missile placement by mid-November. Khrushchev had planned to sign a treaty with Castro in late November, then travel to New York to address the United Nations, and issue an ultimatum for a settlement of the Berlin issue (*Tapes*, 679; *Gamble*, 182), using the missiles in Cuba as a threat for this purpose. Khrushchev thought Kennedy would accept the missile placement as a *fait accompli*. It appears Khrushchev made these plans on his own. Some of his top advisers privately thought them too adventurous, but the top governmental decision-making body of the Soviet Union, the Presidium, supported him, although its response was largely a rubber stamp (*Gamble*, 180). Castro was at first reluctant to accept the missiles, fearing they would trigger a U.S. invasion (*Tapes*, 676–678), but in the end he too accepted them, and the prospect gave him great confidence and lent some swagger to his statements about the United States (*Gamble*, 186–187, 229–230).

In all ExComm meetings up to and including the one on the morning of Thursday, October 18, everyone appears to have assumed that the U.S. response would be purely military. The only options they discussed seriously during this time were (1) an air strike directed exclusively at the missile sites and (probably) the SAM sites nearby, (2) a wider air strike including Soviet and Cuban aircraft parked at airfields, and (3) a full-scale invasion of Cuba. If anything, the attitudes hardened when the evidence of the presence of the longer-range IRBMs arrived. In fact, at the Thursday meeting, Kennedy discussed a timetable for air strikes to commence that weekend (*Tapes*, 148).

McNamara had first mentioned a blockade toward the end of the meeting on Tuesday, October 16, and developed the idea (in a form uncannily close to the course of action actually taken) in a small group after the formal meeting had ended (*Tapes*, 86, 113). Ball argued that an air strike without warning would be a "Pearl Harbor," and that the United States should not do this (*Tapes*, 115); he was most importantly supported by Robert Kennedy (*Tapes,* 149). The civilian members of ExComm further shifted toward the blockade option when they found that what the military Joint Chiefs of Staff wanted was a massive air strike; the military regarded a limited strike aimed at only the missile sites so dangerous and ineffective that "they would prefer taking no military action than to take that limited strike" (*Tapes,* 97).

Between October 18 and Saturday, October 20, the majority opinion within ExComm gradually coalesced around the idea of starting with a blockade, simultaneously issuing an ultimatum with a short deadline (48 to 72 hours was mentioned), and proceeding to military action if necessary after this deadline expired. International law required a declaration of war in order to set up a blockade, but this problem was ingeniously resolved by proposing to call it a "naval quarantine" of Cuba (*Tapes,* 190–196).

Some people held the same positions throughout these discussions (from October 16 to 21)—for example, the military Chiefs of Staff constantly favored a major air strike—but others shifted their views, at times dramatically. Bundy initially favored doing nothing (*Tapes,* 172), then switched toward a preemptive surprise air attack (*Tapes,* 189). President Kennedy's own position also shifted away from an air strike toward a blockade. He wanted the U.S. response to be firm. While his reasons undoubtedly were mainly military and geopolitical, as a good domestic politician he was also fully aware that a weak response would hurt the Democratic party in the imminent Congressional elections. On the other hand, the responsibility of starting an action that might lead to nuclear war weighed very heavily on him. He was impressed by the CIA's assessment that some of the missiles were already operational, which increased the risk that any air strike or invasion could lead to the Soviets' firing these missiles and to large U.S. civilian casualties (*Gamble,* 235). In the second week of the crisis (October 22 through 28), his decisions seemed to consistently favor the lowest-key options discussed by ExComm.

By the end of the first week's discussions, the choice lay between a blockade and an air strike, two position papers were prepared, and in a straw vote on October 20 the blockade won 11 to 6 (*War,* 516). Kennedy made the decision to start by imposing a blockade, and announced it in a television address to the nation on Monday, October 22. He demanded a halt to the shipment of Soviet missiles to Cuba, and a prompt withdrawal of those already there.

Kennedy's speech brought the whole drama and tension into the public arena. The United Nations held several dramatic but unproductive debates. Other world leaders and the usual busybodies of international affairs offered advice and mediation.

Between October 23 and 25 the Soviets at first tried bluster and denial; Khrushchev called the blockade "banditry, a folly of international imperialism" and said his ships would ignore it. The Soviets, in the United Nations and elsewhere, claimed their intentions were purely defensive, and also issued statements of defiance. In secret, they explored ways to end the crisis. This included some direct messages from Khrushchev to Kennedy. It also included some very indirect and lower-level approaches by the Soviets. In fact, as early as Monday, October 22—before Kennedy's TV address—the Soviet Presidium had decided not to let this crisis lead to war. By Thursday, October 25, they had decided they were willing to withdraw from Cuba in exchange for a promise by the United States not to invade Cuba, but they had also agreed to "look around" for better deals (*Gamble*, 241, 259). Of course, the United States did not know any of the Soviet thinking about this.

In public as well as in private communications, the USSR broached the possibility of a deal involving withdrawal of U.S. missiles from Turkey and Soviet ones from Cuba. This possibility had already been discussed by ExComm. The missiles in Turkey were obsolete, so the United States wanted to remove them anyway and replace them with a Polaris submarine stationed in the Mediterranean Sea. But it was thought that the Turks would regard the presence of U.S. missiles as a matter of prestige and so it might be difficult to persuade them to accept the change. (The Turks might also correctly regard missiles, fixed on Turkish soil, as a firmer signal of the U.S. commitment to Turkey's defense than an offshore submarine, which could move away on short notice; see *Tapes*, 568.)

The blockade went into effect on Wednesday, October 24. Despite their public bluster, the Soviets were cautious in testing it. It appears they were surprised that the United States had discovered the missiles in Cuba before the whole installation program was completed; Soviet personnel in Cuba had observed the U-2 overflights but not reported them back to Moscow (*Tapes*, 681). The Soviet Presidium ordered the ships carrying the most sensitive materials (actually the IRBM missiles) to stop or turn around. But it also ordered General Issa Pliyev, the commander of the Soviet troops in Cuba, to get his troops combat-ready, and to use all means except nuclear weapons to meet any attack (*Tapes*, 682). In fact, the Presidium twice prepared, and then canceled without sending, orders authorizing him to use tactical nuclear weapons in the case of a U.S. invasion (*Gamble*, 242–243, 272, 276). Of course, the U.S. side saw only that several Soviet ships (which were actually carrying oil and other nonmilitary cargo) continued to sail toward the blockade zone. The U.S. Navy showed some moderation in its enforcement of the blockade. A tanker was allowed to pass without being boarded; another tramp steamer carrying industrial cargo was boarded but allowed to proceed after only a cursory inspection. But tension was mounting, and neither side's actions were as cautious as the top-level politicians on both sides would have liked.

On the morning of Friday, October 26, Khrushchev sent Kennedy a conciliatory private letter offering to withdraw the missiles in exchange for a U.S. promise not to invade Cuba. But later that day he toughened his stance. It seems he was emboldened by two items of evidence. First, the U.S. Navy was not being excessively aggressive in enforcing the blockade. They had let through some obviously civilian freighters; they boarded only one ship, the *Marucla*, and let it pass after a cursory inspection. Second, some dovish statements had appeared in U.S. newspapers. Most notable among these was an article by the influential and well-connected syndicated columnist Walter Lippmann, who suggested the swap whereby the United States would withdraw its missiles in Turkey in exchange for the USSR's withdrawing its missiles in Cuba (*Gamble*, 275). Khrushchev sent another letter to Kennedy on Saturday, October 26, offering this swap, and this time he made the letter public. The new letter was presumably a part of the Presidium's strategy of "looking around" for the best deal. Members of ExComm concluded that the first letter was Khrushchev's own thoughts, while the second was written under pressure from hard-liners in the Presidium—or even that Khrushchev was no longer in control (*Tapes*, 498, 512–513). In fact, both of Khrushchev's letters were discussed in and approved by the Presidium (*Gamble*, 263, 275).

ExComm continued to meet, and opinions within it hardened. One reason was the growing feeling that the blockade by itself would not work. Kennedy's television speech had imposed no firm deadline, and, as we know, in the absence of a deadline a compellent threat is vulnerable to the opponent's procrastination. Kennedy had seen this quite clearly and as early as Monday, October 22: in the morning ExComm meeting preceding his speech he commented, "I don't think we're gonna be better off if they're just sitting there" (*Tapes*, 216). But a hard, short deadline was presumably thought to be too rigid. By Thursday, others in ExComm were realizing the problem; for example, Bundy said, "A plateau here is the most dangerous thing" (*Tapes*, 423). The hardening of the Soviet position, as shown by the public "Saturday letter" that followed the conciliatory private "Friday letter," was another concern. More ominously, that Friday, U.S. surveillance had discovered there were tactical nuclear weapons (FROGs) in Cuba (*Tapes*, 475). This showed the Soviet presence there to be vastly greater than thought before, but it also made invasion more dangerous to U.S. troops. Also on Saturday, a U.S. U-2 plane was shot down over Cuba. (It now appears this was done by the local commander, who interpreted his orders more broadly than Moscow had intended (*War*, 537; *Tapes*, 682). In addition, Cuban antiaircraft defenses fired at low-level U.S. reconnaissance planes. The grim mood in ExComm throughout that Saturday was well encapsulated by Dillon: "We haven't got but one more day" (*Tapes*, 534).

On Saturday, plans leading to escalation were being put in place. An air strike was planned for the following Monday, or Tuesday at the latest, and Air Force reserves were called up (*Tapes*, 612–613). Invasion was seen as the inevitable culmination of events (*Tapes*, 537–538). A tough private letter to Khrushchev from President Kennedy was drafted, and was handed over by Robert Kennedy to the

Soviet Ambassador in Washington, Anatoly Dobrynin. In it, Kennedy made the following offer: (1) The Soviet Union withdraws its missiles and IL-28 bombers from Cuba with adequate verification (and ships no new ones). (2) The United States promises not to invade Cuba. (3) The U.S. missiles in Turkey will be removed after a few months, but this offer is void if the Soviets mention it in public or link it to the Cuban deal. An answer was required within 12 to 24 hours; otherwise "there would be drastic consequences" (*Tapes*, 605–607).

On the morning of Sunday, October 28, just as prayers and sermons for peace were being offered in many churches in the United States, Soviet radio broadcast the text of a letter that Khrushchev was sending to Kennedy, in which he announced that construction of the missile sites was being halted immediately, and that the missiles already installed would be dismantled and shipped back to the Soviet Union. Kennedy immediately sent a reply welcoming this decision, which was broadcast to Moscow by the Voice of America radio. It now appears that Khrushchev's decision to back down was made before he received Kennedy's letter via Dobrynin, but that the letter only reinforced it (*Tapes*, 689).

That did not quite end the crisis. The U.S. Joint Chiefs of Staff remained skeptical of the Soviets and wanted to go ahead with their air strike (*Tapes*, 635). In fact, the construction activity at the Cuban missile sites continued for a few days. Verification by the United Nations proved problematic. The Soviets tried to make the Turkey part of the deal semipublic. They also tried to keep the IL-28 bombers in Cuba out of the withdrawal. It was not until November 20 that the deal was finally clinched and the withdrawal began (*Tapes*, 663–665; *Gamble*, 298–310).

2 A SIMPLE GAME-THEORETIC EXPLANATION

At first sight the game-theoretic aspect of the crisis looks very simple. The United States wanted the Soviet Union to withdraw its missiles from Cuba; thus the U.S. objective was to achieve compellence. For this purpose, the United States deployed a threat: Soviet failure to comply would eventually lead to a nuclear war between the superpowers. The blockade was a starting point of this inevitable process, and an actual action that demonstrated the credibility of U.S. resolve. In other words, Kennedy took Khrushchev to the brink of disaster. This was sufficiently frightening to Khrushchev that he complied. Of course, the prospect of nuclear annihilation was equally frightening to Kennedy, but that is in the nature of a threat. All that is needed is that the threat be sufficiently costly to the other side to induce them to act in accordance with our wishes; then we don't have to carry out the bad action anyway.

A somewhat more formal statement of this argument proceeds by drawing a game tree like that shown in Figure 13.1. The Soviets have installed the missiles, and now the United States has the first move. It chooses between doing nothing and issuing a threat. If the United States does nothing, this is a major military and political achievement for the Soviets, so we score the payoffs as −2 for the United States and 2 for the Soviets. If the United States issues its threat, the Soviets get to move, and they can either withdraw or defy. Withdrawal is a humiliation (a substantial minus) for the Soviets and a reaffirmation of U.S. military superiority (a small plus), so we score it 1 for the United States and −4 for the Soviets. If the Soviets defy the U.S. threat, there will be a nuclear war. This is terrible for both, but particularly bad for the United States, which as a democracy cares more for its citizens, so we score this −10 for the United States and −8 for the Soviets. This quantification is very rough guesswork, but the conclusions do not depend on the precise numbers we have chosen. If you disagree with our choice, you can substitute other numbers you think to be a more accurate representation; as long as the *relative* ranking of the outcomes is the same, you will get the same subgame-perfect equilibrium.

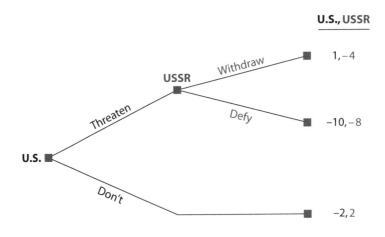

FIGURE 13.1 The Simple Threat Model of the Crisis

Now we can easily find the subgame-perfect equilibrium. If faced with the U.S. threat, the Soviets get −4 from withdrawal and −8 by defiance, so they prefer to withdraw. Looking ahead to this, the United States reckons on getting 1 if it issues the threat and −2 if it does not; therefore it is optimal for the United States to make the threat. The outcome gives payoffs of 1 to the United States and −4 to the Soviets.

But a moment's further thought shows this interpretation to be unsatisfactory. One might start by asking why the Soviets would deploy the missiles in Cuba at all, when they could look ahead to this unfolding of the subsequent game in which they would come out the losers. But more importantly, several facts about the situation and several events during its unfolding do not fit into this picture of a simple threat.

Before explaining the shortcomings of this analysis and developing a better explanation, however, we digress to an interesting episode in the crisis that sheds light on the requirements of successful compellence. As we pointed out in Chapter 9, a compellent threat must have a deadline; otherwise the opponent can nullify it by procrastination. The discussion of the crisis at the U.N. Security Council on Tuesday, October 23, featured a confrontation between U.S. ambassador Adlai Stevenson and Soviet ambassador Valerian Zorin. Stevenson asked Zorin point-blank whether the USSR had placed and was placing nuclear missiles in Cuba. "Yes or no—don't wait for the translation—yes or no?" he insisted. Zorin replied: "I am not in an American courtroom. . . . You will have your answer in due course," to which Stevenson retorted, "I am prepared to wait for my answer until hell freezes over." This was dramatic debating; Kennedy watching the session on live television remarked "Terrific. I never knew Adlai had it in him" (*Profile*, 406). But it was terrible strategy. Nothing would have suited the Soviets better than to keep the Americans "waiting for their answer" while they went on completing the missile sites. "Until hell freezes over" is an unsuitable deadline for compellence.

3 ACCOUNTING FOR ADDITIONAL COMPLEXITIES

Let us return to developing a more satisfactory game-theoretic argument. As we pointed out before, the idea that a threat has only a lower limit on its size, namely that it be large enough to frighten the opponent, is correct only if the threatener can be absolutely sure that everything will go as planned. But almost all games have some element of uncertainty. You cannot know your opponent's value system for sure, and you cannot be completely sure that the players' intended actions will be accurately implemented. Therefore a threat carries a twofold risk. Your opponent may defy it, requiring you to carry out the costly threatened action; or your opponent may comply, but the threatened action may occur by mistake anyway. When such risks exist, the cost of the threatened action to oneself becomes an important consideration.

The Cuban missile crisis was replete with such uncertainties. Neither side could be sure of the other's payoffs—that is, of how seriously the other regarded the relative costs of war and of losing prestige in the world. Also, the choices of "blockade" and "air strike" were much more complex than the simple phrases suggest, and there were many weak links and random effects between an order in Washington or Moscow and its implementation in the Atlantic Ocean or in Cuba.

Graham Allison's excellent book, *Essence of Decision*, brings out all of these complexities and uncertainties. They led him to conclude that the Cuban missile crisis cannot be explained in game-theoretic terms. He considers two alternatives, one explanation based on the fact that bureaucracies have their set rules and procedures, and another based on the internal politics of U.S. and Soviet governance and military apparatuses. He concludes that the political explanation is best.

We broadly agree, but interpret the Cuban missile crisis differently. It is not the case that game theory is inadequate for understanding and explaining the crisis; rather, the crisis was *not a two-person game*—United States versus USSR, or Kennedy versus Khrushchev. Each of these two "sides" was itself a complex coalition of players, with differing objectives, information, actions, and means of communication. The players within each side were engaged in other games, and some members were also directly interacting with their counterparts on the other side. In other words, the crisis can be seen as a complex many-person game with alignments into two broad coalitions. Kennedy and Khrushchev can be regarded as the top-level players in this game, but each was subject to constraints of having to deal with others in his own coalition with divergent views and information, and neither had full control over the actions of these others. We argue that this more subtle game-theoretic perspective is not only a good way to look at the crisis, but also essential in understanding how to practice brinkmanship.

We begin with some items of evidence that Allison emphasizes, and others that emerge from other writings.

First, there are several indications of divisions of opinion on each side. On the U.S. side, as already noted, there were wide differences within ExComm. In addition, Kennedy found it necessary to consult others such as former President Eisenhower and leading members of Congress. Some of them had very different views; for example, Senator Fulbright said in a private meeting that the blockade "seems to me the worst alternative" (*Tapes*, 271). The media and the political opposition would not give the President unquestioning support for too long either. Kennedy could not have continued on a moderate course if the opinion among his advisers and the public became decisively hawkish.

Individual people also *shifted* positions during the two weeks. For example, McNamara was at first quite dovish, arguing that the missiles in Cuba were not a significant increase in the Soviet threat (*Tapes*, 89) and favoring blockade and negotiations (*Tapes*, 191), and ended up more hawkish, arguing that Khrushchev's conciliatory letter of Friday, October 26, was "full of holes" (*Tapes*, 495, 585) and arguing for an invasion (*Tapes*, 537). Most importantly, the U.S. military chiefs always advocated a far more aggressive response. Even after the crisis was over and everyone thought the United States had won a major round in the cold war, Air Force General Curtis Lemay remained dissatisfied and wanted action: "We lost!

We ought to just go in there today and knock 'em off," he said. (*Essence*, 206; *Profile*, 425).

Even though Khrushchev was the dictator of the Soviet Union, he was not in full control of the situation. Differences of opinion on the Soviet side are less well documented, but for what it is worth, later memoirists have claimed that Khrushchev made the decision to install the missiles in Cuba almost unilaterally, and when he informed them, they thought it a reckless gamble (*Tapes*, 674; *Gamble*, 180). There were limits to how far he could count on the Presidium to rubber-stamp his decisions. Indeed, two years later the disastrous Cuban adventure was one of the main charges leveled against Khrushchev when the Presidium dismissed him, (*Gamble*, 353–355). It has also been claimed that Khrushchev wanted to defy the U.S. blockade, and only the insistence of First Deputy Premier Anastas Mikoyan led to the cautious response (*War*, 521). Finally, on Saturday, October 27, Castro ordered his antiaircraft forces to fire on all U.S. planes overflying Cuba, and refused the Soviet ambassador's request to rescind the order (*War*, 544).

Various parties on the U.S. side had very different information and a very different understanding of the situation, and at times this led to actions that were inconsistent with the intentions of the leadership or even against their explicit orders. The concept of an "air strike" to destroy the missiles is a good example. The nonmilitary people in ExComm thought this would be very narrowly targeted and would not cause significant Cuban or Soviet casualties, but the Air Force intended a much broader attack. Luckily this difference came out in the open early, leading ExComm to decide against an air strike and the President to turn down an appeal by the Air Force (*Essence*, 123, 209). As for the blockade, the U.S. Navy had set procedures for this action. The political leadership wanted a different and softer process: form the ring closer to Cuba to give the Soviets more time to reconsider, allow the obviously nonmilitary cargo ships to pass unchallenged, and cripple but not sink the ships that defy challenge. Despite McNamara's explicit instructions, however, the Navy mostly followed its standard procedures (*Essence*, 130–132). The U.S. Air Force created even greater dangers. A U-2 plane drifted "accidentally" into Soviet air space and almost caused a serious setback. General Curtis Lemay, acting without the President's knowledge or authorization, ordered the Strategic Air Command's nuclear bombers to fly past their "turnaround" points and some distance toward Soviet air space to positions where they would be detected by Soviet radar. Fortunately, the Soviets responded calmly; Khrushchev merely protested to Kennedy.[3]

There was similar lack of information and communication, and weakness of

[3] Richard Rhodes, *Dark Sun: The Making of the Hydrogen Bomb* (New York: Simon & Schuster, 1995), pp. 573–575. Lemay, renowned for his extreme views and his constant chewing of large unlit cigars, is supposed to be the original inspiration, in the 1963 movie *Dr. Strangelove*, for General Jack D. Ripper, who orders his bomber wing to launch an unprovoked attack on the Soviet Union.

the chain of command and control, on the Soviet side. For example, the construction of the missiles was left to the standard bureaucratic procedures. The Soviets, used to construction of ICBM sites in their own country where they did not face significant risk of air attack, laid out the sites in Cuba in a similar way, where they would have been much more vulnerable. At the height of the crisis, when the Soviet SA-2 troops saw an overflying U.S. U-2 plane on Friday, October 26, Pliyev was temporarily away from his desk and his deputy gave the order to shoot it down; this created far more risk than Moscow would have wished (*Gamble*, 277–278). And at numerous other points—for example, when the U.S. Navy was trying to get the freighter *Marucla* to stop and be boarded—the people involved might have set off an incident with alarming consequences by taking some action in fear of the immediate situation.

All these factors made the outcome of any decision by the top-level commander on each side somewhat *unpredictable*. This gave rise to a substantial risk of the "threat going wrong." In fact, Kennedy thought that the chances of the blockade leading to war were "between one out of three and even" (*Essence*, 1).

As we pointed out, such uncertainty can make a simple threat too large to be acceptable to the threatener. We will take one particular form of the uncertainty, namely U.S. lack of knowledge of the Soviets' true motives, and analyze its effect formally, but similar conclusions hold for all other forms of uncertainty.

Reconsider the game shown in Figure 13.1. Suppose the Soviet payoffs from withdrawal and defiance are the opposite of what they were before: −8 for withdrawal and −4 for defiance. In this alternative scenario, the Soviets are hardliners. They prefer nuclear annihilation to the prospect of a humiliating withdrawal and the prospect of living in a world dominated by the capitalist United States; their slogan is "Better dead than red-white-and-blue." We show the game tree for this case in Figure 13.2. Now if the United States makes the threat, the Soviets defy it. So the United States stands to get −10 from the threat, and only −2 if it makes no threat and accepts the presence of the missiles in Cuba. It takes the lesser of the two evils. In the subgame-perfect equilibrium of this version of the game, the Soviets "win" and the U.S. threat does not work.

In reality, when the United States makes its move, it does not know whether the Soviets are hard-liners, as in Figure 13.2, or softer, as in Figure 13.1. The United States can try to estimate the probabilities of the two scenarios, for example, by studying past Soviet actions and reactions in different situations. We can regard Kennedy's statement that the probability of the blockade leading to war was between one-third and one-half this estimate of the probability that the Soviets are hard-line. Since the estimate is imprecise over a range, we work with a general symbol, p, for the probability, and examine the consequences of different values of p.

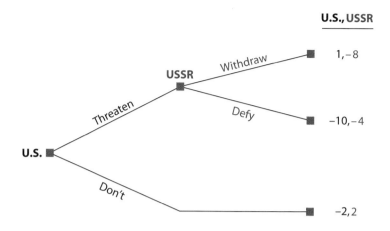

FIGURE 13.2 The Game with Hard-line Soviets

The tree for this more complex game is shown in Figure 13.3. The game starts with an outside force (here labeled "Nature") determining the Soviets' type. Along the upper branch of Nature's choice, the Soviets are hard-line. This leads to the upper node, where the United States makes its decision whether to issue its threat, and the rest of the tree is exactly like the game in Figure 13.2. Along the lower branch of Nature's choice, the Soviets are soft. This leads to the lower node, where the United States makes its decision whether to issue its threat, and the rest of the tree is exactly like the game in Figure 13.1. But the United States does not know from which node it is making its choice. Therefore the two U.S. nodes are enclosed in an "information set." Its significance is that the United States cannot take different actions at the nodes within the set, such as issuing the threat only if the Soviets are soft. It must take the same action at both nodes, either threatening at both nodes or not threatening at both. It must make this decision in the light of the probabilities that the game might in truth be "located"at the one node or the other, that is, by calculating the *expected* payoffs of the two actions.

Of course, the Soviets themselves know what type they are. So we can do some rollback near the end of the game. Along the upper path, the hard-line Soviets will defy a U.S. threat, and along the lower path, the soft Soviets will withdraw in the face of the threat. Therefore the United States can look ahead and calculate that a threat will get a −10 if the game is actually moving along the upper path (a probability of p), and a 1 if it is moving along the lower path (a probability of $1 - p$). The expected U.S. payoff from making the threat is therefore $-10p + (1 - p) = 1 - 11p$.

If the United States does not make the threat, it gets a −2 along either path, so its expected payoff is also −2. Comparing the expected payoffs of the two actions, we see that the United States should make the threat if $1 - 11p > -2$, or $11p < 3$, or $p < 3/11 = 0.27$.

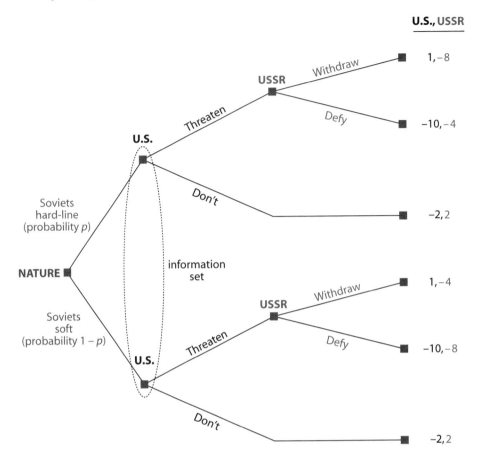

FIGURE 13.3 The Threat with Unknown Soviet Payoffs

If the threat were sure to work, the United States would not care how bad its payoff could be if the Soviets defied it, whether −10 or even far more negative. But the risk that the Soviets might be hard-liners and thus defy a threat makes the −10 relevant in the U.S. calculations. Only if the probability, p, of the Soviets' being hard-line is small enough will the United States find it acceptable to make the threat. Thus the upper limit of 3/11 on p is also the upper limit of this U.S. tolerance, given the specific numbers we have chosen. If we choose different numbers, we will get a different upper limit; for example, if we rate a nuclear war as −100 for the United States, then the upper limit on p will be only 3/101. But the idea of a large threat being "too large to make" if the probability of its going wrong is above a critical limit holds in general.

In this instance, Kennedy's estimate was that p lay somewhere in the range from 1/3 to 1/2. The lower end of this range, 0.33, is unfortunately just above our upper limit 0.27 for the risk the United States is willing to tolerate. Therefore the simple bald threat "if you defy us, there will be nuclear war" is too large, too risky, and too costly for the United States to make.

4 A PROBABILISTIC THREAT

If an outright threat of war is too large to be tolerable, and if you cannot find another, naturally smaller threat, then you can reduce the threat by creating merely a probability rather than a certainty that the dire consequences for the other side will occur if it does not comply. However, this does not mean that you decide after the fact whether to take the drastic action. If you had that freedom, you would shirk from the terrible consequences, your opponents would know or assume this, and so the threat would not be credible in the first place. You must relinquish some freedom of action and make a credible commitment. In this case, you must commit to a probabilistic device.

When making a *simple threat*, one player says to the other player: "If you don't comply, something will *surely* happen that will be very bad for you. By the way, it will also be bad for me, but my threat is credible because of my reputation [or through delegation, or other reasons]." With a **probabilistic threat,** one player says to the other, "If you don't comply, there is a *risk* that something very bad for you will happen. By the way, it will also be very bad for me, but later I will be powerless to reduce that risk."

Metaphorically, a probabilistic threat of war is a kind of Russian roulette (an appropriate name in this context). You load a bullet into one chamber of a revolver and spin the barrel. The bullet acts as a "detonator" of the mutually costly war. When you pull the trigger, you do not know whether the chamber in the firing path is loaded. If it is, you may wish you had not pulled the trigger, but by then it will be too late. Before the fact, you would not pull the trigger if you knew the bullet was in that chamber (that is, if the certainty of the dire action were too costly), but you are willing to pull the trigger knowing that there is only a 1 in 6 chance—in which the threat has been reduced by a factor of 6, to a point where it is now tolerable.

Brinkmanship is the creation and control of a suitable risk of this kind. It requires two apparently inconsistent things: On the one hand, you must let matters get enough out of your control that you will not have full freedom after the fact to refrain from taking the dire action, and so your threat will remain credible. On the other, you must retain sufficient control to keep the risk of the action from becoming too large and so your threat becomes too costly. Such "controlled lack of control" looks difficult to achieve, and it is. We will discuss in Section 5 how the trick can be performed. Just one hint: all the complex differences of judgment, the dispersal of information, and the difficulties of enforcing orders, which made a simple threat too risky, are exactly the forces that make it possible to create a risk of war and therefore make brinkmanship credible. The real difficulty is not how to lose control, but how to do so in a controlled way.

We first focus on the mechanics of brinkmanship. For this purpose we slightly alter the game of Figure 13.3 to get Figure 13.4. Here we introduce a different kind of U.S. threat. It consists of choosing and fixing a probability, q, such that, if the Soviets defy the United States, war will occur with that probability. With the remaining probability, $(1 - q)$, the United States will give up and agree to accept the Soviet missiles in Cuba. Remember that if the game gets to the point where the Soviets defy the United States, the latter does not have a choice in the matter. The Russian-roulette revolver has been set for the probability, q, and chance determines whether the firing pin hits a loaded chamber (that is, whether nuclear war actually happens).

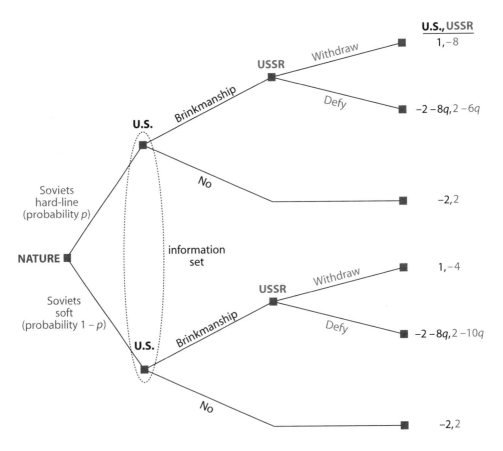

FIGURE 13.4 The Brinkmanship Model of the Crisis

Thus nobody knows the precise outcome and payoffs that will result if the Soviets defy this brinkmanship threat, but they know the probability, q, and can calculate expected values. For the United States, the outcome is -10 with the probability q and -2 with the probability $(1 - q)$, so the expected value is

$$-10q - 2(1 - q) = -2 - 8q.$$

For the Soviets, the expected payoff depends on whether they are hard-line or soft (and only they know their own type). If hard-line, they get a –4 from war, which happens with probability q, and a 2 if the United States gives up, which happens with the probability $(1 - q)$. The Soviets' expected payoff is $-4q + 2(1 - q)$ $= 2 - 6q$. If they were to withdraw, they would get a –8, which is clearly worse no matter what value q takes between 0 and 1. Thus the hard-line Soviets will defy the brinkmanship threat.

The calculation is different if the Soviets are soft. Reasoning as before, we see that they get the expected payoff $-8q + 2(1 - q) = 2 - 10q$ from defiance, and the sure payoff –4 if they withdraw. For them, withdrawal is better if $-4 > 2 - 10q$, or $10q > 6$, or $q > 0.6$. Thus U.S. brinkmanship must contain at least a 60% probability of war, otherwise it will not deter the Soviets, even if they are the soft type. We call this lower bound on the probability q the **effectiveness condition.**

Observe how the expected payoffs for U.S. brinkmanship and Soviet defiance shown in Figure 13.4 relate to the simple-threat model of Figure 13.3; the latter can now be thought of as a special case of the general brinkmanship-threat model of Figure 13.4, corresponding to the extreme value $q = 1$.

We can solve the game shown in Figure 13.4 in the usual way. We have already seen that along the upper path the Soviets, being hard-line, will defy the United States, and that along the lower path the soft Soviets will comply with U.S. demands if the effectiveness condition is satisfied. If this condition is not satisfied, then both types of Soviets will defy the United States, so the latter would do better never to make this threat at all. So let us proceed assuming that the soft Soviets will comply, and look at the U.S. choices. This is basically the question of how risky the U.S. threat can be and still remain tolerable to the United States.

If the United States makes the threat, it runs the risk, p, that it will encounter the hard-line Soviets, who will defy the threat. Then the expected U.S. payoff will be $(-2 - 8q)$, as calculated before. The probability is $(1 - p)$ that the United States will encounter the soft-type Soviets. We are assuming that they comply; then the United States gets a 1. Therefore the expected payoff to the United States from the probabilistic threat, assuming that it is effective against the soft-type Soviets, is

$$(-2 - 8q) \times p + 1 \times (1 - p) = -8pq - 3p + 1.$$

If the United States refrains from making the threat, it gets a –2. Therefore the condition for the United States to make the threat is

$$-8pq - 3p + 1 > -2$$

$$\text{or } q < \frac{3}{8} \frac{1 - p}{p} = 0.375 \, (1 - p)/p.$$

That is, the probability of war must be small enough to satisfy this expression or the United States will not make the threat at all. We call this upper bound on q the **acceptability condition.** Note that p enters the formula for the maximum value of q that will be acceptable to the United States; the larger the chance that the Soviets will not give in, the smaller the risk of mutual disaster that the United States finds acceptable.

If the probabilistic threat is to work, it should satisfy both the effectiveness condition and the acceptability condition. We can determine the appropriate level of the probability of war using Figure 13.5. The horizontal axis is the probability, p, that the Soviets are hard-line, and the vertical axis is the probability, q, that war will occur if they defy the U.S. threat. The horizontal line $q = 0.6$ gives the lower limit of the effectiveness condition; the threat should be such that its associated (p, q) combination is above this line if it is to work even against the soft-type Soviets. The curve $q = 0.375 \, (1 - p)/p$ gives the upper limit of the acceptability condition; the threat should be such that (p, q) is below this curve if it is to be tolerable to the United States even assuming it works against the soft-type Soviets. Therefore an effective and acceptable threat should fall somewhere between these two lines, above and to the left of their point of intersection, at $p = 0.38$ and $q = 0.6$ (shown as a gray "wedge" in Figure 13.5).

The curve reaches $q = 1$ when $p = 0.27$. For values of p less than this, the dire threat (certainty of war) is acceptable to the United States, and effective against the soft-type Soviets. This just confirms our analysis in Section 4.

For values of p in the range from 0.27 to 0.38, the dire threat with $q = 1$ puts (p, q) to the right of the acceptability condition and is too large to be tolerable to the United States But a scaled-down threat can be found. For this range of values of p, there are some values of q low enough to be acceptable to the United States and yet high enough to compel the soft-type Soviets. Brinkmanship (using a probabilistic threat) can do the job in this situation, whereas a simple dire threat would be too risky.

If p exceeds 0.38, then there is no value of q that satisfies both conditions. If the probability that the Soviets will never give in is greater than 0.38, then any threat large enough to work against the soft-type Soviets ($q \geq 0.6$) creates a risk of war too large to be acceptable to the United States. If $p \geq 0.38$, therefore, the United States cannot help itself by using the brinkmanship strategy.

5 PRACTICING BRINKMANSHIP

If Kennedy has a very good estimate of the probability, p, of the Soviets being hard-liners, and is very confident about his ability to control the risk, q, that the blockade will lead to nuclear war, then he can calculate and implement his best strategy. As we saw in Section 4, if $p < 0.27$, the dire threat of a certainty of war is acceptable to Kennedy. (Of course, even then he will prefer to use the smallest effective threat, namely $q = 0.6$.) If p is between 0.27 and 0.38, then he has to use brinkmanship. Such a threat has to have the risk of disaster $0.6 < q < 0.375(1 - p)/p$, and again Kennedy prefers the smallest of this range, namely $q = 0.6$. If $p > 0.38$, then he should give in.

In practice Kennedy does not know p precisely; he only estimates that it lies within the range from 1/3 to 1/2. Similarly, he cannot be confident about the exact location of the critical value of q in the acceptability condition. That depends on the numbers used for the Soviet payoffs in various outcomes, for example, the –8 (for war) versus –4 (for compliance), and Kennedy can only estimate these values. Finally, he may not even be able to control the risk created by his brinkmanship action very precisely. All these ambiguities make it necessary to proceed cautiously.

Suppose Kennedy thinks that $p = 0.35$, and issues a threat backed by an action that carries the risk $q = 0.65$. The risk is greater than what is needed to be effective, namely 0.6. The limit of acceptability is $0.375 \times (1 - 0.35)/0.35 = 0.7$, and the risk $q = 0.65$ is less than this limit. Thus according to Kennedy's calculations, the risk satisfies both of the conditions—effectiveness and acceptability. However, suppose Kennedy is mistaken. For example, if he has not realized that Lemay might actually defy orders and take an excessively aggressive action, then q may in reality be higher than Kennedy thinks it is; for example, q may equal 0.8, which Kennedy would regard as too risky. Or suppose p is actually 0.4; then

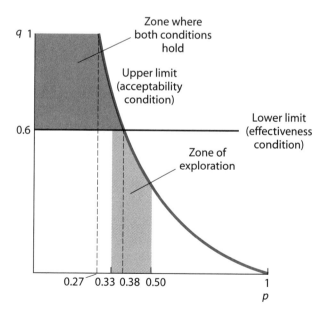

FIGURE 13.5 Conditions for Successful Brinkmanship

Kennedy would regard even $q = 0.65$ as too risky. Or Kennedy's experts may have misestimated the values of the Soviet payoffs. If they rate the humiliation of withdrawal as –5 instead of –4, then the threshold of the effectiveness condition will actually be $q = 0.7$, and Kennedy's threat with $q = 0.65$ will go wrong.

All that Kennedy knows is that the general shape of the effectiveness and acceptability conditions is like that shown in Figure 13.5. He does not know p for sure. Therefore he does not know exactly what value of q to choose in order to fulfill both the effectiveness and acceptability conditions; indeed, he does not even know if such a range exists for the unknown true value of p: it might be greater than or less than the borderline value of 0.38 that divides the two cases. And he is not able to fix q very precisely; therefore even if he knew p, he would not be able to act confident of his willingness to tolerate the resulting risk.

With such hazy information, imprecise control, and large risks, what is Kennedy to do? He has to *explore* the boundaries of the Soviets' risk tolerance as well as his own. It would not do to start the exploration with a value of q that might turn out to be too high. Instead, Kennedy must explore the boundaries "from below"; he must start with something quite safe, and gradually increase the level of risk to see "who blinks first." That is exactly how brinkmanship is practiced in reality.

We explain this with the aid of Figure 13.5. Observe the blue-shaded area. Its left and right boundaries, $p = 1/3$ and $p = 1/2$, correspond to the limits of Kennedy's estimated range of p. The lower boundary is the horizontal axis ($q = 0$). The upper boundary is composed of two segments. For $p < 0.38$ this segment corresponds to the effectiveness condition, and for $p > 0.38$ it corresponds to the acceptability condition. Remember that Kennedy does not know the precise positions of these boundaries, but must grope toward them from below. Therefore the blue-shaded region is where he must start the process.

Suppose Kennedy starts with a very safe action, say q equaling approximately 0.01 (1%). In our context of the Cuban missile crisis we can think of this as his television speech, which announced that a quarantine would soon go into effect. At this juncture the point with coordinates (p, q) lies somewhere near the bottom edge of the shaded region. Kennedy does not know exactly where, because he does not know p for sure. But the overwhelming likelihood is that at this point the threat is quite safe but also ineffective. Therefore Kennedy escalates it a little bit. That is, he moves the point (p, q) in a vertically upward direction from wherever it was initially. This could be the actual start of the quarantine. If that proves to be still safe but ineffective, he jacks up the risk one more notch. This could be the leaking of information about bombing plans.

As he proceeds in this way, eventually his exploration will encounter one of the boundaries of the blue-shaded area in Figure 13.5, and which boundary this is depends on the value of p. One of two things comes to pass. Either the threat becomes serious enough to deter the Soviets; this happens if the true value of p is less than its true critical value, here 0.38. On the diagram, we see this as a movement out of the blue-shaded area and into the area in which the threat is both acceptable *and* effective. Then the Soviets concede and Kennedy has won. Or the threat becomes too risky for the Unites States; this happens if $p > 0.38$. Kennedy's exploration in this case pushes him above the acceptability condition. Then Kennedy decides to concede, and Khrushchev has won. Again we point out that since Kennedy is not sure of the true value of p, he does not know in advance which of these two outcomes will prevail. As he gradually escalates the risk, he may get some clues from Soviet behavior that enable him to make his estimate of p somewhat more precise. Eventually he will reach sufficient precision to know which part of the boundary he is headed toward, and therefore whether the Soviets will concede or the United States must be the player to do so.

Actually, there are two possible outcomes only so long as the ever-present and steadily increasing mutual risk of disaster does not come to pass as Kennedy is groping through the range of ever more risky military options. Therefore there is a third possibility, namely that the explosion occurs before either side recognizes that it has reached its limit of tolerance of risk and climbs down. This continuing and rising risk of a very bad outcome is what makes brinkmanship such a delicate and dangerous strategy.

Thus brinkmanship in practice is the **gradual escalation of the risk of mutual harm.** It can be visualized vividly as **chicken in real time.** In our analysis of chicken in Chapter 4, we gave each player a simple binary choice: either go straight or swerve. In reality, the choice is usually one of timing. The two cars are rushing toward each other, and either player can choose to swerve at any time. When the cars are very far apart, swerving ensures safety. As they get closer together, they face an ever-increasing risk that they will collide anyway, that even swerving will not avoid a collision. As the two players continue to drive toward each other, each is exploring the limit of the other's willingness to take this risk, and perhaps at the same time exploring his own limit. The one who hits that limit first swerves. But there is always the risk that they have left it long enough and are close enough that even after choosing Swerve, they can no longer avoid the collision.

Now we see why, in the Cuban missile crisis, the very features that make it inaccurate to regard it as a two-person game make it easier to practice such brinkmanship. The blockade was a relatively small action, unlikely to start a nuclear war at once. But once Kennedy set the blockade in motion, its operation, escalation, and other features were not totally under his control. So Kennedy was not saying to Khrushchev, "If you defy me (cross a sharp brink), I will coolly and deliberately launch a nuclear war that will destroy both our peoples." Rather, he was saying, "The wheels of the blockade have started to turn and are gathering their own momentum. The more or longer you defy me, the more likely it is that some operating procedure will slip up, or the political pressure on me will rise to a point where I must give in, or some hawk will run amok. If this risk comes to pass, I will be unable to prevent nuclear war, no matter how much I may regret it at that point. Only you can now defuse the tension by complying with my demand to withdraw the missiles."

We believe that this perspective gives a much better and deeper understanding of the crisis than can most analyses based on simple threats. It tells us why the *risk* of war played such an important role in all discussions. It even makes Allison's compelling arguments about bureaucratic procedures and internal divisions on both sides an integral part of the picture: these features allow the top-level players on both sides to credibly lose some control, that is, to practice brinkmanship.

One important condition remains to be discussed. In Chapter 9 we saw that every threat has an associated implicit promise, namely that the bad consequence will not occur if your opponent complies with your wishes. The same is required for brinkmanship. If, as you are increasing the level of risk, your opponent does comply, you must be able to "go into reverse"—begin reducing the risk immediately and quite quickly remove it from the picture. Otherwise the opponent would not gain anything by compliance. This may have been a problem in the Cuban missile crisis. If the Soviets feared that Kennedy could not control the hawks like Lemay ("We ought to just go in there today and knock 'em off"), they would gain nothing by giving in.

To reemphasize and sum up, brinkmanship is the strategy of exposing your rival and yourself to a gradually increasing risk of mutual harm. The actual occurrence of the harmful outcome is not totally within the threatener's control.

Viewed in this way, brinkmanship is everywhere. In most confrontations—for example, between a company and a labor union, a husband and a wife, a parent and a child, and the President and Congress—one player cannot be sure of the other party's objectives and capabilities. Therefore most threats carry a risk of error, and every threat must contain an element of brinkmanship. We hope we have given you some understanding of this strategy, and that we have impressed upon you the risks it carries. You will have to face up to brinkmanship, or conduct it yourself, on many occasions in your personal and professional lives. Please do so carefully, and with a clear understanding of its potentialities and risks.

To help you do so, we now recapitulate the important lessons learned from the handling of the Cuban missile crisis, reinterpreted as a labor union leadership contemplating a strike in pursuit of its wage demand, unsure whether this will result in the whole firm shutting down:

1. Start small and safe. Your first step should not be an immediate walkout; it should be to schedule a membership meeting at a date a few days or weeks hence, while negotiations continue.

2. Raise the risks gradually. Your public and private statements, and stirring up

of the sentiments of the membership, should induce management to believe that acceptance of its current low-wage offer is becoming less and less likely. If possible, stage small incidents, for example, a few one-day strikes or local walkouts.

3. As this process continues, read and interpret signals in management's actions to figure out whether the firm has enough profit potential to afford the union's high-wage demand.

4. Retain enough control over the situation; that is, retain the power to induce your membership to ratify the agreement you will reach with management; otherwise management will think that the risk will not deescalate even if it concedes your demands.

SUMMARY

In some game situations, the risk of error in the presence of a threat may call for the use of as small a threat as possible. When a large threat cannot be reduced in other ways, it can be scaled down by making its fulfillment probabilistic. Strategic use of a *probabilistic threat,* in which you expose your rival and yourself to an increasing risk of harm, is called *brinkmanship.*

Brinkmanship requires a player to relinquish control over the outcome of the game without completely losing control. You must create a threat with a risk level that is both large enough to be effective in compelling or deterring your rival and small enough to be acceptable to you. To do so, you must determine the levels of risk tolerance of both players through a *gradual escalation of the risk of mutual harm.*

The Cuban missile crisis of 1962 serves as a case study in the use of brinkmanship on the part of President Kennedy. Analyzing the crisis as an example of a simple threat, with the U.S. blockade of Cuba establishing credibility, is inadequate. A better analysis accounts for the many complexities and uncertainties inherent in the situation and the likelihood that a simple threat was too risky. Because the actual crisis involved numerous political and military players, Kennedy was able to achieve "controlled loss of control" by ordering the blockade and gradually letting incidents and tension escalate, until Khrushchev yielded in the face of the rising risk of nuclear war.

KEY TERMS

acceptability condition (454)
brinkmanship (451)
chicken in real time (457)
effectiveness condition (453)

gradual escalation of the
  risk of mutual harm (457)
probabilistic threat (451)

EXERCISES

1. Examine the following situations of confrontation, in which one side or the other used brinkmanship. Some are successes in the sense that a mutually acceptable deal was reached, and the others failures in the sense that the mutually bad outcome (disaster) occurred.

   In each case, do the following: (i) identify the interests of the parties, (ii) describe the nature of the uncertainty inherent in the situation, and (iii) give the strategies the parties used in order to escalate the risk of disaster. (iv) Discuss whether the strategies were good ones; in particular, in the cases where disaster occurred, discuss how one or both parties could have played the game better. (v) [Optional] If you can, set up a small mathematical model of the kind you saw in the text of this chapter.

   In each case we provide a few readings to get you started; you should locate more using the resources of your library and resources on the World Wide Web such as Lexis-Nexis.

   Successes:
   (a) The Uruguay Round of international trade negotiations that started in 1986 and led to the formation of the World Trade Organization in 1994. *Reading:* John H. Jackson, *The World Trading System*, 2nd ed. (Cambridge, Mass.: MIT Press, 1997), pp. 44–49, and chaps. 12 and 13.
   (b) The Camp David accords between Israel and Egypt in 1978. *Reading:* William B. Quandt, *Camp David: Peacemaking and Politics* (Washington, D.C.: Brookings Institution, 1986).
   (c) The negotiations between the South African apartheid regime and the African National Congress to establish a new constitution with majority rule, 1989 to 1994. *Reading:* Allister Sparks, *Tomorrow Is Another Country* (New York: Hill and Wang, 1995).

   Failures:
   (d) The confrontation between the regime and the student pro-democracy demonstrators in Beijing, 1989. *Readings: Massacre in Beijing: China's Struggle for Democracy*, ed. Donald Morrison (New York: Time Magazine Publications, 1989); *China's Search for Democracy: The Student and Mass Movement of 1989*, ed. Suzanne Ogden, Kathleen Hartford, L. Sullivan, D. Zweig (Armonk, N.Y.: M. E. Sharpe, 1992).
   (e) The U.S. budget confrontation between President Clinton and the Republican-controlled Congress in 1995. *Readings:* Sheldon Wolin, "Democracy and Counterrevolution," *Nation*, April 22, 1996; David Bowermaster, "Meet the Mavericks," *U.S. News and World Report*, December 25, 1995/January 1, 1996; "A Fight that Never Seems to End," *Economist*, December 16, 1995; and several other articles available for reading on Lexis-Nexis.

(f) The Caterpillar strike, 1991 to 1998. *Readings:* "The Caterpillar Strike: Not Over Till It's Over," *Economist,* February 28, 1998; "Caterpillar's Comeback," *Economist,* June 20, 1998; Aaron Bernstein, "Why Workers Still Hold a Weak Hand," *BusinessWeek* March 2, 1998.

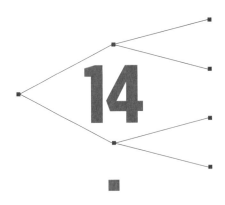

14

Strategy and Voting

V OTING IN ELECTIONS may seem straightforward at first glance—nothing that game theory could say anything new or surprising about. Citizens of the United States decide who they want for President and vote for that person; faculty members decide whether they want to require every student to take a course in statistics and vote yes or no; students decide who would make the best college government president and vote for her. Where is the strategy in this?

It turns out that a lot more is involved in participating in an election than deciding what you like and voting accordingly. There are different types of voting procedures used to determine winners in different types of elections. Different procedures can lead to different outcomes, and one particular procedure may lead to an outcome you prefer. Then you can try to make use of your strategic abilities in order to see that this procedure is used. Or you may find that the given procedure for a specific election is open to manipulation by voters who alter their preferences in order to alter the outcome of the election. If other voters behave strategically in this way, then it is likely to be in your best interests to do so as well. But maybe not. Again, knowledge of game theory will help you determine your best course of action.

In the following sections we consider each of these important issues, first looking at the different types of voting rules and procedures and then at some nonintuitive or paradoxical results that can arise under certain rules. We then discuss some of the criteria used to judge the performance of the various voting methods. We also address strategic behavior of voters and the scope for outcome manipulation. Finally, we present two different versions of a famous result known as the *median voter theorem*—as a two-person zero-sum game with discrete strategies and with continuous ones.

1 VOTING RULES AND PROCEDURES

There are many ways in which elections can be held, especially when there are numerous alternatives (that is, candidates or issues) among which people can choose. Just increasing the number of alternatives to anything larger than two is enough to make election design issues interesting. We describe in this section a variety of procedures from three broad classes of voting, or vote aggregation, methods. The number of possible voting procedures is enormous, and even the simple taxonomy that we provide here can be complicated by allowing elections based on a combination of procedures; there is a considerable literature in both economics and political science devoted to just this topic. We have not attempted to provide an exhaustive survey but rather give a flavor of that literature. If you are interested, we suggest you consult the broader literature for more details on the subject.[1]

A. Binary Methods

Vote aggregation methods can be classified according to the number of options or candidates considered by the voters at any given time. **Binary methods** require that voters be asked to choose between only two alternatives at a time. In elections involving exactly two candidates, votes can be aggregated using the well known principle of **majority rule,** which simply requires that the alternative with a majority of votes wins. When dealing with a slate of more than two alternatives, **pairwise voting**—a method involving a repetition of binary votes—can be used. Pairwise procedures are **multistage;** they entail voting on pairs of alternatives in a series of majority votes to determine which is most preferred.

A number of pairwise procedures exist, distinguished by the order of votes or the method of pairing alternatives. A complete round-robin of majority votes—pitting each alternative against all of the others—is often called the **Condorcet method,** after the 18th-century French theorist Jean Antoine Nicholas Caritat, Marquis de Condorcet. He suggested that the candidate who defeats each of the

---

[1] For an introduction to voting issues, see the symposium on "Economics of Voting," *Journal of Economic Perspectives,* vol. 9, no. 1, (Winter 1995); William Riker, *Liberalism Against Populism* (San Francisco: W. H. Freeman, 1982); Michael Dummett, *Voting Procedures* (Oxford: Clarendon Press, 1984); or L. F. Cranor, "Declared-Strategy Voting: An Instrument for Group Decision-Making" (Ph.D. dissertation, Department of Engineering and Policy, Washington University, 1996).

others in such a series of one-on-one contests should win the entire election; such a candidate, or alternative, is now termed a **Condorcet winner.** Other pairwise procedures successively eliminate candidates or produce "scores" like the **Copeland index**, which measures an alternative's win-loss record in a round-robin of contests; the first round of the World Cup soccer tournament uses a type of Copeland index to determine which teams from each group move on to the second round of play.[2]

Another well known pairwise procedure, used when there are three possible alternatives, is the **amendment procedure,** required by the parliamentary rules of the U.S. Congress when legislation is brought to a vote. When a bill is brought before Congress, any amended version of the bill must first win a vote against the original version of the bill. The winner of that vote is then paired against the status quo and members vote on whether to adopt the version of the bill that won the first round; majority rule can then be used to determine the winner. The amendment procedure can be used to consider any three alternatives by pairing two in a first-round election and then putting the third up against the winner in a second-round vote.

Pairwise voting is not the only way to handle an election involving multiple candidates. A large slate can also be reduced to two alternatives using some plurative method like those considered next in Section 1.B. Once the number of alternatives has been reduced to two, a binary method can be used to determine the final winner. Such **mixed methods** are described in Section 1.C.

B. Plurative Methods

Plurative methods allow voters to consider simultaneously any number of alternatives beyond two and generally use information from voter preference orderings in aggregating votes. Some of these procedures require knowledge of a voter's entire preference ordering; others use only a part of that ordering. One plurative method, the familiar **plurality rule,** uses information only on each voter's most preferred alternative. With plurality rule, all possible alternatives are put to a vote simultaneously. Each voter casts a single vote in favor of one particular alternative, and the alternative with the most votes wins. Thus, for instance, the 1994 Maine gubernatorial election saw independent candidate Angus King capture 36% of the vote, Democrat Joseph Brennan 34%, Republican Susan Collins 23%, and Green Party candidate Jonathan Carter 6%. King gained the largest portion of the vote, and this plurality won him the governorship. Note that, as in King's case, a plurality need *not* be a majority; that is, a multiple alter-

---

[2] Note that such indices, or scores, must have precise mechanisms in place in order to deal with ties; World Cup soccer uses a system that relatively undervalues a tie in order to encourage more aggressive play. See Barry Nalebuff and Jonathan Levin, "An Introduction to Vote Counting Schemes," *Journal of Economic Perspectives*, vol. 9, no. 1, (Winter 1995), pp. 3–26.

native election can be won with a plurality of less than 50% of the vote. Plurality rule is the plurative generalization of majority rule.

Of the plurative voting procedures that make use of voter preference orderings, one of the most well known is the **Borda count,** named after Jean-Charles de Borda, a fellow countryman and contemporary of Condorcet. Borda described the new procedure as an alternative to (or an improvement on, from his perspective) plurality rule. The Borda count requires voters to rank-order all of the possible alternatives in an election and indicate their rankings on their ballot cards. Then points are assigned to each alternative, based on its position on each voter's ballot. In a three-person election, the candidate at the top of your ballot would get three points, the next candidate two points, and the bottom candidate one point. After the ballots are collected, each candidate's points are summed, and the one with the most points wins the election. Some variations on this method allow for relatively larger numbers of points to be awarded to a voter's most preferred alternative. A Borda count procedure is used in a number of sports-related elections, including professional baseball's Cy Young Award and college football's championship elections.

A relatively recently designed plurative method that is also gaining support as an alternative to existing procedures is **approval voting,** which allows voters to cast a single vote for each alternative of which they "approve." The alternative that receives the most approvals wins; in elections in which more than one winner can be selected (if electing a school board, for instance), a threshold level of approvals is set in advance and alternatives with more than the required minimum approvals are elected.

In a three-candidate election between alternatives A, B, and C, if one group of voters slightly prefers A to B but vehemently dislikes C, approval voting can allow that group of voters to cast votes showing their approval of *both* A and B; that is, they could vote for both A and B. This procedure makes it less likely for a candidate ranked last by those not voting for her—like C—to win an election. It also favors relatively moderate alternatives over those at either end of the spectrum. Several professional societies have adopted approval voting in order to elect their officers, and some states have used or are considering using this method for public elections.

C. Mixed Methods

Some multistage voting procedures combine plurative and binary methods. The **majority runoff** procedure, for instance, is a two-stage method used to decrease a large group of possibilities to a binary decision. In a first-stage election, voters indicate their most preferred alternative, and these votes are tallied. If one candidate receives a majority of votes in the first stage, she wins. However, if there is no majority choice, a second-stage election pits the two most preferred alternatives against each other. Majority rule chooses the winner in the second stage.

Another such procedure involves voting in successive **rounds.** Voters may be asked to consider a number of alternatives in a first round of voting, through either a single vote or a preference ranking, such as the Borda count. At the end of each round, the worst-performing alternative is eliminated, and voters are asked to consider the remaining alternatives in a next round. The elimination continues until only two alternatives remain; at that stage the method becomes binary, and a final majority runoff determines a winner. A similar procedure is used to choose sites for the Olympic Games.

The next method we consider uses information on a voter's full preference ranking and provides for **proportional representation.** Proportional representation implies, for example, that a state electorate consisting of 50% Republicans, 30% Democrats, and 20% Independents would yield a body of representatives mirroring the party affiliations of that electorate: in other words, 50% of the U.S. Representatives from such a state would be Republican, and so on. Note that in a plurality rule system, in which voting is done by district, if the voter mix in all districts exactly mirrors the overall voter mix in the state, then, using this example, *all* of the U.S. Representatives from this state—100%—would be Republican; the Republican candidate would get 50% of the vote and win in each and every district.

The method of the **single transferable vote,** also known as the **Hare procedure,** after Englishman Thomas Hare, produces proportional representation based on voter rankings of the available alternatives.[3] To continue with our election example, in the Hare procedure, all voters in the state consider a list of all possible candidates. Each voter gets a single vote but indicates her preference ranking over all candidates. Then, depending on the exact specifications of the procedure, candidates who attain a certain quota of votes are elected and others who fall below a certain level (or sometimes the single candidate with the fewest votes) are eliminated. Votes for those candidates who are eliminated are "transferred," using the voters' preference orderings, to the next highest ranking candidate. This procedure continues until an appropriate number of candidates are elected.

Clearly, there is room for considerable strategic thinking in the choice of a vote aggregation method, and strategy is also important, even after the rule has been chosen. A pairwise election, for example, can be strategically manipulated by the person or persons in charge of setting the order of the votes. Such agenda setting has important implications for election outcomes, just as rule setting does. We will examine some of the issues involved in rule making as well as agenda setting in Section 2. Furthermore, strategic behavior on the part of vot-

---

[3] For an overview of this system, see Nicolaus Tideman, "The Single Transferable Vote," *Journal of Economic Perspectives*, vol. 9, no. 1 (Winter 1995), pp. 27–38. The term *single-transferable vote* refers to a class of voting rules having the same basic structure.

ers, often called **strategic voting** or **strategic misrepresentation of preferences,** can also alter election outcomes under any set of rules. While some procedures, including the Borda count in particular, are especially vulnerable to such behavior, any of the rules described here can fall prey to some type of manipulation. We will look more closely at issues of strategic voting in Section 4, after describing in Section 3 some of the desirable properties one might like a vote aggregation method to have.

2 VOTING PARADOXES

Even when people vote according to their true preferences, certain voting procedures under specific conditions can give rise to curious outcomes. In addition, election outcomes can depend critically on the type of procedure used to aggregate votes. This section describes some of the most famous of the curious outcomes—the so-called voting paradoxes—as well as some examples of how election results can change under different vote aggregation methods with no change in voter preferences and no strategic voting.

A. The Condorcet Paradox

The **Condorcet paradox** is one of the most famous and important of the voting paradoxes.[4] As mentioned earlier, the Condorcet method calls for the winner to be the candidate who had gained a majority of votes in each round of the round-robin of pairwise comparisons. The paradox arises when no clear winner—no Condorcet winner—emerges from this process.

   To illustrate the paradox, we construct an example in which three individuals vote on three alternative outcomes using the Condorcet method. Consider three city councillors (Left, Center, and Right) who are asked to rank their preferences for three alternative welfare policies, one that extends the welfare benefits currently available (call this one Generous), one that cuts back on available benefits (Limited), and one that maintains the status quo (Average). They are then asked to vote on each pair of policies to establish a council ranking, or a **social ranking.** This ranking is meant to describe how the council as a whole judges the merits of the possible welfare systems.

   Suppose Councillor Left's preferences are such that she prefers the Generous policy most and the Limited policy least, with the Average policy in between. Councillor Center most prefers the Average policy, but she has qualms about the

---

[4]It is so famous that economists have been known to refer to it as *the* voting paradox. Political scientists appear to know better, since they are far more likely to use its formal name. As we will see, there are any number of possible voting paradoxes, not just the one named for Condorcet.

state of the city budget, so she prefers the Limited policy to the Generous one. Finally, Councillor Right most prefers Limited but places Generous above Average; she believes the Generous policy will soon cause a serious budget crisis and turn public opinion so much against benefits that a more permanent state of Limited benefits will result, whereas Average benefits could go on indefinitely. We can rewrite these preference orderings using the shorthand P to indicate that one alternative is preferred to another. (Technically, P is referred to as a *binary ordering relation.*) Thus the preferences are for Councillor Left, Generous P Average P Limited; for Councillor Center, Average P Limited P Generous; and for Councillor Right, Limited P Generous P Average. The three preference orderings are shown in Figure 14.1, with the most preferred alternative at the top and the least preferred at the bottom.

We can now look at the results. In the pairing of Generous against Average, Left and Right would vote for Generous, so that policy would beat Average. In the next pairing, of Average against Limited, Left and Center would vote for Average, so it would beat Limited. And in the final pairing of Generous against Limited, the vote would again be 2 to 1, this time for Limited over Generous. If you look closely at the set of winners here, you will see that when the council votes on alternative pairs of policies, a majority prefer Generous over Average, Average over Limited, *and* Limited over Generous. No one policy has a majority over both of the others. The group's preferences are cyclical: Generous P Average P Limited P Generous.

This cycle of preferences is an example of an **intransitive ordering** of preferences. The concept of rationality is usually taken to mean that individual preference orderings are **transitive** (the opposite of intransitive). If someone is given choices A, B, and C, and you know she prefers A to B and B to C, then transitivity implies she also prefers A to C. (The terminology comes from the transitivity of numbers in mathematics; for instance, if $3 > 2$ and $2 > 1$, then we know that $3 > 1$.) A transitive preference ordering will not cycle as does the social ordering derived in our city council example; hence, we say that such an ordering is intransitive.

Notice that all of the *councillors* have transitive preferences over the three welfare policy alternatives but the *council* does not. This important distinction is

LEFT	CENTER	RIGHT
Generous	Average	Limited
Average	Limited	Generous
Limited	Generous	Average

FIGURE 14.1   Councillor Preferences Over Welfare Policies

an example of the Condorcet paradox: even if all individual preference orderings are transitive, there is no guarantee that the social preference ordering induced by Condorcet's voting procedure will also be transitive. The result has far-reaching implications for public servants, as well as for the general public. It calls into question, for instance, the notion of the "public interest" when we know that such interests may not be easily defined or may not even exist. Our city council does not have any well defined set of group preferences over the welfare policies. The lesson is that societies, institutions, or other large groups of people should not always be analyzed as if they acted like individuals.

The Condorcet paradox can even arise more generally. There is no guarantee that the social ordering induced by *any* formal group voting process will be transitive just because individual preferences are. However, some estimates have shown that the paradox is most likely to arise when large groups of people are considering large numbers of alternatives. Smaller groups considering smaller numbers of alternatives are more likely to have similar preferences over those alternatives; in such situations, the paradox is much less likely to appear.[5] In fact, the paradox arose in our example because the council completely disagreed not only about which alternative was best but also about which was worst. The smaller the group, the less likely such outcomes are to occur.

B. The Reversal Paradox

The Borda count voting procedure can give rise to a number of peculiar outcomes when the slate of candidates open to voters changes. One of the most well known of these outcomes is the **reversal paradox.** It arises in an election with at least four alternatives when one of these is removed from consideration after votes have been submitted, making recalculation necessary.

Suppose there are four candidates for a (hypothetical) special commemorative Cy Young Award to be given to a retired major-league baseball pitcher. The candidates are Steve Carlton (SC), Sandy Koufax (SK), Robin Roberts (RR), and Tom Seaver (TS). Seven prominent sportswriters are asked to rank these pitchers on their ballot cards. The top-ranked candidate on each card will get 4 points; decreasing numbers of points will be allotted to candidates ranked second, third, and fourth.

The completed ballot cards for the seven voters are shown in Figure 14.2. When the votes are tallied, Seaver gets $3 + 2 + 3 + 4 + 2 + 2 + 4 = 20$ points, Koufax gets $4 + 3 + 4 + 1 + 3 + 3 + 1 = 19$ points, Carlton gets $1 + 4 + 1 + 2 + 4 + 4 + 2 = 19$ points, and Roberts gets $2 + 1 + 2 + 3 + 1 + 1 + 3 = 13$ points. Seaver wins the election, followed by Koufax and Carlton tied, and Roberts in last place.

[5] See Peter Ordeshook, *Game Theory and Political Theory* (Cambridge: Cambridge University Press, 1986), p. 58.

1	2	3	4	5	6	7
SK	SC	SK	TS	SC	SC	TS
TS	SK	TS	RR	SK	SK	RR
RR	TS	RR	SC	TS	TS	SC
SC	RR	SC	SK	RR	RR	SK

FIGURE 14.2 Sportswriter Preferences for Award Candidates

Now suppose it is discovered that Roberts is not really eligible for the commemorative award because he never actually won a Cy Young Award, reaching the pinnacle of his career in the years just prior to the institution of the award in 1956. This discovery invalidates the original election, so the sportswriters are asked to complete new ballot cards, considering only the three remaining eligible candidates. In the new election, the cards look like those in Figure 14.2, except that Roberts has been eliminated from each ballot; the new ballots are illustrated in Figure 14.3. The top spot on each card now gets three points, while the second and third spots receive two and one, respectively.

Adding votes in this second election shows that Carlton receives 15 points, Koufax receives 14 points, and Seaver receives 13 points. Winner has turned loser as the new results reverse the standings found in the first election. No change in preference orderings accompanies this strange result. The only difference in the two elections is the number of candidates being considered.

C. The Agenda Paradox

The third paradox that we consider involves the ordering of alternatives in a binary voting procedure. In a parliamentary setting with a committee chair who determines the specific order of voting for a three-alternative election, substantial power over the final outcome lies with the chair. In fact, the chair can take

1	2	3	4	5	6	7
SK	SC	SK	TS	SC	SC	TS
TS	SK	TS	SC	SK	SK	SC
SC	TS	SC	SK	TS	TS	SK

FIGURE 14.3 Sportswriter Preferences Over a Narrowed Field

advantage of the intransitive social preference ordering that arises from some sets of individual preferences and, by selecting an appropriate agenda, manipulate the outcome of the election in any manner she desires.

Consider again the city councillors Left, Center, and Right, who must decide among Generous, Average, and Limited welfare policies. The councillors' preferences over the alternatives were shown in Figure 14.1. Let us now suppose that one of the councillors has been appointed chair of the council by the mayor and the chair is given the right to decide which two welfare policies get voted on first, and which goes up against the winner of that initial vote. With the given set of councillor preferences and common knowledge of the preference orderings, the chair can get any outcome she wants; this result is known as the **agenda paradox.**

If the chair wants the Generous policy to win, she sets the Average policy against the Limited policy in the first round and has the winner go up against the Generous policy. Under this set of rules, Average beats Limited (2 to 1) but is then beaten by Generous (2 to 1). Similarly, if the chair wants the Average policy to win, she pits Generous against Limited in the first round and the winner (Limited) against Average in the second. Finally, if she wants to have the Limited policy win, she sets Limited against the winner of a first round between Generous and Average.

With the set of preferences illustrated in Figure 14.1, then, we get not only the intransitive social preference ordering of the Condorcet paradox but also the result that the outcome of a binary procedure could be any of the possible alternatives. The only determinant of the outcome in such a case is the ordering of the agenda. This result implies that setting the agenda is the real game here and, because the chair sets the agenda, the appointment or election of the chair is the true outlet for strategic behavior. Proponents of the various bills need to focus their efforts on ensuring that the councillor who holds their views is made chair. Here, as in many other strategic situations, what appears to be the game (in this case, choice of a welfare policy) is not the true game at all; rather, those involved in the game engage in strategic play at an earlier point (deciding the identity of the chair) and vote according to set preferences in the eventual election.

D. Change the Voting Method, Change the Outcome

Strategic thinking can be used in a number of ways when dealing with voting rules and procedures. If you have some control over the voting rule that will be used in a particular election, you can use this power to your advantage. As should be clear from the preceding discussion, election outcomes are likely to differ under different sets of voting rules.

As an example, consider 100 voters who can be broken down into three groups based on their preferences over three candidates (A, B, and C). There are

40 voters in the first group; these voters like candidate A best, B next, and C least. The second group consists of 25 voters, each of whom prefers candidates B, C, and A (in that order). Finally, in the third group are 35 voters who prefer C the most, then B, and A the least. Depending on the vote-aggregation method used, any of these three candidates could win the election.

With simple plurality rule, candidate A gets 40% of the vote while C gets 35% and B gets only 25%. Candidate A thus wins a plurality rule election even though 60% of the voters rank her lowest of the three. Supporters of candidate A would obviously prefer this type of election. If they had the power to choose the voting method, then plurality rule, a seemingly "fair" procedure, would win the election for A in spite of the majority's strong dislike for that candidate.

A Borda count, however, would produce a different outcome. In a Borda system with three points going to the most-preferred candidate, two points to the middle candidate, and one to the least-preferred candidate, A gets 40 first-place votes and 60 third-place votes, for a total of $40(3) + 60(1) = 180$ points. B gets 25 first-place votes and 75 second-place votes, for a total of $25(3) + 75(2) = 225$ points; and C gets 35 first-place votes, 25 second-place votes, and 40 third-place votes, for a total of $35(3) + 25(2) + 40(1) = 195$ points. With this procedure, B wins, with C in second place and A last. B would also win with an approval voting system in which candidates cast votes in favor of their top two preferred candidates.

And what about candidate C? She can win the election if a majority runoff system is used; the top two vote getters in a first-round election move on to face each other in a majority rule runoff. Then A and C, with 40 and 35 votes in the first round, survive to face each other in the runoff. And, because A is the least-preferred alternative for 60 of the 100 voters, candidate C wins the runoff election 60 to 40.

Another example of how different procedures can lead to different outcomes can be seen in the voting for the site of the 1996 (Summer) Olympics. The actual voting procedure for the Olympic-site selection is a mixed method involving multiple rounds of plurality rule with elimination, and a final majority rule vote. Each country representative participating in the election casts one vote, and the candidate city with the lowest vote total is eliminated after each round. Majority rule is used to choose between the final two candidate cities. When the vote for the 1996 site was taken, Athens (Greece) led the early rounds due in part to the fact that 1996 marked the centenary of the modern games. Thus, had a single round of plurality rule been used, Athens would have been selected. The use of the mixed procedure, however, led to a situation in which the plurality rule winner (Athens) gained little additional support as other cities were eliminated. Athens actually retained a plurality through the next-to-last round of voting, in which the other contenders were Atlanta (Georgia) and Sydney (Australia). After Sydney was eliminated in that round, most of its votes went to Atlanta rather than Athens, and Atlanta won in the final runoff.

3 EVALUATING VOTE AGGREGATION METHODS

Our discussion of the various voting paradoxes in Section 2 suggests that voting methods can suffer from a number of faults that lead to unusual, unexpected, or even unfair outcomes. In addition, it leads us to ask: Is there one voting system that satisfies certain regularity conditions, including transitivity, and is the most "fair"—that is, most accurately captures the preferences of the electorate? Kenneth Arrow's **impossibility theorem** tells us the answer to this question is a firm no.[6]

The technical content of Arrow's theorem makes it beyond our scope to prove completely. But the sense of the theorem is relatively straightforward. Arrow argued that no preference aggregation method could satisfy all six of the critical principles he identified:

1. The social or group ranking must rank all alternatives (be complete).

2. It must be transitive.

3. It should satisfy a condition known as *positive responsiveness*, or the Pareto property—in which unanimous preferences within the electorate are reflected in the aggregate ranking.

4. The ranking must not be imposed by external considerations (such as customs) independent of the preferences of individuals in the society.

5. It must not be dictatorial—no single voter should determine the group ranking.

6. And it should be independent of irrelevant alternatives—that is, no change in the set of candidates (addition to or subtraction from) should change the rankings of the unaffected candidates.

Often the theorem is abbreviated by imposing the first four conditions and focusing on the difficulty of simultaneously obtaining the last two; the simplified form states that we cannot have independence of irrelevant alternatives without dictatorship.[7]

Arrow's theorem has provoked extensive research into the robustness of his conclusion to changes in the underlying assumptions. No minimal reduction in the number of criteria nor modest relaxation of Arrow's principles has led to anything other than Arrow's conclusion; no truly fair voting method can be found without violating some apparently reasonable aggregation requirement. Most economic and political theorists now accept that some form of compromise is necessary when choosing a vote, or preference, aggregation method. There have also been additional suggestions made regarding criteria that a good aggregation

---

[6]A full description of this theorem, often called *Arrow's General Possibility Theorem*, can be found in Kenneth Arrow, *Social Choice and Individual Values*, 2nd ed. (New York: Wiley, 1963).

[7]See Walter Nicholson's treatment of Arrow's impossibility theorem in his *Microeconomic Theory* 7th ed. (New York: Dryden Press, 1998) pp. 764–766, for more detail at a level appropriate for intermediate-level economics students.

system should satisfy. Some of these include the *Condorcet criterion* (that a Condorcet winner should be selected by a voting system if such a winner exists), the *consistency criterion* (that an election including all voters should elect the same alternative as two elections held for an arbitrary division of the entire set of voters), and lack of manipulability (a voting system should not encourage manipulability—strategic voting—on the part of voters).

We can see immediately that some of the voting methods considered earlier do not satisfy all of Arrow's principles, and some do not satisfy the Condorcet or consistency criteria. The Borda count, for example, most obviously violates Arrow's requirement of the independence of irrelevant alternatives; our example of the reversal paradox shows this vividly. The Borda method also does not satisfy the Condorcet criterion, although it is nondictatorial, consistent, and satisfies the Pareto property. Similarly, the single transferable vote method also violates independence of irrelevant alternatives. (You should be able to prove this to yourself relatively easily.) All of the other systems we have considered satisfy independence, but break down on one of the other principles.

The manipulability criterion is an interesting one. The literature on vote aggregation tells us that all nondictatorial systems are manipulable,[8] and some theorists have argued that voting systems should be evaluated on their tendency to encourage manipulation. The relative manipulability of a voting system can be determined by the amount of information required by voters to successfully manipulate an election. Some research based on this criterion suggests that, of the procedures so far discussed, plurality rule is the most manipulable (that is, requires the least information). In decreasing order of manipulability are approval voting, the Borda count, the amendment procedure, majority rule, and the Hare procedure (single transferable vote).[9] Note that this classification depends only on the amount of information necessary to manipulate a voting system, and is not based on the ease of putting such information to good use or whether manipulation is most easily achieved by individuals or groups; in practice, the manipulation of plurality rule by *individuals* is quite difficult. The potential for manipulability in each of our procedures, however, leads us to ask exactly how such actions are taken. We examine exactly this issue in the next section.

4 STRATEGIC VOTING

Several of the voting systems we have considered yield considerable scope for strategic misrepresentation of preferences on the part of voters. The process of agenda manipulation found in the agenda paradox, for instance, is based on the

---

[8]For the theoretical details on this result, see A. Gibbard, "Manipulation of Voting Schemes: A

assumption that councillors did not vote strategically. At least one of the councillors, though, might have good reason to strategically misrepresent her preferences. In our discussion in Section 2.C, we showed that an agenda-setting chair could engineer a win for the Generous policy by setting the Average policy against the Limited policy in a first round of voting, with the winner facing the Generous policy in a second round. In that case, Councillor Center's least-preferred policy is chosen in the election. She should anticipate this outcome and alter her vote in the first round if by doing so she can alter the final outcome; that is, she should vote strategically. Thus, voters can choose to vote for candidates, issues, or policies that are not actually their most-preferred outcomes if they believe that such behavior in an early voting round can alter the final election results in their favor; similarly, voters might vote for something other than their most preferred alternative in order to prevent their least-favorite alternative from winning. In this section, we consider a number of ways in which strategic voting behavior can affect elections.

A. Plurality Rule

Plurality rule elections, often perceived as the most fair by many voters, still provide opportunities for strategic behavior. In Presidential elections, for instance, there are generally two major candidates in contention. When such a race is relatively close, there is potential for a third candidate to enter the race and divert votes away from the leading candidate; if the entry of this third player truly threatens the chances of the leader winning the election, the late entrant is called a **spoiler.**

Spoilers are generally believed to have little chance to win the whole election, but their role in changing the election outcome is undisputed. In elections with a spoiler candidate, those who prefer the spoiler to the leading major candidate but least prefer the trailing major candidate may do best to strategically misrepresent their preferences in order to prevent the election of their least-favorite candidate. That is, you should vote for the leader in such a case even though you would prefer the spoiler because the spoiler is unlikely to garner a plurality; voting for the leader then prevents the trailing candidate, your least favorite, from winning.[10]

Some evidence shows there was a great deal of strategic misrepresentation of voter preferences during the 1992 Presidential election. Ross Perot ran a strong third-party campaign for much of the primary season, only to drop out of the

---

General Result," *Econometrica*, vol. 41, no. 4 (July 1973), pp. 587–601, and M. A. Satterthwaite, "Strategy-Proofness and Arrow's Conditions," *Journal of Economic Theory*, vol. 10 (1975), pp. 187–217.

[9] H. Nurmi's classification can be found in his *Comparing Voting Systems* (Norwell, Mass.: D. Reidel, 1987); Cranor, 1996, "Declared-Strategy Voting," provides an overview.

[10] Note that an approval voting method would not suffer from this same problem.

race at midsummer and reappear in the fall. His reappearance at such a late date cast him as a spoiler in the then close race between the incumbent, President Bush, and his Democratic challenger, Bill Clinton. Perot's role in dividing the conservative vote and clinching Clinton's victory that year has been debated, but it is certainly possible that Clinton would not have won without Perot on the ballot. Similarly, it has been argued that Maine's Angus King would not have won his 1994 election if the Green Party had not diverted some of the Democratic votes away from that party's candidate, Joseph Brennan.

A more interesting interpretation of the Bush-Clinton-Perot contest, however, claims there was significant misrepresentation of preferences by voters at the polls that November. According to *Newsweek*, "exit polls on election day [in 1992] asked voters how they would have cast their ballots if they 'believed Ross Perot had a chance to win.' In this hypothetical contest, Perot had won 40 percent of the vote, against 31 percent for Clinton and 27 percent for Bush."[11] That is, voters chose rational and optimizing voting strategies given their (perhaps inaccurate) understanding of the chances of the three candidates. It seems, however, that if everyone had believed Perot had a chance to win, he might have done so, but since no one thought he had a chance, he did not.

In elections for legislatures, where many candidates are chosen, the performance of third parties is very different under a system of proportional representation of the whole population in the whole legislature than under a system of plurality in separate constituencies. Britain has the constituency and plurality system. In the last 50 years, the Labor and Conservative Parties have shared power. The Liberal Party, despite a sizable third-place support in the electorate, has suffered from strategic voting and therefore has had disproportionately few seats in Parliament. Italy has had the nationwide list and proportional representation system; there is no need to vote strategically in such a system and even small parties can have significant presence in the legislature. Often no party has a clear majority of seats, and small parties can affect policy through the bargaining for alliances.

Of course, a party cannot flourish if it is largely ineffective in influencing the country's political choices. Therefore we tend to see just two major parties in countries with the plurality system, and several parties in those with the proportional representation system. Political scientists call this observation *Duverger's law*.

In the legislature, the constituency system tends to produce only two major parties, often one with a clear majority of seats, and therefore more decisive government. But it runs the risk that the minority's interests will be overlooked, that is, of producing a "tyranny of the majority." A proportional representation system gives better voice to minority views. But it can produce inconclusive bar-

---

[11] "Ross Reruns," *Newsweek.* Special Election Recap Issue, November 18, 1996, p. 104.

gaining for power and legislative gridlock. Interestingly, each country seems to believe that its system performs worse and considers switching to the other—in Britain there are strong voices calling for proportional representation, and Italy has been seriously considering a constituency system.

B. Pairwise Voting

When you know you are bound by a pairwise method like the amendment procedure, you can use your prediction of the second-round outcome to determine your optimal voting strategy in the first round. It may be in your interest to appear committed to a particular candidate or policy in the first round, even if it is not your most-preferred alternative, so that your least-favorite alternative cannot win the entire election in the second round.

We return here to our example of the city council with an agenda-setting chair who determines the order of the votes to be taken on three possible welfare policies; again, all three preference rankings are assumed to be known to the entire council. Suppose Councillor Left, who most prefers the Generous welfare package, is appointed chair and sets the Average and Limited policies against each other in a first vote, with the winner facing off against the Generous policy in the second round. If the three councillors vote strictly according to their preferences, shown in Figure 14.1, Average will beat Limited in the first vote and Generous will then beat Average in the second vote; the chair's preferred outcome will be chosen. The city councillors are likely to be well-trained strategists, however, who can look ahead to the final round of voting and use rollback to determine which way to vote in the opening round.

In the scenario just described, Councillor Center's least-preferred policy will be chosen in the election. Realizing this to be the likely outcome, she might consider the possibility of voting strategically in the first round in order to alter the election's outcome. If Center votes for her most-preferred policy in the first round, she could vote for the Average policy, which would then beat Limited in that round and lose to Generous in round 2. However, she could instead vote strategically for the Limited policy in the first round, which would be enough to help Limited beat Average on the first vote. Then, when Limited is set up against Generous in the second round, Generous will lose to Limited. Councillor Center's misrepresentation of her preference ordering with respect to Average and Limited will help her to change the outcome of the election from Generous to Limited. While Limited is not her most-preferred outcome, it is better than Generous from her perspective.

This strategy works well for Center if she can be sure that no other strategic votes will be cast in the election. Thus we need to fully analyze both rounds of voting, using our theory from Chapters 3 and 4, in order to verify the Nash equilibrium strategies for the three councillors. We do this by using rollback on the two simultaneous-vote rounds of the election. In the analysis, we abbreviate the names of the policies to G, A, and L (for Generous, Average, and Limited).

Under the agenda set by Left, the councillors vote on A versus L in the first round and then on the winner versus G in the second round. The second vote, then, could be between either A and G or L and G. To determine equilibrium strategies in that round, we have to look at both of these possibilities.

Figure 14.4 illustrates the outcomes that arise under different vote configurations in each of the possible second-round elections. The two tables at the top of the figure show the winning policy (not payoffs to the players) when A has won the first round and is pitted against G; the tables on the bottom show the winning policy when L has won the first round. In both cases, we use the convention developed in Chapter 4 for illustrating three-player games in strategic form. Councillor Left chooses the row of the final outcome, Center chooses the column, and Right chooses the actual table (left or right).

You should be able to establish that each councillor has a dominant strategy in both the A-versus-G and the L-versus-G elections. In the A-versus-G election, Left's dominant strategy is to vote for G, Center's dominant strategy is to vote for A, and Right's dominant strategy is to vote for G; G will win this election. When

(a) A versus G election

Right votes

for A:

		CENTER	
		A	G
LEFT	A	A	A
	G	A	G

for G:

		CENTER	
		A	G
LEFT	A	A	G
	G	G	G

(b) L versus G election

Right votes

for L:

		CENTER	
		L	G
LEFT	L	L	L
	G	L	G

for G:

		CENTER	
		L	G
LEFT	L	L	G
	G	G	G

FIGURE 14.4 Election Outcomes in Two Possible Second-Round Votes

the councillors consider L versus G, Left's dominant strategy is still to vote for G, and Right and Center both have a dominant strategy to vote for L; in this vote, L wins.

We can express the councillors' second-round strategies as contingent on the winner of the first-round election. That is, Right's dominant strategy in the second round can be described as "vote for G if A has won the first round, and vote for L if L has won the first round," and similarly for Left and Center. However, in this particular case, we can actually describe the second-round voting strategies more easily. A quick check shows that all of the councillors vote according to their true preferences in this round. Thus their dominant strategies are all the same: "vote for the alternative I prefer." Because there is no future to consider in the second-round vote, the councillors simply vote for whichever policy ranks higher in their preference ordering.

Given the outcomes of the second-round election shown in Figure 14.4, we now step back to consider optimal voting strategies in the first round. There is only one possible configuration of alternatives in that round, A versus L. And since we know how the councillors will vote in the next round given the winner here, we show the outcome of the entire election in the tables in Figure 14.5.

As an example of how we arrived at these outcomes, consider the G in the upper-left-hand cell of the right-hand table in Figure 14.5. The outcome in that cell occurs when Left and Center both vote for A in the first round while Right votes for L. Thus A and G are paired in the second round and, as we saw in Figure 14.4, G wins. The other outcomes are derived in similar fashion.

Look carefully now at Figure 14.5. Councillor Left (who is the chair and has set the agenda) has a dominant strategy to vote for A in this round. Her vote is not strategic; she prefers A to L. Councillor Right has a dominant strategy to vote for L; this vote is not strategic either, because Right prefers L to A. Councillor Center, however, has a dominant strategy to vote for L *even though* she strictly prefers A to L. As our intuitive discussion noted, she has a strong incentive to misrepresent her preferences in the first round of voting; it turns out she is the

Right votes

for A:

		CENTER	
		A	L
LEFT	A	G	G
	L	G	L

for L:

		CENTER	
		A	L
LEFT	A	G	L
	L	L	L

FIGURE 14.5 Election Outcomes Based on First-Round Votes

only one who votes strategically. And her behavior changes the outcome of the election from G (the result in the absence of strategic voting) to L.

Note that the chair, Councillor Left, set the agenda in the hopes of having her most preferred alternative chosen. Instead, her *least*-preferred alternative has prevailed. It appears that the power to set the agenda may not be so beneficial after all. But of course, Councillor Left should anticipate the strategic behavior. Then she can choose the agenda so as to take advantage of her understanding of games of strategy. In fact, if she sets L against G in the first round and then the winner against A, the Nash equilibrium outcome is G, the chair's most-preferred outcome; Right misrepresents her preferences in the first round to vote for G over L in order to prevent A, her least-preferred outcome, from winning. You should verify that this is Councillor Left's best agenda-setting strategy. In the full voting game where setting the agenda is considered an initial, prevoting round, we should expect to see the Generous welfare policy adopted when Councillor Left is chair.

We can also see an interesting pattern emerge when looking more closely at the voting behavior of our councillors in the strategic version of the election. There are pairs of councillors who vote "together" (the same as each other) in both rounds. Under the original agenda, Right and Center vote together in both rounds, and in the suggested alternative (L versus G in the first round), Right and Left vote together in both rounds. In other words, a sort of long-lasting coalition has formed between two councillors in each case.

Strategic voting of this type appears to have occurred in Congress. One example pertains to a federal school construction funding bill considered in 1956.[12] Before being brought to a vote against the status quo of no funding, the bill was amended in the House of Representatives to require that aid be offered only to states with no racially segregated schools. Under the parliamentary voting rules of Congress, a vote on whether to accept the so-called Powell Amendment was taken first, with the winning version of the bill considered afterward. Political scientists who have studied the history of this bill argue that opponents of school funding strategically misrepresented their preferences regarding the amendment in order to defeat the original bill. A key group of Representatives voted for the amendment but then joined opponents of racial integration in voting against the full bill in the final vote; the bill was defeated. Voting records of this group indicate that many of them had voted against racial integration issues in other circumstances, implying that their vote for integration in this case was merely an instance of strategic voting and not an indication of their true feelings regarding school integration.

---

[12] Ordeshook, *Game Theory and Political Theory*, mentions this example on p. 83. A more complete analysis of the case can be found in Riker, *Liberalism Against Populism*.

C. The Borda Count

We showed earlier that the Borda count procedure can give rise to intriguing paradoxical results when an alternative is removed from contention in an election. The Borda count can also be strategically manipulated by voters to guarantee that a certain alternative is chosen, or that another is definitely *not* chosen.

Suppose you are a football coach participating in a coaches' poll to choose the national collegiate football champion. Coaches are asked to list teams on the ballot in the order they wish to rank them; thus, a coach's top-ranked team appears at the top of the ballot, and so on. This particular season has seen no single team dominate the early-season polls; rather, two excellent teams have been trading the number 1 position back and forth all year. In the days before the final vote, it has become clear that the coaches are divided nearly 50-50 over the two top candidates. All the coaches seem to agree, however, on which two teams are at the top.

Under the circumstances, you might want to adopt some strategic voting behavior. In particular, if you are not certain whether the team of your choice has a majority, you can significantly improve their chances of coming out on top in the poll by putting their leading opposition at the bottom of your ballot. You do this even though you really believe that team is the second best in the nation because, by doing so, you significantly reduce the number of points it earns and you significantly increase your choice's chances of winning. Even if your choice is a few votes short of a majority, your behavior can easily send it on to victory—assuming the rest of the coaches do not also make use of similar strategies.

This particular example has the trappings of a multi-person prisoners' dilemma game, such as those discussed in Section 2 of Chapter 11. Each individual coach has an incentive to put her own top choice first and her second choice *last*. But if every coach does this, then the winner is the same team that would have won in the absence of strategic voting. Not only that, but when the ballots are published, all of the coaches are criticized for unsportsman-like conduct. It is for just this reason that ballots are open to public scrutiny in sports award elections using any variant on the Borda count. The fear of public reprisal is generally enough to keep voters "honest" in the elections. The temptation to misrepresent preferences still exists, however, and does lead to oddly ordered ballots in the occasional election.

5 THE MEDIAN VOTER THEOREM

All of the preceding sections have focused on the behavior, strategic and otherwise, of voters in multiple alternative elections. However, strategic analysis can also be applied to *candidate* behavior in such elections. Given a particular distribution of voters and voter preferences, candidates will, for instance, need to determine optimal strategies in building their political platforms. When there are just two candidates in an election, when voters are distributed in a "reasonable" way along the political spectrum, and when each voter has "reasonably" consistent preferences (later in this section we will define what we mean by "reasonable"), the **median voter theorem** tells us that both candidates will position themselves on the political spectrum at the same place as the median voter. The **median voter** is the "middle" voter in that distribution, more precisely, the one at the 50th percentile.

The full game here has two stages. In the first stage, candidates choose their locations on the political spectrum. In the second stage, voters elect one of the candidates. The general second-stage game is open to all of the varieties of strategic misrepresentation of preferences that we discussed earlier. Hence we have reduced the choice of candidates to two for our analysis in order to prevent such behavior from arising in equilibrium. With only two candidates, second-stage votes will directly reflect voter preferences and the first-stage location decision of the candidates remains the only truly interesting part of the larger game. It is in that stage that the median voter theorem defines Nash equilibrium behavior.

A. Discrete Political Spectrum

Let us first consider a population of 9 million voters, each of whom has a preferred position on a five-point political spectrum: Far Left (FL), Left (L), Center (C), Right (R), and Far Right (FR). Voters will choose the candidate who publicly identifies herself as being closer to their own position on the spectrum in an election. If both candidates are politically equidistant from a group of like-minded voters, each voter flips a coin to decide which candidate to choose; this process gives each candidate one-half of the voters in that group.

In any given election, candidates choose political positions on the spectrum to maximize the number of votes they receive (and thus increase the chances of winning). We can use strategic analysis to determine the optimal location strategies of the two candidates if we know how voters are distributed among the different possible positions. So, let us suppose that the 9 million voters are spread symmetrically around the center of the political spectrum, with 3 million located at C, 2 million each at L and R, and 1 million each at FL and FR. We illustrate this **discrete distribution** of voters in Figure 14.6, using a **histogram,** or bar chart. The height of each bar indicates the number of voters located at that position.

Let us further suppose there is an upcoming Presidential election between

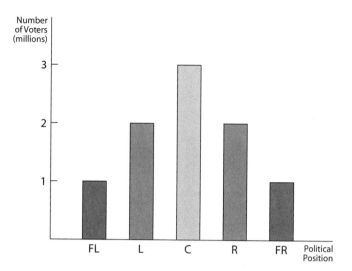

FIGURE 14.6   Discrete Distribution of Voters

an Ex-Governor and an Ex-Actor.[13] Under the configuration of voters illustrated in Figure 14.6, we can construct a payoff table for the two candidates showing the number of votes each can expect to receive under all of the different combinations of political platform choices. This five-by-five table is shown in Figure 14.7, with totals denoted in millions of votes.

Here is how the votes are allocated. Consider first what happens when both candidates choose the *same* position (the results shown in the five cells along the top-left-to-bottom-right diagonal of the table). Each candidate gets exactly one-half of the votes—since all voters are equidistant from each candidate, all of them flip coins to decide their choices, and each candidate garners 4.5 million votes. When the two candidates choose *different* positions, the more-left candidate gets all the votes at or to the left of her position while the more-right candidate gets all the votes at or to the right of her position. In addition, each candidate gets the votes in central positions closer to her than to her rival, and the two of them split the votes from any voters in a central position equidistant between them. Thus, if Ex-Governor locates herself at L while Ex-Actor locates herself at FR, Ex-Governor gets the 2 million votes at L as well as the 1 million at FL *and* the 3 million at C (since C is closer to L than to FR). Ex-Actor gets only the 1 million votes at FR and the 2 million at R. The payoff is (6,3). Similar calculations determine the outcomes in the rest of the table.

---

[13]Any resemblance between our hypothetical candidates and actual past candidates in the United States is not meant to imply an analysis of their performances relative to the Nash equilibrium. Nor is our distribution of voters meant to typify U.S. voter preferences.

		EX-ACTOR				
		FL	L	C	R	FR
EX-GOVERNOR	FL	4.5, 4.5	1, 8	2, 7	3, 6	4.5, 4.5
	L	8, 1	4.5, 4.5	3, 6	4.5, 4.5	6, 3
	C	7, 2	6, 3	4.5, 4.5	6, 3	7, 2
	R	6, 3	4.5, 4.5	3, 6	4.5, 4.5	8, 1
	FR	4.5, 4.5	3, 6	2, 7	1, 8	4.5, 4.5

FIGURE 14.7 Election Results: Symmetric Voter Distribution

The table in Figure 14.7 is large, but the game can be solved very quickly. We begin with the now familiar search for dominant, or dominated, strategies for the two players. Immediately we see that, for Ex-Governor, FL is dominated by L and FR is dominated by R; the same is true for Ex-Actor (since the game is symmetric). Furthermore, L and R are dominated by C for both players. The only remaining cell in the table is (C, C); this is the Nash equilibrium.

We now take note of two important characteristics of the equilibrium in the candidate location game. First, both candidates locate at the *same* position in equilibrium. This illustrates the **principle of minimum differentiation,** a general result in all two-player location models, whether they explain political platform choice by presidential candidates, hotdog-cart location choices by street vendors, or product feature choices by electronics manufacturing firms.[14] When the individuals who vote for or buy from you can be arranged on a well-defined spectrum of preferences, you do best by looking as much like your rival as possible. This explains a diverse collection of behaviors on the part of political candidates and businesses. It may help you understand, for example, why there is never just one gas station at a heavily traveled intersection, or why all brands of four-door sedans (or minivans or sport utility vehicles) seem to look the same even though every brand claims to be coming out continually with a "new" look.

Second, and perhaps more crucial, both candidates locate at the position of the median voter in the population. In our example, the median position is occupied by the middle 1 million voters when you count in from either side of the political spectrum—right at the center of the spectrum. This is the result predicted by the median voter theorem.

The median voter theorem can be expressed in different ways. One version

---

[14]Economists learn this result within the context of Hotelling's model of spatial location. See Harold Hotelling, "Stability in Competition," *Economic Journal*, vol. 39, no. 1 (March 1929), pp. 41–57.

states simply that the position of the median voter is the equilibrium location position of the candidates in a two-candidate election. Another version says that the position that the median voter most prefers will be the Condorcet winner in a round-robin where every position is pitted against every other . This position will defeat every other position in a pairwise contest. For example, if M is this median position and L is any position to the left of M, then M will get all the votes of people who most prefer a position at or to the right of M, plus some to the left of M but closer to M than to L. Thus M will get more than 50 percent of the votes. The two versions amount to the same thing, because in a two-candidate election, both seeking to win a majority will adopt the Condorcet-winner position. These interpretations are identical. In addition, to guarantee that the result holds for a particular population of voters, the theorem (in either form) requires that voters' preferences be "reasonable" as suggested earlier. *Reasonable* here means "single-peaked." Each voter has a unique most-preferred position on the political spectrum, and her utility (or payoff) decreases away from that position in either direction. In actual U.S. Presidential elections, the theorem is borne out by the tendency for the main candidates to make very similar promises to the electorate.

That candidates locate at the position of the median voter is a general proposition. But it does not require that the electorate be distributed symmetrically around the center of the political spectrum, as was the case in our hypothetical example. Suppose the 9 million voters are instead divided into 4 million at L, 2 million at FR, and 1 million each at FL, C, and R. Then Ex-Governor and Ex-Actor face the payoff table shown in Figure 14.8.

You should practice your skills in successive elimination of dominated strategies to convince yourself that the equilibrium does indeed entail both candidates locating at L, the new position of the median voter. Notice that this result helps to explain why state political candidates in Massachusetts, for example,

		EX-ACTOR				
		FL	L	C	R	FR
EX-GOVERNOR	FL	4.5, 4.5	1, 8	3, 6	5, 4	5.5, 3.5
	L	8, 1	4.5, 4.5	5, 4	5.5, 3.5	6, 3
	C	6, 3	4, 5	4.5, 4.5	6, 3	6.5, 2.5
	R	4, 5	3.5, 5.5	3, 6	4.5, 4.5	7, 2
	FR	3.5, 5.5	3, 6	2.5, 6.5	2, 7	4.5, 4.5

FIGURE 14.8   Election Results: Asymmetric Voter Distribution

tend *all* to be more liberal than candidates for similar positions in Texas or South Carolina.

B. Continuous Political Spectrum

The median voter theorem can also be proven for the case of a continuous distribution of political positions. Rather than having five, three, or any finite number of positions from which to choose, a **continuous distribution** assumes there are effectively an infinite number of political positions. These political positions are then associated with locations along the real number line between 0 and 1.[15] Voters are still distributed along the political spectrum as before, but because the distribution is now continuous rather than discrete, we use a voter **distribution function** rather than a histogram to illustrate voter locations. Two common functions—the **uniform distribution** and the **normal distribution**—are illustrated in Figure 14.9.[16] The area under each curve represents the total number of votes available; at any given point along the interval from 0 to 1, such as x in Figure 14.9a, the number of votes up to that point is determined by finding the area under the distribution function from 0 to x. It should be clear that the median voter in each of these distributions is located at the center of the spectrum, at position 0.5.

It is not feasible to construct a payoff table for our two candidates in the continuous-spectrum case; such tables must necessarily be finitely dimensioned and thus cannot accommodate an infinite number of possible strategies for players. We can, however, solve the game by applying the same strategic logic that we used for the discrete (finite) case discussed in Section 5.A.

Consider the options of Ex-Governor and Ex-Actor as they contemplate the possible political positions open to them. Each knows she must find her Nash equilibrium strategy—her best response to the equilibrium strategy of her rival. We can define a set of strategies that are best responses quite easily in this game, even though the complete set of possible strategies is impossible to delineate.

Suppose Ex-Actor locates at a random position on the political spectrum, such as x in Figure 14.9a. Ex-Governor can then calculate how the votes will be split for all possible positions she might choose. If she chooses a position to the left of x, she gets all the votes to her left and half of the votes lying between her and Ex-Actor's position. If she locates to the right of x, she gets all the votes to her right and half of the votes lying between her and x. Finally, if she, too, locates at x, she and Ex-Actor split the votes 50-50. These three possibilities effectively sum-

---

[15]This construction is the same one used in Chapters 10 and 11 for analyzing large populations of individuals.

[16]We do not delve deeply into the mechanics underlying distribution theory nor the mathematics required to calculate the exact proportion of the voting population lying to the left or right of any particular position on the continuous political spectrum. Here we present only enough information to convince you that the median voter theorem continues to hold in the continuous case.

(a) Uniform distribution

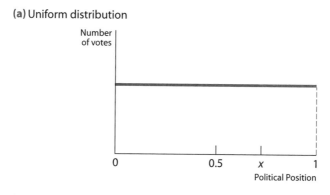

(b) Normal distribution

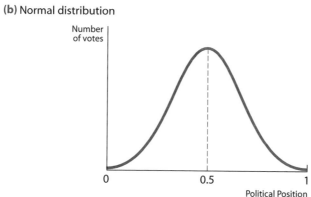

FIGURE 14.9   Continuous Voter Distributions

marize all of Ex-Governor's location choices, given that Ex-Actor has chosen to locate at x.

But which of the response strategies just outlined is Ex-Governor's "best" response? The answer depends on the location of x relative to the median voter. If x is to the right of the median, then Ex-Governor knows that her best response will be to maximize the number of votes she gains, which she can do by locating an infinitely small bit to the left of x.[17] In that case, she effectively gets all the votes from 0 to x while Ex-Actor gets those from x to 1. When x is to the right of the median, as in Figure 14.9a, then the number of voters represented by the area under the distribution curve from 0 to x is by definition larger than the number of voters from x to 1, so Ex-Governor would win the election. Similarly, if x is to the left of the median, Ex-Governor's best response will be to locate an infinitely small bit to the right of x and thus gain all the votes from x to 1. When x is exactly at the median, Ex-Governor does best by choosing to locate at x also. The best-response strategies for Ex-Actor are constructed exactly the same way and, given the location of her rival, are exactly the same as those described for Ex-Governor.

[17] Such a location, infinitesimally removed from x to the left, is feasible in the continuous case. In our discrete example, candidates had to locate at exactly the same position.

We now have complete descriptions of the best-response strategies for both candidates. Which exact strategies constitute the Nash equilibrium here? This is easy to determine if we pick any starting location for one of the candidates and apply the best-response strategies over and over until each candidate is located at a position that represents her best response to the position chosen by her rival. If Ex-Actor were contemplating locating at x in Figure 14.9a, Ex-Governor would want to locate just to the left of x, but then Ex-Actor would want to locate just to the left of that, and so on. Only when the two candidates locate exactly at the median of the distribution (whether the distribution is uniform or normal or some other kind), do they find that their decisions are best responses to each other. Again we see that the Nash equilibrium is for both candidates to locate at the position of the median voter.

More complex mathematics are needed to prove the continuous version of the median voter theorem to the satisfaction of a true mathematician. For our purposes, however, the discussion given here should convince you of the *validity* of the theorem in both its discrete and continuous forms. The most important *limitation* of the median voter theorem is that it applies when there is just one issue, or a one-dimensional spectrum of political differences. If there are two or more dimensions—for example, if being conservative versus liberal on social issues does not coincide with being conservative versus liberal on economic issues—then the population is spread out in a two-dimensional "issue space" and the median voter theorem no longer holds. The preferences of every individual can be single-peaked, in the sense that the individual has a most-preferred point and her payoff value drops away from this point in all directions, like the height going away from the peak of a hill. But we cannot identify a median voter, in two dimensions, such that exactly the same number of voters have their most-preferred point to the one side of her as to the other side. In two dimensions there is no unique sense of side, and the numbers of voters to the two sides can vary, depending on just how we define "side."

SUMMARY

Elections can be held using a variety of different voting procedures that alter the order in which issues are considered or the manner in which votes are tallied. Voting procedures are classified as *binary, plurative,* or *mixed methods.* Binary methods include *majority rule,* as well as *pairwise* procedures like the *Condorcet method,* and the *amendment procedure. Plurality rule,* the *Borda count,* and *approval voting* are plurative methods. And *majority runoffs, rounds,* and *proportional representation* are mixed methods.

Voting paradoxes (like the *Condorcet,* the *agenda,* and the *reversal paradox*) show how counterintuitive results can arise due to difficulties associated with aggregating preferences or due to small changes in the list of issues being considered. Another paradoxical result is that outcomes in any given election under a given set of voter preferences can change depending on the voting procedure used. Certain principles for evaluating voting methods can be described, although Arrow's *impossibility theorem* shows that no one system satisfies all of the criteria at the same time.

Voters have scope for strategic behavior in the game that chooses the voting procedure or in an election itself through the *misrepresentation of their own preferences*. Preferences may be strategically misrepresented in order to achieve the voter's most-preferred, or to avoid the least-preferred, outcome.

Candidates may also behave strategically in building a political platform. A general result known as the *median voter theorem* shows that in elections with only two candidates, both locate at the preference position of the *median voter*. This result holds when voters are distributed along the preference spectrum either *discretely* or *continuously*.

KEY TERMS

agenda paradox (471)
amendment procedure (464)
approval voting (465)
binary method (463)
Borda count (465)
Condorcet method (463)
Condorcet paradox (467)
Condorcet winner (464)
Copeland index (464)
continuous distribution (486)
discrete distribution (482)
distribution function (486)
Hare procedure (466)
histogram (482)
impossibility theorem (473)
intransitive ordering (468)
majority rule (463)
majority runoff (465)
median voter (482)
median voter theorem (482)

mixed method (464)
multistage procedure (463)
normal distribution (486)
pairwise voting (463)
plurality rule (464)
plurative method (464)
principle of minimum
  differentiation (484)
proportional
  representation (466)
reversal paradox (469)
rounds (466)
single transferable vote (466)
social ranking (467)
spoiler (475)
strategic misrepresentation
  of preferences (467)
strategic voting (467)
transitive ordering (468)
uniform distribution (486)

EXERCISES

1. Consider a vote being taken by three roommates, A, B, and C, who share a triple-dorm room. They are trying to decide which of three elective courses to take together this term. (Each roommate has a different major and is taking required courses in her major for the rest of her courses.) Their choices are Philosophy, Geology, and Anthropology, and their preferences for the three courses are as shown here:

A	B	C
Philosophy	Anthropology	Geology
Geology	Philosophy	Anthropology
Anthropology	Geology	Philosophy

They have decided to have a two-round vote and will draw straws to determine who sets the agenda.

Suppose A sets the agenda and wants the Philosophy course to be chosen. How should she set the agenda to achieve this outcome if she knows that everyone will vote truthfully in all rounds? What agenda should she use if she knows that they will all vote strategically?

2. Repeat Exercise 1 for the situation in which B sets the agenda and wants to ensure that Anthropology wins.

3. Suppose that voters 1 through 4 are being asked to consider three different candidates—A, B, and C—in a Borda count election. Their preferences are:

1	2	3	4
A	A	B	C
B	B	C	B
C	C	A	A

Assume that voters will cast their votes truthfully (no strategic voting).

Find a Borda weighting system—a number of points to be allotted to first, second, and third preferences—in which candidate A wins.

4. Repeat Exercise 3 to find a Borda weighting system in which candidate B wins.

5. Consider a group of 50 residents attending a town meeting in Massachusetts. They must choose between three proposals for dealing with town garbage.

Proposal 1 asks the town to provide garbage collection as one of its services, Proposal 2 calls for the town to hire a private garbage collector to provide collection services, and Proposal 3 calls for residents to be responsible for getting their own garbage. There are three types of voters. The first type prefers Proposal 1 to Proposal 2 and Proposal 2 to Proposal 3; there are 20 of these voters. The second type prefers Proposal 2 to Proposal 3 and Proposal 3 to Proposal 1; there are 15 of these voters. The third type prefers Proposal 3 to Proposal 1 and Proposal 1 to Proposal 2; there are 15 of these.

(a) Under a plurality voting system, which proposal wins?

(b) Suppose voting proceeds using a Borda count, in which voters list the proposals, in order of preference, on their ballot. The proposal listed first (or at the top) on a ballot gets three points, the proposal listed second gets two points and the proposal listed last gets one point. In this situation, with no strategic voting, how many points are gained by each proposal? Which proposal wins?

(c) What strategy can the second and third types of voters use to alter the outcome of the Borda count vote in part (b) to one that *both* types prefer? If they use this strategy, how many points does each proposal get, and which wins?

6. During the Cuban missile crisis, serious differences of opinion arose within the ExComm group advising President John Kennedy, which we summarize here. There were three options: Soft (a blockade), Medium (a limited air strike), and Hard (a massive air strike or invasion). There were also three groups in ExComm. The civilian doves ranked the alternatives Soft best, Medium next, and Hard last. The civilian hawks preferred Medium best, Hard next, and Soft last. The military preferred Hard best, but they felt "so strongly about the dangers inherent in the limited strike that they would prefer taking no military action rather than to take that limited strike."[18] In other words, they ranked Soft second and Medium last.

If the matter were to be decided by majority vote in ExComm, and if each group was about one-third the total membership of ExComm, which alternative, if any, would win?

7. John Paulos, in his book *A Mathematician Reads the Newspaper*, gives the following caricature based on the 1992 Democratic Presidential primary caucuses. There are five candidates: Jerry Brown, Bill Clinton, Tom Harkin, Bob Kerrey, and Paul Tsongas. There are 55 voters, with different preference orderings concerning the candidates. There are six different orderings, which we

---

[18] *The Kennedy Tapes: Inside the White House During the Cuban Missile Crisis*, ed. Ernest R. May and Philip D. Zelikow (Cambridge, Mass.: Harvard University Press, 1997), p.97.

label I through VI. The preference orderings (1 for best to 5 for worst) along with the numbers of voters with each ordering, are shown in the following table; the candidates are identified by the first letters of their last names.[19]

RANKING	GROUPS (AND THEIR SIZES)					
	I (18)	II (12)	III (10)	IV (9)	V (4)	VI (2)
1	T	C	B	K	H	H
2	K	H	C	B	C	B
3	H	K	H	H	K	K
4	B	B	K	C	B	C
5	C	T	T	T	T	T

(a) First suppose that all voters vote sincerely, and consider the outcomes of each of several different election rules. Show each of the following outcomes: (i) Under the plurality method (the one with the most first preferences), Tsongas wins. (ii) Under the runoff method (the top two first preferences go into a second round), Clinton wins. (iii) Under the elimination method (at each round, the one with the least first preferences at that round is eliminated, and the rest go into the next round), Brown wins. (iv) Under the Borda count method (five points for first preference, four for second, and so on; the candidate with the most points wins), Kerrey wins. (v) Under the Condorcet method (pairwise comparisons), Harkin wins.

(b) Suppose that you are a Brown, Kerrey, or Harkin supporter. Under the plurality method you would get your worst outcome. Can you benefit by voting strategically? If so, how?

(c) Are there opportunities for strategic voting under each of the other methods as well? If so, explain who benefits from voting strategically and how they can do so.

8. An election has three candidates and takes place under the plurality rule. There are numerous voters, spread along an ideological spectrum from left to right. Represent this by a horizontal straight line whose extreme points are 0 (left) and 1 (right). Voters are uniformly distributed along this spectrum; so the number of voters in any segment of the line is proportional to the length of that segment. Thus a third of the voters are in the segment from 0 to 1/3, a quarter in the segment from 1/2 to 3/4, and so on. Each voter votes for the candidate whose declared position is closest to the voter's own position. The can-

---

[19] John Allen Paulos, *A Mathematician Reads the Newspaper* (New York: Basic Books, 1995), pp.104–106.

didates have no ideological attachment, and take up any position along the line, each seeking only to maximize her share of votes.

(a) Suppose you are one of the three candidates. The left-most of the other two is at point x, and the right-most is at the point $(1 - y)$, where $x + y < 1$ (so the right-most candidate is a distance y from 1). Show that your best response is to take up the following positions under the given conditions:

(i) just slightly to the left of x if $x > y$ and $3x + y > 1$,

(ii) just slightly to the right of $(1-y)$ if $y > x$ and $x + 3y > 1$,

(iii) and exactly halfway between the other candidates   if $3x + y < 1$ and $x + 3y < 1$.

(b) In a graph with x and y along the axes, show the areas (the combination of x and y values) where each of the response rules i to iii in part (a) is best for you.

(c) From your analysis, what can you conclude about the Nash equilibrium of the game where the three candidates each choose positions?

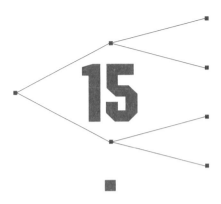

Bidding Strategy and Auction Design

W E HAVE ARGUED REPEATEDLY in earlier chapters that the strategic topics we are discussing—mixed strategies, threats and promises, screening and signaling, even brinkmanship—are important and relevant to all of you in your daily lives. In this chapter, we continue that theme in our discussion of strategies in auctions. Now, you might argue that you have never been, nor do you anticipate ever being, at Christie's or Sotheby's, or an antique auction in the wilds of Vermont, or a livestock auction in Illinois. Nevertheless, you do probably participate in more auctions than you imagine.

Auctions entail the transfer of a particular object from a seller to a buyer (or bidder) for a certain price (or bid). Considered in this simple form, many market transactions resemble auctions. For example, stores like Filene's Basement in Boston use a clever pricing strategy to keep customers coming back for more: they reduce the prices on items remaining on the racks successively each week until either the goods are purchased or the price gets so low that they donate the items to charity. Shoppers love this. Little do they realize that they are participating in what is known as a *descending*, or *Dutch, auction.* This is one of the types of auctions described in detail in this chapter.

Even if you do not personally participate in many auctions, your life is profoundly influenced by auctions. The Federal Communications Commission (FCC) in recent years has auctioned off large parts of the electromagnetic broadcasting spectrum. The kind of television you watch for the next few decades and the kinds of cellular phones you use will be affected by this process and its outcomes. These auctions have already raised some $23 billion in government revenues, of which approximately $12 billion were collected as of October 1997. Because these revenues have made significant contributions to the federal budget, they have affected important macroeconomic magnitudes, such as interest rates. Understanding how auctions work will help you understand these important events and their implications.

This chapter explains the different types of auctions you might encounter, as well as the strategies you might want to employ as either the buyer or the seller in such situations. While the formal theory of auctions relies on advanced calculus to derive its results, we eschew most of this difficult mathematics in favor of more intuitive descriptions of optimal behavior and strategy choice.[1] As an example of how auction design and bidding theory has been used in reality, we provide a detailed discussion of the recent airwave spectrum auctions held by the FCC.

1 TYPES OF AUCTIONS

Auctions come in a large variety of forms involving different methods for the submission of bids, for the determination of the identity of the winner, and of the final price paid by the winner, and for valuing the object being auctioned. We start our study of auctions by simply running through a number of the possibilities, describing their characteristics and mechanics.

There are two methods for submitting bids at an auction. In a **sealed bid** auction, buyers secretly write their bid on a piece of paper and seal it in an envelope before handing it in. (Sealed-bid auctions do not require that bids literally be sealed in envelopes—only that no bids be visible to other bidders before the auction is closed.) Thus each bidder gets one bid in this type of auction.[2] In **open-outcry bidding,** on the other hand, bidders call out their bids in public; this method perhaps fits best the popular vision of the way auctions work—an image that includes feverish bidders and an auctioneer. Not all auctions have auctioneers, however, and even those that do may not be characterized by fever-

---

[1] Our sources include Paul Milgrom, "Auctions and Bidding: A Primer," Orley Ashenfelter, "How Auctions Work for Wine and Art," and John G. Riley, "Expected Revenues from Open and Sealed Bid Auctions," all in the *Journal of Economic Perspectives*, vol. 3, no. 3 (Summer 1989), pp. 3–50, and readable by those of you with a reasonably strong calculus background. More complex information on the subject can be found in R. Preston McAfee and John McMillan, "Auctions and Bidding," *Journal of Economic Literature*, vol. 25 (June 1987), pp. 699–738, and Paul Milgrom, "Auction Theory," in *Advances in Economic Theory*, ed. Truman Bewley (Cambridge: Cambridge University Press, 1987), pp. 1–32.

[2] Technically, bidders could submit as many bids as they wanted, but only the highest one would have any potential impact on the outcome of the auction, so why bother to outbid yourself? (Unless you plan to renege on your high bid—more on this later.)

ish bidding. Sealed-bid auctions have no need for an auctioneer; they require only an overseer who opens the bids and determines the winner. And open-outcry auctions may entail bidding that occurs with the mere nod of a head or the flick of a wrist.

Two other well known types of auctions do require auctioneers, but only one of these—the **English auction,** also known as an **ascending auction**—conforms to the popular impression of auctions. Such auctions were, and are, the norm at English auction houses like Christie's and Sotheby's. The auction houses have a conventional auctioneer who starts at a low price and calls out successively higher prices for an item, waiting to receive a bid at each price before going on; when no further bids can be obtained, the item goes to the most recent—highest—bidder. There are also now Internet sites running what are essentially English auctions for virtually any item imaginable. ONSALE, Inc., (at http://www.onsale.com) is a publicly traded company that conducts real-time auctions in three main categories (computers, home office, and sports/fitness); eBay Auction Classifieds (at http://www.ebay.com) offers virtually anything you might want from a very long list of categories.[3] English auctions can involve any number of bidders, although only the top bidder gains the item.

The other type of auctioneer-led auction is the **Dutch,** or **descending, auction**. Dutch auctions, which get their name from the way tulips are auctioned in the Netherlands, work in the opposite "direction" from English auctions: The auctioneer starts at an extremely high price and calls out successively *lower* prices until one of the assembled potential bidders accepts the price, makes a bid, and takes the item. The Dutch tulip auctions use a "clock" that ticks down to ever lower prices until one bidder "stops the clock" and collects her tulips.

There are also several methods by which auction prices can be determined. The most common is for the highest bidder to pay her bid and take her item. An auction using this pricing scheme is known as a **first-price** auction. In a **second-price** auction, the highest bidder wins the item but pays a price equal to the bid of the *second-highest* bidder. A second-price structure can be extremely useful for eliciting truthful bids, as we will see in Section 4. (Third-, fourth-, fifth-, and lower-price auctions also exist, although these are not commonly used.)

Finally, there are two ways in which bidders may value an item up for auction. In **common-value,** or **objective-value,** auctions, the value of the object is the same for all the bidders, but each bidder generally knows only an imprecise estimate of it. For example, an oil drilling tract has a given amount of oil that should entail the same revenue for all companies, but each company has only its

---

[3]ONSALE offers what it calls "Yankee auctions," which are similar to English auctions except that they offer multiple (identical) items for sale; winning bids are classified on the basis of price *and* quantity. eBay offers what it calls a Dutch auction, although its version does not conform strictly to the definition given here. Sellers at eBay can choose the type of auction they desire for the sale of their particular product.

own expert's estimate of the amount of oil contained under the tract. Similarly, each bond trader has only an estimate of the future course of interest rates. In such auctions, each bidder should be aware of the fact that other bidders possess some (however sketchy) information about an object's value, and she should attempt to infer the contents of that information from the actions of rival bidders.

The second type of valuation process gives us **private-value,** or **subjective-value,** auctions. In this case, each bidder places different values on an object. For example, a gown worn by Princess Diana or a necklace worn by Jacqueline Bouvier Kennedy Onassis may have sentimental value to some bidders. Bidders do not know each other's valuations of an object, and the seller does not know any of the bidders' valuations either, although it is possible that each may have some rough estimates. This information problem affects not only bidding strategies in private-value auctions, but also the seller's strategy in designing the form of auction to identify the highest valuation and to extract the best price.

In setting up an auction, various combinations of bidding type (sealed or open), setting (English or Dutch), pricing (first or second), and valuation method (private or common) are possible. Other, odder configurations can also be used to sell goods at auction. For example, you could set up an auction in which the highest bidder wins but the *top two bidders* pay their bids, or one in which the high bidder wins but *all bidders* pay their bids, a procedure that we discuss in Section 5. We do not attempt to consider all such combinations, however. Rather, we analyze several of the most common auction schemes using examples that bring out important strategic concepts.

2 THE WINNER'S CURSE

A standard but often ignored outcome arises in common-value auctions. Recall that such auctions entail the sale of an object whose value is fixed and identical for all bidders, although each bidder can only estimate it. The **winner's curse** is a warning to bidders that if they win the object in the auction, they are likely to have paid more than it is worth.

Suppose you are a corporate raider bidding for Targetco. Your experts have studied this company and produced estimates that, in the hands of the current management, it is worth somewhere between 0 and $10 billion, all values in this range being equally likely. The current management knows the precise figure, but of course they are not telling you. You believe that, whatever Targetco is worth under existing management, it will be worth 50% more under your control. What should you bid?

You might be inclined to think that, on average, Targetco is worth $5 billion under existing management and thus $7.5 billion, on average, under yours. If so, then a bid somewhere between \$5 and \$7.5 billion should be profitable. But such a bidding strategy reckons without the response of the existing management to your bid. If Targetco is actually worth more than your bid, the current owners are not going to accept the bid. You are going to get the company only if its true worth is toward the lower end of the range.

Suppose you bid amount b. Your bid will be accepted and you will take over the management of Targetco if it is worth somewhere between 0 and b under the current management; on average, you can expect the company to be currently worth $b/2$ if your bid is accepted. In your hands, the average worth will be 50% more than the current worth, or $(1.5)(b/2) = 0.75 \ b$. Because this value is always less than b, you would win the takeover battle only when it was not worth winning! It seems that many raiders have discovered this fact too late.

But it is not only corporate raiders, often involved in one-on-one negotiations with target firms resembling auctions with only one bidder, who are affected by the winner's curse. Similar problems arise when you are competing with other bidders in a common-value auction and all of you have separate estimates for the object's value.

Consider the case of a tract of land (or sea) being offered for oil exploration.[4] You win only when your rivals make estimates of the value of the tract that are lower than your estimate. You should recognize this and try to learn from it.

Suppose the true value of the tract, unknown to any of the bidders, is \$1 billion. (In this case, the seller probably does not know the true value of the tract either.) Suppose there are 10 oil companies in the bidding. Each one's experts estimate the value of the tract with an error of ±\$100 million, all numbers in this range being equally likely. If all 10 of the estimates could be pooled, their arithmetic average would be an unbiased and much more accurate indicator of the true value than any single estimate. But when each bidder sees only one estimate, the largest of these estimates is biased: on average, it will be \$1.08 billion, right near the upper end of the range.[5] Thus, the winning company is likely to pay too much, unless it recognizes the problem and adjusts its bid downward to compensate for this bias. The exact calculation required to determine how far to shade down your bid without losing the auction is difficult, however, because you must also recognize that all the other bidders will be making the same adjustment.

We do not pursue the advanced mathematics required to create an optimal bidding strategy in the common-value auction. However, we can provide you with some general advice. If you are bidding on an item, the question "Would I be willing to purchase the tract of land for \$1.08 billion, given what I know before

---

[4] The United States auctions leases for offshore oil drilling rights.

[5] The 10 estimates will, on average, range from \$0.9 billion to \$1.1 billion (\$100 million on either side of \$1 billion). The low and high estimates will, on average, be at the extremes of the distribution.

submitting my bid?" is very different from the question, "Would I still be willing to purchase the tract of land for $1.08 billion, given what I know before submitting my bid *and* given the knowledge that I will be able to purchase the land only if no one else is willing to bid $1.08 billion for it?"[6] Even in a sealed-bid auction, it is the second question that reveals correct strategic thinking, since you win with any given bid only when all others bid less—only when all other bidders are not willing to pay what you have bid.

If you do not take the winner's curse into account in your bidding behavior, you should expect to lose substantial amounts. The numerical calculations we did earlier for bidding on the hypothetical Targetco showed that this should be so. How real is this danger in practice? Richard Thaler has marshaled a great deal of evidence to show that the danger is very real indeed.[7]

The simplest experiment to test the winner's curse is to auction a jar of pennies. The prize is objective, but each bidder forms a subjective estimate of how many pennies there are in the jar and therefore of the size of the prize; this is a pure example of a common-value auction. Most teachers have conducted such experiments with students and found significant overbidding. In one similar but related experiment, M.B.A. students were asked to bid for a hypothetical company instead of a penny jar. The game was repeated, with feedback after each round on the true value of the company. Only five of 69 students learned to bid less over time; the average bid actually went up in the later rounds.

Observations of reality confirm these findings. There is evidence that winners of oil and gas drilling leases at auctions take substantial losses on their leases. Baseball players who as free agents went to new teams were found to be overpaid by comparison with those who re-signed with their old teams.

We repeat: the precise calculations that show how much you should *shade down* your bidding to take into account the winner's curse are beyond the scope of this text; the articles cited in footnote 1 contain the necessary mathematical analysis. Here we merely wish to point out the problem and emphasize the need for caution. When your willingness to pay depends on your anticipated ability to make a profit from your purchase, or on the expected resale value of the item, be wary.

This analysis shows the importance of the prescriptive role of game theory. From observational and experimental evidence, we know that many people fall prey to the winner's curse. By doing so, they lose a lot of money. Learning the basics of game theory would help them anticipate the winner's curse and avoid attendant losses.

[6] Steven Landsburg, *The Armchair Economist* (New York: Free Press, 1993), p. 175.

[7] Richard Thaler, "Anomalies: The Winner's Curse," *Journal of Economic Perspectives,* vol. 2, no. 1 (Winter 1988), pp. 191–201.

3 BIDDING STRATEGIES

We turn now to private-value auctions and a discussion of optimal bidding strategies. Suppose you are interested in purchasing a particular lot of Chateau Margaux 1952 Bordeaux wine.[8] Consider some of the different possible auction procedures that could be used to sell the wine.

Suppose first that you are participating in a standard, open-outcry English auction. Your optimal bidding strategy is straightforward, given that you know your valuation V. Start at any step of the bidding process. If the last bid made by a rival bidder, call it r, is at or above V, you are certainly not willing to bid higher, so you need not concern yourself with any further bids. Only if the last bid is still below V do you bid at all. In that case, you can add a penny (or the smallest increment allowed by the auction house) and bid r plus one cent. If the bidding ends there, you get the wine for r (or virtually r), and you make an effective profit of $V - r$. If the bidding continues, you repeat the process, substituting the value of the new last bid for r. In this type of auction, the high bidder gets the wine for (virtually) the valuation of the second-highest bidder.

Now suppose the wine auction is first-price sealed-bid and you suspect that you are a very high value bidder. You need to decide whether to bid V or something other than V. Should you put in a bid equal to the full value V that you place on the object?

Remember that the high bidder in this auction will be required to pay her bid. In that case, you should not in fact bid V. Such a bid would be sure to give you zero profit, and you could do better by reducing your bid somewhat. If you bid a little less than V, you run the risk of losing the object should a rival bidder make a bid above yours but below V. But as long as you do not bid so low that this outcome is guaranteed, you have a positive probability of making a positive profit. Your optimal bidding strategy entails **shading** your bid. Calculus would be required to describe the actual strategy required here, but an intuitive understanding of the result is simple. An increase in shading (a lowering of your bid from V) provides both an advantage and a disadvantage to you; it increases your profit margin if you obtain the wine, but it also lowers your chances of being the high bidder and therefore of actually obtaining the wine. Your bid is optimal when the last bit of shading just balances these two effects.

What about a Dutch auction? Your bidding strategy in this case is similar to that for the first-price sealed-bid auction. Consider your bidding possibilities. When the price called out by the auctioneer is above V, you choose not to bid. If no one has bid by the time the price gets down to V, you may choose to do so.

---

[8] For details on the workings of actual wine auctions, see Ashenfelter, "How Auctions Work for Wine and Art," *Journal of Economic Perspectives*, vol. 3, no. 3 (Summer 1989), pp. 23–36.

But, again, as in the sealed-bid case, you have two options. You can bid now and get zero profit, or wait for the price to drop lower. Waiting a bit longer will increase the profit that you take from the sale, but it also increases your risk of losing the wine to a rival bidder. Thus shading is in your interest here as well, and the precise amount of shading depends on the same cost-benefit analysis described in the previous paragraph.

We have not yet considered the possibility of a second-price auction. There is, of course, no such thing as a second-price Dutch auction because there is no second-highest bidder in such an auction. A second-price English auction, however, does exist, but optimal bidding strategies are essentially the same as in the first-price situation. The only difference is that now it is no longer crucial for your final bid to be only pennies above the previous highest bid. Because you pay a price equal only to the second-highest bid, you need not concern yourself with the distance between the last two bids. Second-price sealed-bid auctions are more interesting from a strategic perspective. We pursue the analysis of bidding strategies in such auctions in the following section.

4 VICKREY'S TRUTH SERUM

We have just seen that bidding strategies in private-value open-outcry English auctions differ from those for first-price sealed-bid auctions. The high bidder in the English auction can get the object for essentially the valuation of the second-highest bidder. In the sealed-bid auction, the high bidder pays her bid, regardless of the distance between it and the next highest bid. Strategic bidders in a sealed-bid auction recognize this fact and attempt to retain some profit (surplus) for themselves by shading their bids. Of course, all else being equal, sellers would prefer bids that were not shaded downward. Let us see how sellers might induce bidders to reveal their true valuations with their bids.

Rather than using a first-price auction when selling a subjectively (private) valued object with sealed bids, William Vickrey showed that truthful revelation of valuations from bidders would arise if the seller used a modified version of the sealed-bid scheme; his suggestion was to modify the sealed-bid auction so that it more closely resembles its open-outcry counterpart.[9] That is, the highest bidder should get the object for a price equal to the second-highest bid—a second-price sealed-bid auction. Vickrey showed that with these rules, every bidder has a dominant bidding strategy to bid her true valuation. Thus we facetiously dub it **Vickrey's truth serum.**

---

[9] Vickrey was one of the most original minds in economics during the last four decades. In 1996 he won the Nobel Prize for his work on auctions and truth-revealing procedures. Sadly, he died just three days after the prize was announced.

However, we saw in Chapter 12 that there is a cost to extracting information. Auctions are no exception. Buyers reveal the truth about their valuations in an auction using Vickrey's scheme only because it gives them some profit from doing so. This reduces the profit for the seller, just as the shading of bids does in a first-price auction. The relative merit of the two procedures from the seller's point of view therefore depends on which one entails a greater reduction in her profit. We discuss this later in Section 6; but first we explain how Vickrey's scheme works.

Suppose you are one of the growing horde of Beanie Baby collectors bidding on the Princess Bear beanie in a local auction. The beanie is being sold in a sealed-bid second-price auction. You value the Princess beanie at $100, but you do not know the valuations of the other bidders. The rules of the auction allow you to bid any real-dollar value for the beanie; we will call your bid b and consider all of its possible values. Because you are not constrained to a small specific set of bids, we cannot draw a finite payoff matrix for this bidding game, but we can logically deduce the optimal bid.

The success of your bid will obviously depend on the bids submitted by others interested in Princess, primarily because you need to consider whether your bid will win. The outcome thus depends on all rival bids, but only the largest bid among them will affect your outcome. We call this largest bid r and disregard all bids below r.

What is your optimal value of b? We will look at bids both above and below $100 to determine whether any option other than exactly $100 can yield you a better outcome than bidding your true valuation.

We start with $b > 100$. There are three cases to consider. (1) If your rival bids less than $100 (so $r < 100$), then you get Princess at the price r. Your profit, which depends only on what you pay relative to your true valuation, is $(100 - r)$, which is what it would have been had you simply bid $100. (2) If your rival's bid falls between your actual bid and your true valuation (so $100 < r < b$), then you are forced to take Princess for more than she is worth to you. Here you would have done better to bid $100; you would not have gotten the Beanie Baby, but you would not have given up the $(r - 100)$ in lost profit either. (3) Your rival bids even more than you do (so $b < r$). You still do not get the Beanie Baby, but you would not have gotten it even had you bid your true valuation. Bidding your true valuation is never worse, and sometimes better, than bidding something higher.

What about the possibility of shading your bid slightly and bidding $b < 100$? Again, there are three situations. (1) If your rival's bid is lower than yours (so $r < b$), then you are the high bidder, and you get Princess for r. Here you could have gotten the same result by bidding $100. (2) If your rival's bid falls between 100 and your actual bid (so $b < r < 100$), your rival gets the Beanie Baby. If you had bid $100 in this case, you would have gotten Princess, paid r, and still made a profit of $(100 - r)$. (3) Your rival's bid could have been higher than \$100 (so $100 < r$). Again, you do not get the Beanie Baby, but if you had bid \$100, you still would not have gotten it, so there would have been no harm in doing so. Bidding your true valuation, then, is no worse, and sometimes better, than bidding something lower.

If truthful bidding is never worse and sometimes better than bidding either above or below your true valuation, then you do best to bid truthfully. That is, no matter what your rival bids, it is always in your best interest to be truthful. Put another way, bidding your true valuation is your *dominant strategy* whether you are allowed discrete or continuous bids.

Vickrey's remarkable result that truthful bidding is a dominant strategy in second-price sealed-bid auctions has many other applications. For example, if each member of a group is asked what she would be willing to pay for a public project that will benefit the whole group, each has an incentive to understate her own contribution—to become a "free rider" on the contributions of the rest; we have already seen examples of such effects in the collective-action games of Chapter 11. A variant of the Vickrey scheme can elicit the truth in such games as well.

5 ALL-PAY AUCTIONS

We have considered most of the standard auction types discussed in Section 1 but none of the more creative configurations that might arise. Here we consider a common-value sealed-bid first-price auction in which every bidder, win or lose, pays to the auctioneer the amount of her bid. An auction where the losers also pay may seem strange. But, in fact, many contests result in this type of outcome. In political contests, all candidates spend a lot of their own money and a lot of time and effort for fund raising and campaigning. The losers do not get any refunds on all their expenditures. Similarly, hundreds of competitors spend four years of their lives preparing for an event at the next Olympic games. Only one wins the gold medal and the attendant fame and endorsements; two others win the far less valuable silver and bronze medals; the efforts of the rest are wasted. Once you start thinking along these lines, you will realize that such **all-pay auctions** are, if anything, more frequent in real life than situations resembling the standard formal auctions where only the winner pays.

How should you bid (that is, what should your strategy be for expenditure of time, effort, and money) in an all-pay auction? Once you decide to participate, your bid is wasted unless you win, so you have a strong incentive to bid very aggressively. In experiments, the sum of all the bids often exceeds the value of the prize by a large amount, and the auctioneer makes a handsome profit.[10] In that case, everyone submitting extremely aggressive bids cannot be the equilibrium outcome; it seems wiser to stay out of such destructive competition altogether. But if everyone else did that, then one bidder could walk away with the prize for next to nothing; thus, not bidding cannot be an equilibrium strategy either. This analysis suggests that the equilibrium lies in mixed strategies.

Consider a specific auction with n bidders. To keep the notation simple, we choose units of measurement so that the common-value object (prize) is worth 1. Bidding more than 1 is sure to bring a loss, so we restrict bids to those between 0 and 1. It is easier to let the bid be a continuous variable x, where x can take on any (real) value in the interval $(0,1)$. Since the equilibrium will be in mixed strategies, each person's bid, x, will be a continuous random variable. Because you win the object only if all other bidders submit bids below yours, we can express your equilibrium mixed strategy as $P(x)$, the probability that your bid takes on a value less than x; for example, $P(1/2) = 0.25$ would mean that your equilibrium strategy entailed bids below $1/2$ one-quarter of the time (and bids above $1/2$ three-quarters of the time).[11]

As usual, we can find the mixed-strategy equilibrium by using an indifference condition. Each bidder must be indifferent about the choice of any particular value of x, given that the others are playing their equilibrium mix. Suppose you, as one of the n bidders, bid x. You win if all of the remaining $(n-1)$ are bidding less than x. The probability of anyone else bidding less than x is $P(x)$; the probability of two others bidding less than x is $P(x) \times P(x)$, or $[P(x)]^2$; the probability of all $(n-1)$ of them bidding less than x is $P(x) \times P(x) \times P(x) \ldots$ multiplied $(n-1)$ times, or $[P(x)]^{n-1}$. Thus with a probability of $[P(x)]^{n-1}$, you win 1. Remember that you pay x no matter what happens. Therefore, your net expected payoff for any bid of x is $[P(x)]^{n-1} - x$. But you could get 0 for sure by bidding 0. Thus, because you must be indifferent about the choice of any particular x, *including 0*, the condition that defines the equilibrium is $[P(x)]^{n-1} - x = 0$. In a full mixed-strategy equilibrium, this condition must be true for all x. Therefore the equilibrium-mixed strategy bid is

$$P(x) = x^{1/(n-1)}.$$

A couple of sample calculations will illustrate what this implies. First con-

[10] One of us (Dixit) has auctioned $10 bills to his Games of Strategy class, and made a profit of as much as $60 from a 20-student section. At Princeton there is the tradition of giving the professor a polite round of applause at the end of the semester. Once Dixit offered $20 to the student who kept applauding continuously the longest. This is an open-outcry all-pay auction with payments in kind (applause). While most students dropped out between five and 20 minutes, three went on for four and a half hours!

[11] $P(x)$ is called the *cumulative probability distribution function* for the random variable x. The more familiar probability density function for x is its derivative, $P'(x) = p(x)$. Then $p(x)\,dx$ denotes the probability that the variable takes on a value in a small interval from x to $x + dx$.

sider the case of $n = 2$; then $P(x) = x$ for all x. Therefore the probability of bidding a number between two given levels x_1 and x_2 is $P(x_2) - P(x_1) = x_2 - x_1$. Since the probability that the bid lies in any range is simply the length of that range, any one bid must be just as likely as any other bid. That is, your equilibrium mixed-strategy bid should be random and uniformly distributed over the whole range from 0 to 1.

Next let $n = 3$. Then $P(x) = \sqrt{x}$. For $x = 1/4$, $P(x) = 1/2$, so the probability of bidding 1/4 or less is 1/2. The bids are no longer uniformly distributed over the range from 0 to 1; they are more likely to be in the lower end of the range.

Further increases in n reinforce this tendency. For example, if $n = 10$, then $P(x) = x^{1/9}$, and $P(x)$ equals 1/2 when $x = 1/2^9 = 1/512 = 0.00195$. In this situation, your bid is as likely to be smaller than 0.00195 as it is to be anywhere within the whole range from 0.00195 to 1. Thus your bids are likely to be very close to 0.

Your average bid should correspondingly be smaller the larger the number n. In fact a more precise mathematical calculation shows that if everyone bids according to this strategy, the average or expected bid of any one player will be just $(1/n)$.[12] With n players bidding, on average, $1/n$ each, the total expected bid is 1, and the auctioneer makes zero expected profit. This provides more precise confirmation that the equilibrium strategy eliminates overbidding.

The idea that your bid should be much more likely to be close to 0 when the total number of bidders is large makes excellent intuitive sense, and the finding that equilibrium bidding eliminates overbidding lends further confidence to the theoretical analysis. Unfortunately, many people in actual all-pay auctions either do not know or forget this theory, and bid to excess.

6 HOW TO SELL AT AUCTION

Bidders are not the only auction participants who need to carefully consider their optimal strategies. An auction is really a sequential-play game in which the first move is the setting of the rules and bidding starts only in the second round of moves. It falls to the sellers, then, to determine the path that later bidding will follow by choosing a particular auction structure.

---

[12] The expected bid of any one player is calculated as the expected value of x, using the probability density function, $p(x)$. In this case,

$$p(x) = P'(x) = \frac{1}{n-1} x^{(2-n)/(n-1)},$$

and the expected value of x is

$$\int_0^1 xp(x)\,dx = \frac{1}{n}.$$

As a seller interested in auctioning off your prized art collection, or even your home, say, you must decide on the best type of auction to use. Obviously, to guarantee yourself the greatest profit from your sale, you must look ahead to the predicted outcome of the different types of auctions before making a choice. One concern of many sellers is that an item will go to a bidder for a price lower than the value the seller places on the object. To counter this, most sellers insist on setting a **reserve price** for auctioned objects; they reserve the right to withdraw the object from the sale if no bid higher than the reserve price is obtained.

Beyond this, however, what can sellers do to determine the type of auction that might net them the most profit possible? One possibility is to use Vickrey's suggested scheme, a second-price sealed-bid auction. According to him, this kind of auction elicits truthful bidding from potential buyers. Does this effect make it a good auction type from the seller's perspective?

In a sense, the seller in such a second-price auction is giving the bidder a profit margin in order to counter the temptation to shade down the bid in the hope of a larger profit. But this outcome then reduces the seller's revenue, just as shading down in a first-price sealed-bid auction would. Which type of auction is ultimately better for the seller actually turns out to depend on the bidders' attitudes toward risk and their beliefs about the value of the object for sale.

A. Risk-Neutral Bidders and Independent Estimates

The least-complex configuration of bidder risk attitudes and beliefs occurs when there is risk neutrality (no risk aversion) and when bidder estimates about the value of the object for sale remain independent of each other. As we said in the Appendix to Chapter 5, *risk-neutral* individuals care only about the expected monetary value of their outcomes, regardless of the level of uncertainty associated with those outcomes. *Independence in estimates* means a bidder is not influenced by the estimates of other bidders when determining how much an object is worth to her; the bidder has decided independently exactly how much the object is worth to her. In this case, there can be no winner's curse. If these conditions for bidders hold, sellers can anticipate the same expected average revenue (over a large number of trials) from any of the four primary types of auction: English, Dutch, and first- and second-price sealed-bid.

This **revenue equivalence** result does not imply that all of the auctions will yield the same revenue for every item sold, but that the auctions will yield the same selling price *on average* over the course of numerous auctions.[13] We can see

---

[13] Vickrey, in his "Counterspeculation, Auctions, and Competitive Sealed Tenders," *Journal of Finance,* vol. 16, no. 1, (March 1961), pp. 8–37, was one of the first to note the existence of revenue equivalence. A more recent study gathering a number of the results on revenue outcomes for various auction types is J. G. Riley and W. F. Samuelson, "Optimal Auctions," *American Economic Review,* vol. 71, no. 3, (June 1981), pp. 381–392.

the equivalence quite easily between second-price sealed-bid auctions and English auctions. We have already seen that, in the second-price auction, each bidder's dominant strategy is to bid her true valuation. The highest bidder gets the object for the second-highest bid, and the seller gets a price equal to the valuation of the second-highest bidder. Similarly, in an English auction, bidders drop out as the price increases beyond their valuations, until only the first- and second-highest-valuation bidders remain. Once the price reaches the valuation of the second-highest bidder, that bidder will also drop out, and the remaining (highest-valuation) bidder will take the object for just a cent more than the second-highest bid. Again, the seller gets a price (essentially) equivalent to the valuation of the second-highest bidder.

More advanced mathematical techniques are needed to prove that revenue equivalence can be extended to Dutch and first-price sealed-bid auctions as well, but the intuition should be clear. In all four types of auctions, the highest-valuation bidder should win the auction and pay on average a price equal to the second-highest valuation. If the seller is likely to use a particular type of auction repeatedly, she need not be overly concerned about her choice of auction structure; all four would yield her the same expected price.

B. Correlated Estimates

Suppose, however, that in determining their own valuations of an object, bidders are influenced by the estimates (or their beliefs about the estimates), of other bidders. Such a situation is relevant for common-value auctions, like those for oil exploration considered in Section 1. Suppose your experts have not presented a glowing picture of the future profits to be gleaned from a specific tract of land. You are therefore pessimistic about its potential benefits, and you have constructed an estimate V of its value that you believe reflects your pessimism.

Under the circumstances, you may be concerned that your rival bidders have also received negative reports from their experts. When bidders believe their valuations are all likely to be similar, either all relatively low or all relatively high, for example, we say those beliefs or estimates of value are *positively correlated.* Recall that we introduced the concept of correlation in Chapter 12, where we discussed correlated risks and insurance. In the current context, the likelihood that your rivals' estimates are also unfavorable may magnify the effect of your pessimism on your own valuation. If you are participating in a first-price sealed-bid auction, you may be tempted to shade down your bid even more than you would in the absence of correlated beliefs. Of course, if bidders are optimistic and valuations generally high, correlated estimates may lead to less shading than when estimates are independent.

However, the increase in the shading of bids that accompanies correlated low (or pessimistic) bids in a first-price auction should be a warning to sellers. With positively correlated bidder beliefs, the seller may want to avoid the first-price auction and take advantage of Vickrey's recommendation to use a second-price structure. We have just seen that this auction type encourages truthful revelation and, when correlated estimates are possible, the seller does even better to avoid auctions in which any additional shading of bids might occur.

A first-price open-outcry English auction will have the same ultimate outcome as the second-price sealed-bid auction, while an open-outcry Dutch auction will have the same outcome as a first-price sealed-bid auction. Thus a seller facing bidders with correlated estimates of an object's value should also prefer the English to the Dutch versions of the open-outcry auction. If you are bidding on the oil land in an open-outcry English auction, and the price is nearing your estimate of the land's value but your rivals are still bidding feverishly, you can infer that their estimates are at least as high as yours—perhaps significantly higher. The information you obtain from observing the bidding behavior of your rivals may convince you that your estimate is too low. You might even increase your own estimate of the land's value as a result of the bidding process. Your continuing to bid may provide an impetus for further bidding by other bidders, and the process may continue for a while. If so, the seller reaps the benefits. More generally, the seller can expect a higher selling price in an English auction than in a first-price sealed-bid auction when bidder estimates are correlated. For the bidders, however, the effect of the open bidding is to disperse additional information and to reduce the effect of the winner's curse.

Our discussion of correlated estimates assumes that there are a fairly large number of bidders involved in the auction. But an open-outcry English auction can be beneficial to the seller if there are only two bidders, both of whom are particularly enthusiastic about the object for sale. They would bid against each other as long as possible, pushing the price up to the lower of the valuations, both of which were relatively high from the start. The same auction could be disastrous for the seller, however, if one of the bidders has a very low valuation; the other is then quite likely to have a valuation considerably higher than the first. In this case, we say that bidder valuations are *negatively correlated*. We encourage any seller facing a small number of bidders with potentially very different valuations to choose a Dutch or first-price sealed-bid structure. Either of these would reduce the possibility of the high-estimate bidder gaining the object for well under her true valuation; that is, these would transfer the available profit from the buyer to the seller.

C. Risk-Averse Bidders

Here we return to assuming that bids and beliefs are uncorrelated but consider instead the possibility that auction outcomes could be affected by bidders' attitudes toward risk. In particular, suppose bidders are *risk-averse*. They may be much more concerned, for example, about the losses caused by underbidding— losing the object—than the costs associated with bidding at or close to their true valuations. Thus, risk-averse bidders generally want to win if possible without ever overbidding.

What does this preference structure do to the types of bids they submit in first-price versus second-price (sealed-bid) auctions? Again, think of the first-price auction as being equivalent to the Dutch auction. Here risk aversion leads bidders to bid earlier rather than later. As the price drops to the bidder's valuation and beyond, there is greater and greater risk in waiting to bid. With risk-averse bidders, we expect them to bid quickly, not to wait just a little bit longer in the hopes of gaining those extra few pennies of profit. Applying this reasoning to the first-price sealed-bid auction, we expect bidders to shade down their bids by *less* than they would if they were not risk-averse: too much shading actually increases the risk of not gaining the object, which risk-averse bidders would want to avoid.

Compare this outcome to that of the second-price auction, where bidders pay a price equal to the second-highest bid. Bidders bid their true valuations in such an auction, but pay a price less than that. If they shade their bids only slightly in the first-price auction, then those bids will tend to be close to the bidders' true valuations—and bidders pay their bids in such auctions. Thus bids will be shaded somewhat, but the price ultimately paid in the first-price auction will probably exceed what would be paid in the second-price auction. When bidders are risk-averse, the seller then does better to choose a first-price auction than a second-price auction.

The seller does better with the first-price auction in the presence of risk aversion only in the sealed-bid case. If the auction were open-outcry (English), the bidders' attitude toward risk is irrelevant to the outcome. Thus, risk aversion does not alter the outcome for the seller in these auctions.

7 SOME ADDED TWISTS TO CONSIDER

A. Multiple Objects

When you think about an auction of a group of items, like a bank's auctioning repossessed vehicles or estate sales auctioning the contents of a home, you probably envision the auctioneer bringing each item to the podium individually and selling it to the highest bidder. This process is appropriate when each bidder has independent valuations for each item. However, independent valuations may not always be an appropriate way to model bidder estimates. Then, if bidders value specific groups or whole packages of items higher than the sum of their values for the component items, the choice of auctioning the lots separately or together makes a big difference to bidding strategies as well as to outcomes.

Consider a real-estate developer named Red who is interested in buying a very large parcel of land in order to build a town house community for professionals. Two townships, Cottage and Mansion, are each auctioning a land parcel big enough to suit her needs. Both parcels are essentially square in shape and encompass four square acres. The mayor of Cottage has directed that the auctioneer sell the land as quarter-acre blocks, one at a time, starting with the perimeter of the land and working inward, selling the corner lots first and then the lots on the north, south, east, and west borders in that order. The mayor of Mansion has at the same time directed that the auctioneer attempt to sell the land in her town first as a full four-acre block, then as two individual two-acre lots, and then as four one-acre lots after that, if no bids exceed the set reserve prices.

Red has determined by virtue of extensive market analyses that the blocks of land in Cottage and Mansion would provide the same value to her. However, she has to obtain the full 4 acres of land in either town in order to have enough room for her planned development. The auctions are being held on the same day at the same time. Which should she attend?

It should be clear that her chances of acquiring a four-acre block of land for a reasonable price—less than or equal to her valuation—are much better in Mansion than in Cottage. In the Mansion auction, she would simply wait to see how bidding proceeded, submitting a final high bid if the second-highest offer fell below her valuation of the property. In the Cottage auction, she would need to win each and every one of the 16 parcels up for sale. Under the circumstances, she should expect rival bidders interested in owning land in Cottage to become more intent on their goal—perhaps even joining forces—as the number of available parcels decreases during the course of the auction. Red would have to bid aggressively enough to win parcels in the early rounds while being conservative enough to ensure that she did not exceed her total valuation by the end of the auction. The difficulties in crafting a bidding strategy for such an auction are numerous and the probability of being unable to profitably obtain every parcel quite large; hence Red's preference for the Mansion auction.

Note that from the seller's point of view, the Cottage auction is likely to bring in greater revenue than the Mansion auction if there are an adequate number of bidders interested in small pieces of land. If the only bidders were all developers like Red, however, they might be hesitant to even participate in the Cottage auction for fear of being beaten in just one round. In that case, the Mansion-type auction is better for the seller.

B. Defeating the System

We saw earlier which auction structure is best for the seller, given different assumptions about how bidders felt toward risk and whether their estimates were correlated. There is always an incentive for bidders, though, to come up with a bidding strategy that defeats the seller's efforts. The best-laid plans for a profitable auction can almost always be defeated by an appropriately clever bidder or, more often, group of bidders.

Even the Vickrey second-price sealed-bid auction can be defeated if there are only a few bidders in the auction, all of whom can collude among themselves. By submitting one high bid and a lowball second-highest bid, collusive bidders can obtain an object for the second-bid price. This outcome relies on the fact that no other bidders submit intermediate bids, or that the collusive group is able to prevent such an occurrence. The possibility of collusion highlights the need for the seller's reserve prices, although they only partially offset the problem in this case.

First-price sealed-bid auctions are less vulnerable to bidder collusion for two reasons. The potential collusive group engages in a multiperson prisoners' dilemma game in which each bidder has a temptation to cheat. Such cheating might entail an individual submitting her own high bid in order to win the object for herself, reneging on any obligation to share profits with group members. Collusion among bidders in this type of auction is also difficult to sustain because cheating (that is, making a different bid from that agreed to within the collusive group) is easy to do but difficult for other buyers to detect. Thus the sealed-bid nature of the auction prevents detection of a cheater's behavior, and hence punishment, until the bids are opened and the auction results announced; at that point, it is simply too late. However, there may be more scope for sustaining collusion if a particular group of bidders participates in a number of similar auctions over time, so that they engage in the equivalent of a repeated game.

Other tricky bidding schemes can be created to meet the needs of specific individuals, or groups, in any particular type of auction. One very clever and quite recent example has arisen in the auction of the U.S. airwave spectrum, as we will show later in Section 8.C. In addition, there is always the specter of outright fraud on the part of bidders. One bidder might submit several bids, planning to default on all but a very low—but still winning—bid after all the rival bids have become known. Penalties for default can counter such problems but must be quite stringent to work in all cases.

C. Information Disclosure

Finally, we consider the possibility that the seller has some private information about an object that might affect the bidders' valuations of that object. Such a situation arises when the quality or durability of a particular object, like an automobile, a house, or a piece of electronic equipment, is of great importance to the buyers. Then the seller's experience with the object in the past may be a good predictor of the future benefits that will accrue to the winning bidder.

Under such circumstances, the seller must carefully consider any temptation to conceal information. If the bidders know that the seller has some private information, they are likely to interpret any failure to disclose that information as a signal that the information is unfavorable. Even if the seller's information *is* unfavorable, she may be better off to reveal it; bidders' beliefs might be worse than the actual information. Thus honesty is often the best policy.

Honesty can also be in the seller's interests for another reason. When she has private information about a common-value object, she should disclose that information in order to sharpen the bidders' estimates of the value of the object. The more confident the bidders are that their valuations are correct, the more likely they are to bid up to those valuations. Thus disclosure of private seller information in a common-value auction can help not only the seller by reducing the amount of shading done by bidders but also the bidders by reducing the effects of the winner's curse.

8 THE AIRWAVE SPECTRUM AUCTIONS

Probably no set of auctions in history has been the subject of such widespread public, commercial, and academic interest and analysis as those held recently by the FCC of the electromagnetic airwave spectrum, more commonly known as the *PCS (personal communication systems) auctions*, or, simply, *airwave auctions*. The broadcast licenses being sold gave buyers the rights to provide a variety of wireless communication services (depending on the type of license purchased). These include mobile or portable radio services, (such as telephone, fax, paging, and so on); digital voice, video, and data transmission (satellite service); interactive video and data services (used for short-distance services, like meter reading, for example); and wireless cable services (via microwave) for video programming, teleconferencing, and high-speed data transmission (for Internet access). Such licenses had been allocated after judicial hearings or lengthy systems until the Omnibus Budget Reconciliation Act of 1993 authorized provision using a competitive bidding process. Numerous economists were consulted both by the FCC during the auction design stage and by the firms engaged in bidding before and during the auction process; their analyses necessarily made use of many of the topics we have discussed in preceding sections. Our goal here is to give you a flavor for the ways in which strategic auction theory was used in writing the rules for the airwave auctions and in the bidding strategies actually employed; in between these two discussions, we provide some specifics on the actual auction outcomes.[14]

A. Designing the Rules

The importance of the right set of rules should be clear, especially if you have read the earlier sections of this chapter. Although Congress did not push the revenue aspect of the auctions, choosing rather to emphasize their efficiency aspects, the economists working to design the rules knew that bad choices could lead not only to inefficient outcomes, like mismatches of licenses to firms, but also to fiscally impractical ones that reduced the government's potential revenue. The economists were also concerned about the possibility of the auctions simply breaking down in some unforeseen way.[15]

The design process, which took place over a several-month period from fall 1993 through most of spring 1994, benefitted from observing the results of spectrum auctions in other countries, as well as from auction theory. Auctions in New Zealand (the first country to use them for broadcast licenses) and Australia had led to difficulties attributed to faulty design procedures, so the Americans wanted to learn from the mistakes of their counterparts "down under."[16] Auction theory worked well for explaining why problems had arisen in these earlier auctions and for providing ways to prevent such problems in the FCC auctions. In addition, this theory and analysis helped the U.S. government achieve specific goals, including the prevention of monopoly and the promotion of small businesses.

In announcing the final rules for the auctions, the FCC left itself considerable leeway to pick and choose from a variety of schemes, depending on the specific circumstances of each auction. (In fact, the rules have been updated several times since the first auction in 1994.) But the main characteristics of the mecha-

---

[14] In keeping with the technological level of the auctions themselves, we refer you to the FCC Auctions web page *(http://www.fcc.gov/wtb/auctions/)* for additional information. We also consulted a variety of other Internet sources in the preparation of this section, including CNN's financial newswire service *(http://www.cnnfn.com/index.html)*, *Wireless Week (http://www.wirelessweek.com /News/)*, and the FCC's online news release directory *(http://www.fcc.gov/Bureaus /Wireless/News_Releases/199x/)*. More conventional sources include, "The FCC Report to Congress on Spectrum Auctions," WT Docket No. 97-150, October 1997 (available also at *http://www.fcc.gov/wtb/auctions/ papers/fc970353.pdf)*; Peter Spiegel, "Hollow Victory," *Forbes*, January 27, 1997; "Learning to Play the Game," *Economist*, May 17, 1997, p. 86; John McMillan, "Selling Spectrum Rights," *Journal of Economic Perspectives*, vol. 8, no. 3, (Summer 1994), pp. 145–162; and Peter Cramton, "The FCC Spectrum Auctions: An Early Assessment," *Journal of Economics and Management Strategy*, vol. 6, no. 3, (Fall 1997), pp. 431–495.

[15] McMillan, "Selling Spectrum Rights," p. 148.

[16] A particularly comic example of such problems arose in New Zealand. Second-price auctions were used, presumably for their revelation features, but no minimum bids or reserve prices were set. Thus a university student in Dunedin won a television license with a bid of N.Z.$1 for which he ultimately paid zero because no one else had bid against him (ibid., p. 148).

nisms chosen remain in place and each relies on a specific piece of the theory to explain its role.

It was decided first that the auctions would not be sealed-bid but open. True, sealed-bid auctions are less susceptible to collusive behavior on the part of bidders, since bidders cannot directly observe each other's behavior. However, it was assumed that bidder estimates of the license values would be positively correlated; an open auction would allow bidders to glean as much information as possible from rival bids, which would partially alleviate the problems associated with the winner's curse. The benefits associated with improving the efficiency of the bidding were believed to outweigh the costs associated with any increase in collusion.

In addition, rather than the traditional open-outcry method, the FCC decided to use a previously untried auction method for most of the auctions: simultaneous multiple rounds of bidding. This choice was based on the belief that there would be significant interdependence among the estimated values (that is, correlated estimates) of the licenses being auctioned and that auctioning individual licenses one at a time might make them less attractive to bidders. In this, the FCC followed the logic of our "multiple objects" discussion in Section 7.A. The simultaneous multiple-round structure put all of the licenses in a particular portion of the spectrum up for auction at the same time. In each round, bidders could submit bids for any number of licenses at the same time. Following each round, bids were published so that all participants could follow the progress of the auction. Rounds of bidding continued until no further bids were received for any of the licenses, at which point the high bidders would be awarded their winnings. This scheme allowed large and small firms to compete for the same types of licenses on an essentially equal footing.

The FCC did not, however, permit packaged bids in its auction rules. *Package bidding* (also called *combinatorial bidding*) occurs when a bidder submits a bid for a *group* of objects, licenses in this case, and wins or loses the entire group. There was some discussion during the auction-design stage that disallowing package bids was inefficient, but there were offsetting concerns that packaging would favor large bidders and be inordinately complex to implement, given the FCC's time pressure.[17] Congress has asked the FCC to consider the possibility of allowing combinatorial bidding in future auctions, and the FCC is working toward implementing such a system.[18]

The simultaneous multiple-round structure required several special safety features to make it work as planned. In particular, the FCC learned from the New Zealand auction that reserve prices should be set and from the Australia auction that down payments along with bid withdrawal and default penalties should be

---

[17] Cramton, "FCC Spectrum Auctions," pp. 437–439.
[18] "FCC Report to Congress," p. 36.

used. Bidders would be required to pay a deposit in order to be able to bid (20% of their bid as a down payment within five days of the auction) and to incur significant penalties for defaulting on a winning bid. Reserve prices, which, according to theory, are required only when bidding is weak (as with the small number of bidders in New Zealand), would be set only when a small number of bidders applied to participate in a particular auction.

Interdependent values were expected to be important in the geographically based auctions of licenses for particular regions of the country, and for the nationwide license auctions as well. These were thus held using the simultaneous multiple-round scheme. The one license type for which bidder valuations were expected to be independent was interactive video and data service (IVDS); such service is local and for short-distance communications only. The FCC chose to sell all of the available IVDS licenses in a single open-outcry auction. Again, theory showed that an open auction would be a sensible choice in that particular context.

Finally, the FCC needed a mechanism that would favor small businesses and previously unrepresented groups—together comprising a subset of bidders known as *designated entities*—without damaging the efficiency of the auction. Their ultimate decision was to offer a combination of installment payment plans, "bidding credits" of 25% to 35% for small and very small businesses (defined on the basis of gross annual revenues), and "entrepreneurs' blocks" that were held for bidders under a given financial level.

B. How the Auctions Have Fared

As of the end of 1998, the FCC had run a total of 18 different auctions—including one reauction of 18 licenses for which high bids were withdrawn—for portions of the airwave spectrum. The most recent took place during December 1998, and the FCC has plans for at least eight more auctions, four of which have already been scheduled for the first half of 1999. As discussed at the beginning of this section, the FCC has sold the rights to a wide variety of different wireless communication services at both national and regional levels. The total revenue generated for the government is now over $23 billion, although roughly half of this amount is still to be collected.

According to the FCC, the first auction sold 10 nationwide *narrowband* licenses for a total of $650 million; such licenses allow the provision of *nationwide* mobile radio, and data and paging services. There were 29 eligible bidders in that auction, each of whom paid a refundable $350,000 deposit for the right to participate. Four bidders did not appear at the auction, and eight others subsequently dropped out, leaving only 17 after the first round. The auction consisted of simultaneous multiple rounds (the untried procedure described in Section 8.A), and lasted for four days and 47 rounds before bidding was halted on July 29,

1994. This was relatively short in comparison with the first *broadband* auction, which covered *regional* mobile radio services and took a full three months to close (running from December 5, 1994, until March 13, 1995), through 112 rounds of bidding, and the second broadband auction, which ultimately closed in May 1996 after five months and 184 rounds of bidding. These two auctions brought in $7.7 and $10.2 billion respectively, for a total of 592 different licenses (99 in the first auction, the rest in the second).

The second broadband auction won notoriety for some of its high bidders, including one relatively small firm that, with the promise of foreign backing, bid $4.2 billion for 56 different licenses.[19] Not surprisingly, this firm has experienced difficulties in meeting its financial obligations. We will discuss this case from the bidder's perspective shortly.

Later broadband PCS auctions netted significantly smaller revenues. The third auction brought in $2.5 billion, only about one-quarter as much as the second, and the April 1997 auction yielded only $13.6 million, far less than was anticipated. Analyses indicate that the bids at the second auction may have been out of line with the actual profit potential of the licenses (hence the winning bidder's difficulties already mentioned) and that later auctions suffered from a rush to meet a starting deadline, possibly leaving interested bidders unable to participate. Other more sinister possibilities have also been hypothesized. Again, we provide more on this issue shortly.

Overall, the auctions seem to have been well designed and well played by bidders and sellers alike. The mere numbers, however, do not tell us much about the actual bidding or participation strategies employed by firms interested in obtaining licenses. Section 8.B considers two examples that speak to these issues: defaulting on bids and charges of bid rigging under investigation by the Department of Justice.

C. Two Specific Bidding Issues

DEFAULT   As mentioned earlier, the second broadband PCS auction received significant attention because of the enormity of some of the winning bids. The most eye-popping was a $4.2 billion bid submitted by a relatively small, foreign-backed firm. Their expectations of significant profit from the broadcast opportunities afforded by their 56 licenses pushed their bid to a point where they ended up paying about twice as much per expected customer as any other winner had paid up to that point.

Prices *were* generally higher in this auction than in previous ones, a fact that has been attributed both to increased competition for the licenses and to the attractive payment terms made available to small bidders by the FCC.[20] However, some bidders obviously overstepped themselves.

[19] See *http://www.cnnfn.com/news/9605/06/fcc_auction_pkg/index.htm.*
[20] Cramton, "FCC Spectrum Auctions," pp. 473–475.

The small firm in question failed to make its initial deposit, defaulted on 17 licenses that it had won, and went on to seek help in meeting its additional financial obligation to the FCC. The FCC was quick to handle the difficulties brought about by the defaults, however, arranging for a reauctioning of these 17 licenses within a few months after the close of the original auction. In addition, it was forced to reconsider its installment plan allowances and to remove them in later auctions.

BID RIGGING   The final and most interesting example of strategic behavior in the airwave auctions has to do with the claims of bid rigging in the April 1997 auction. After watching prices soar in some of the earlier auctions, bidders were apparently anxious to come up with a way to reduce the price of the winning bids. Because outright collusion among firms was difficult, they searched for a way to use the auction structure itself to help them achieve beneficial outcomes.

The result of their search? It has been alleged that certain firms engaged in *code bidding*; that is, they signaled their intentions to go after licenses for certain geographic locations by using the FCC codes or telephone area codes for those areas as the last three digits of their bids. The FCC has taken these allegations quite seriously. The case is now pending at the Department of Justice.

Other signaling devices were apparently used in at least one of the early broadband auctions, as well. While some firms literally announced their intentions to win a particular license, others used a variety of strategic bidding techniques to signal their interest in specific licenses or to dissuade rivals from horning in on their territories. In the first broadband auction, for example, GTE and other firms apparently used the code-bidding technique of ending their bids with the numbers that spelled out their names on a touch-tone telephone keypad.[21]

Some changes to the bidding rules are being discussed as a response to the signaling behavior that has been observed. In particular, the FCC is considering a new bidding system that will reduce the flexibility bidders have in specifying their bid amounts; an electronic system is in the works that would allow bidders merely to click on a key or screen icon to submit a minimum acceptable bid or bid increment. This would prevent the difficulties associated with allowing bidders to use any numbers they choose in their bids—in the last three digits in particular.[22]

The FCC auctions are very recent, and their experience is still in the process of being analyzed and discussed by experts. That, and limitations of space, mean we cannot provide an exhaustive analysis. But we hope you begin to see that the theory of auction design and equilibrium bidding strategies can be use-

---

[21] Ibid., p. 467.
[22] "FCC Report to Congress," p. 15.

ful to both sellers and buyers in actual auctions. Both the FCC and the bidders in the airwave auctions needed, and used, strategic analysis to their benefit. Perhaps you will have the opportunity to use these concepts for your own benefit as well.

SUMMARY

In addition to the standard *first-price open-outcry ascending* variety, auctions may be *second-price, sealed-bid,* or *descending*. Objects for bid may have a single *common value* or many *private values* specific to each bidder. With common-value auctions, bidders often win only when they have overbid, falling prey to the *winner's curse*. In private-value auctions, optimal bidding strategies, including decisions about when to *shade down* bids from your true valuation, depend on the auction type used. In the familiar first-price auction, there is a strategic incentive to underbid.

Vickrey showed that sellers can elicit true valuations from bidders by using a second-price sealed-bid auction. Generally, sellers will choose the mechanism that guarantees them the most profit; this choice will depend on bidder risk attitudes and bidder beliefs about an object's value. If bidders are risk-neutral and have independent valuation estimates, all auction types will yield the same outcome. Decisions regarding how to auction a large number of objects, individually or as a group, and whether to disclose information are nontrivial. Sellers must also be wary of bidder collusion or fraud.

The FCC made extensive use of auction theory when designing the rules for its recent airwave auctions. These simultaneous multiple-round auctions have incorporated many safety features and led to significant revenues for the government. These auctions seem to have been well designed and well played, despite some concerns regarding bidder default and bid-rigging schemes.

KEY TERMS

all-pay auction (503)
ascending auction (496)
common value (496)
descending auction (496)
Dutch auction (496)
English auction (496)
first-price auction (496)
objective value (496)
open outcry (495)

private value (497)
reserve price (506)
revenue equivalence (506)
sealed bid (495)
second-price auction (496)
shading (500)
subjective value (497)
Vickrey's truth serum (501)
winner's curse (497)

EXERCISES

1. A house painter has a regular contract to work for a builder. On these jobs, her cost estimates are generally right: sometimes a little high, sometimes a little low, but correct on average. When the regular work is slack, she bids competitively for other jobs. "Those are different," she says. "They almost always end up costing more than I estimate." Assuming her estimating skills do not differ between the jobs, what can explain the difference?

2. Consider an auction where n identical objects are on offer, and there are $(n + 1)$ bidders. The actual value of an object is the same for all bidders and equal for all objects, but each bidder gets only an independent estimate, subject to error, of this common value. The bidders put in sealed bids. The top n bidders get one object each, and each pays what she had bid. What considerations will affect your bidding strategy and how?

3. The idea of the winner's curse can be expressed slightly differently from its usage in the text: "The only time your bid matters is when you win, which happens when your estimate is higher than the estimates of all the other bidders. Therefore you should focus on this case. That is, you should always act as if all the others have received estimates lower than yours, and use this 'information' to revise your own estimate." Here we ask you to apply this idea to a very different situation.

   A jury consists of 12 people who hear and see the same evidence presented, but each of them interprets this evidence using her own thinking and experience, and arrives at an estimate of the guilt or the innocence of the accused. Then each is asked to vote—Guilty or Not guilty. The accused is convicted if all 12 vote Guilty and acquitted if one or more vote Not guilty. Each juror's objective is to arrive at a verdict that is the most accurate verdict in light of the evidence. Each juror votes strategically, seeking to maximize this objective and using all the devices of information inference that we have studied. Will the equilibrium outcome of their voting be optimal?

4. You are in the market for a used car, and see an ad for the model you like. The owner has not set a price, but invites potential buyers to make offers. Your pre-purchase inspection gives you only a very rough idea of the value of the car; you think it is equally likely to be anywhere in the range of $1,000 to $5,000 (so your calculation of the average of this value is $3,000). The current owner knows the exact value, and will accept your offer if it exceeds that value. If your offer is accepted and you get the car, then you will find out the truth. But you have some special repair skills, and know that once you own the car, you will be able to work on it and increase its value by a third (33.3 ... %) of whatever it is worth when it comes into your hands.

 (a) What is your expected profit if you offer $3,000? Should you make such an offer?

 (b) What is the highest offer you can make without losing money on the deal?

5. **[Optional]** The mathematical analysis of bidding strategies and equilibrium is unavoidably difficult. If you know some simple calculus, we lead you here through a very simple example.

 Consider a first-price sealed-bid private-values auction. There are two bidders. Each knows her own value but not that of the other. The two values are equally likely to lie anywhere between 0 and 1; that is, each bidder's value is uniformly distributed over the interval $[0,1]$. The strategy of each can be expressed as a complete plan of action drawn up in advance: "If my value is v, I will bid b." In other words, the strategy expresses the bid as a function of value, $b = B(v)$. We construct an equilibrium in such strategies. The players are identical, and we look for a symmetric equilibrium in which their strategies are the same function B. Of course, we must use the equilibrium conditions to solve for the function B.

 Suppose the other player is using the function B. Your value is v. You are contemplating your best response. You can always act as if your value were x, that is, bid $B(x)$. With this bid you will win if the other player's value is less than x. Since all values on the interval $[0,1]$ are equally likely, the probability of finding a value less than x is just x. If you win, your surplus is your true value minus your bid, that is, $v - B(x)$. Thus your expected payoff from the strategy of bidding as if your value were x is given by $x[v - B(x)]$.

 (a) Write down the first-order condition for x to maximize your expected payoff.

 (b) In equilibrium, you will follow the strategy of the function B; that is, you will choose x equal to your true value v. Rewrite the first-order condition of part (a) using this.

 (c) Substitute $B(v) = v/2$ into the equation you obtained in part (b), and see that it holds. Assuming that the equation in (b) has a unique solution, this is the equilibrium strategy we seek. Interpret it intuitively.

6. **[Optional]** Repeat Exercise 5 but with n bidders instead of two, and verify that the function $B(v) = v(n-1)/n$ gives a common equilibrium strategy for all bidders. Explain intuitively why $B(v)$ increases for any v as n increases.

16

Bargaining

P EOPLE ENGAGE IN BARGAINING throughout their lives. Children start by negotiating to share toys and to play games with other children. Couples bargain about matters of housing, child rearing, and the adjustments each must make for the other's career. Buyers and sellers bargain over price, workers and bosses over wages. Countries bargain over policies of mutual trade liberalization; superpowers negotiate mutual arms reduction. And the two authors of this book had to bargain with each other—generally very amicably—about what to include or exclude, how to structure the exposition, and so forth. To get a good result from such bargaining, the participants must devise good strategies. In this chapter we raise and explicate some of these basic issues and strategies.

All bargaining situations have two things in common. First, the total payoff the parties to the negotiation are capable of creating and enjoying as a result of reaching an agreement should be greater than the sum of the individual payoffs they could achieve separately—the whole must be greater than the sum of the parts. Without the possibility of this excess value, or "surplus," the negotiation would be pointless. If two children considering whether to play together cannot see a net gain from having access to a larger total stock of toys or from each other's company in play, then it is better for each to "take his toys and play by himself." Of course, the world is full of uncertainty and the anticipated benefits may not materialize. But when engaged in bargaining, the parties must at least perceive some gain therefrom: Faust, when he agreed to sell his soul to the Devil, thought the benefits of knowledge and power he gained were worth the price he would eventually have to pay.

The second important general point about bargaining follows from the first: it is not a zero-sum game. When a surplus exists, the negotiation is over how to divide it up. Each bargainer tries to get more for himself and leave less for the others. This may appear to be zero-sum, but behind it lies the danger that if the agreement is not reached, no one will get any surplus at all. This mutually harmful alternative, and *both* parties' desire to avoid it, is what creates the potential for the threats—explicit and implicit—that make bargaining such a strategic issue.

Before the advent of game theory, one-on-one bargaining was generally thought to be a difficult and even indeterminate problem. Observation of widely different outcomes in otherwise similar-looking situations lent support to this view. Management-union bargaining over wages yields different outcomes in different contexts; different couples make different choices. Theorists were not able to achieve any systematic understanding of why one party gets more than another, and attributed this result to vague and inexplicable differences in "bargaining power."

Even the simple theory of Nash equilibrium does not take us any farther. Suppose two people are to split \$1. Let us construct a game in which each is asked to announce what he would want. The moves are simultaneous. If their announcements x and y add up to 1 or less, each gets what he announced. If they add up to more than 1, neither gets anything. Then *any* pair (x,y) adding to 1 constitutes a Nash equilibrium in this game: *given* the announcement of the other, each player cannot do better than to stick to his own announcement.[1]

Further advances in game theory have brought progress along two quite different lines, each using a distinct mode of game-theoretic reasoning. In Chapter 2 we distinguished between cooperative game theory, in which the players decide and implement their actions jointly, and noncooperative game theory, in which the players decide and take their actions separately. Each of the two lines of advance in bargaining theory uses one of these two approaches. One views bargaining as a *cooperative* game, in which the parties find and implement a solution jointly, perhaps using a neutral third party like an arbitrator for enforcement. The other views bargaining as a *noncooperative* game, in which the parties choose strategies separately and we look for an equilibrium. However, unlike our earlier simple game of simultaneous announcements, whose equilibrium was indeterminate, here we impose more structure and specify a sequential-move game of offers and counteroffers, which leads to a determinate equilibrium. As in Chapter 2, we emphasize that the labels "cooperative" and "noncooperative" refer to joint versus separate actions, not to nice versus nasty behavior nor to

[1]As we saw in Chapter 7, this game can be used as an example to bolster the critique that the Nash equilibrium concept is too imprecise. In the bargaining context, we might say the multiplicity of equilibria is just a formal way of showing the indeterminacy that previous analysts had claimed.

compromise versus breakdown. The equilibria of noncooperative bargaining games can entail a lot of compromise.

1 NASH'S COOPERATIVE SOLUTION

In this section we present Nash's cooperative-game approach to bargaining. First we present the idea in a simple numerical example; then we develop the more general algebra.[2]

A. Numerical Example

Imagine two Silicon Valley entrepreneurs, Andy and Bill. Andy produces a microchip set which he can sell to any computer manufacturer for $900. Bill has a software package that can retail for $100. The two meet, and realize that their products are ideally suited for each other, and that with a bit of trivial tinkering they can produce a combined system of hardware and software worth $3,000 in each computer. Thus together they can produce an extra value of $2,000 per unit, and they expect to sell millions of these units each year. The only obstacle that remains on this path to fortune is to agree to a division of the spoils. Of the $3,000 revenue from each unit, how much should go to Andy and how much to Bill?

Bill's starting position is that without his software, Andy's chip set is just so much metal and sand, so Andy should get only the $900, and Bill himself should get $2,100. Andy counters that without his hardware, Bill's programs are just symbols on paper or magnetic signals on a diskette, so Bill should get only $100, and $2,900 should go to him, Andy.

Watching them argue, you might suggest they "split the difference." But that is not an unambiguous recipe for agreement. Bill might offer to split the profit on each unit equally with Andy. Under this scheme each will get a profit of $1,000, meaning that $1,100 of the revenue goes to Bill and $1,900 to Andy. Andy's response might be that they should have an equal percentage of profit on their contribution to the joint enterprise. Thus Andy should get $2,700 and Bill $300.

The final agreement depends on their stubbornness or patience if they negotiate directly with each other. If they try to have the dispute arbitrated by a third party, the arbitrator's decision depends on his sense of the relative value of hardware and software, and on the rhetorical skills of the two principals as they present their arguments before the arbitrator. For the sake of definiteness, suppose the arbitrator decides that the division of the profit should be 4 :1 in favor of

---

[2] John F. Nash, Jr., "The Bargaining Problem," *Econometrica*, vol. 18, no. 2 (1950), pp. 155–162.

Andy; that is, Andy should get four-fifths of the surplus while Bill gets one-fifth, or Andy should get four times as much as Bill. What is the actual division of revenue under this scheme? Suppose Andy gets a total of x and Bill gets a total of y; thus Andy's profit is $(x - 900)$ and Bill's is $(y - 100)$. The arbitrator's decision implies that Andy's profit should be four times larger than Bill's, so that $x - 900 = 4(y - 100)$, or $x = 4y + 500$. Of course, the total revenue available to both is \$3,000, so it must also be true that $x + y = 3,000$, or $x = 3,000-y$. Then $x = 4y + 500 = 3,000 - y$, or $5y = 2,500$, or $y = 500$, and thus $x = 2,500$. This division mechanism leaves Andy with a profit of $2,500 - 900 = \$1,600$, and Bill with $500 - 100 = \$400$. This is the 4:1 split in favor of Andy that the arbitrator wants.

We now develop this simple idea into a general algebraic formula that you will find useful in many practical applications. Then we go on to examine more specifics of what determines the ratio in which the profits in a bargaining game get split.

B. General Theory

Suppose two bargainers, A and B, seek to split a total value v, which they can achieve if and only if they agree on a specific division. If no agreement is reached, A will get a and B will get b, each by acting alone or in some other way acting outside of this relationship. Call these their *backstop* payoffs or, in the jargon of the Harvard Negotiation Project, their **BATNAs (best alternative to a negotiated agreement)**.[3] Often a and b are both zero, but more generally, we only need to assume that $a + b < v$, so that there is a positive **surplus** $(v - a - b)$ from agreement; if this were not the case, the whole bargaining would be moot because each side would just take up its outside opportunity and get its BATNA.

Consider the following rule: each player is to be given his BATNA plus a share of the surplus, a fraction h of the surplus for A and a fraction k for B, such that $h + k = 1$. Writing x for the amount that A finally ends up with, and similarly y for B, we translate these statements as

$$x = a + h(v - a - b) = a(1 - h) + h(v - b) \quad \text{and} \quad y = b + k(v - a - b) = b(1 - k) + k(v - a)$$
$$x - a = h(v - a - b) \qquad\qquad\qquad\qquad y - b = k(v - a - b)$$

We call these expressions the **Nash formulas.** Another way of looking at them is to say that the surplus $(v - a - b)$ gets divided between the two bargainers in the proportions of $h : k$, or

$$\frac{y - b}{x - a} = \frac{k}{h}$$

or, in slope-intercept form,

---

[3] See Roger Fisher and William Ury, *Getting to Yes*, 2nd ed. (New York: Houghton Mifflin, 1991).

$$y = b + \frac{k}{h}(x - a) = \left(b - \frac{ak}{h}\right) + \frac{k}{h}x$$

To use up the whole surplus, x and y must also satisfy $x + y = v$. The formulas for x and y displayed above are actually the solutions to these last two simultaneous equations.

A geometric representation of the **Nash cooperative solution** is shown in Figure 16.1. The backstop or BATNA is the point P, with coordinates (a, b). All points (x, y) that divide the gains in proportions $h : k$ between the two players lie along the straight line passing through P and having slope k/h; this is just the line $y = b + (k/h)(x - a)$ that we derived earlier. All points (x, y) that use up the whole pie lie along the straight line joining $(v, 0)$ and $(0, v)$; this line is the second equation we derived, namely $x + y = v$. The Nash solution occurs at the intersection of the lines, at the point Q. The coordinates of this point are the parties' payoffs after the agreement.

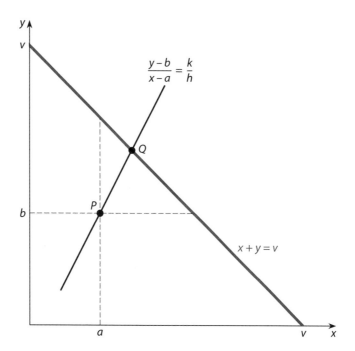

FIGURE 16.1 The Nash Bargaining Solution in the Simplest Case

Of course, the Nash formula says nothing about how or why such a solution might come about. And this vagueness is its merit—it can be used to encapsulate the results of many different theories taking many different perspectives.

At the simplest, you might think of the Nash formula as a shorthand description of the outcome of a bargaining process that we have not specified in detail. Then h and k can stand for the two parties' relative bargaining strengths. Of course, this is a cop-out; a more complete theory should explain where these bargaining strengths come from, and why one party might have more than the other. We do this in a particular context later in the chapter. In the meantime, by summarizing any and all of the sources of bargaining strength in these numbers h and k, the formula has given us a good tool.

Nash's own approach was quite different, and indeed different from the whole approach to game theory we have taken thus far in this book. Therefore it deserves more careful explanation. In all the games we have studied so far, the players chose and played their strategies separately from each other. We have looked for equilibria in which each player's strategy was in his own best interests, given the strategies of the others. Some such outcomes were very bad for some or even all of the players, the prisoners' dilemma being the most prominent example. In such situations, there was scope for the players to get together and agree that they would all follow some particular strategy. But in our framework, there was no way they could be sure that the agreement would hold. After reaching an agreement, the players would disperse, and when it was each player's turn to act, he would actually take the action that served his own best interests. The agreement for joint action would unravel in the face of such separate temptations. True, in discussing repeated games in Chapter 8, we found that the implicit threat of the collapse of an ongoing relationship might sustain an agreement, and in Chapter 12, we did allow for communication by signals. But individual action was of the essence, and any mutual benefit could be achieved only if it did not fall prey to the selfishness of separate individual actions. In Chapter 2, we called this approach to game theory *noncooperative*, emphasizing that the term signified how actions are taken, not whether outcomes are jointly good. The important point, again, is that any joint good has to be an equilibrium outcome of separate action in such games.

What if joint action *is* possible? For example, the players might take all their actions immediately after the agreement is reached, in one another's presence. Or they might delegate the implementation of their joint agreement to a neutral third party, or to an arbitrator. In other words, the game might be *cooperative* (again in the sense of joint action). Nash modeled bargaining as a cooperative game.

The thinking of a collective group that is going to implement a joint agreement by joint action can be quite different from that of a set of individuals who know they are *interacting* strategically but are *acting* noncooperatively. Whereas the latter set will think in terms of an equilibrium, and then delight or grieve, depending on whether or not they like the results, the former can think first of what is a good outcome and then see how to implement it. In other words, the theory defines the outcome of a cooperative game in terms of some general principles or properties that seem reasonable to the theorist.

Nash formulated a set of such principles for bargaining and proved that they implied a unique outcome. His principles are roughly as follows: (1) The outcome should be invariant if the scale in which the payoffs are measured changes linearly. (2) The outcome should be **efficient.** (3) If the set of possibilities is reduced by removing some that are irrelevant, in the sense that they would not be chosen anyway, then the outcome should not be affected.

The first of these principles conforms to the theory of expected utility, which we discussed briefly in the Appendix to Chapter 5. We saw there that a nonlinear rescaling of payoffs represents a change in a player's attitude toward risk and a real change in behavior; a concave rescaling implies risk aversion, and a convex rescaling implies risk preference. A linear rescaling, being the intermediate case between these two, represents no change in the attitude toward risk. Therefore it should have no effect on expected payoff calculations and no effect on outcomes.

The second principle simply means that no available mutual gain should go unexploited. In our simple example of A and B splitting a total value v, it would mean that x and y has to sum to the full amount of v available, and not to any smaller amount; in other words, the solution has to lie on the $x + y = v$ line in Figure 16.1. More generally, the complete set of logically conceivable agreements to a bargaining game, when plotted on a graph as was done in Figure 16.1, will be bounded above and to the right by the subset of agreements that leave no mutual gain unexploited. This subset need not lie along a straight line like $x + y = v$ (or $y = v - x$); it could lie along any curve of the form $y = f(x)$.

In Figure 16.2, all of the points on and below (that is, "south" and to the "west" of) the thick blue curve labeled $y = f(x)$ constitute the complete set of conceivable outcomes. The curve itself consists of the efficient outcomes; there are no conceivable outcomes that include more of both x and y than the outcomes on $y = f(x)$, so there are no unexploited mutual gains left. Therefore we call the curve $y = f(x)$ the **efficient frontier** of the bargaining problem. As an example, if the frontier is a quarter circle, $x^2 + y^2 = v^2$, then $y = \sqrt{v^2 - x^2}$, and the function $f(x)$ is defined by $f(x) = \sqrt{v^2 - x^2}$.

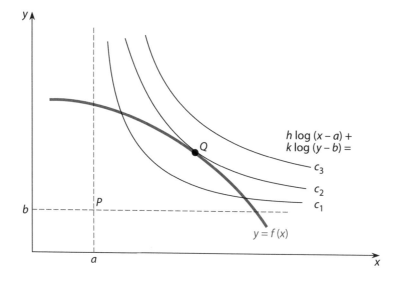

FIGURE 16.2 The General Form of the Nash Bargaining Solution

The third principle also seems appealing. If an outcome that a bargainer wouldn't have chosen anyway drops out of the picture, what should it matter? This assumption is closely connected to the "independence of irrelevant alternatives" assumption of Arrow's impossibility theorem, which we met in Chapter 14, Section 3, but we must leave the development of this connection to more advanced treatments of the subject.

Nash proved that the cooperative outcome that satisfied all three of these assumptions could be characterized by the mathematical maximization problem: Choose x and y to

$$\text{maximize } (x-a)^h (y-b)^k \text{ subject to } y = f(x).$$

Here x and y are the outcomes, a and b the backstops, and h and k two positive numbers summing to 1, which are like the bargaining strengths of the Nash formula. The values for h and k cannot be determined by Nash's three assumptions alone; thus they leave a degree of freedom in the theory and in the outcome. Nash actually imposed a fourth assumption on the problem, that of symmetry between the two players; this additional assumption led to the outcome $h = k = 1/2$ and fixed a unique solution. We have given the more general formulation that subsequently became common in game theory and economics.

Figure 16.2 also gives a geometric representation of the objective of the maximization. The black curves labeled c_1, c_2, and c_3 are the level curves, or contours, of the function being maximized; along each such curve, $(x-a)^h(y-b)^k$ is constant and equals c_1, c_2, or c_3 (with $c_1 < c_2 < c_3$) as indicated. The whole space could be filled with such curves, each with its own value of the constant, and curves farther to the "northeast" would have higher values of the constant.

It is immediately apparent that the highest-possible value of the function occurs at that point of tangency, Q, between the efficient frontier and one of the level curves.[4] The location of Q is defined by the property that the contour passing through Q is tangent to the efficient frontier. This tangency is the usual way to illustrate the Nash cooperative solution geometrically.[5]

In our example of Figure 16.1, we can also derive the Nash solution mathematically; to do so requires calculus, but the ends here are more important—at least to the study of games of strategy—than the means. For the solution, it helps to write $X = x - a$ and $Y = y - b$; thus X is the amount of the surplus that goes to A, and Y is the amount of the surplus that goes to B. The efficiency of the outcome

[4]One and only one of the (convex) level curves can be tangential to the (concave) efficient frontier; in Figure 6.2, this level curve is labeled c_2. All lower level curves (like c_1) cut the frontier in two points; all higher level curves (like c_3) do not meet the frontier at all.

[5]If you have taken an elementary microeconomics course, you will have encountered the concept of social optimality, illustrated graphically by the tangent point between the production possibility frontier of an economy and a social indifference curve. Our Figure 16.2 is similar in spirit: the efficient frontier in bargaining is like the production possibility frontier, and the level curves of the objective in cooperative bargaining are like social indifference curves.

guarantees that $X + Y = x + y - a - b = v - a - b$, which is just the total surplus, which we will write as S. Then $Y = X - S$, and

$$(x - a)^h (y - b)^k = X^h Y^k = X^h (S - X)^k.$$

In the Nash solution, X takes on the value that maximizes this function. Elementary calculus tells us that the way to find X is to take the derivative of this expression with respect to X and set it equal to zero. Using the rules of calculus for taking the derivatives of powers of X and of the product of two functions of X, we get

$$h X^{h-1} (S - X)^k - X^h k (S - X)^{k-1} = 0.$$

When we cancel the common factor $X^{h-1} (S - X)^{k-1}$, this becomes

$$h(S - X) - kX = 0$$
$$hY - kX = 0$$
$$kX = hY$$
$$\frac{X}{h} = \frac{Y}{k}.$$

Finally, expressing the equation in terms of the original variables x and y, we have $(x - a)/h = (y - b)/k$, which is just the Nash formula. The punch line: Nash's three conditions lead to the formula we originally stated as a simple way of splitting the bargaining surplus.

The three principles, or desired properties, that determine the Nash cooperative bargaining solution are simple and even appealing. But in the absence of a good mechanism to make sure the parties take the actions that are stipulated by the agreement, these principles may come to nothing. A player who can do better by strategizing on his own than by using the Nash solution may simply reject the principles. If an arbitrator can enforce a solution, the player may simply refuse to go to arbitration. Therefore Nash's cooperative solution will seem more compelling if it can be given an alternative interpretation, namely as the Nash equilibrium of a noncooperative game played by the bargainers. This can indeed be done, and we will develop an important special case of it in Section 5.

2 VARIABLE-THREAT BARGAINING

In this section, we embed the Nash cooperative solution within a specific game, namely as the second stage of a sequential-play game. We assumed in Section 1 that the players' backstops (BATNAs) a and b were fixed. But suppose there is a first stage to the bargaining game in which the players can make strategic moves to manipulate their BATNAs within certain limits. After they have done this, the

Nash cooperative outcome starting from those BATNAs will emerge in a second stage of the game. This type of game is called **variable-threat bargaining.** What kind of manipulation of the BATNAs is in a player's interest in this type of game?

We show the possible outcomes from a process of manipulating BATNAs in Figure 16.3. The originally given backstops (*a* and *b*) are the coordinates for the game's backstop point *P*; the Nash solution to a bargaining game with these backstops occurs at the outcome *Q*. If player A can increase his BATNA in order to move the game's backstop point to P_1, then the Nash solution starting there leads to the outcome *Q'*, which is better for player A (and worse for B). Thus a strategic move that improves one's own BATNA is desirable. For example, if you have a good job offer in your pocket—a higher BATNA—when you go for an interview at another company, you are likely to get a better offer from that employer than you would if you did not have the first alternative.

The result that improving your own BATNA can improve your ultimate outcome is quite obvious, but the next step in the analysis is less so. It turns out that if player A can make a strategic move that *reduces* player B's BATNA and moves the game's backstop point to P_2, the Nash solution starting there leads to the *same* outcome *Q'* that was achieved after A increased his own BATNA to get to the backstop point P_1. Therefore this alternative kind of manipulation is equally in player A's interest. As an example of decreasing your opponent's BATNA, think of a situation in which you are already working and want to get a raise. Your chances are better if you can make yourself indispensable to your employer so

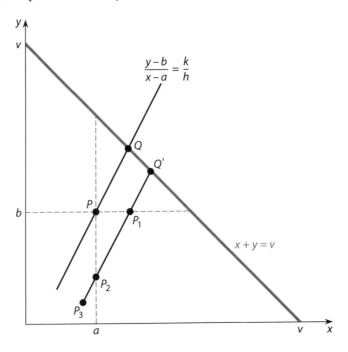

FIGURE 16.3 Bargaining Game of Manipulating BATNAs

that without you his business has much worse prospects; his low outcome in the absence of an agreement—not offering you a raise and your leaving the firm—may make him more likely to accede to your wishes.

Finally, and even more dramatically, if player A can make a strategic move that lowers *both players' BATNAs* so that the game's backstop point moves to P_3, that again has the same result as each of the previous manipulations. This particular move is like using a threat that says, "This will hurt you more than it hurts me."

In general, the key for player A is to shift the game's BATNA point to somewhere below the line PQ. The farther southeast the BATNA point is moved, the better it is for player A in the eventual outcome. As is usual with threats, the idea is not to actually suffer the low payoff, but merely to use its prospect as a lever to get a better outcome.

The possibility of manipulating BATNAs in this way depends on the context. We offer just one illustration. In 1980 there was a baseball players' strike. It took a very complicated form. The players struck during spring training, then resumed working (playing, really) when the regular season began in April, and went on strike again starting on Memorial Day. A strike is costly to both sides, employers and employees, but the costs differ. During spring training the players do not have salaries, but the owners make some money from vacationing spectators. At the start of the regular season, in April and May, the players get salaries but the weather is cold and the season is not yet exciting; therefore the crowds are small, so the cost of a strike to the owners is low. The crowds start to build up from Memorial Day onward, which raises the cost of a strike to the owners, while the salaries the players stand to lose stay the same. So we see that the two-piece strike was very cleverly designed to lower the BATNA of the owners *relative to* that of the players as much as possible.[6]

One puzzle remains: why did the strike occur at all? According to the theory, everyone should have seen what was coming; a settlement more favorable to the players should have been reached so that the strike would have been unnecessary. A strike that actually happens is a threat that has "gone wrong." Some kind of uncertainty—asymmetric information or brinkmanship—must be responsible.

3 ALTERNATING-OFFERS MODEL I: TOTAL VALUE DECAYS

Here we move back to the more realistic noncooperative game theory, and think about the process of individual strategizing that may produce an equilibrium in

---

[6] See Larry DeBrock and Alvin Roth, "Strike Two: Labor-Management Negotiations in Major League Baseball," *Bell Journal of Economics* vol. 12, no. 2 (Autumn 1981), 413–425.

a bargaining game. Our standard picture of this process is one of **alternating offers.** One player, say A, makes an offer. The other, say B, either accepts it or makes a counteroffer. If he does the latter, then A can either accept that or come back with another offer of his own. And so on. Thus we have a sequential-move game, and look for its rollback equilibrium.

To find a rollback equilibrium we must start at the end and work backward. But where is the end point? Why should the process of offers and counteroffers ever terminate? Perhaps more drastically, why would it ever start? Why would the two bargainers not stick to their original positions and refuse to budge? Of course, it is costly to both if they fail to agree at all, but the benefit of an agreement is likely to be smaller to the one who makes the first or the larger concession. The reason anyone concedes must be that continuing to stand firm would cause an even greater loss of benefit. This loss takes one of two broad forms. The available pie or surplus may **decay** (shrink) with each offer, a possibility we consider in this section. The alternative possibility is that time has value, and **impatience** is important, so a delayed agreement is worth less; we examine this possibility in Section 5.

Consider the following story of bargaining over a shrinking pie. A fan arrives at a professional football (or basketball) game without a ticket. He is willing to pay up to $25 to watch each quarter of the game. He finds a scalper who states a price. If the fan is not willing to pay this price, he goes to a nearby bar to watch the first quarter on the big-screen TV. At the end of the quarter he comes out, finds the scalper still there, and makes a counteroffer for the ticket. If the scalper does not agree, the fan goes back to the bar. He comes out again at the end of the second quarter, when the scalper makes him yet another offer. If that offer is not acceptable to the fan, he goes back into the bar, emerging at the end of the third quarter to make yet another counteroffer. Of course, the value of watching the rest of the game is declining as the quarters go by.[7]

Rollback analysis enables us to predict the outcome of this alternating-offers bargaining process. At the end of the third quarter, the fan knows that if he does not buy the ticket then, the scalper will be left with a small piece of paper of no value. So the fan will be able to make a very small offer that for the scalper will still be better than nothing. Thus, on his last offer, the fan can get the ticket almost for free. Backing up one period, at the end of the second quarter, the scalper has the initiative in making the offer. But he must look ahead and recognize that he cannot hope to extract the whole of the remaining two quarters' value from the fan. If the scalper asks for more than $25—the value of the *third* quarter to the fan—the fan will turn down the offer because he knows he can get the fourth quarter later for almost nothing; so the scalper can at most ask for $25.

---

[7]Just to keep the argument simple, we imagine this as a process of one-on-one bargaining. Actually there may be several fans and several scalpers, turning the situation into a *market*. We analyze interactions in markets in detail in Chapter 17.

Now consider the situation at the end of the first quarter. The fan knows that if he does not buy the ticket now, the scalper can expect to get only $25 later, so $25 is all the fan needs to offer now to secure the ticket. Finally, before the game even begins, the scalper can look ahead to all this and ask for $50; this $50 represents the $25 value of the *first* quarter to the fan, plus the $25 for which the fan can get the remaining three quarters' worth. Thus the two will strike an immediate agreement, and the ticket will change hands for $50, but the price is determined by the full forward-looking rollback reasoning.[8]

This story can easily be turned into a more general argument for two bargainers, A and B. Suppose A makes the first offer to split the total surplus, which we call v (in some currency, say dollars). If B refuses the offer, the total available drops by x_1 to $(v - x_1)$; B offers a split of this. If A refuses B's offer, the total drops by a further amount x_2 down to $(v - x_1 - x_2)$; A offers a split of this. This offer and counteroffer process continues until finally, say after 10 rounds, $v - x_1 - x_2 - \cdots - x_{10} = 0$, so the game ends. As usual with sequential-play games, we begin our analysis at the end.

If the game has gone to the point where only x_{10} is left, B can make a final offer whereby he gets to keep "almost all" of this, leaving a measly cent or so to A. Left with the choice of that or absolutely nothing, A should accept the offer. To avoid the finicky complexity of keeping track of tiny cents, let us call this outcome "x_{10} to B, 0 to A." We will do the same at the other (earlier) rounds.

Knowing what is going to happen in round 10, we turn to round 9. Here A is to make the offer, and $(x_9 + x_{10})$ is left. A knows he must offer at least x_{10} to B, or else B will refuse the offer and take the game to round 10, where he can get that much. Of course, A does not want to offer any more to B. So on round 9, A will offer a split where he keeps x_9 and leaves x_{10} to B.

Then on the round before, when $x_8 + x_9 + x_{10}$ is left, B will offer a split where he gives x_9 to A and keeps $(x_8 + x_{10})$. Working backward, on the very first round A will offer a split where he keeps $(x_1 + x_3 + x_5 + x_7 + x_9)$ and gives $(x_2 + x_4 + x_6 + x_8 + x_{10})$ to B. This offer will be accepted.

You can remember this by means of a simple trick. *Hypothesize* a sequence in which all offers are refused. (This is, of course, *not* what actually happens.) Then add up the amounts that would be destroyed by the refusals of one player. This is what the other player gets in the actual equilibrium. For example, when B refused A's first offer above, the total available surplus dropped by x_1, and x_1 became part of what went to A in the equilibrium of the game.

If each player has a positive BATNA, the analysis must be modified somewhat to take these into account. At the last round, B must offer A at least the

---

[8] To keep the analysis simple, we omitted the possibility that the game might get exciting, so the value of the ticket might actually increase as the quarters go by. The uncertainty makes the problem much more complex but also more interesting. The ability to deal with such possibilities should serve as an incentive for you to go beyond this book or course to study more advanced game theory.

BATNA *a*. If x_{10} is greater than *a*, this leaves B with $(x_{10} - a)$; if not, the game must terminate before this round is reached. Now at round 9, A must offer B the larger of the two amounts—the $(x_{10} - a)$ that B can get in round 10, and the BATNA *b* that B can get outside this agreement. The analysis can proceed all the way back to round 1 in this way; we leave it to you to complete the rollback reasoning for this case.

We have found the rollback equilibrium of the alternating-offers bargaining game, and in the process of deriving the outcome we have also described the full strategies—complete contingent plans of action—behind the equilibrium, namely what each player *would* do if the game reached some later stage. In fact, actual agreement is immediate on the very first offer. The later stages are not reached; they are off-equilibrium nodes and paths. But as usual with rollback reasoning, the foresight about what rational players would do at those nodes if they were reached is what informs the initial action.

The other important point to note is that *gradual decay* (several potential rounds of offers), leads to a more even or fairer split of the total than *sudden decay* (only one round of bargaining permitted). In the latter, no agreement would result if B turned down A's very first offer; then in a rollback equilibrium A would get to keep (almost) the whole surplus, giving B an "ultimatum" to accept a measly cent or else get nothing at all. The subsequent rounds give B the credible ability to refuse a very uneven first offer.

4 EXPERIMENTAL EVIDENCE

The theory of this particular type of bargaining process is fairly simple, and many people have staged laboratory or classroom experiments that create such conditions of decaying totals, in order to observe what the experimental subjects actually do. We mentioned some of these briefly in Chapter 7 when discussing the validity of rollback reasoning; now we examine them in more detail in the context of bargaining.[9]

The simplest bargaining experiment is the **ultimatum game,** where there is only one round: player A makes an offer, and if B does not accept it, the bargaining ends and both get nothing. The general structure of these games is as follows. A pool of players is brought together, either in the same room or at computer terminals in a network. They are paired; one person in the pair is designated to be the *proposer* (the one who makes the offer or is the seller who posts a price), and

[9]For more details, see Douglas D. Davis and Charles A. Holt, *Experimental Economics* (Princeton, N.J.: Princeton University Press, 1993), pp. 263–269, and *The Handbook of Experimental Economics*, ed. John H. Kagel and Alvin E. Roth, (Princeton, N.J.: Princeton University Press, 1995), pp. 255–274.

the other to be the *chooser* (the one who accepts or refuses the offer, or is the customer who decides whether to buy at that price). The pair is given a fixed surplus, usually $1 or some other sum of money, to split.

Rollback reasoning suggests that A should offer B the minimal unit, say 1 cent out of a dollar, and that B should accept such an offer. Actual results are dramatically different. In the case in which the subjects are together in a room, and the assignment of the role of proposer is made randomly, the most common offer is a 50:50 split. Very few offers worse than 75:25 are made (where the proposer to keep 75% and the chooser to get 25%), and if made, they are often rejected.

This finding can be interpreted in one of two ways. Either the players cannot or do not perform the calculation required for rollback, or the payoffs of the players include something other than what they get out of this round of bargaining. Surely the calculation in the ultimatum game is simple enough that anyone should be able to do it, and the subjects in most of these experiments are college students. A more likely explanation is the one we put forth in Chapters 3 (Section 9) and 7 (Section 1)—that the theory, which assumed payoffs to consist only of the sum earned in this one round of bargaining, is too simplistic.

Participants can have payoffs that include other things. They may have self-esteem or pride that prevents them from accepting a very unequal split. Even if the proposer A does not include this consideration in his own payoff, if he thinks that B might, then it is a good strategy for A to offer enough to make it likely that B will accept. A balances his higher payoff with a smaller offer to B against the risk of getting nothing if B rejects an offer deemed too unequal.

A second possibility is that, when the participants in the experiment are gathered in a room, the anonymity of pairing cannot be guaranteed. If the participants come from a group like college students or villagers who have ongoing relationships outside of this game, then they may value those relationships. Then the proposers fear that if they offer too unequal a split in this game, those relationships may suffer. Therefore they would be more generous in their offers than the simplistic theory suggests. If this is the explanation, then ensuring greater anonymity should enable the proposers to make more unequal offers, and experiments do find this to be the case.

Finally, people may have a sense of fairness drilled into them during their nurture and education. This may have evolutionary value for society as a whole, and may therefore have become a social norm. Whatever its origin, it may lead the proposers to be relatively generous in their offers, quite irrespective of the fear of rejection. One of us (Skeath) has conducted classroom experiments involving several different ultimatum games. Students who had known partners with whom to bargain were noticeably "fairer" in their split of the pie. In addition, several students cited specific cultural backgrounds as explanations for behavior that was inconsistent with theoretical predictions.

Experimenters have tried variants of the basic game to differentiate between these explanations. The point about ongoing relationships can be handled by stricter procedures that visibly guarantee anonymity. Doing this by itself has some effect on the outcomes but still does not produce offers as extreme as those predicted by the purely selfish rollback argument of the theory. The remaining explanations—namely "fear of rejection" and the "ingrained sense of fairness"—remain to be sorted out.

The fear of rejection can be removed by considering a variant called the *dictator game*. Again the participants are matched in pairs. One person, say A, is designated to determine the split, and the other, say B, is simply a passive recipient of what A decides. Now the split becomes decidedly more uneven, but even here a majority of As choose to keep no more than 70%. This result suggests a role for an ingrained sense of fairness.

But such a sense has its limits. In some experiments, a sense of fairness was created simply when the experimenter randomly assigned roles of proposer and chooser. In one variant, the participants were given a simple quiz, and those who performed best were made proposers. This created a sense that the role of proposer had been earned, and the outcomes did show more unequal splits. When the dictator game was played with earned rights and with stricter anonymity conditions, most As kept everything, but some (about 5%) still offered a 50:50 split.

One of us (Dixit) carried out a classroom experiment in which students in groups of 20 were gathered together in a computer cluster. They were matched randomly and anonymously in pairs, and each pair tried to agree on how to split 100 points. Roles of proposer and chooser were not assigned; either could make an offer or accept the other's offer. Offers could be made and changed at any time. The pairs could exchange messages instantly with their matched opponent on their computer screens. The bargaining round ended at a random time between three and five minutes; if agreement was not reached in time by a pair, both got zero. There were 10 such rounds with different random opponents each time. Thus the game itself offered no scope for cooperation through repetition. In a classroom context the students had ongoing relationships outside the game, but they did not generally know or guess who they were playing in any round even though no great attempt was made to enforce anonymity. Each student's score for the whole game was the sum of his point score for the 10 rounds. The stakes were quite high, as this counted for 5% of the course grade!

The highest total of points achieved was 515. Those who quickly agreed on 50:50 splits did the best, and those who tried to hold out for very uneven scores, or refused to split a difference of 10 points or so between the offers and ran the risk of time running out on them, did poorly.[10] It seems that moderation and fairness do get rewarded, even as measured in terms of one's own payoff.

---

[10] Those who were best at the mathematical aspects of game theory, such as problem sets, did a little worse than the average, probably because they tried too hard to eke out an extra advantage and met resistance. And women did slightly better than men.

5 ALTERNATING-OFFERS MODEL II: IMPATIENCE

Now we consider a different kind of cost of delay in reaching an agreement. Suppose the actual monetary value of the total available for splitting does not decay, but players have a "time value of money" and therefore prefer early agreement to later agreement. They make offers alternately as described in Section 3, but their time preferences are such that money now is better than money later. For concreteness, we will say that both bargainers believe that having only 95 cents right now is as good as having $1 one round later.

A player who prefers having something right away to having the same thing later is impatient; he attaches less importance to the future relative to the present. We came across this idea in Chapter 8 (Section 2), and saw two reasons for it. First, player A may be able to invest his money, say $1 now, and get his principal back along with interest and capital gains at a rate of return r, for a total of $(1 + r)$ during the next period (tomorrow, or next week, or next year, or whatever is the length of the period). Second, there may be some risk that the game will end between now and the next offer (like the sudden end at a time between three and five minutes in the classroom game described earlier). If p is the probability that the game continues, then the chance of getting a dollar next period has an expected value of only p now.

Suppose we consider a bargaining process between two players with zero BATNAs. Let us start the process with one of the two bargainers, say A, making an offer to split $1. If the other player, B, rejects A's offer, then B will have the opportunity to make his own offer one round later. The two bargainers are in identical situations when each makes his offer, because the amount to be split is always $1. Thus in equilibrium the amount that goes to the person currently in charge of making the offer (call it x) is the same, regardless of whether that person is A or B. We can use rollback reasoning to find an equation that can be solved for x.

Suppose A starts the alternating offer process. He knows that B can get x in the next round when it is B's turn to make the offer. Therefore, A must give B at least an amount that is equivalent, in B's eyes, to getting x in the next round; A must give B at least $0.95x$ now. (Remember that for B, 95 cents received now is equivalent to $1 received in the next round, so $0.95x$ now is as good as x in the next round.) Of course, A will not give B any more than is required to induce B's acceptance. Thus A offers B exactly $0.95x$ and is left with $(1 - 0.95x)$. But the amount that A gets when making the offer is just what we called x. Therefore $x = 1 - 0.95\ x$, or $(1 + 0.95)x = 1$, or $x = 1/1.95 = 0.512$.

Two things about this calculation should be noted. First, even though the process allows for an unlimited sequence of alternating offers and counteroffers, in the equilibrium the very first offer A makes gets accepted and the bargaining ends. Since time has value, this is an efficient outcome. The cost of delay governs how much A must offer B to induce acceptance; it thus affects A's rollback reasoning. Second, the player who gets to make the first offer gets more than half of the pie, namely 0.512 as opposed to 0.488. Thus each player gets more when he makes the first offer than when the other player makes the first offer. But this advantage is far smaller than that in an ultimatum game with no future rounds of counteroffers.

Now suppose the two players are not equally patient (or impatient, as the case may be). B still regards $1 in the next round as being equivalent to 95 cents now, but A regards it as being equivalent to only 90 cents now. Thus A is willing to accept a smaller amount in order to get it sooner; in other words, A is more impatient. This inequality in rates of impatience can translate into unequal equilibrium payoffs from the bargaining process. To find the equilibrium for this example, we write x for the amount that A gets when he starts the process, and y for what B gets when he starts the process.

A knows that he must give B at least $0.95y$; otherwise B will reject the offer in favor of the y he knows he can get when it becomes his turn to make the offer. Thus the amount that A gets, x, must be $1 - 0.95y$; $x = 1 - 0.95y$. Similarly, when B starts the process, he knows he must offer A at least $0.90x$, and then $y = 1 - 0.90x$. These two equations can be solved for x and y:

$$x = 1 - 0.95(1 - 0.9\,x)$$
$$[1 - 0.95(0.9)]x = 1 - 0.95$$
$$0.145x = 0.05$$
$$x = 0.345$$

and

$$y = 1 - 0.9(1 - 0.95\,y)$$
$$[1 - 0.9(0.95)]\,y = 1 - 0.9$$
$$0.145\,y = 0.10$$
$$y = 0.690$$

Note that x and y do not add up to 1 because each of these amounts is the payoff to a given player when he makes the first offer. Thus when A makes the first offer, A gets 0.345 and B gets 0.655; when B makes the first offer, B gets 0.69 and A gets 0.31. Once again, each player does better when he makes the first offer than when the other player makes the first offer, and once again the difference is small.

The outcome of this case with unequal rates of impatience differs from that of the previous case with equal rates of impatience in a major way. With unequal rates of impatience, the more impatient player, A, gets a lot less than B even when he is able to make the first offer. We expect that the person who is willing to accept less in order to get it sooner ends up getting less, but the difference is very dramatic. The proportions of A's and B's shares are almost 1:2.

As usual, we can now build on these examples to develop the more general algebra. Suppose A regards $1 immediately as being equivalent to $(1 + r) one offer later, or equivalently, A regards $1/(1 + r) immediately as being equivalent to $1 one offer later. For brevity, we substitute a for $1/(1 + r)$ in the calculations that follow. Likewise, suppose player B regards $1 today as being equivalent to

$(1 + s)$ one offer later; we use b for $1/(1 + s)$. If r is high (or equivalently, if a is low), then player A is very impatient. Similarly, B is impatient if s is high (or if b is low).

Here we look at bargaining that occurs in alternating rounds, with a total of $1 to be divided between two players, both of whom have zero BATNAs. (You can do the even more general case easily once you understand this one.) What is the rollback equilibrium?

We can find the payoffs in such an equilibrium by extending the simple argument used earlier. Suppose A's payoff in the rollback equilibrium is x when he makes the first offer; B's payoff in the rollback equilibrium is y when he makes the first offer. We look for a pair of equations linking the values x and y and then solve these equations to determine the equilibrium payoffs.[11]

When A is making the offer, he knows he must give B an amount that B regards as being equivalent to y one period later. This amount is $by = y/(1 + s)$. Then, after making the offer to B, A can keep only what is left: $x = 1 - by$.

Similarly, when B is making the offer, he must give A the equivalent of x one period later, namely ax. Therefore $y = 1 - ax$. Solving these two equations is now a simple matter. We have $x = 1 - b(1 - ax)$, or $(1 - ab) x = 1 - b$. Expressed in terms of r and s, this becomes

$$x = \frac{1 - b}{1 - ab} = \frac{s + rs}{r + s + rs}.$$

Similarly, $y = 1 - a(1 - by)$, or $(1 - ab)y = 1 - a$. This becomes

$$y = \frac{1 - a}{1 - ab} = \frac{r + rs}{r + s + rs}.$$

Although this quick solution might seem a sleight of hand, it follows the same steps used earlier, and we soon give a different reasoning yielding exactly the same answer. First let us examine some features of the answer.

First note that, as in our simple unequal impatience example, the two magnitudes x and y add up to more than 1:

$$x + y = \frac{r + s + 2rs}{r + s + rs} > 1$$

Remember that x is what A gets when he has the right to make the first offer, and y is what B gets when he has the right to make the first offer. When A makes the

[11] We are taking a shortcut; we have simply assumed that such an equilibrium exists and that the payoffs are uniquely determined. More rigorous theory proves these things. For a step in this direction, see John Sutton, "Non-Cooperative Bargaining: An Introduction," *Review of Economic Studies*, vol. 53, no. 5 (October 1986), pp. 709–724. The fully rigorous (and quite difficult) theory is given in Ariel Rubinstein, "Perfect Equilibrium in a Bargaining Model," *Econometrica* vol. 50, no. 1 (January 1982), pp. 97–109.

first offer, B gets $(1 - x)$, which is less than y; this just shows A's advantage from being the first proposer. Similarly, when B makes the first offer, B gets y and A gets $(1 - y)$, which is less than x.

However, usually r and s are small numbers. When offers can be made at short intervals like a week or a day or an hour, the interest your money can earn from one offer to the next, or the probability that the game ends precisely within the next interval, is quite small. For example, if r is 1% (or 0.01) and s is 2% (0.02), then the formulas yield $x = 0.668$ and $y = 0.337$, so the advantage of making the first offer is only 0.005. (A gets 0.668 when making the first offer, but $1 - 0.337 = 0.663$ when B makes the first offer; the difference is only 0.005.) More formally, when r and s are each small compared with 1, then their product rs is very small indeed; thus we can ignore rs in order to write an approximate solution for the split that does not depend on which player makes the first offer:

$$x = \frac{s}{r + s} \quad \text{and} \quad y = \frac{r}{r + s}.$$

Now $x + y$ is approximately equal to 1.

Most importantly, x and y in the approximate solution are the shares of the surplus that go to the two players, and $y/x = r/s$; that is, the shares of the players are inversely proportional to their rates of impatience as measured by r and s. If B is twice as impatient as A, then A gets twice as much as B, so the shares are 1/3 and 2/3, or 0.333 and 0.667, respectively. Thus we see that patience is an important advantage in bargaining. Our formal analysis supports the intuition that if you are very impatient, the other player can offer you a quick but poor deal, knowing you will accept it.

This effect of impatience hurts the United States in numerous negotiations that our government agencies and diplomats conduct with other countries. The American political process puts a great premium on speed. The media, interest groups, and rival politicians all demand results, and are quick to criticize the administration or the diplomats for any delay. With this pressure to deliver, the negotiators are always tempted to come back with results of any kind. Such results are often poor from the long-run U.S. perspective; the other countries' concessions often have loopholes, and their promises are less than credible. Of course, the U.S. administration hails the deals as great victories, but they usually unravel after a few years. A similar example, perhaps closer to people's everyday lives, involves insurance companies, which often make lowball offers of settlement to people who have suffered a major loss; the insurers know that their clients urgently want to make a fresh start and are therefore very impatient.

As a conceptual matter, the formula $y/x = r/s$ ties our noncooperative game approach to bargaining to the cooperative approach of the Nash solution discussed in Section 1. The formula for shares of the available surplus that we derived there becomes, with zero BATNAs, $y/x = k/h$. In the cooperative approach, the shares of the two players stood in the same proportions as their bargaining strengths, but these strengths were assumed to be given somehow from the outside. Now we have an explanation for the bargaining strengths in terms of some more basic characteristics of the players—h and k are inversely proportional to the players' rates of impatience r and s. In other words, Nash's cooperative solution can also be given an alternative and perhaps more satisfactory interpretation as the rollback equilibrium of a noncooperative game of offers and counteroffers, if we interpret the abstract bargaining-strength parameters in the cooperative solution correctly in terms of the players' characteristics like impatience.

Finally, note that agreement is once again immediate—the very first offer is accepted. The whole rollback analysis as usual serves the function of disciplining the first offer, by making the first proposer recognize that the other would credibly reject a less adequate offer.

To conclude this section, we offer an alternative derivation of the same (precise) formula for the equilibrium offers that we derived earlier. Suppose this time that there are 100 rounds; A is the first proposer and B the last. Start the backward induction in the 100th round; B will keep the whole dollar. Therefore in the 99th round, A will have to offer B the equivalent of $1 in the 100th round, namely b, and A will keep $(1 - b)$. Then proceed backwards:

In round 98, B offers $a(1 - b)$ to A, and keeps $1 - a(1 - b) = 1 - a + ab$.
In round 97, A offers $b(1 - a + ab)$ to B, and keeps $1 - b(1 - a + ab) = 1 - b + ab - ab^2$.
In round 96, B offers $a(1 - b + ab - ab^2)$ to A, and keeps $1 - a + ab - a^2b + a^2b^2$.
In round 95, A offers $b(1 - a + ab - a^2b + a^2b^2)$ to B, and keeps $1 - b + ab - ab^2 + a^2b^2 - a^2b^3$.

Proceeding in this way and following the established pattern, we see that in round 1, A gets to keep

$$1 - b + ab - ab^2 + a^2b^2 - a^2b^3 + \cdots + a^{49}b^{49} - a^{49}b^{50} = (1 - b)\,[1 + ab + (ab)^2 + \cdots + (ab)^{49}]$$

The consequence of allowing more and more rounds is now clear. We just get more and more of these terms, growing geometrically by the factor ab for every two offers. To find A's payoff when he is the first proposer in an infinitely long sequence of offers and counteroffers, we have to find the limit of the infinite geometric sum. In the Appendix to Chapter 8 we saw how to sum such series. Using the formula obtained there, we get the answer

$$(1 - b)\,[1 + ab + (ab)^2 + \cdots + (ab)^{49} + \cdots = \frac{1 - b}{1 - ab}$$

This is exactly the solution for x that we obtained before. By a similar argument, you can find B's payoff when he is the proposer, and in doing so, improve your understanding and technical skills at the same time.

6 MANIPULATING INFORMATION IN BARGAINING

We have seen that the outcomes of a bargain depend crucially on various characteristics of the parties to the bargain, most importantly their BATNAs and their impatience. We proceeded thus far assuming that the players knew each other's characteristics as well as their own. In fact, we assumed each player knew that the other knew, and so on; that is, the characteristics were common knowledge. In reality we often engage in bargaining without knowing the other side's BATNA or degree of impatience; sometimes we do not even know our own BATNA very precisely.

As we saw in Chapter 12, a game with such uncertainty or informational asymmetry has associated with it an important game of signaling and screening of strategies for manipulating information. Bargaining is replete with such strategies. A player with a good BATNA or a high degree of patience wants to signal this fact to the other. However, since someone without these good attributes will want to imitate them, the other party will be skeptical and will examine the signals critically for their credibility. And each side will also try screening, using strategies that induce the other to take actions that will reveal its characteristics truthfully.

In this section we look at some such strategies used by buyers and sellers in the housing market. Most Americans are active in the housing market several times in their lives, and many people are professional estate agents or brokers who have even more extensive experience in the matter. Moreover, housing is one of the few markets in the United States where haggling or bargaining over price is accepted and even expected. Therefore considerable experience of strategies is available. We draw on this experience for many of our examples, and interpret it in the light of our game-theoretic ideas and insights.[12]

When you contemplate buying a house in a new neighborhood, you are unlikely to know the general range of prices for the particular type of house you are interested in. Your first step should be to find this out so you can then determine your BATNA. And that does not mean looking at newspaper ads or realtors' listings, which indicate only asking prices. Local newspapers and some Internet sites list recent actual transactions and the actual prices; you should check these against the asking prices of the same houses to get an idea of the state of the market and the range of bargaining that might be possible.

Next comes finding out (screening) the other side's BATNA and level of impatience. If you are a buyer, you can find out why the house is being sold, and how long it has been on the market. If it is empty, why, and how long has it been that

---

[12]We have taken the insights of practitioners from Andrée Brooks, "Honing Haggling Skills," *New York Times*, December 5, 1993.

way? If the owners are getting divorced, or have moved elsewhere and are financing another house on an expensive bridge loan, it is likely that they have a low BATNA or are rather impatient.

You should also find out other relevant things about the other side's preferences, even though these may seem irrational to you. For example, some people regard an offer too far below the asking price as an insult, and will not sell at any price to someone who makes such an offer. There are norms of this kind that vary across regions and times. It pays to find out what the common practices are.

Most importantly, the *acceptance* of an offer more accurately reveals a player's true willingness to pay than anything else and therefore is open to exploitation by the other player. A brilliant game-theorist friend of ours tried just such a ploy. He was bargaining for a floor lamp. Starting with the seller's asking price of $100, the negotiation proceeded to a point where our friend made an offer to buy the lamp for $60. The seller said yes, at which point our friend thought: "This guy is willing to sell it for $60, so his true rock-bottom price must be even lower. Let me try to find out whether it is." So our friend said, "How about $55?" The seller got very upset, refused to sell for any price, and asked our friend to leave the store and never come back.

The seller's behavior conformed to the norm that it is utmost bad faith in bargaining to renege on an offer once it is accepted. This makes good sense as a norm in the whole context of all bargaining games that occur in society. If an offer on the table cannot be accepted in good faith by the other player without fear of the kind of exploitation attempted by our friend, then each bargainer will wait to get the other to accept an offer, thereby revealing the limit of his true rock-bottom acceptance level, and the whole process of bargains will grind to a halt. Therefore such behavior has to be disallowed, and making it a social norm to which people adhere instinctively, as the seller in the example did, is a good way for society to achieve this aim.

Of course, the offer may explicitly say that it is open only for a specified and limited time; this stipulation can be part of the offer itself. Job offers usually specify a deadline for acceptance; stores have sales for limited periods. But in that case the offer is truly a *package* of price and time, and reneging on either dimension provokes a similar instinctive anger. For example, customers get quite mad if they arrive at a store during the sale period and find an advertised item unavailable. The store must offer a rain check, which allows the customer to buy the item at its sale price when next available at the regular price; even this offer causes some inconvenience to the customer and risks some loss of goodwill. The store can specify "limited quantities, no rain checks" very clearly in its advertising of the sale; even then, many customers get upset if they find that the store has run out of the item.

Next on our list of strategies to use in one-on-one bargaining, like the housing market, comes signaling your own high BATNA or patience. The best way to signal patience is to *be* patient. Do not come back with counteroffers too quickly.

"Let the sellers think they've lost you." This is a credible signal because someone not genuinely patient would find it too costly to mimic the leisurely approach. Similarly, you can signal a high BATNA by starting to walk away, a tactic that is common in negotiations at bazaars in other countries and some flea markets and tag sales in the United States.

Even if your BATNA is low, you may commit yourself to not accepting an offer below a certain level. This constraint acts just like a high BATNA, because the other side cannot hope to get you to accept anything less. In the housing context, you can claim your inability to concede any further by inventing (or creating) a tightwad parent who is providing the down payment, or a spouse who does not really like the house and will not let you offer any more. Sellers can try similar tactics. A parallel in wage negotiations is the *mandate*. A meeting is convened of all the workers who pass a resolution—the mandate—authorizing the union leaders to represent them at the negotiation, but with the constraint that the negotiators must not accept an offer below a certain level specified in the resolution. Then at the meeting with the management, the union leaders can say that their hands are tied, there is no time to go back to the membership to get their approval for any lower offer.

Most of these strategies involve some risk. While you are signaling patience by waiting, the seller of the house may find another willing buyer. As employer and union wait for each other to concede, tensions may mount so high that a strike that is costly to both sides nevertheless cannot be prevented. In other words, many strategies of information manipulation are instances of brinkmanship. We saw in Chapter 13 how such games can have an outcome that is bad for both parties. The same is true in bargaining. *Threats* of breakdown of negotiations or of strikes are strategic moves intended to achieve quicker agreement or a better deal for the player making the move; however, an *actual* breakdown or strike is an instance of the threat "gone wrong." Of course, the player making the threat—initiating the brinkmanship—must assess the risk and the potential rewards when deciding whether and how far to proceed down this path.

7 BARGAINING WITH MANY PARTIES AND ISSUES

Our discussion thus far has been confined to the classic situation where two parties are bargaining about the split of a given total surplus. But many real-life negotiations involve several parties or several issues simultaneously. Although the games get more complicated, often the enlargement of the group or the set of issues actually makes it easier to arrive at a mutually satisfactory agreement. In this section we take a brief look at such matters.[13]

[13] For a more thorough treatment, see Howard Raiffa, *The Art and Science of Negotiation* (Cambridge, Mass.: Harvard University Press, 1982), parts III and IV.

A. Multi-Issue Bargaining

In a sense we have already considered multi-issue bargaining. The negotiation over price between a seller and a buyer always involves *two* things: (1) the object offered for sale or considered for purchase and (2) money. The potential for mutual benefit arises when the buyer values the object more than the seller does, that is, when the buyer is willing to give up more money in return for getting the object than the seller is willing to accept in return for giving up the object. Both players can be better off as a result of their bargaining agreement.

The same principle applies more generally. International trade is the classic example. Consider two hypothetical countries, Freedonia and Ilyria. If Freedonia can convert one loaf of bread into two bottles of wine (by using less of its resources like labor and land in the production of bread and using them to produce more wine instead), while Ilyria can convert one bottle of wine into one loaf of bread (by switching its resources in the opposite direction), then between them they can create more goods "out of nothing." For example, suppose that Freedonia can produce 200 more bottles of wine if it produces 100 fewer loaves of bread, and Ilyria can produce 150 more loaves of bread if it produces 150 fewer bottles of wine. These switches in resource utilization create an extra 50 loaves of bread and 50 bottles of wine relative to what the two countries produced originally. This extra bread and wine is the "surplus" they can create, if they can agree how to divide it between them. For example, suppose Freedonia gives 175 bottles of wine to Ilyria and gets 125 loaves of bread. Then each country will have 25 more loaves of bread and 25 more bottles of wine than it did before. But there is a whole range of possible exchanges corresponding to different divisions of the gain. At one extreme, Freedonia may give up all the 200 extra bottles of wine it has produced in exchange for 101 loaves of bread from Ilyria, in which case Ilyria gets almost all the gain from trade. At the other extreme, Freedonia may give up only 151 bottles of wine in exchange for 150 loaves of bread from Ilyria, so that Freedonia gets almost all the gain from trade.[14] Between these limits lies the frontier where the two can bargain over the division of the gains from trade.

The general principle should now be clear. When two or more issues are on the bargaining table at the same time, and the two parties are willing to trade more of one against less of the other at different rates, then a mutually beneficial deal exists. The mutual benefit can be realized by trading at a rate somewhere between the two parties' different rates of willingness to trade. The division of

---

[14]Economics uses the concept *ratio of exchange*, or price, which here is expressed in terms of the number of bottles of wine that trade for each loaf of bread. The crucial point is that the the possibility of gain for both countries occurs with any ratio that lies between the 2:1 at which Freedonia can just convert bread into wine, and the 1:1 at which Ilyria can do so. At a ratio close to 2:1, Freedonia gives up almost all of its 200 extra bottles of wine and gets little more than the 100 loaves of bread that it sacrificed to produce the extra wine; thus Ilyria has almost all of the gain. Conversely, at a ratio close to 1:1, Freedonia has almost all of the gain. The issue in the bargaining is the division of gain, and therefore the ratio or the price at which the two should trade.

gains depends on the choice of the rate of trade. The closer it is to one side's willingness ratio, the less that side gains from the deal.

Now you can also see how the possibilities for mutually beneficial deals can be expanded by bringing more issues to the table at the same time. With more issues, you are more likely to find divergences in the ratios of valuation between the two parties, and thereby more likely to locate possibilities for mutual gain. In the case of a house, for example, many of the fittings or furnishings may be of little use to the seller in the new house to which he is moving, but they may be of sufficiently good fit and taste that the buyer values having them. Then, if the seller cannot be induced to lower the price, he may be amenable to including these items in the original price in order to close the deal.

However, the expansion of issues is not an unmixed blessing. If you value something greatly, you may fear putting it on the bargaining table; you may worry that the other side will extract big concessions from you, knowing that you want to protect that one item of great value. At the worst, a new issue on the table may make it possible for one side to deploy threats that lower the other side's BATNA. For example, a country engaged in diplomatic negotiations may be vulnerable to an economic embargo; then it would much prefer to keep the political and economic issues distinct.

B. Multiparty Bargaining

Having many parties simultaneously engaged in bargaining may also facilitate agreement, because instead of having to look for pairwise deals, the parties can seek a circle of concessions. International trade is again the prime example. Suppose the United States can produce wheat very efficiently but is less productive in cars, Japan is very good at producing cars but has no oil, and Saudi Arabia has a lot of oil but cannot grow wheat. In pairs they can achieve little, but the three together have the potential for a mutually beneficial deal.

As with multiple issues, expanding the bargaining to multiple parties is not simple. In our example, the deal would be that the United States would send an agreed amount of wheat to Saudi Arabia, which would send its agreed amount of oil to Japan, which would in turn ship its agreed number of cars to the United States. But suppose that the latter reneges on its part of the deal. Saudi Arabia cannot retaliate against the United States because, in this scenario, it is not offering anything to the United States that it can potentially withhold. Saudi Arabia can only break its deal to send oil to Japan, an innocent party. Thus enforcement of multilateral agreements may be problematic. The General Agreement on Tariffs and Trade (GATT) between 1946 and 1994, and the World Trade Organization (WTO) since then, have indeed found it difficult to enforce their agreements and to levy punishments on countries that violate the rules.

SUMMARY

Bargaining negotiations attempt to divide the *surplus* (excess value) that is available to the parties if an agreement can be reached. Bargaining can be analyzed as a *cooperative* game in which parties find and implement a solution jointly or as a (structured) *noncooperative* game in which parties choose strategies separately and attempt to reach an equilibrium.

Nash's cooperative solution is based on three principles of the outcomes' invariance to linear changes in the payoff scale, *efficiency*, and invariance to removal of irrelevant outcomes. The solution is a rule that states the proportions of division of surplus, beyond the backstop payoff levels (also called *BATNAs*, or *best alternatives to a negotiated agreement*) available to each party, based on relative bargaining strengths. Strategic manipulation of the backstops can be used to increase a party's payoff.

In a noncooperative setting of *alternating offer and counteroffer*, rollback reasoning is used to find an equilibrium; this generally involves a first-round offer that is immediately accepted. If the surplus value *decays* with refusals, the sum of the (hypothetical) amounts destroyed due to the refusals of a single player is the payoff to the other player in equilibrium. If delay in agreement is costly due to *impatience*, the equilibrium offer shares the surplus roughly in inverse proportion to the parties' rates of *impatience*. Experimental evidence indicates that players often offer more than is necessary to reach an agreement in such games; this behavior is thought to be related to issues of player anonymity as well as beliefs about fairness.

The presence of information asymmetries in bargaining games makes signaling and screening important. Some parties will wish to signal their high BATNA levels or extreme patience while others will want to screen to obtain truthful revelation of such characteristics. When more issues are on the table or more parties are participating, agreements may be easier to reach but bargaining may be riskier or the agreements more difficult to enforce.

KEY TERMS

alternating offers (532)
best alternative to a negotiated
   agreement (BATNA) (524)
decay (532)
efficient frontier (527)
efficient outcome (526)
impatience (532)

Nash (cooperative)
   solution (525)
Nash formulas (524)
surplus (524)
ultimatum game (534)
variable-threat
   bargaining (530)

EXERCISES

1. Consider the bargaining situation between Compaq Computer Corporation and the California businessman who owned the Internet address *www. altavista.com*.[15] Compaq, which had recently taken over Digital Equipment Corporation, wanted to use this man's Web address for Digital's Internet search engine, which was then accessed at *www.altavista.digital.com*. Compaq and the businessman apparently negotiated long and hard during the summer of 1998 over a selling price for the latter's address. Although the businessman was the "smaller" player in this game, it appears that the final agreement entailed a $3.35 *million* price tag for the Web address in question. Compaq confirmed the purchase in August, and began using the address in September, but refused to divulge any of the financial details of the settlement. Given this information, provide some comment on the likely values of the BATNAs for these two players, their bargaining strengths or levels of impatience, and whether a cooperative outcome appears to have been attained in this game.

2. Consider a two-person bargaining game in which the total value V of the available surplus decays by a constant c after each refused offer. If players A and B have positive BATNAs of a and b, respectively, what is the equilibrium of the bargaining game. (*Hint:* Use a rollback analysis like the one in the text. First find how many rounds are necessary to make the surplus vanish.)

3. Recall the variant of the pizza pricing game from Chapter 8, in which one store (Donna's Deep Dish) was much larger than the other (Pierce's Pizza Pies). The payoff table (Figure 8.6) is reproduced here:

		PIERCE'S PIZZA PIES	
		High	Medium
DONNA'S DEEP DISH	High	156,60	132,70
	Medium	150,36	130,50

The noncooperative dominant strategy equilibrium is (High, Medium), yielding profits of 132 to Donna's and 70 to Pierce's, for a total of 202. If the two could achieve (High, High), their total profit would be 156 + 60 = 216, but Pierce's would not agree to this. Suppose the two stores can reach an enforce-

---

[15] Details regarding this bargaining game were reported in "A Web Site by Any Other Name Would Probably Be Cheaper," *Boston Globe*, July 29, 1998, and in Hiawatha Bray's "Compaq Acknowledges Purchase of Web Site," *Boston Globe*, August 12, 1998.

able agreement where both charge High, and Donna's pays Pierce's a sum of money. The alternative to this agreement is simply the noncooperative dominant strategy equilibrium. They bargain over this agreement, and Donna's has 2.5 times as much bargaining power as Pierce's. In the resulting agreement, what sum will Donna's pay to Pierce's?

4. Two hypothetical countries, Euphoria and Militia, are holding negotiations to settle a dispute. They meet once a month starting in January, and take turns making offers. Suppose the total at stake is 100 points. The government of Euphoria is facing reelection in November. Unless the government produces an agreement at the October meeting, it will lose the election, which it regards as being just as bad as getting zero points from an agreement. The government of Militia does not really care about reaching an agreement; it is just as happy to prolong the negotiations or even to fight, as it would be settling for anything significantly less than 100. What will be the outcome of the negotiations? What difference will it make who moves first?

5. In light of the answer to Exercise 4, discuss why actual negotiations often continue right down to the deadline.

6. Let x be the amount that player A asks for, and y be the amount B asks for, when making the first offer in an alternating-offers bargaining game with impatience. Their rates of impatience are r and s, respectively..

   (a) If we use the approximate formulas $x = s/(r + s)$ for x and $y = r/(r + s)$ for y, and if B is twice as impatient as A, then A gets two-thirds of the surplus and B gets one-third. Verify that this result is correct.

   (b) Let $r = 0.01$ and $s = 0.02$, and compare the x and y values found using the approximation method with the more exact solutions for x and y found using the formulas $x = (s + rs)/(r + s + rs)$ and $y = (r + rs)/(r + s + rs)$ derived in the text.

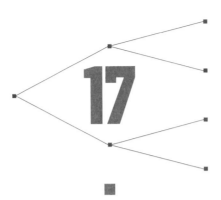

Markets and Competition

A N ECONOMY IS A SYSTEM ENGAGED in production and consumption. Its primary resources, such as labor, land, and raw materials, are used in the production of goods and services, including some that are then used in the production of other goods and services. The endpoints of this chain are the goods and services that go to the ultimate consumers. The system has evolved differently in different places and times, but always and everywhere it must create methods of linking its component parts, that is to say, a set of institutions and arrangements that enable various suppliers, producers, and buyers to deal with one another.

In the previous two chapters we examined two such institutions, auctions and bargaining. In auctions, one seller generally deals with several actual or potential buyers, although the reverse arrangement—in which one buyer deals with several actual or potential sellers—also exist; an example is a construction or supply project being offered for tender. Bargaining generally confronts one buyer with one seller.

That leaves perhaps the most ubiquitous economic institution of all, namely the *market*, where several buyers and several sellers can deal simultaneously. A market can conjure up the image of a bazaar in a foreign country, where several sellers have their stalls, several buyers come, and much bargaining is going on at the same time between pairs of buyers and sellers. But most markets operate differently. A town typically has two or more supermarkets, and the households in the town decide to shop at one (or perhaps buy some items at one and some at another) by comparing availability, price, quality, and so on. Supermarkets are generally not located next to each other; nevertheless, they know they have to compete with each other for the townspeople's patronage, and this affects their decisions on what to stock, what prices to charge, and so forth. There is no overt bargaining between a store and a customer; prices are posted, and one can either pay that price or go elsewhere. But the existence of the other store limits the price each store can charge.

In this chapter, we will briefly examine markets from a strategic viewpoint. How should, and how do, a group of sellers and buyers act in this environment? What is the equilibrium of their interaction? Are buyers and sellers efficiently matched to one another in a market equilibrium?

We have already seen a few examples of strategies in markets. Remember the pricing game of the two pizza stores introduced in Chapter 4? We considered two versions: one with discrete prices (High, Low, Medium), and one with a continuous range of prices. Because the strategic interaction pits two sellers against each other, we call it a *duopoly* (from the Greek *duo,* meaning "two," and *polein,* "to sell"). We found that the equilibrium was a prisoners' dilemma, in which both players charge prices below the level that would maximize their joint profit. In Chapter 8, we saw some ways, most prominently through the use of a repeated relationship, whereby the two stores can resolve the prisoners' dilemma and sustain tacit collusion with high prices; they thus become a *cartel,* or in effect a *monopoly.*

But even when the pizza stores do not collude, their prices are still higher than their costs because each store has a somewhat captive clientele whom it can exploit and profit from. When there are more than two sellers in a market—and as the number of sellers increases—we would expect the prisoners' dilemma to worsen, or the competition among the sellers to intensify, to the point where prices are as low as they can be. If there are hundreds or thousands of sellers, each might be too small to have a truly strategic role, and the game-theoretic approach would have to be replaced with one more suitable for dealing with many individually small buyers and sellers. In fact, that is what the traditional economic "apparatus" of *supply and demand* does—assumes that each such small buyer or seller is unable to affect the market price. Each party simply decides how much to buy or sell at the prevailing price. Then the price adjusts, through some invisible and impersonal market, to the level that equates supply and demand and thus "clears" the market. Remember the distinction we drew in Chapter 2 between an individual decision-making situation and a game of strategic interaction. The traditional economic approach regards the market as a collection of individual decision makers who interact not strategically but through the medium of the price set by the market.

We can now go beyond this cursory treatment, to see precisely how an increase in the number of sellers and buyers intensifies the competition between them, and leads to an outcome that looks like an impersonal market in which the price adjusts to equate demand and supply. That is to say, we can derive our supply-and-demand analysis from the more basic strategic considerations of bargaining and competition. This will deepen your understanding of the market mechanism even if you have already studied it in elementary economics courses and textbooks. If you have not, this chapter will allow you to leapfrog ahead of others who have studied economics but not game theory.

In addition to our discussion of the market mechanism, we will show when and how competition yields an outcome that is socially desirable; we also consider what is meant by the phrase "socially desirable." Finally, in Sections 4 and 5 we examine some alternative mechanisms that attempt to achieve efficient or fair allocations where markets cannot or do not function well.

1 A SIMPLE TRADING GAME

We begin with a situation that involves pure bargaining between two people, and then we introduce additional bargainers and competition. Imagine two people, a seller S and a buyer B, negotiating over the price of a house. S may be the current owner who gets a payoff with a monetary equivalent of $100,000 from living in the house. Or S may be a builder whose cost of constructing the house is $100,000. In either interpretation, this amount represents the seller's BATNA, or in the language of auctions, his reserve price. From living in the same house, B can get a payoff of $300,000, which represents the maximum he would be willing to pay. Right now, however, B does not own a house, so his BATNA is 0. Throughout this section, we assume that the sellers' reserve prices and the buyers' willingness to pay are common knowledge for all the participants, so the negotiation games are not plagued by the additional difficulty of information manipulation. We consider the latter in connection with the market mechanism later in the chapter.

If the house is sold for a price p, the seller gets p but gives up the house, so his surplus, measured in thousands of dollars, is $(p - 100)$. The buyer gets to live in the house, which gives him a benefit of 300, for which he pays p, so his surplus is $(300 - p)$. The two surpluses add up to

$$(300 - p) + (p - 100) = 300 - 100 = 200$$

regardless of the specific value of p. Thus 200 is the total surplus that the two can realize by striking a deal—the extra value that exists when S and B reach an agreement with each other.

The function of the price in this agreement is to divide up the available surplus between the two. For S to agree to the deal, he must get a nonnegative surplus from it, therefore p must be at least 100. Similarly, for B to agree to the deal, p cannot exceed 300. The full range of the negotiation for p lies between these two limits, but we cannot predict p precisely without specifying more details of the bargaining process. In Chapter 16 we saw two ways to model this process. The Nash cooperative model determined an outcome, given the relative bargaining strengths of S and B; the noncooperative game of alternating offers and counteroffers fixed an outcome, given the relative patience of the two.

Here we develop a different graphical illustration of the BATNAs, the surplus, and the range of negotiation from the approach used in Chapter 16. This device will help us to relate ideas based on bargaining to those of trading in markets. Figure 17.1 shows quantity on the horizontal axis and money measured in thousands of dollars on the vertical axis. The curve S, which represents the seller's behavior, consists of three vertical and horizontal blue lines that together are shaped like a rising step.[1] The curve B, which represents buyer's behavior, consists of three vertical and horizontal gray lines, shaped like a falling step.

Here's how to construct the seller's curve. First locate the point on the vertical axis corresponding to the seller's reserve price, here 100. Then draw a line going up from 0 to 100 on the vertical axis, indicating that no house is available for prices below 100. From there, draw a horizontal line (shown blue in the figure) of length 1; this shows that the quantity the seller wants to sell is exactly one house. Finally, from the right-hand end of this line, draw a line vertically upward to the region of very high money amounts. This emphasizes that, in this example, there is no other seller, so no quantity greater than 1 is available for sale, no matter how much money is offered for it.

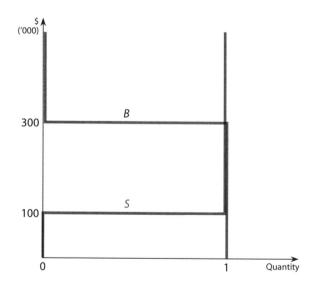

FIGURE 17.1   One Buyer, One Seller

[1] The word *curve* is more commonly used for a smoothly curving shape; therefore you may think it strange to have a "curve" consisting of three (or later, more) straight line segments joined end-to-end at right angles. But we use it, for want of a better word.

In a similar manner, the buyer's bargaining curve B is constructed. As before, first locate the point on the vertical axis representing the buyer's willingness to pay, here 300. From there, draw a vertical line going upward along the axis, indicating that, for prices above 300, the buyer is not willing to purchase any house at all. Next, draw a horizontal line (shown gray in Figure 17.1) of length 1, representing the quantity (one house) that the buyer wants to buy. Finally, from the right end of this horizontal line, draw a line going vertically downward to the horizontal axis, suggesting that, in this example, there is no other buyer, so no quantity greater than 1 finds any customers, no matter how low the asking price.

Note that the two step-shaped curves S and B overlap along their vertical segments, at quantity 1 and money amounts ranging between 100 and 300. The overlap indicates the range of negotiation for bargaining between buyer and seller. The horizontal coordinate of the overlapping segment (here 1) is the amount traded; the vertical segment of overlap (here from 100 to 300) represents the range of prices at which the final bargaining agreement can occur.

Those of you who have had an introductory course in economics will recognize the S as the seller's **supply curve** and B as the buyer's **demand curve.** For those unfamiliar with these concepts, we can explain them here briefly. Suppose the trades are initiated and negotiated not by the buyers and sellers themselves, as in the two-person game described in Figure 17.1, but in an organized market. A neutral "market maker," whose job it is to bring buyers and sellers together by determining a price at which they can trade, finds out how much will be on offer for sale and how much will be demanded at any price, and then adjusts the price so that the market clears, that is, the quantity demanded equals the quantity supplied. This market maker is sometimes a real person, for example, the manager of the trading desk for a particular security in financial markets; at other times he is a hypothetical construct in economic theorizing. The same conclusions follow in either case.

Let us see how we can apply the market concept to the outcome in our bargaining example. First consider the supply side. Curve S in Figure 17.1 tells us that for any price below 100, nothing is offered for sale, while for any price above 100, exactly one house is offered. This curve also shows that when the price is exactly 100, the seller would be just as happy selling the house as he would be remaining in it as the original owner (or not building it at all if he is the builder), so the quantity offered for sale could be 0 or 1. Thus this curve indicates the quantity offered for sale at each possible price. In our example, the fractional quantities between 0 and 1 do not have any significance, but for other objects of exchange such as wheat or oil, it is natural to think of the supply curve as showing continuously varying quantities as well.

Next consider the demand side. The graph of the buyer's willingness to pay, labeled B in Figure 17.1, indicates that for any price above 300, there is no willing buyer, while for any price below 300, there is a buyer willing to buy exactly one house. At a price of exactly 300, the buyer is just as happy buying as not, so he may purchase either 0 or 1 item.

The demand curve and the supply curve coincide at quantity 1 over the range of prices from 100 to 300. When the market maker chooses any price in this range, exactly one buyer and exactly one seller will appear. The market maker can then bring them together and consummate the trade; the market will have cleared. We explained this outcome earlier as arising from a bargaining agreement, but economists would call it a **market equilibrium.** Thus the market is an institution or arrangement that brings about the same outcome as would direct bargaining between the buyer and the seller. Moreover, the range of prices that clears the market is the same as the range of negotiation in bargaining.

The institution of a market, where each buyer and seller responds passively to a price set by a market maker, seems strange when there is just one buyer and one seller. Indeed, the very essence of markets is competition, which demands the presence of several sellers and several buyers. Therefore we now gradually extend the scope of our analysis by introducing more of each type of agent, and at each step we examine the relationship between direct negotiation and market equilibrium.

Let us consider a situation in which there is just the one original buyer, B, who is still willing to pay a maximum of $300,000 for a house. But now we introduce a second seller, S_2, who has a house for sale identical to that offered by the first seller, whom we now call S_1. S_2 has a BATNA of $150,000. It may be higher than S_1's because S_2 is an owner who places a higher subjective value on living in his house than S_1 does, or a builder with a higher cost of construction than S_1.

The existence of S_2 means that S_1 cannot hope to strike a deal with B for any price higher than 150. If S_1 and B try to make a deal at a price of even 152, for example, S_2 could undercut this price with an offer of 151 to the buyer. S_2 would still get a positive surplus instead of being left out of the bargain and getting 0. Similarly, S_2 cannot hope to charge more than 150, because S_1 could undercut that price even more eagerly. Thus the presence of the competing seller narrows the range of bargaining in this game from (100, 300) to (100, 150). The equilibrium trade is still between the original two players, S_1 and B, because it will occur at a price somewhere *between* 100 and 150, which cuts S_2 out of the trade. The presence of the second seller thus drives down the price that the original seller can get. Where the price settles between 100 and 150 depends on the relative bargaining powers of B_1 and S_1. Even though S_2 drops out of the picture at any price below 150, if B_1 has a lot of bargaining power for other reasons not mentioned here (for example, relative patience), then he may refuse an offer from S_1 that is close to 150, and hold out for a figure much closer to 100. If S_1 has relatively little bargaining power, he may have to accept such a counteroffer.

We show the demand and supply curves for this case in Figure 17.2a. The demand curve is the same as before, but the supply curve has two steps. For any price below 100, neither seller is willing to sell, so the quantity is zero; this is the vertical line along the axis from 0 to 100. For any price from 100 to 150, S_1 is willing to sell but S_2 is not. Thus the quantity supplied to the market is 1, and the vertical line at this quantity extends from 100 to 150. For any price above 150, both sellers are willing to sell and the quantity is 2; the vertical line at this quantity, from the price of 150 upward, is the last segment of the supply curve. Again we draw horizontal segments from quantity 0 to quantity 1 at price 100, and from quantity 1 to quantity 2 at price 150, to show the supply curve as an unbroken entity.

The overlap of the demand and supply curves is now a vertical segment representing a quantity of 1 and prices ranging from 100 to 150. This is exactly the range of negotiation we found earlier when considering bargaining instead of a market. Thus the two approaches predict the same range of outcomes: one house changes hands, the seller with the lower reserve price makes the sale, and the final price is somewhere between $100,000 and $150,000.

What if both sellers had the same reserve price of $100,000? Then neither would be able to get any price above 100. For example, if S_1 asked for $101,000, then S_2, facing the prospect of making no sale at all, could counter by asking only $100,500. The range of negotiation would be eliminated entirely. One seller would succeed in selling his house for $100,000. The other would be left out, but at this price he is just as happy not selling the house as he is selling it.

We show this case in the context of market supply and demand in Figure 17.2b. Neither seller offers his house for sale for any price below 100, and both

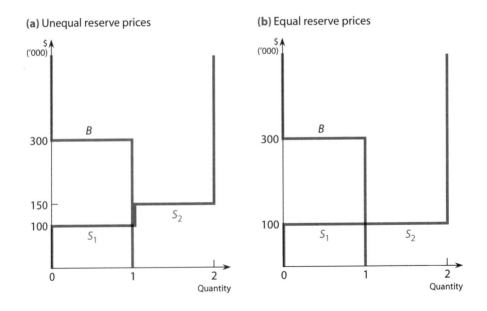

(a) Unequal reserve prices

(b) Equal reserve prices

FIGURE 17.2 One Buyer, Two Sellers

sellers offer houses as soon as the price rises above 100. Therefore the supply curve has a horizontal segment of length 2 at height 100; at its right-hand end starts the terminal vertical upward line of indefinite length. The demand curve has not changed, so now the two curves overlap only at the point (1,100). The range of prices is now reduced to one value, and this price represents the market equilibrium. One house changes hands, one of the two sellers (it does not matter which) makes the sale, and the price is exactly 100. The competition between the two sellers with identical reserve prices prevents either from getting more than the reserve price; the buyer reaps the benefit of the competition between the sellers in the form of a large surplus, namely $300 - 100 = 200$.

Of course, the situation in the market could be reversed, so that there is just one seller, S, but a second buyer, B_2, who is willing to pay \$200,000 for a house. In this case, the first buyer, B_1, cannot hope to acquire the house for any less than 200. You should be able to figure out why—B_2 will bid the price up to at least 200, so if B_1 is to get the house, he will have to offer more than that amount. The range of possible prices on which S and B_1 could agree now extends from 200 to 300. This range is again narrower than the original 100 to 300 that we had with only one buyer and one seller. We leave you to do the explicit analysis using the negotiation approach and the market supply and demand curves.

Proceeding with our pattern of introducing more buyers and sellers into our bargaining or market, we next consider a situation with two of each type of trader. Suppose the sellers, S_1 and S_2, have reserve prices of 100 and 150 respectively, and the buyers, B_1 and B_2, have a maximum willingness to pay of 300 and 200 respectively. We should expect the range of negotiation in the bargaining game to extend from 150 to 200, and can prove this as follows.

Each buyer and each seller is looking for the best possible deal for himself. They can make and receive tentative proposals while continuing to look around for better ones, until no one can find anything better, at which point the whole set of deals is finalized. Consider one such tentative proposal, in which one house, say S_1, would sell at a price $p_1 < 150$ and the other, S_2, at $p_2 > 150$. The buyer who would pay the higher price, say B_1, can then approach S_1, and offer him $(p_1 + p_2)/2$. This offer is greater than p_1 and so is better for S_1 than the original tentative proposal, and less than p_2 and so is better for B_1 as well. Of course, this would be worse for the other two traders, but since B_1 and S_1 are free to choose their partners and trades, B_2 and S_2 can do nothing about it. Thus any price less than 150 cannot prevail. Similarly you can check that any price higher than 200 cannot prevail either.

All of this is a fairly straightforward continuation of our reasoning in the earlier cases with fewer traders. But with two of each type of trader, an interesting new feature emerges: the two houses (assumed to be identical) must both sell for the same price. If they did not, the buyer who was scheduled to pay more in the original proposal could strike a mutually better deal with the seller who was getting less.

Note the essential difference between competition and bargaining here. You *bargain* with someone on the "other side" of the deal; you *compete* with someone on the "same side" of the deal for the opportunity to strike a deal with someone on the other side. Buyers *bargain* with sellers; buyers *compete* with other buyers, and sellers *compete* with other sellers.

We draw the supply and demand curves for the two-seller, two-buyer case in Figure 17.3a. Now each curve has three vertical segments and two horizontal lines joining these at the reserve prices of the two traders on that side of the market. The curves overlap at the quantity 2 and along the range of prices from 150 to 200. This is where the market equilibrium must lie. The two houses will change hands at the same price; it does not matter whether S_1 sells to B_1 and S_2 to B_2, or S_1 sells to B_2 and S_2 to B_1.

We move finally to the case of three buyers and three sellers. Suppose the first two sellers, S_1 and S_2, have the same respective reserve prices of 100 and 150 as before, but the third, S_3, has a higher reserve price of 220. Similarly, the first two buyers B_1 and B_2 have the same willingness to pay as before, 300 and 200 respectively, while the third buyer's willingness to pay is lower, only 140. The new buyer and seller do not alter the negotiation or its range at all; only two houses get traded, between sellers S_1 and S_2 and buyers B_1 and B_2, for a price somewhere between 150 and 200 as before. B_3 and S_3 do not get to make a trade.

To see why B_3 and S_3 are left out of the trading, suppose S_3 does come to an agreement to sell his house. Only B_1 could be the buyer because neither B_2 nor B_3 is willing to pay S_3's reserve price. Suppose the agreed-on price is 221, so B_1 gets a surplus of 79 and S_3 gets only 1. Then consider how the others must pair off. S_1,

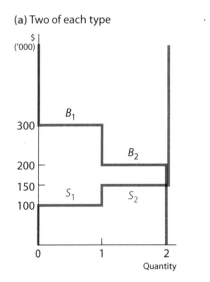
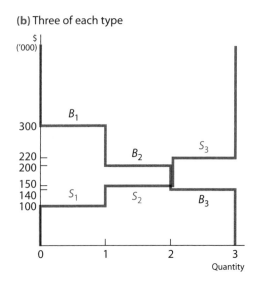

FIGURE 17.3 More Buyers and Sellers

with his reserve price of 100, must sell to B_3, with his willingness to pay of 140; and S_2, with his reserve price of 150, must sell to B_2, with his willingness to pay of 200. The surplus from the S_1 to B_3 sale is 40, and the surplus from the S_2 to B_2 sale is 50. Therefore S_1 can get at most almost 40 in surplus from his sale, and S_2 can get at most 50 from his; either seller could benefit by approaching B_1 and striking a mutually beneficial alternative deal. For example, S_2 could approach B_1 and say, "With your willingness to pay of 300 and my reserve price of 150, we have a total available surplus of 150 between us. You are getting 79 in your deal with S_3; I am getting less than 50 in my deal with B_2. Let us get together; you could have 90 and I could have 60, and we both would be better off." B_1 would accept this deal, and his tentative arrangement with S_3 would fall through.

The numbers we have derived here are specific to our example, but the argument itself is perfectly general. For any number of buyers and sellers, first rank sellers by increasing reserve prices and buyers by decreasing willingness to pay. Then pair buyers and sellers of the same rank. When you find the pairing number n such that the nth buyer's willingness to pay is less than the nth seller's reserve price, you know that all buyers and sellers ranked n or more will not get to make a trade.

We can see this argument even more clearly using market supply-and-demand-curve analysis as in Figure 17.3b. Construction of the supply and demand curves follows the same steps as before, so we do not describe it in detail here. The two curves overlap at the quantity 2, and the prices in the overlap range from 150 to 200. Therefore, in the market equilibrium, two houses change hands at a price somewhere between these two limits. Given the range of possible prices, we know that B_3's willingness to pay, 140, will be less than the market price, while S_3's reserve price, 220, will be higher than the market price. Therefore, just as we could say in the general case that the nth buyer and seller do not get to trade, we can assert that these two agents will not be active in the market equilibrium. Once again, it does not matter whether S_1 sells to B_1 and S_2 to B_2, or S_1 sells to B_2 and S_2 to B_1.

Suppose for the sake of definiteness that the market price in this case is 175. Then the surpluses for the traders are 75 for S_1, 25 for S_2, 125 for B_1, 25 for B_2, and 0 for S_3 and B_3; the total surplus is 250. You can check that the total surplus will be the same no matter where the price ends up within the range from 150 to 200.

But couldn't we pair people differently and do better—especially if all three houses get sold? The answer is no. The negotiation process, or the market mechanism, produces the largest possible surplus. As before, to come up with an alternate solution we have to pair B_1 with S_3, B_3 with S_1, and B_2 with S_2. Then the B_1 and S_3 pairing generates a surplus of $300 - 200 = 80$ and the B_3 and S_1 pairing generates $140 - 100 = 40$, while the remaining pairing between B_2 and S_2 generates $200 - 150 = 50$. The total surplus in this alternative scenario comes only to 170, which is 80 less than the surplus of the 250 associated with the negotiation outcome or the market equilibrium. The problem is that B_3 is willing to pay only 140, a price below the equilibrium range, while S_3 has a reserve price of 220, above the equilibrium price range; between them they bring to the game a negative surplus of $140 - 220 = -80$. If they get the opportunity to trade, there is no escaping the fact that the total surplus will go down by 80.

Although the outcome of the negotiation process or the market equilibrium yields the highest total surplus, it does not necessarily guarantee the most equal distribution of surplus: B_1, B_2, S_1, and S_2 share the whole 250, while B_3 and S_3 get nothing. In our alternate outcome, everyone got something, but the pairings could not survive the process of renegotiation. If you are sufficiently concerned about equity, you might prefer the alternative outcome despite the level of surplus generated. You could then restrict the process of renegotiation to bring about your preferred outcome. Better still, you could let the negotiation or market mechanisms go their way and generate the maximum surplus; then you could take some of that surplus away from those who enjoy it and redistribute it to the others using a tax-and-transfer policy. The subject of public economics considers such issues in depth.

Let us pause here to sum up what we have accomplished in this section. We have developed a method of thinking about how negotiations between buyers and sellers proceed to secure mutually advantageous trades. We also linked the outcomes of this bargaining process with those of the market equilibrium based on supply-and-demand analysis. Although our analysis involved simple numerical examples with small numbers of buyers and sellers, the ideas have broad-ranging implications, which we develop in a more general way in the sections that follow. In the process, we also take the negotiation idea beyond that of the trade context. As we saw in Section 7 of Chapter 16, with many agents and many goods and services available to trade, the best deals may be multilateral. A good general theory must allow not just pairs, but also triplets or more general groupings of the participants. Such groups—sometimes called *coalitions*—can get together to work out tentative deals as the individuals and groups continue the search for better alternatives. The process of deal making stops only when no person or coalition can negotiate anything better for itself. Then the tentative deals become final and are consummated. We take up the subject of creating mutually advantageous deals in the next section.

2 THE CORE

We now introduce some notation and a general theory for analyzing a process of negotiation and renegotiation for mutually advantageous deals. Then we apply it in a simple context. We will find that, as the number of participants becomes large, the complex process of coalition formation, re-formation, and deal making produces an outcome very close to what a traditional market would produce. Thus we will see that the market is an institution that can simplify the complex process of deal making among large numbers of traders.

Consider a general game with n players, labeled simply 1, 2, . . . n, such that the set of all players is $N = \{1, 2, 3, \ldots, n\}$. Any subset of players in N is a **coalition** C. When it is necessary to specify in full the members of a coalition, we list them in braces; $C = \{1, 2, 7\}$ might be one such coalition. We allow the N to be a subset of itself, and call it the **grand coalition** G; thus $G = N$. A simple formula in combinatorial mathematics gives the total number of subsets of N that include one or two or three or . . . n people out of the n members of N. This number of possible subsets is $(2^n - 1)$; there are thus $(2^n - 1)$ possible coalitions altogether.

For the simple two-person trading game discussed in Section 1, $n = 2$, so we can imagine three coalitions: two single-member ones of one buyer and one seller respectively, and the grand coalition of the two together. When there are two sellers and one buyer, we have seven ($2^3 - 1$) coalitions: three trivial ones consisting of one player each; three pairs, $\{S_1, B\}$, $\{S_2, B\}$, and $\{S_1, S_2\}$; and the grand coalition, $\{S_1, S_2, B\}$.

A coalition can strike deals among its own members to exploit all available mutual advantage. For any coalition C, we let the function $v(C)$ be the total surplus that its members can achieve on their own, regardless of what the players not in the coalition (that is, in the set $N - C$) do. This surplus amount is called the **security level** of the coalition. When we say "regardless of what the players not in the coalition do," we visualize the worst-case scenario, namely what would happen to members of C if the others (those not in C) took the action that was the worst from C's perspective. This idea is the analogue of an individual player's minimax, which we studied in Chapters 4 and 5.

The calculation of $v(C)$ comes from the specific details of a game. As an example, consider our simple game of house trading from Section 1. No individual can achieve any surplus on his own without some cooperation from a player on the other side, so the security levels of all single-member coalitions are zero; $v(\{S_1\}) = v(\{S_2\}) = v(\{B\}) = 0$. The coalition $\{S_1, B\}$ can realize a total surplus of $300 - 100 = 200$, and S_2 on his own cannot do anything to reduce the surplus the other two can enjoy (arson being outside the rules of the game), so $v(\{S_1, B\}) = 200$. Similarly, $v(\{S_2, B\}) = 300 - 150 = 150$. The two sellers on their own cannot achieve any extra payoff without involving the buyer, so $v(\{S_1, S_2\}) = 0$.

In a particular game then, for any coalition C, we can find its security level, $v(C)$; this is also called the **characteristic function** of the game. We assume that the characteristic function is common knowledge, that there is no asymmetry of information.

Next we consider possible outcomes of the game. An **allocation** is a list of surplus amounts, $(x_1, x_2, \ldots x_n)$, for the players. An allocation is **feasible** if it is logically conceivable within the rules of the game, that is, if it arises from any deal that could in principle be made. Any feasible allocation may be proposed as a tentative deal, but the deal can be upset if some coalition can break away and do better for itself. We say that a feasible allocation is **blocked** by a coalition C if

$$v(C) > \sum_{i \text{ in } C} x_i.$$

This says that the security level of the coalition—the total surplus it can minimally guarantee for its members—exceeds the sum of the surpluses that its members get in the proposed tentative allocation. If this is true, then the coalition can form and will strike a bargain among its members that leaves all of them better off than in the proposed allocation. Therefore the implicit threat by the coalition to break away (back out of the deal if the surplus gained is not high enough) is credible; that is how the coalition can upset (block) the proposed allocation.

The set of allocations that cannot be blocked by any coalition contains all of the possible stable deals, or the bargaining range of the game. This set cannot be reduced further by any groups searching for better deals; we call this set of allocations the **core** of the game.

A. Numerical Example

We have already analyzed the blocking in the simple two-person trading game in Section 1, although we did not use this terminology. However, we can now see that what we called the "range of negotiation" corresponds exactly to the core. So we can translate, or restate, our results as follows: With one buyer and one seller, the range of negotiation is from 100 to 300. This outcome corresponds to a core consisting of all allocations ranging from the one (with price 100) where the buyer gets surplus 200 and the seller gets 0, to the one (with price 300) where the buyer gets 0 and the seller gets 200.

Now we use our new analytical tools to consider the case in which one buyer, willing to pay up to 300, meets two sellers having equal reserve prices of 100. What is the core of the game among these three participants?

Consider an allocation of surpluses, say x_1 and x_2 to the two sellers and y to the buyer. For this allocation to be in the core, no coalition should be able to achieve more for its members than what they receive here. We know that each single-member coalition can achieve only zero; therefore a core allocation must satisfy the conditions

$$x_1 \geq 0, \quad x_2 \geq 0, \quad \text{and} \quad y \geq 0.$$

Next, the coalitions of one buyer and one seller can each achieve a total of $300 - 100 = 200$, while the coalition of the two sellers can achieve only zero. Therefore for the proposed allocation to survive being blocked by a two-person coalition, it must satisfy

$$x_1 + y \geq 200, \quad x_2 + y \geq 200, \quad \text{and} \quad x_1 + x_2 \geq 0.$$

Finally, all three participants together can achieve a surplus of 200; therefore if the proposed allocation is not to be blocked by the grand coalition, we must have

$$x_1 + x_2 + y \geq 200.$$

But since the total surplus available is only 200, $x_1 + x_2 + y$ cannot be greater than 200. Therefore we get the equation $x_1 + x_2 + y = 200$.

Given this last equation, we can now argue that x_1 must be zero. If x_1 were strictly positive, and we combined the equation $x_1 > 0$ with one of the two-person inequalities, $x_2 + y \geq 200$, then we would get $x_1 + x_2 + y > 200$, which is impossible. Similarly, we find $x_2 = 0$ also. Then it follows that $y = 200$. In other words, the sellers get no surplus—it all goes to the buyer. To achieve such an outcome, one of the houses must sell for its reserve price $p = 100$; it does not matter which.

We conducted the analysis in the previous section using our intuition about competition. When both sellers have the same reserve price, the competition between them becomes very fierce. If one has negotiated a deal to sell at a price of 102, the other is ready to undercut and offer the buyer a deal at 101 rather than go without a trade. In this process the sellers "beat each other down" to 100. In this section, we have developed some general theory about the core, and applied it in a mechanical and mathematical way to get a result that leads back to the same intuition. We hope this two-way traffic between mathematical theory and intuitive application reinforces both in your mind. The purely intuitive thinking becomes harder as the problem grows more complex, with many objects, many buyers, and many sellers. The mathematics provides a general algorithm we can apply to all of these problems; you will have ample opportunities to use it. But, having solved the more complex problems using the math, you should also pause and relate the results to the intuition, as we have done here.

B. Some Properties of the Core

EFFICIENCY   We saw earlier that an allocation is in the core if it cannot be blocked by *any* coalition—including the grand coalition. Therefore, if a core allocation gives the n players surpluses of $(x_1, x_2, \ldots x_n)$, the sum of these surpluses must be at least as high as the security level that the grand coalition can achieve:

$$x_1 + x_2 + \cdots + x_n \geq v(\{1, 2, \ldots, n\})$$

But the grand coalition can achieve anything that is feasible. Therefore for *any* feasible allocation of surpluses, say $(z_1, z_2, \ldots z_n)$, it must be true that

$$v(\{1, 2, \ldots, n\}) \geq z_1 + z_2 + \cdots + z_n.$$

These two inequalities together imply that it is impossible to find another feasible allocation that will give everyone a higher payoff than a core allocation; we *cannot* have

$$z_1 > x_1, \quad z_2 > x_2, \quad \ldots, \quad \text{and} \quad z_n > x_n$$

simultaneously. For this reason the core allocation is said to be **efficient.** (In the jargon of economists, it is called *Pareto efficient,* after Wilfredo Pareto, who first introduced this concept of efficiency.)

Efficiency sounds good, but it is only one of a number of good properties we might want a core allocation to have. Efficiency simply says that no other allocation will improve everyone's payoff simultaneously; the core allocation could still leave some people with very low payoffs and some with very high payoffs. That is, efficiency says nothing about the equity or fairness or distributive justice of a core allocation. In our trading game, a seller got zero surplus if there was another seller with the same reserve price; otherwise he had hope of some positive surplus. This is a matter of luck, not fairness. And we saw that some sellers with high reserve prices and some buyers with a low willingness to pay could get cut out of the deals altogether because the others could get greater total surplus; this outcome also sacrifices equity for the sake of efficiency.

A core allocation will always yield the maximum total surplus, but it does not regulate the distribution of that surplus. In our two-buyer, two-seller example, the buyer with the higher willingness to pay (B_1) gets more surplus because competition from the other buyer (B_2) means that the two pay the same price. Similarly, the lower-reserve-price seller (S_1) gets the most surplus—or, equivalently, the lower-cost builder (S_1) makes the most profit.

In our simple trading game, the core was generally a range rather than a single point. Similarly, in other games there may be many allocations in the core. They will all be efficient, but will entail different distributions of the total payoff among the players, some fairer than others. The concept of efficiency says nothing about how we might choose one particular core allocation when there are many. Some other theory or mode of analysis is needed for that. Later in the chapter we take a brief look at other mechanisms that do pay attention to the distribution of payoffs among the players.

EXISTENCE Is there a guarantee that all games will have a core? Alas, no. In games where the players have negative spillover effects (externalities) on one another, it may be possible for every tentatively proposed allocation to be upset by a coalition of players who gang together to harm the others. Then the core may be empty.

An extreme example of this is the *garbage game.* There are n players, each with one bag of garbage. Each can dump it in his own yard or a neighbor's yard; he can even divide it up and dump some in one yard and some in others. The payoff from having b bags of garbage in your yard is $-b$, where b does not have to be an integer. Then a coalition of m people has the security level $-(n - m)$ as long as $m < n$, because they get rid of their m bags in nonmembers' yards but, in the worst-case scenario that underlies the calculation of the security level, all $(n - m)$ nonmembers dump their bags in members' yards. Therefore the characteristic function of the game is given by $v(m) = -(n - m) = m - n$ so long as $m < n$. But the grand coalition has no nonmembers on whom to dump their garbage, so $v(n) = -n$.

Now it is easy to see that any proposed allocation can be blocked. The most severe test occurs for a coalition of $(n - 1)$ people, which has a very high security level of -1 because they are dumping all their bags on the one unfortunate person excluded from the coalition, and suffer only his one bag among the $(n - 1)$ of them. Suppose person n is currently the excluded one. Then he can get together with any $(n - 2)$ members of the coalition, say the last $(n - 2)$, and offer them an even better deal. The new coalition of $(n - 1)$, namely 2 through n, should dump their bags on 1. Person n will accept the whole of 1's bag. Thus persons 2 through $(n - 1)$ will be even better off than in the original coalition of the first $(n - 1)$ because they have 0 surplus to share among them rather than -1. And person n will be dramatically better off with 1 bag on his lawn than in the original situation in which he was the excluded person and suffered $(n - 1)$ bags. Thus the original allocation can be blocked. The same is true of every allocation; therefore the core is nonexistent, or empty.

The problem arises because of the nature of "property rights," which is implicit in the game. People have the right to dump their garbage on others; much as in some places and at some times firms have a right to pollute the air or the water, or people have the right to smoke. In such a regime of "polluter's rights," negative externalities go unchecked and an unstable process of coalition formation and reformation can result, leading to an empty core. In an alternative system where people have the right to remain free from the pollution caused by others, this problem does not arise. In our game, if the property right is changed so that no one can dump garbage on anyone else without their consent, then the core is not empty; the allocation in which every person keeps his own bag in his own yard and gets a payoff of -1 cannot be blocked.

C. Discussion

COOPERATIVE OR NONCOOPERATIVE GAME?   The concept of the core was developed for an abstract formulation of a game, starting with the characteristic function. No details are provided as to just *how* a coalition C can achieve total surplus equal to its security level $v(C)$. Those details are specific to each application. In the same way, when we find a core allocation, nothing is said about how the individual actions that give rise to that allocation are implemented. Implicitly, it is assumed that some external referee sees to it that all players carry out their assigned roles and actions. Therefore the core must be understood as a concept within the realm of cooperative game theory, where actions are jointly chosen and implemented.

But the freedom of individual players to join coalitions, or to break up existing coalitions and form new ones, introduces a strong noncooperative element into the picture. That is why the core is often a good way to think about competition. However, for competition to occur freely, it is important that all buyers and sellers have equal and complete freedom to form and re-form coalitions. What if only the sellers are active players in the game, and the buyers must passively accept their stated prices and cannot search for better deals? Then the core of the sellers' game that maximizes their total profit entails forming the grand coalition of sellers, which is a cartel or a monopoly. This problem is often observed in practice. Because buyers are numerous small individuals who are in the market infrequently, they find it difficult to get together or to negotiate with sellers for a better deal. Sellers are fewer in number and have repeated relationships among themselves; they are more likely to form coalitions. That is why antitrust laws try to prevent cartels and to preserve competition, whereas similar laws to prevent combination on the buyers' side are almost nonexistent.[2]

ADDITIVE AND NONADDITIVE PAYOFFS   We defined the security level of a coalition as the minimum sum total of payoffs that could be guaranteed for its members. When we add individuals' payoffs in this way, we are making two assumptions: (1) that the payoffs are measured in some comparable units, money being the most frequently used yardstick of this kind; and (2) that the coalition somehow solves its own internal bargaining problem. Even when it has a high total payoff, a coalition can break up if some members get a lot while others get very little—unless the former can compensate the latter in a way that keeps everyone better off by being within the coalition than outside it. Such compensation must take the form of transfers of the unit in which payoffs are measured, again typically money.

If money (or something else that is desired by all and measured in comparable units for all), is not transferable, or if people's payoffs are nonlinear rescalings of money amounts—for example, because of risk aversion—then a coalition's security level cannot be described by its total payoff alone. We then have to keep track of the separate payoffs of all individual members. In other words, the characteristic function $v(C)$ is no longer a number but a set of vectors listing all members' payoff combinations that are possible for this coalition regardless of what others do. The theory of the core for this situation, technically called the case of

---

[2] In labor markets in some European countries such as Sweden, the coalition of all buyers (owners or management) deals with the coalition of all sellers (unions), but that is not the way the concept of core assumes the process will work. The core would allow coalitions of smaller subsets of companies and workers to form and re-form, constantly seeking a better deal for themselves. In practice there is pure bargaining between two fixed coalitions.

nontransferable utility, can be developed, and you will find it in more advanced treatments of cooperative game theory.[3]

3 THE MARKET MECHANISM

We have already looked at supply and demand curves for markets with small numbers of traders. Generalizing them for more realistic markets is straightforward, as we see in Figure 17.4. There can be hundreds or thousands of potential buyers and sellers of each commodity; on each side of the market they span a range of reserve prices and willingness to pay. For each price p, measured on the vertical axis, the line labeled S shows the quantity of all sellers whose reserve price is below p—those who are willing to sell at this price. When we increase p enough that it exceeds a new seller's reserve price, that seller's quantity comes onto the market, so the quantity supplied increases; we show this as a horizontal segment of the curve at the relevant value of p. The rising-staircase form that results from this procedure is the supply curve. Similarly, the descending staircase graph on the buyers' side shows, for each price, the quantity the buyers are willing to purchase (demand); therefore it is called the demand curve.

If all the sellers have access to the same inputs to production and the same technology of production, they may have the same costs and therefore the same reserve price. In that case, the supply curve may be horizontal rather than stepped. However, among the buyers there are generally idiosyncratic differences of taste, so the demand curve is always a step-shaped curve.

Market equilibrium occurs where the two curves meet, which can happen in three possible ways. One is a vertical overlap at one quantity. This indicates that exactly that quantity is the amount traded, and the price (common to all the trades) can be anything in the range of overlap. The second possibility is that the two curves overlap on a horizontal segment. Then the price is uniquely determined, but the quantity can vary over the range shown by the overlap. A horizontal overlap further means that the unique price is simultaneously the reserve price of one or more sellers and the willingness to pay of one or more buyers; at this price the two types are equally happy making or not making a trade, so it does not matter where along the overlap the quantity traded "settles." The third possibility is that a horizontal segment of one curve cuts a vertical segment of the other curve. Then there is only one point common to both curves, and the price and the quantity are both uniquely determined.

[3]R. Duncan Luce and Howard Raiffa, *Games and Decisions* (New York: Wiley, 1957), especially chaps. 8 and 11, remains a classic treatment of cooperative games with transferable utility. An excellent modern general treatment that also allows nontransferable utility is Roger B. Myerson, *Game Theory* (Cambridge, Mass.: Harvard University Press, 1991), chap. 9.

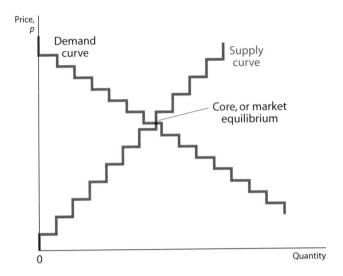

FIGURE 17.4 Market Equilibrium with Many Buyers and Sellers

In each case, the price (even if not uniquely determined) is one at which, if our neutral market maker were to call it out, the quantity demanded would equal the quantity supplied, and the market would clear. Any such price at which the market clears is called a market equilibrium price. It is the market maker's purpose to find such a price, and he does so by a process of trial and error. He calls out a price and asks for offers to buy and sell. All of these are tentative at first. If the quantity demanded at the current price is greater than the quantity supplied, the market maker tries out a slightly higher price; if demand is less than supply, he tries a slightly lower price. The market maker adjusts the trial price in this way until the market clears. Only then are the tentative offers to buy and sell made actual, and they are put into effect at that market equilibrium price.

When thousands of buyers are ranked according to their willingness to pay, and hundreds or thousands of sellers according to their reserve prices, the differences between two successive traders on each side are quite small. Therefore each step of the demand curve falls only a little, and that of the supply curve rises only a little. With this type of supply and demand curves, even when there is a vertical overlap, the range of indeterminacy of the price in the market equilibrium is small, and the market-clearing price will be almost uniquely determined.

We showed graphically in simple examples that the core of our trading game, where each seller has one unit to offer and each buyer wants to buy just one unit, coincides with the market equilibrium. Similar results are true for any number of buyers and sellers, each of whom may want to buy or sell multiple units at different prices, but the algebra is more complex than is worthwhile, so we will omit it. If some sellers or some buyers want to trade more than one unit each, then the link between the core and the market equilibrium is slightly weaker. A market equilibrium always lies in the core, but the core may have other allocations that are not market equilibria. However, in large markets, with numerous buyers and sellers, there are relatively few such extra allocations, and the core and the market outcomes virtually coincide. Because this result comes from rather advanced microeconomic theory, we must leave it as an assertion, hoping that the simple examples worked out here give you confidence in its truth.[4]

A. Properties of the Market Mechanism

EFFICIENCY   The correspondence between the outcomes of the core and of the market equilibrium has an immediate and important implication. We saw in Section 2 that a core allocation was efficient; it maximized the total possible surplus. Therefore the market equilibrium must also be efficient, a fact that accounts in part for the conceptual appeal of the market.

As with the core, the market does not guarantee a fair or just distribution of income or wealth. A seller's profit depends on the accident of how far the market price happens to be above his cost, and a buyer's surplus on how far the market price happens to be below his willingness to pay.

INFORMATION AND MANIPULABILITY   The market maker typically does not know the reserve price of any seller or the willingness to pay of any buyer. He sets a price, and asks traders to disclose some of this information through their offers to sell and buy. Do they have the incentive to reply truthfully? In other words, is the market mechanism manipulable? This question is quite central to the strategic perspective of this book, and relates to our discussion of information in Chapter 12.

First consider the case in which each seller and each buyer wants to trade just one unit. For simplicity, consider the two-buyer, two-seller example from Section 1, in which seller S_1 has a reserve price of 100. He could pretend to have a higher reserve price (by not offering to sell his house until the market maker calls out a price appropriately higher than 100) or a lower one. That is, he could pretend that his supply curve was different than it is. Would that be in his interest?

Just as the market maker does not know anyone's reserve price or willingness to pay, S_1 does not know this information about anyone else. So he does not know what the market-clearing price will be. Suppose he pretends that his reserve price is 120. It might be that the market-clearing price exceeds 120, in which case he would get to make the sale, as he would have anyway, and would get the market price as he would have had he shown his reserve price to be 100; his exaggeration

---

[4] For a recent discussion of the general theory about the relationship between the core and market equilibrium, see Andreu Mas-Colell, Michael D. Whinston, and Jerry R. Green, *Microeconomic Theory* (New York: Oxford University Press, 1995), pp. 652–660.

would make no difference. But the market-clearing price might be somewhere between 100 and 120, in which case the market maker would not take up S_1's offer to sell, and S_1 would lose a deal that would have been profitable. In this instance, exaggeration can only hurt him. Similarly, he might understate his reserve price and offer to sell when the market maker calls out a price of 80. If the market-clearing price ended up below 80, the understatement would make no difference. But if it ended up somewhere between 80 and 100, then the market maker would hold S_1 to his offer, and the sale would net a loss to S_1. Thus understatement as with exaggeration can only hurt him. In other words, truthful revelation is the dominant strategy for S_1. The same is true for all buyers and sellers.

It is no coincidence that this assertion is similar to the argument in Chapter 15 that truthful bidding is the dominant strategy in a private-value second-price auction. A crucial thing about the second-price auction is that what you pay when you win depends not on what you bid, but on what someone else bid. Over- or underbidding can only affect your chances of winning in an adverse way; you might fail to win when it would have been good, and might win when it would be bad. Similarly, in the market, if you get to make a sale or a purchase, the price depends not on your reserve price or your willingness to pay, but on that of someone else. Misstating your values can only affect your chances of making a trade in an adverse way.

This argument does not work if some individual has several units to trade, and is a large trader who can affect the price. For example, a seller might have a reserve price of 100 and 15 units for sale. At a price of 100, there might be 15 willing buyers, but the seller might pretend to have only 10 units. Then the market maker would believe there was some scarcity, and he would set a higher market-clearing price at which the number of willing buyers matched the number of units the seller chose to put up for sale. If the buyers' willingness to pay rises sufficiently rapidly as the quantity goes down (that is, if the demand curve is steep), then the seller in question might make more profit by selling 10 units at the higher price than he would by selling 15 units at a lower price. Then the misstatement would work to his advantage, and the market mechanism would be manipulable. In the jargon of economics, this is an instance of the seller exercising some **monopoly power.** By offering less than he actually has (or can produce) at a particular price, the individual is driving up the market equilibrium price. Similarly, a multiunit buyer can understate his demand to drive the price down and thereby get greater surplus on the units he does buy; this is called the exercise of **monopsony power.**

However, if the market is so large that any one seller's attempt to exercise monopoly power can alter the price only very slightly, then such actions will not be profitable to the seller. When each seller and each buyer is small in relation to the market, each has negligible monopoly and monopsony power. In the jargon of economics, competition becomes *perfect*; in our jargon of game theory, the market mechanism becomes **nonmanipulable,** or **incentive-compatible.**

B. Experimental Evidence

We have told a "story" of the process by which a market equilibrium is reached. However, as noted earlier, the market maker who played the central role in that story, calling out prices, collecting offers to buy and sell, and adjusting the price until the market cleared, does not actually exist in most markets. Instead, the buyers and sellers have to search for trading partners, and attempt to strike deals, on their own. The process is somewhat like the formation and reformation of coalitions in the theory of the core, but even more haphazard and unorganized. Can such a process lead to a market-clearing outcome?

Through observation of actual markets we can readily find many instances where markets seem to function well, and many where they fail. But situations of the former kind differ in many ways from those of the latter, and many of these differences are not easily observable. Therefore the evidence does not help us understand whether and how markets can solve the problems of coordinating the actions of many buyers and sellers to arrive at an equilibrium price. Laboratory experiments, on the other hand, allow us to observe outcomes under controlled conditions, varying just those conditions whose effects we want to study. Therefore they are a useful approach and a valuable item of evidence for judging the efficacy of the market mechanism.

Over the last half-century, many such experiments have been carried out. Their objectives and methods vary, but in some form they all bring together a group of subjects who are asked to trade a tangible or intangible "object." Each is informed of his own valuation and given real incentives to act upon this valuation. For example, would-be sellers who actually consummate a trade are given real monetary rewards equal to their profit or surplus, which is equivalent to the price they succeed in obtaining *minus* the cost they are assigned.

The results are generally encouraging for the theory. In experiments with only two or three traders on one side of the market, attempts to exercise monopoly or monopsony power are observed. But otherwise, even with relatively small numbers of traders (say, five or six on each side), trading prices generally converge quite quickly to the market equilibrium that the experimenter has calculated, knowing the valuations he has assigned to the traders. If anything, the experimental results are "too good." Even when the traders know nothing about the valuations of other traders, have no understanding of the theory of market equilibrium, and have no prior experience in trading, they arrive at the equilibrium configuration relatively quickly. In recent work, the focus has shifted away from asking *whether* the market mechanism works, to asking *why* it works so much better than we have any reason to expect.[5]

There is one significant exception, namely markets for assets. The decision

---

[5] Douglas D. Davis and Charles A. Holt, *Experimental Economics* (Princeton, N.J.: Princeton University Press, 1993), Chap. 3.

to buy or sell a long-lived asset must take into account not merely the relation between the price and your own valuation, but also the expected movement of the price in the future. Even if you derive no benefit from owning a tulip bulb, you might buy and hold it for a while if you believed that you could then sell it to someone else for a sufficiently higher price, thereby making a capital gain. You might come to this belief because you have observed rising prices for a while. But prices might be rising because others are buying tulip bulbs too, similarly expecting to resell them with a capital gain. If many people simultaneously try to sell the bulbs to obtain those expected gains, the price will fall, the gains will fail to be realized, and the expectations will collapse. Then we would say that the tulip bulb market has experienced a *speculative bubble* that had burst. Such bubbles are observed in experimental markets, as they are in reality.[6]

4 THE SHAPLEY VALUE

The core has an important desirable property of efficiency, but it also has some undesirable ones. Most basically, there are games that have no core; many others have a very large core consisting of a whole range of outcomes, so the concept does not determine an outcome uniquely. Other concepts that do better in these respects have been constructed. The best known of these is the Shapley value, named after Lloyd Shapley of UCLA.

Like Nash's bargaining solution, which we discussed in Chapter 16, the Shapley value is a cooperative concept. It is similarly grounded in a set of axioms or assumptions that, it is argued, a solution should satisfy. And it is the unique solution that conforms to all these axioms. Of course, the desirability of the properties is a matter of judgment; other game theorists have specified other combinations of properties, each of which leads to its own unique solution. But the solutions for different combinations of properties are different, and not all the properties can be equally desirable. Perhaps more importantly, each player will judge the properties not by their innate attractiveness, but by what the outcome gives him. If an outcome is not satisfactory for some players, and they believe they can do better by insisting on a different game, the proposed solution may be a nonstarter.

Therefore any cooperative solution should not be accepted as the likely actual outcome of the game based solely on the properties that seem desirable to some game theorists or in the abstract. Rather, it should be judged by its actual performance for prediction or analysis. From this practical point of view, the

---

[6] Shyam Sunder, "Experimental Asset Markets: A Survey," *Handbook of Experimental Economics*, ed. John H. Kagel and Alvin E. Roth (Princeton, N.J.: Princeton University Press, 1995), pp. 468–474.

Shapley value turns out to do rather well. We do not go into the axiomatic basis of the concept here. We simply state its formula, briefly interpret and motivate it, and demonstrate it in action.

Suppose a game has n players and its characteristic function is $v(C)$, which is the minimum total payoff that coalition C can guarantee to its members. The **Shapley value** is an allocation of payoffs u_i to each player i as defined by

$$u_i = \sum_C \frac{(n-k)!\,(k-1)!}{n!}[v(C) - v(C - \{i\})], \qquad (17.1)$$

where $n!$ denotes the product $1 \times 2 \times \cdots \times n$; where the sum is taken over all the coalitions C that have i as a member, and where, in each term of the sum, k is the size of the coalition C.

The idea is that each player should be given a payoff equal to the average of the "contribution" he makes to each coalition to which he could belong, where all coalitions are regarded as equally likely in a suitable sense. First consider the size k of the coalition—from 1 to n. Since all sizes are equally likely, a particular size coalition occurs with probability $1/n$. Then the $(k-1)$ partners of i in a coalition of size k can be chosen from among the remaining $(n-1)$ players in any of

$$\frac{(n-1)!}{[(n-1)-(k-1)]!\,(k-1)!} = \frac{(n-1)!}{(n-k)!\,(k-1)!}$$

ways. The reciprocal of this expression is the probability of any one such choice. Combining that reciprocal with $(1/n)$ gives us the probability of a particular coalition C of size k containing member i, and that is what appears as the built-up fraction in each term of the sum on the right-hand side of the formula for u_i. What multiplies the fraction is the difference between the security level of the coalition C, namely $v(C)$, and the security level the remaining people would have if i were removed from the coalition, namely $v(C - \{i\})$. This term measures the contribution i makes to C.

The idea that each player's payoff should be commensurate with his contribution has considerable appeal. Most importantly, if the player must make some effort in order to generate this contribution, then such a payment scheme gives him the correct incentive to make that effort. In the jargon of economics, each person's incremental contribution to the economy is called his *marginal product*; therefore the concept of tying payoffs to contributions is called the *marginal productivity theory of distribution*. A market mechanism would automatically reward each participant for his contribution. The Shapley value can be understood as a way of implementing this principle—of achieving a marketlike outcome—at least approximately when an actual market cannot be arranged, such as within units of a single larger organization.

Of course, the formula embodies much more than the general principle of marginal productivity payment. Averaging over all coalitions to which a player could belong, and regarding all coalitions as equally likely, are very specific procedures that may or may not have any counterpart in reality. Nevertheless, the Shapley value often produces outcomes that have some realistic features. We now examine two specific examples of the Shapley value in action, one from politics and one from economics.

A. Power in Legislatures and Committees

Suppose a 100-member legislature consists of four parties, Red, Blue, Green, and Brown. The Reds have 43 seats, the Blues 33, the Greens 16, and the Browns 8. Each party is a coherent block that votes together, so each can be regarded as one player. A majority is needed to pass any legislation; therefore no party can get legislation through without the help of another block and so no party can govern on its own. In this situation the power of a party will depend on how crucial that party is to the formation of a majority coalition. The Shapley value provides a measure of just that power. Therefore in this context it is called the **power index,** and is attributed to Shapley, Martin Shubik, and John Banzhaf.

Give the value 1 to any coalition (including the grand coalition) that has a majority and 0 to any coalition without a majority. Then in the formula for the Shapley value of party i (Eq. 17.1), the contribution it makes to a coalition C, namely $v(C) - v(C - \{i\})$, is 1 if the coalition C has a majority but $C - \{i\}$ if it does not. When party i's contribution to C is 1, we say that party i is a *pivotal member* of the majority coalition C. In all other cases—namely if coalition C does not have a majority at all, or if C would have a majority even without i, the contribution of i to C is zero.

We can now list all the majority coalitions and which party is pivotal in each. None of the four possible (albeit trivial) one-party "coalitions" has a majority. Three of the six possible two-party coalitions have a majority, namely {Red, Blue}, {Red, Green}, and {Red, Brown}, and in each case both members are pivotal because the loss of either would mean the loss of the majority. Of the four possible three-party coalitions, in three of them—namely {Red, Blue, Green}, {Red, Blue, Brown}, and {Red, Green, Brown}—only Red is pivotal. In the fourth three-party coalition, namely {Blue, Green, Brown}, all three parties are pivotal. In the grand coalition, no party is pivotal.

In the Shapley value formula, the term corresponding to each two-party coalition takes the value $(4 - 2)!(2 - 1)!/4! = 2! \times 1!/4! = 2/24 = 1/12$, and each three-party coalition gets $(4 - 3)!(3 - 1)! = 1! \times 2!/4! = 1/12$ also.

With this information we can calculate the Shapley value of each party. We have

$$u_{\text{Red}} = \frac{1}{12} \times 3 + \frac{1}{12} \times 3 = \frac{1}{2}$$

because the Red party is pivotal in three two-party coalitions and three three-party ones and each such coalition gets a weight of 1/12. Similarly

$$u_{\text{Blue}} = \frac{1}{12} \times 1 + \frac{1}{12} \times 1 = \frac{1}{6}$$

and $u_{\text{Green}} = u_{\text{Brown}} = 1/6$ likewise, because Blue, Green, and Brown are each pivotal in exactly one two-party and one three-party coalition.

Even though the Blues have almost twice as many members as the Greens, who in turn have twice as many as the Browns, the three parties have equal power. The reason is that they are all equally crucial in the formation of a majority coalition, either one at a time with Red, or the three of them joined together against Red.

We do observe in reality that small parties in multiparty legislatures enjoy far more power than their number proportions in the legislature might lead us to believe. The Shapley value index shows this in a dramatic way. Of course, it does make some unappealing assumptions; for example, it takes all coalitions to be equally likely, and all contributions to have the same value 1 if they create a majority and the same value 0 if they do not. Therefore we should not interpret the power index as a literal or precise quantitative measure, rather as a rough indication of political power levels. More realistic analysis can sometimes show that small parties have even greater power than the Shapley measure might suggest. For example, the two largest parties often have ideological antagonisms that rule out a big coalition between them. If coalitions that include both Red and Blue together are not possible, then Green and Brown will have even greater power because of their ability to construct a majority—one of them together with Red, or both of them together with Blue.

B. Allocation of Joint Costs

Probably the most important practical use of the Shapley value is for allocating the costs of a joint project among its several constituents. Suppose a town is contemplating the construction of a multipurpose auditorium building. The possible uses are as a lecture hall, a theater, a concert hall, and an opera house. A single-purpose lecture hall would cost $1 million. A theater would need more sophisticated stage and backstage facilities, and would cost $4 million. A concert hall would need better acoustics than the lecture hall but less sophisticated staging than the theater, so the cost of a pure concert hall would be $3 million. An opera house would need both staging and acoustics, and the cost of the two together, whether for a theater-concert combination or for opera, would be $6 million. If the opera house is built, it can also be used for any of the three other purposes, and a theater or a concert hall can also be used for lectures.

The town council would like to recoup the construction costs by charging the users. The tickets for each type of activity will therefore include an amount that reflects an appropriate share of the building cost attributable to that activity. How are these shares to be apportioned? This is a mirror image of the problem of distributing payoffs according to contributions—we want to charge each use for the extra cost it entails. And the Shapley value offers an answer to the problem. For each activity, say the theater, we list all the combinations in which it could occur, and what it would cost to construct the facility for that combination, and for that combination minus the theater. This analysis tells us the contribution of the theater to the cost of building for that combination. Assuming all combinations involving this activity to be equally likely, and averaging, we get the overall cost share of the theater.

First we calculate the probabilities of each combination of uses. The theater could be on its own with a probability of 1/4. It could be in a combination of two activities with a probability of 1/4, and there are three ways to combine the theater with another activity, so the probability of each such combination is 1/12. Similarly, the probability of each three-activity combination including the theater is also 1/12. Finally, the probability of the theater being in the grand coalition of all four activities is 1/4.

Next we calculate the contributions the theater makes to the cost of each combination. Here we just give an example. The theater-concert combination costs $6 million, and the concert hall alone would cost $3 million; therefore $6 - 3 = 3$ is the contribution of the theater to this combination. The complete list of combinations and contributions is shown in Figure 17.5. Each row focuses on one activity. The left-most column simply labels the activity: L for lecture, T for theater, C for concert, and O for opera. The successive columns to the right show one-, two-, three-, and four-activity combinations. The heading of each column shows the size and the probability of each combination of that size. The actual cells for each row in these columns list the combinations of that size that include the activity of that row (when there is only one activity we omit the listing), followed by the contribution of the particular activity to that combination. The right-most column gives the average of all these contributions using the appropriate probabilities as weights, that is, the Shapley value.

It is interesting to note that while the cost of building an opera facility is *six* times that of building a lecture hall (6 versus 1 in the bottom and top cells of the "Size 1" column), the cost share of opera in the combined facility is *13* times that of lectures (2.75 versus 0.25 in the "Shapley Value" column). This is because providing for lectures adds nothing to the cost of building for any other use, so the contribution of lectures to all other combinations is zero. On the other hand, providing for opera raises the cost of *all* combinations except those that also involve both theater and concert, of which there is only one.

Note also that the cost shares add up to 6, the total cost of building the combined facility. This is another useful property of the Shapley value. If you have

ACTIVITY		COMBINATIONS				Shapley Value
		Size 1 Prob. = 1/4	Size 2 Prob. = 1/12	Size 3 Prob. = 1/12	Size 4 Prob. = 1/4	
L	1		LT: $4 - 4 = 0$ LC: $3 - 3 = 0$ LO: $6 - 6 = 0$	LTC: $6 - 6 = 0$ LTO: $6 - 6 = 0$ LCO: $6 - 6 = 0$	$6 - 6 = 0$	0.25
T	4		LT: $4 - 1 = 3$ TC: $6 - 3 = 3$ TO: $6 - 6 = 0$	TLC: $6 - 3 = 3$ TLO: $6 - 6 = 0$ TCO: $6 - 6 = 0$	$6 - 6 = 0$	1.75
C	3		CT: $3 - 1 = 2$ CT: $6 - 4 = 2$ CO: $6 - 6 = 0$	CLT: $6 - 4 = 2$ CLO: $6 - 6 = 0$ CTO: $6 - 6 = 0$	$6 - 6 = 0$	1.25
O	6		OL: $6 - 1 = 5$ OT: $6 - 4 = 2$ OC: $6 - 3 = 3$	OLT: $6 - 4 = 2$ OLC: $6 - 3 = 3$ OTC: $6 - 6 = 0$	$6 - 6 = 0$	2.75

FIGURE 17.5 Cost Allocation for Auditorium Complex

sufficient mathematical facility, you can use Eq. (17.1) to show that the payoffs u_i of all the players sum to the characteristic function v of the grand coalition.

To sum up, the Shapley value allocates the costs of a joint project by calculating what the presence of each participant adds to the total cost. Each participant is then charged this addition to the cost (or in the jargon of economics his *marginal cost*).

The principle can be applied in reverse. Suppose some firms could create a larger total profit by cooperating than they could by acting separately. They could bargain over the division of the surplus (as we saw in Chapter 16), but the Shapley value provides a useful alternative, in which each firm receives its marginal contribution to the joint enterprise. Adam Brandenburger and Barry Nalebuff, in their book *Co-opetition*, develop this idea in detail.[7]

5 FAIR DIVISION MECHANISMS

Neither the core nor the market mechanism nor Shapley value guarantee any fairness of payoffs. Game theorists have examined other mechanisms that focus on the fairness issue, and have developed a rich line of theory and applications. Here we give you just a brief taste.[8]

---

[7] Adam Brandenburger and Barry Nalebuff, *Co-opetition* (New York: Doubleday, 1996).

[8] For a thorough treatment, see Steven J. Brams and Alan D. Taylor, *Fair Division: From Cake-cutting to Dispute Resolution* (New York: Cambridge University Press, 1996).

The oldest and best-known **fair division mechanism** is "one divides, the other chooses." If A divides the cake into two pieces, or the collection of objects available into two piles, and B chooses one of the pieces or piles, the outcome will be fair. The idea is that, if A were to produce an unequal division, then B would choose the larger. Knowing this, A will divide equally. Such behavior then constitutes a rollback equilibrium of this sequential-move game.

A related mechanism has a referee who slowly moves a knife across the face of the cake, say from left to right. At any point, either player can say "stop." The referee cuts the cake at that point and gives the piece to the left of the knife to the player who spoke.

We do not require the cake to be homogeneous—one side may have more raisins and another more walnuts—nor that all objects in the pile to be identical. The participants can differ in their preferences for different parts of the cake or different objects, and therefore they may have different perceptions about the relative values of portions and about who got a better deal. Then we can distinguish two concepts of fairness. First, the division should be *proportional*, meaning that each of n people believes that his share is at least $1/n$ of the total. Second, the division should be **envy-free,** which means that no participant believes someone else got a better deal. With two people the two concepts are equivalent, but with more, envy-freeness is a stronger requirement, because each may think he got $1/n$ while also thinking that someone else got more than he did.

With more than two people, simple extensions of the "one cuts, the other chooses" procedure ensure proportionality. Suppose there are three people, A, B and C. Let any one, say A, divide the total into three parts. Ask B and C which portions they regard as acceptable, that is, of size at least 1/3. If there is only one piece that both reject, then give that to A, leaving each of B and C a piece he regards as acceptable. If they reject two pieces, give one of the rejected pieces to A, and reassemble the remaining two pieces. Both B and C must regard this total as being of size at least 2/3 in size. Then use the "one cuts, the other chooses" procedure to divide this between them. Each gets what he regards as at least 1/2 of this, which is at least 1/3 of the initial whole. And A will make the initial cuts to create three pieces that are equal in his own judgment, to ensure himself of getting 1/3. With more people, the same idea can be extended by using mathematical induction. Similar, and quite complex, generalizations exist for the moving-knife procedure.

Ensuring envy-freeness with more than two people is harder. Here is how it can be done with three. Let A cut the cake into three pieces. If B regards one of the pieces as being larger than the other two, he "trims" the piece just enough so that there is at least a two-way tie for largest. The trimming is set aside and C gets to choose from the three pieces. If C chooses the trimmed piece, then B can choose either of the other two, and A gets the third. If C chooses an untrimmed piece, then B must take the piece he trimmed, and A gets the third.

Leaving the trimming out of consideration for the moment, this division is envy-free. Since C gets to choose any of the three pieces, he does not envy anyone. B created at least a two-way tie for what he believed to be the largest, and gets one of the two (the trimmed piece if C does not choose it), or the opportunity to choose one of the two, so B does not envy anyone. Finally, A will create an initial cut he regards as equal, so he will not envy anyone.

Then whoever of B or C did *not* get the trimmed piece divides the trimming into three portions. Whoever of B or C *did* get the trimmed piece takes the first pick, A gets the second pick, and the divider of the trimming gets the residual piece of it. A similar argument shows that the division of the whole cake is envy-free. For cases of even more people, and for many other fascinating details and applications, read Brams and Taylor's *Fair Division* (cited earlier).

Of course, we do not get fairness without paying a price in the form of the loss of some other desirable property of the mechanism. In realistic environments with heterogeneous cakes or other objects, and with diverse preferences, these fair division mechanisms generally are not efficient; they are also manipulable.

SUMMARY

We consider a market where each buyer and each seller attempts to make a deal with someone on the other side, in competition with others on his own side looking for similar deals. This is an example of a general game where *coalitions* of players can form to make tentative agreements for joint action. An outcome can be prevented or blocked if a coalition can form and guarantee its members better payoffs. The *core* consists of outcomes that cannot be blocked by any coalition.

In the trading game with identical objects for exchange, the core can be interpreted as a *market equilibrium* of *supply* and *demand*. When there are several buyers and sellers with slight differences in the prices at which they are willing to trade, the core shrinks and a competitive market equilibrium is determinate. This outcome is *efficient*. With few sellers or buyers, preference manipulation can occur and amounts to the exercise of *monopoly* or *monopsony power*. Laboratory experiments find that with as few as five or six participants, trading converges quickly to the market equilibrium.

The *Shapley value* is a mechanism for assigning payoffs to players based on an average of their contribution to possible coalitions. It is useful for allocating costs of a joint project, and gives insight into the relative power of parties in a legislature.

Neither the core nor the Shapley value guarantees fair outcomes. Other *fair division mechanisms* exist; they are often variants or generalizations of the "one cuts, the other chooses" idea. But these are generally inefficient and manipulable.

KEY TERMS

allocation (561)
blocking (562)
characteristic function (561)
coalition (561)
core (562)
demand curve (554)
efficient allocation (564)
envy-free allocation (578)
fair division mechanism (578)
feasible allocation (561)
grand coalition (561)

incentive-compatible
  market mechanism (570)
market equilibrium (555)
monopoly power (570)
monopsony power (570)
nonmanipulable
  market mechanism (570)
power index (574)
security level (561)
Shapley value (572)
supply curve (554)

EXERCISES

1. Consider the case of two buyers, B_1 and B_2, and one seller, S, from Section 1. The seller's reserve price is 100; B_1 has a willingness to pay of 300, and B_2 a willingness to pay of 200. (All numbers are dollar amounts in the thousands.)
   (a) Use the negotiation approach to describe the equilibrium range of negotiation and the trade that occurs in this situation.
   (b) Construct and illustrate the market supply and demand curves for this case. Show on your diagram the equilibrium range of prices and the equilibrium quantity traded.

2. In the case of two sellers with unequal reserve prices, 100 for S_1 and 150 for S_2, and one buyer B with a willingness to pay of 300, show that in a core allocation, the respective surpluses x_1, x_2, and y must satisfy $x_1 \leq 50$, $x_2 = 0$, and $y \geq 150$. Verify that this outcome means that S_1 sells his house for a price between 100 and 150.

3. There are four sellers and three buyers. Each seller has one unit to sell and a reserve price of 100. Each buyer wishes to buy one unit. One is willing to pay up to 400; each of the other two, up to 300. Find the market equilibrium by drawing a figure. Find the core by setting up and solving all the no-blocking inequalities.

4. In Exercise 3, suppose the four sellers get together and decide that only two of them will go to the market, and all four will share any surplus they get there.

Can they benefit by doing so? Can they benefit even more if one or three of them go to the market? What is the intuition for these results?

5. An airport runway is going to be used by four types of planes: small corporate jets and commercial jets of the narrow-body, wide-body, and jumbo varieties. A corporate jet needs a runway 2,000 feet long, a narrow-body 6,000 feet, a wide-body 8,000 feet, and a jumbo 10,000 feet. The cost of constructing a runway is proportional to its length. Considering each *type* of aircraft as a player, use Shapley value to allocate the costs of constructing the 10,000-foot runway among the four types. Is this a reasonable way to allocate the costs?

Index

acceptability condition, 454–57
actions, cross-effects of, 16–18
addition rule, 165–66, 176
additive payoffs, 566–67
adverse selection, 399, 401, 427
agenda paradox, 470–71, 474–75, 488–89
agreements, 23–24
  See also BATNAs; contracts
agriculture, strategic aspects of, 16–17
airwave spectrum auctions, 494–95, 511, 512–18
Allison, Graham, 446, 458
allocation,
  blocked, 562–64, 565, 579
  efficient, 564
  feasible, 561–62
  of joint costs, 575–77, 579
  in markets, 561–67, 575–77, 579
all-pay auctions, 503–5
amendment voting procedure, 464, 474, 477, 488
ant colonies, 350, 370
approval voting, 465, 472, 474, 488
arbitration, 276–77, 523–24, 529
arms race, 16, 107–10
  See also Cuban missile crisis
Arrow, Kenneth, 473–74, 489, 527
artificial intelligence, 68
ascending auctions. *See* English auctions
Ashenfelter, O., 276–77
asset markets, 571–72
assurance games,
  as collective-action, 357, 361, 362, 366–67, 368, 369–70, 371, 393
  as evolutionary games, 334–38, 352
  and expected payoffs, 335–36
  focal points in, 335
  history of ideas influencing, 368
  mixed strategies in, 124, 335, 336
  multiple equilibria in, 124
  Nash equilibrium in, 212–13, 366, 371
  payoffs in, 335, 336
  probability in, 335–36
  pure strategies in, 335, 336, 337, 361
  and risk, 336
  as simultaneous-move games, 107–10, 118, 124
  See also specific topic

auctions,
  all-pay, 503–5
  ascending. *See* English
  cheating at, 511
  common value, 496–9, 503–5, 507, 512, 518
  and defeating the system, 511
  definition of, 494
  descending. *See* Dutch
  and dominant strategies, 503, 507, 508
  Dutch, 496–7, 500–501, 506–9, 518
  English, 496–7, 500–501, 506–9, 518
  estimates for, 506–8, 511
  first move at, 505
  first-price, 496–7, 500–509, 511, 518
  free riders at, 503
  and incentives, 511, 518
  independent valuations at, 509–10
  information for, 502, 508, 511–12, 514
  as markets, 550, 570
  multiple objects at, 509–10
  objective value, *see* common value
  open outcry, 495–7, 500–501, 508–9, 514, 518
  and penalties/punishments, 511, 514–15
  private value, 497, 500–501, 518, 570
  repetition of, 511
  reserve prices at, 511
  rules for, 513–15
  sealed bid, 495–7, 499–509, 511, 514, 518
  second price, 496–7, 501–3, 506–11, 518, 570
  sellers at, 505–9, 511–12, 518
  as sequential-move games, 505
  strategic moves at, 289–90
  and type of games, 18
  types of, 495–97, 518
  winner's curse at, 497–99, 512, 514, 518
  See also bidding; *type of auction or bidding*
audits, 124–25
automatic fulfillment, 308
Axelrod, Robert, 272–74, 375

backstop payoffs. *See* BATNAs
backward induction. *See* rollback
bargaining,
  alternating offers in, 531–34, 537–41, 547
  arbitration in, 276–77, 523–24, 529